Near-Infrared Spectroscopy

Yukihiro Ozaki · Christian Huck ·
Satoru Tsuchikawa · Søren Balling Engelsen
Editors

Near-Infrared Spectroscopy

Theory, Spectral Analysis, Instrumentation, and Applications

 Springer

Editors
Yukihiro Ozaki
School of Science and Technology
Kwansei Gakuin University
Sanda, Japan

Satoru Tsuchikawa
Nagoya University
Nagoya, Japan

Christian Huck
Institut für Analytische Chemie
und Radiochemie
University of Innsbruck
Innsbruck, Tirol, Austria

Søren Balling Engelsen
University of Copenhagen
Copenhagen, Denmark

ISBN 978-981-15-8650-7   ISBN 978-981-15-8648-4   (eBook)
https://doi.org/10.1007/978-981-15-8648-4

This Springer imprint is published by the registered company Springer Nature Singapore Pte Ltd.
The registered company address is: 152 Beach Road, #21-01/04 Gateway East, Singapore 189721, Singapore

Preface

Despite being a well-established and very mature technique, near-infrared (NIR) spectroscopy continues to demonstrate remarkable progress. New principles for instrumentation have provided cutting-edge developments within NIR imaging, handheld instruments and laser-based techniques. As with the field of data analysis in general, NIR spectral analysis and data treatments also continue to demonstrate prominent advancements that only accelerate in its scope and accomplishments. We now have instruments available that are capable of generating very high volumes of high-quality spectral data, in breath-taking speeds, perhaps even distributed over several online measurement points, and terms such as artificial intelligence, big data and deep learning are more and more commonly seen as the tools used to decipher the hidden information in the spectral data. All these advances open the pathways for the introduction of quantum chemistry to NIR spectroscopy—a decoding and understanding for modelling chemical systems based on quantum theory. It is now possible to imagine applications that use a combination of NIR spectroscopy with information and communication technology, applications that uses low-power and remote-sensing laser NIR spectroscopy, or devices incorporated in handheld instruments—maybe even embedded in our mobile phones—all of which is contributing to some artificial intelligent-assisted model to improve, for instance, food quality and safety for the individual consumer. The possibilities of NIRS technology seem to be endless!

Several textbooks and handbooks on NIR spectroscopy are currently available; they are all important books, but some of them are not up to date. We thus thought there was an increasing demand for a new state-of-the-art textbook on NIR spectroscopy. The present book, we hope, will fill the gap and find a wide audience of newcomers as well as experts to the exciting and versatile NIR spectroscopy technology. The book is intended as a go-to-book for background theory, applications and tutorial work. It consists of four major parts: Introduction and Principles, Spectral Analysis and Data Treatments, Instrumentation, and Applications. We attempted to prepare a well-balanced book with emphasis on underlying principles, spectral analysis and data treatments and at the same time cover almost all areas to which NIR spectroscopy is currently applied.

The contributors, from many countries, are all front-runners in modern vibrational spectroscopy, data analysis, instrumentation and/or applications. Another purpose of this book is making a strong bridge between molecular spectroscopists and researchers and engineers in various fields such as agriculture and food engineering, pharmaceutical engineering, polymer engineering, process engineering and biomedical sciences. We intend the book to become a valuable source for molecular spectroscopists who are interested in new applications of NIR spectroscopy and for researchers and engineers in a variety of fields who would like to study basic principles of molecular spectroscopy and NIR spectroscopy in particular.

Finally, it is our strong hope that this book will be useful for graduate science and engineering students where it can serve as inspiration as textbook for courses and seminars of graduate schools.

Last but not least, we would like to thank Dr. Shinichi Koizumi, Mr. Tony Pressler Sekar and Ms. Taeko Sato of Springer Nature, for their continuous efforts in publishing this exciting book.

Sanda, Japan   Yukihiro Ozaki
Innsbruck, Austria   Christian Huck
Nagoya, Japan   Satoru Tsuchikawa
Copenhagen, Denmark   Søren Balling Engelsen
July 2020

Contents

Part I
Introduction and Principles

Chapter 1
Introduction

Yukihiro Ozaki and Christian Huck

Abstract This chapter describes the introduction to NIR spectroscopy. The discovery of infrared region is mentioned first, and then, the definition of NIR region and characteristics of NIR spectroscopy are explained. Finally, the brief history of NIR spectroscopy is outlined.

Keywords Near-infrared · NIR · Vibrational spectroscopy · Electronic spectroscopy

1.1 Discovery of Infrared (IR) Region

The discovery of an invisible component beyond the red end of the solar spectrum (modern meaning-infrared (IR) region in a broad sense) is ascribed to William Herschel, a German-born British astronomer, who is famous for Herschel telescope. In 1800 one day, he investigated the effect of sunlight divided from violet to red by a prism on temperature increase. He used just sunlight, a prism, and thermometers. Figure 1.1 shows his portrait and the experimental set up he employed. He happened to find that the significant temperature increase occurred even outside of red. He thought there was a different kind of invisible radiation from visible light beyond the red end of sunlight and named this radiation "heat ray." This was really a great discovery in science, but even he could not imagine that this is light. He was 62 years old when he discovered "heat ray." Sixty-two years old in 1800 probably corresponds to today's 80 years old or so. Thus, his discovery demonstrated that even very senior scientist could have intensive serendipity.

Y. Ozaki (✉)
School of Science and Technology, Kwansei Gakuin University, Sanda, Japan
e-mail: yukiz89016@gmail.com

Toyota Physical and Chemical Research Institute, Nagakute, Japan

C. Huck
Institute of Analytical Chemistry and Radiochemistry, CCB-Center for Chemistry and Biomedicine, Leopold-Franzens University, Innsbruck, Austria

© Springer Nature Singapore Pte Ltd. 2021
Y. Ozaki et al. (eds.), *Near-Infrared Spectroscopy*,
https://doi.org/10.1007/978-981-15-8648-4_1

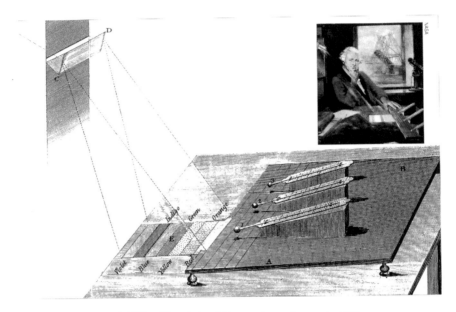

Fig. 1.1 The portrait of William Herschel and his experimental set up in 1800

Interestingly enough, just one year after the discovery of "heat ray," Johann Ritter, a German scientist found another invisible component beyond the violet end of the solar spectrum based on an experiment of blackening of silver chloride. In this way, a new era of light was opened at the turning point from the eighteenth century to the nineteenth century. After the discovery of "heat ray," many scientists investigated it. In 1835, it was confirmed that "heat ray" is invisible light which has longer wavelength than visible light. He named this light "infrared (IR)." Maxwell elucidated theoretically in 1864 that ultraviolet, visible, and IR light are all electromagnetic wave. In 1888, Hz proved it experimentally.

1.2 Introduction to NIR Spectroscopy

IR region is so wide energetically, ranging from 150 to 0.12 kJ mol^{-1} (12,500–10 cm^{-1}). If one compares the energy of the highest wavenumber end of NIR region with that of the lowest wavenumber end of far-IR (FIR) region, one can find that the difference in the energy between the two ends is more than 1000 times. Therefore, nowadays, the IR region is divided into three regions, NIR region (12,500–4000 cm^{-1}; 800–2500 nm), the IR region (mid-IR; 4000–400 cm^{-1}; 2500–25,000 nm), and the far-IR region (FIR; 400–10 cm^{-1}; 25 μm-1 mm) [1–7]. NIR spectroscopy is spectroscopy in the region of 12,500–4000 cm^{-1}, where bands arising from electronic transitions as well as those due to overtones and combinations of normal vibrational modes are expected to appear [1–7]. Therefore, NIR spectroscopy is electronic spectroscopy as well as vibrational spectroscopy. Ultraviolet (UV)-visible (Vis) spectroscopy is electronic spectroscopy while infrared (IR) spectroscopy is vibrational spectroscopy, so that NIR spectroscopy is something special. It lies in between electronic spectroscopy region and vibrational spectroscopy region.

Figure 1.2 shows chemical structure of immobilized metal affinity chromatography (IMAC) material and NIR spectra in the region of 10,000–4000 cm^{-1} of 32 kinds of IMAC materials [8]. Broad features in the 10,000–7500 cm^{-1} region are due to the d-d transition of Ni coordination compound and bands in the 7500–4000 cm^{-1} region arise from overtones and combinations. The spectra in Fig. 1.2 are very interesting examples, demonstrating that in the NIR region, one can observe both bands assigned to electronic transition and those originating from vibrational transitions. Most of the electronic transitions appearing in the NIR region are the d-d transitions, charge-transfer (CT) transitions, and π-π* transitions of large, or long, conjugated systems [1, 3, 7]. NIR spectroscopy involves absorption, emission, scattering, reflection, and diffuse-reflection of light [1–7].

NIR spectroscopy together with Raman, IR, and Terahertz/FIR spectroscopy forms "four sisters of vibrational spectroscopy." Since NIR spectroscopy is concerned

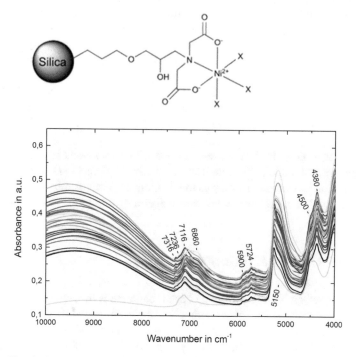

Fig. 1.2 Chemical structure of immobilized metal affinity chromatography (IMAC) material and NIR spectra in the region of 10,000–4000 cm^{-1} of 32 kinds of IMAC materials. Reproduced from Ref. [8] with permission

only with overtones and combination modes as a vibrational spectroscopy, it is very much unique compared with Raman, IR, and Terahertz/FIR spectroscopy.

The NIR region can be divided into three regions; Region I (800–1200 nm; 12,500–8500 cm^{-1}), Region II (1200–1800 nm; 8500–5500 cm^{-1}), and Region III (1800–2500 nm; 5500–4000 cm^{-1}) [1–7]. The boarders of the three regions are not rigorous. In Region I, bands arising from electronic transitions and those due to higher-order overtones and various types combination modes are expected to appear. Region I shows high transparency because all the bands appearing in this region are very weak, allowing biomedical applications and applications to agricultural products. Region I is the region where CCD cameras work very well, and this region is called "window of biological materials" because of high transparency to body. It has also a few more nick names: "the short-wave NIR (SWNIR) region," "near NIR (NNIR) region," or "the Herschel region."

Region II is a region for the first and second overtones of the XH (X = C, O, N) stretching vibrations and various types of combination modes. Region III contains mainly bands attributed to the combination modes except for the second overtone of the C = O stretching vibrational mode. It shows poorer "permeability."

NIR spectroscopy, particularly vibrational NIR spectroscopy, is spectroscopy of anharmonicity [1–7]. The overtones and combination modes are the so-called forbidden transitions for a harmonic potential, yielding very weak bands. Both the frequencies and intensities of NIR bands are controlled by anharmonicity. Therefore, investigations on overtones and combination modes, anharmonicity, vibrational potential, and dipole moment function regarding NIR spectroscopy are important. However, these studies have been far behind applications of NIR spectroscopy probably because until 1990s, it was difficult to obtain accurate NIR spectra and to make reliable band assignments. It is only recent that quantum chemistry has been introduced to studies of frequencies and intensities of overtones and combination bands (Chap. 5).

The fact that bands in the NIR region are weak or very weak is what makes this region unique and markedly different from the other regions [1–7]. The reason why the NIR region is valuable in various applications is because only the NIR region serves as a highly transmitting window to radiation thanks to anharmonicity.

1.3 Brief History of NIR Spectroscopy

It is uncertain when NIR spectroscopy began, but there is the report that Abney and Festing measured the spectra of some simple organic compounds in the 700–1200 nm region as well as in the Vis and IR regions. In the beginning of the twentieth century, main concerns of molecular spectroscopy were UV–Vis and IR spectroscopy. It was 1920s and 30s that systematic measurements of NIR spectra were carried out. A chance came from the development of a spectrometer by Brackett. In 1930s, spectroscopists already recognized that NIR spectra arise from overtones and combination modes [9, 10]. In 1950s, NIR spectroscopy received considerable interest for hydrogen bonding studies and studies on anharmonicity [11, 12]. The development of an innovative spectrometer, Carey 14 Applied Physics in 1954, pushed NIR spectroscopy research as well as UV–Vis studies [1]. However, the development of basic studies of NIR spectroscopy was still rather slow because even development of new spectrometers was insufficient for NIR spectroscopy to observe weak NIR bands accurately, and also systematic analysis of NIR bands was very difficult. Moreover, NIR spectroscopy could not find application fields not only in basic science but also in practical applications [1–7]. Until 1960s, NIR spectroscopy was a "sleeping giant" in terms of both basic science and applications.

It was not a spectroscopist but an engineer in an agricultural field who woke up the sleeping giant. He was Karl Norris (Fig. 1.3a) of the US Department of Agriculture (USDA). Norris learnt electrical engineering as well as agricultural engineering at universities, and thus, he had good background for developing spectrometers and systems with computers. Norris was involved in a research of drying grain by use of infrared technology. He happened to find that the grain had absorbances in the NIR region. He focused on the fact that NIR spectroscopy is suitable for nondestructive analysis, and he and his colleagues tried to use NIR spectroscopy for quality assessments of agricultural products. Norris proposed to use statistical methods to build calibration models from NIR data [13, 14]. First, he employed simple linear regression and then multiple regression. His idea realized the advantages of NIR spectroscopy in practical applications. Thus, Norris is called "Father of NIR spectroscopy." Norris, Phil Williams, Fred McClure, and other engineers applied NIR spectroscopy to many applications in agriculture and then foods. Beltsville (Maryland, USA) was a place for assemblage for the bold and ambitious. Their great success partly came from the strong request of quality assessment from consumers which already started in North America from 1960s and partly from the development of spectrometers and computers.

However, many conventional spectroscopists did not accept adamantly the eccentric idea of utilizing statistical methods such as multiple regression analysis to develop calibration models of NIR data. After rather long dispute, some traditional spectroscopists started to recognize the usefulness of the statistical methods to analyze NIR spectra. Particularly, Tomas Herschfeld played a very important role in making a bridge between the spectroscopists and agricultural engineers.

Of note is that in 1960s, there was significant progress also in the applications of NIR spectroscopy to basic studies [15–17]. For examples, in 1963, Bujis and Choppin [17] measured NIR spectra of pure water and investigated water structure in relation to hydrogen bonds. Late 1960s and early 70s, a few research groups including Camille Sandorfy (Fig. 1.3b) group [16] found very interesting fact concerning with the relative intensities of free and association bands of the OH and NH stretching bands compared for the fundamentals and overtones. The relative intensity of the free band is much greater for the overtones than fundamentals. One can say Sandorfy is a pioneer in basic studies of NIR spectroscopy. He is famous particularly in the research on relation between anharmonicity and hydrogen bondings.

It is also very important to know that there is another great scientist who advanced the practical application of NIR spectroscopy. He was Frans, F. Jobsis (Fig. 1.3c),

Fig. 1.3 Portrait photos of **a** Karl H. Norris, **b** Camille Sandorfy, and **c** Frans F. Jöbsis

(a)

(b)

(c)

who carried out the in vivo monitoring of the redox behavior of cytochrome c oxidase (or cytochrome aa_3) (Chap. 20) [18]. Since his pioneering study medical application of NIR spectroscopy has shown distinctive growth as described later [19].

It is not clear when chemometrics was born, but it is clear that the use of statistical methods by Norris was one of the initiations of the development of chemometrics [1, 2, 4, 6]. Among various molecular spectroscopy, NIR spectroscopy was the first in using chemometrics. For the last half century or so, chemometrics developed NIR spectroscopy and NIR spectroscopy developed chemometrics. Nowadays, chemometrics is used in almost all kinds of spectroscopy including IR, Raman, far-infrared (FIR)/Terahertz, UV–Vis, fluorescence, and NMR spectroscopy.

In 1980s, NIR spectroscopy was used mainly for agriculture and food engineering fields, but applications to polymers and on-line analysis started in those days. After entering 1990s, application of NIR spectroscopy made remarkable progress thanks to the development of spectrometers, detectors, computers, and chemometrics. It has expanded to chemical, polymer and petroleum industries, pharmaceutical industry, biomedical sciences, environmental analysis, and even analysis of cultural resources. In the last ten years or so, development of NIR imaging and portable and handheld instruments has been a matter of big attention. Besides progresses in NIR imaging, and portable and hand-held spectrometers, those in on-line monitoring, process analysis technology (PAT), sensing for security and safety, and medical diagnosis have been particularly noted [1–6]. NIR world is stretching strongly over a huge area of science and technology.

Medical application of NIR spectroscopy is nowadays called functional NIR (fNIR) spectroscopy [20]. It uses mainly electronic NIR spectroscopy in Region 1, the region of "window of biological materials." fNIR is applied not only to medical applications but also to brain research.

Basic studies of NIR spectroscopy such as overtones, combination modes, anharmonicity, and vibrational potential, and application of NIR spectroscopy to basic science like studies of hydrogen bondings, intermolecular interactions, and solution chemistry experienced "renaissance" in the 1990s due to rapid progress in NIR spectrometers particularly FT-NIR spectrometers and spectral analysis methods like two-dimensional correlation analysis.(Chap. 13) [1, 2, 3, 5, 7] Quantum chemical calculations have realized simulations of NIR spectra not only of simple compounds but also of rather complicated molecules such as long chain fatty acids, caffeine, nucleic acid bases, and rosemaric acid (Chap. 5). They also enable one to make band assignments of NIR spectra [20]. It is noted that quantum chemical calculations are useful for both basic studies and applications of NIR spectroscopy.

NIR spectroscopy is expanding markedly to a variety of fields such as astronomy, security and safety sensing, forensic science, building site, paleocultural property science and brain science.

References

1. H.W. Siesler, Y. Ozaki, S. Kawata, H.M. Heise (eds.), *Near-Infrared Spectroscopy*, Principles, Instruments, Applications, Wiley-VCH (2002).
2. Y. Ozaki, W. F. McClure, A. A. Christy, eds., *Near-Infrared Spectroscopy in Food Science and Technology*, Wiley-Interscience (2007).
3. Y. Ozaki, Near-infrared spectroscopy-Its versatility in analytical chemistry. Anal. Sci. **28**, 545–563 (2012)
4. D. A. Burns, E. W. Ciurczak eds., *Handbook of Near-Infrared Analysis*, Third Edition (Practical Spectroscopy), CRC Press (2007).
5. Y. Ozaki, C. W. Huck, K. B. Beć, "Near-IR spectroscopy and its applications" in *Molecular and Laser Spectroscopy: Advances and Applications*, V.P. Gupta ed. 2017, Elsevier, pp.11–38.
6. J. Workman, Jr., L. Weyer, *Practical Guide and Spectral Atlas for Interpretive Near-Infrared Spectroscopy*, 2nd Edition, CRC Press (2012).
7. M.A. Czarnecki, Y. Morisawa, Y. Futami, Y. Ozaki, Advances in molecular structure and interaction studies using near-infrared spectroscopy. Chem. Rev. **115**, 9707–9744 (2015).
8. C.G. Kirchler, R. Henn, J. Modl, F. Munzker, T.H. Baumgartner, F. Meisch, A. Kehle, G.K. Bonn, C.W. Huck, Direct determination of Ni^{2+}-capacity of IMAC materials using near-infrared spectroscopy. Molecules **23**, 3072–3081 (2018).
9. J.W. Ellis, Trans Faraday Soc. **25**, 888 (1928).
10. R. B. Barnes R. R. Brattain, J. Chem. Phys. 3, 446 (1935).
11. W. Kaye, Near-infrared spectroscopy: II. Instrumentation and Technique, Spectrochim Acta **7**, 181–204 (1955).
12. O.H. Wheeler, Near infrared spectra of organic compounds. Chem. Rev. **59**, 629–666 (1959).
13. J.R. Hart, K.H. Norris, C. Golumbic, Determination of the moisture content of seeds by near-infrared spectrophotometry of their methanol extracts. Cereal Chem. **39**, 94–99 (1962).
14. D.R. Massie, K.H. Norris, Trans. Of ASAE **8**, 598 (1965).
15. M.C. Bernard-Houplain, C. Sandorfy, On the similarity of the relative intensities of Raman fundamentals and infrared overtones of free and hydrogen bonded X-H stretching vibrations. Chem. Phys. Lett. **27**, 154–156 (1974).
16. C. Sandorfy, R. Buchet, G. Lachenal: Principles of molecular vibrations for near-infrared spectroscopy, in ref. 2, pp.11–46.
17. K. Bujis, G.R. Choppin, J. Chem. Phys. **39**, 2935 (1963).
18. F.F. Jöbsis, Noninvasive, infrared monitoring of cerebral and myocardial oxygen sufficiency and circulatory parameters. Science **198**, 1264–1267 (1977).
19. G. Margaret, B. Clare, E. E. Tachtsidis: From Jöbsis to the present day: a review of clinical near-infrared spectroscopy measurements of cerebral cytochrome-c-oxidase. J. Biomed. Opt.,**21**, 091307 (2016).
20. Z. Yuan, J. Zhang, X. Lin, Technological advances and prospects in multimodal neuroimaging: Fusions of fNIRS-fMRI, fNIRS-EEG and fMRI-EEG, LAP LAMBERT Academic Publishing (2014).

Chapter 2
Principles and Characteristics of NIR Spectroscopy

Yukihiro Ozaki and Yusuke Morisawa

Abstract This chapter describes the principles and characteristics of NIR spectroscopy. It is divided into two subchapters, 2–1. Characteristics and advantages of NIR spectroscopy: In this subchapter some emphasis is put on the versatility of NIR spectroscopy. Some examples of NIR spectra are explained 2–2. Principles of NIR spectroscopy based on quantum mechanics: To understand principles of NIR spectroscopy, principles of IR spectroscopy are described using quantum mechanics first, and then detailed explanation about molecular vibrations-fundamentals, overtones and combinations is given. Anharmonicity is mentioned briefly.

Keywords Molecular vibrations · Vibrational spectroscopy · Overtones · Combinations · Anharmonicity

NIR spectroscopy has very unique characteristics and advantages in comparison with other spectroscopy like ultraviolet-visible (UV-Vis), IR, and Raman spectroscopy. Those characteristics and advantages of NIR spectroscopy all come from anharmonicity of vibrational modes. Hence, it is important to learn the characteristics and advantages of NIR spectroscopy in relation with anharmonicity, whose description needs fundamental knowledge of quantum mechanics.

Y. Ozaki (✉)
School of Science and Technology, Kwansei Gakuin University, Sanda, Japan
e-mail: yukiz89016@gmail.com

Toyota Physical and Chemical Research Institute, Nagakute, Japan

Y. Morisawa
School of Science and Engineering, Kindai University, Higashiosaka, Japan

© Springer Nature Singapore Pte Ltd. 2021
Y. Ozaki et al. (eds.), *Near-Infrared Spectroscopy*,
https://doi.org/10.1007/978-981-15-8648-4_2

2.1 Characteristics and Advantages of NIR Spectroscopy

2.1.1 Characteristics of NIR Spectroscopy

NIR spectroscopy is concerned with both electronic transitions and vibrational transitions [1–7]. However, as the electronic spectroscopy, it is not easy or almost meaningless to discriminate the NIR region from the visible region. The two regions are seamless in the electronic spectra. In contrast, it is quite straightforward to distinguish the vibrational spectroscopy in the NIR region from that in the IR region because NIR spectroscopy deals only with bands arising from overtones and combination modes, while IR spectroscopy involves mainly bands due to fundamentals, although those originating from overtones and combinations also appear relatively weakly in the IR region.

One of the most characteristic features of NIR spectroscopy come from the fact that bands in the NIR region are weak or very weak. Both bands due to electronic transitions and those originating from vibrational transitions are weak. The overtones and combination modes arise from so-called forbidden transitions [1–7]. The reason why the NIR region is valuable from the point of applications is since only the NIR region offers as a highly transmitting window to radiation.

2.1.2 Characteristics of NIR Bands

Characteristics of bands appearing in the NIR region can be summarized as follows. Here, we consider NIR vibrational bands, overtones and combinations.

(1) Bands observed in the NIR region are all due to overtones and combinations; Not only simple combination bands such as $v_1 + v_2$ but also second order and third order combination bands such as $v_1 + 2v_2$ appear. The NIR region contains many overlapping bands; NIR bands show strong multicolinearlity. Therefore, assignment of the NIR bands is generally not easy.

(2) The NIR bands become weaker and weaker as the wavelength becomes shorter since bands due to higher order overtones and the second and third order combinations appear in the shorter wavelength region. Table 2.1 tabulates the wavelength, wavenumber, and relative intensity of bands due to the fundamental, first, second, and third overtones of CH stretching mode of chloroform. It is noted that the overtone bands become weak abruptly with the increase in the order and that the third overtone bands is located in the Vis region.

(3) Most of the bands in the NIR region originate from functional groups containing a hydrogen atom (*e.g.*, OH, CH, NH). This is partly due to the fact that an anharmonic constant of an XH bond is large, and partly due to the fact that an XH stretching vibration has its fundamental in a high frequency region (3800-2800 cm^{-1}). Hence, NIR spectroscopy is often called "an XH spectroscopic

Table 2.1 The wavelength, wavenumber, and relative intensity of bands due to the fundamental, first, second, and third overtones of CH stretching mode of chloroform	Band position/nm	Band position/cm^{-1}	Intensity/ cm^2 mol^{-1}
υ	3290	3040	25000
2υ	1693	5907	1620
3υ	1154	8666	48
4υ	882	11338	1.7
5υ	724	13831	0.15

method." Besides XH vibrational bands, those arising from the second overtones of $C = O$ stretching modes appear in the NIR region. Recently, bands due to the first and second overtones of $C \equiv N$ stretching modes of acetonitrile are also observed [8]. The $C = O$ and $C \equiv N$ stretching bands are two of the most intense bands in the IR region because of their large transition dipole moments, and thus, even second overtones can be observed in the NIR region.

(4) The first overtones of XH stretching bands give a lower frequency shift upon the formation of a hydrogen bond or an inter- or intra-molecular interaction as in the cases of the corresponding fundamental bands in IR spectra. The shift of the first overtones is almost double of that for the corresponding IR bands.

(5) In the NIR region OH and NH stretching bands of monomeric and polymeric species are much better separated compared with the IR region. It is also possible to distinguish bands ascribed to terminal free OH or NH groups of the polymeric species from those due to their free OH or NH groups in the NIR region.

(6) Because of the larger anharmonicity, the first overtone bands of OH and NH stretching modes of monomeric species are much more enhanced compared with the corresponding bands of polymeric species. On the other hand, fundamental bands originating from the OH and NH stretching modes of polymeric species are much more enhanced than those of the monomeric species due to a larger charge separation of X–H ($^{\delta-}$X–H$^{\delta+}$—:Y) in a hydrogen bonding. In Fig. 4.8 you can see good example for these phenomena. In the IR spectrum a broad feature due to polymeric species is much stronger than a band originating from monometic species while in the NIR spectrum a truly opposite result is observed (Sect. 4.1.2). Therefore, one can monitor more easily the dissociation process from polymeric species into monomeric ones in the NIR region rather than in the IR region by the first overtone of the OH or NH stretching mode of the monomeric species.

Note that almost all of the above characteristics come from the fact that NIR spectroscopy is concerned with forbidden transitions within the harmonic-oscillator approximation.

2.1.3 Advantages of NIR Spectroscopy

Now, let us discuss the advantages of NIR spectroscopy from the point of applications [1–7]. First of all, NIR spectroscopy is a powerful non-destructive and in situ analysis method. One can explore even inside of a material using NIR spectroscopy. Second, it permits non-contact analysis, and analysis using an optical fiber. Third, it is possible to apply NIR spectroscopy to samples in various states, shapes, and thickness. As for the advantages of NIR spectroscopy for fundamental studies we discuss in Chap. 13.

One can discuss the advantages of NIR spectroscopy in comparison with IR spectroscopy.

(1) NIR spectroscopy allows in situ analysis with a sample as it originally is. While one can employ attenuated total reflection (ATR) or photoacoustic spectroscopy (PAS) for in situ analysis in IR spectroscopy, there is no other choice than NIR spectroscopy if one wishes to collect an absorption spectrum on the whole of an apple or a human head. It is also suitable for nondestructive of thick samples.

(2) In general, NIR spectroscopy is more suitable in the analysis of aqueous solutions than IR spectroscopy since the intensity of water bands is much weaker in the NIR spectrum than in the IR spectrum. ATR-IR spectroscopy permits one to examine aqueous solutions, but NIR spectroscopy can probe those in more various manners.

(3) A light-fiber probe can be set in a dangerous environment, and be remotely manipulated. This is one of the reasons why NIR spectroscopy is suitable for on-line analysis. IR Light-fibers are much less robust and more expensive.

(4) In NIR spectroscopy one can select a light path length very freely. In contrast, IR spectroscopy usually requests a cell having a very short path length. NIR spectroscopy allows one to use a 0.1-mm cell, a 1-cm cell, or even a 10-cm cell.

(5) Optical materials used in the NIR region are cheaper than those in the IR region. One can use glass cells, for example.

2.1.4 Versatility of NIR Spectroscopy

NIR spectroscopy holds marked versatility in many aspects. First of all, it has huge versatility in its applications [1–7]. NIR spectroscopy can be used in a laboratory, a factory, a hospital, a field and a museum, at a building site, on a road and in the atmosphere. It may be applied to solids, crystals, fibers, powders, liquids, solutions, and gases. Almost all kinds of materials, from purified samples to bulk materials, can be subjected to NIR measurements.

Another versatility in NIR spectroscopy is the versatility in spectrometers and instruments. (Chap. 9) In the IR region, most of the spectrometers employed are FT spectrometers, while in the NIR region both FT spectrometers and dispersive spectrometers are employed, and dispersive spectrometers with a CCD detector play important roles in the short-wave NIR (SWNIR) region. NIR spectrometers with an acoustic optic tunable filter (AOTF) are also useful. Portable spectrometers and hand-held spectrometers are prety popular in the NIR region. Many kinds of special-purpose instruments are commercially available.

Spectral analysis gives yet another diversity of NIR spectroscopy. Compared with other spectroscopy, chemometrics is quite often used for NIR spectral analysis. Various kinds of chemometrics methods such as PCA (principal components analysis) and PLS (partial least squares) are employed extensively (Chap. 7). A variety of spectral pretreatments are employed in NIR spectroscopy, since it treats various kinds of bulk materials, which yield noise and baseline fluctuations (Sect. 4.1). Nowadays, even quantum chemical calculations are utilized in the spectral analysis in NIR spectroscopy (Sect. 5.2).

2.1.5   Some Examples of NIR Spectra

To understand the characteristics of NIR spectroscopy it is important to study some examples of NIR spectra.

(a)   Water

Figure 2.1 depicts NIR spectra of water in the region of 12000–4000 cm^{-1} obtained using three kinds of cells with different path lengths (1, 0.1, and 0.01 cm). Band intensities vary markedly with the path lengths. In all the spectra an intense foot due to the fundamentals of OH stretching modes can be observed near 4000 cm^{-1}. Water bands become weaker and weaker stepwisely as the wavelength goes to a shorter wavelength. Two strong bands at 5235 and 6900 cm^{-1} are due to the combination of H-O-H antisymmetric stretching mode (ν_3) and bending mode (ν_2) and that of H-O-H symmetric (ν_1) and antisymmetric (ν_3) stretching modes, respectively (these vibrational modes, see Fig. 2.9). These two bands are very useful for investigating

Fig. 2.1 NIR spectra of water in the region of 12000–4000 cm^{-1} obtained by three kinds of cells with different path lengths (0.01, 0.1, and 1 cm)

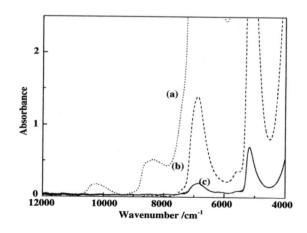

water structure and water contents in various materials. Bands at 10613, 8807, and 8762 cm^{-1} are assigned to $2v_1 + v_3$, $v_1 + v_2 + v_3$, and $2v_1 + v_2$, respectively. The band at 10613 cm^- [1] is valuable for estimating water contents in foods and materials. As you can see here, several bands attributed to the second and third order combination modes appear in the short wavelength region. More detailed analysis of water spectra will be discussed in Sect. 4.1.2.

(b) Methanol

Figure 2.2 depicts an NIR spectrum in the 7700–3700 cm^{-1} region of low concentration (0.005 M, in CCl_4) methanol. In this concentration it is very unlikely that methanol forms hydrogen bonds. Methanol is a very simple molecule, however, note that it gives so many bands in this region. One can easily assign a band at 7130 cm^{-1} to the first overtone of the OH stretching mode of free methanol. Bands in the region of 6100–5600 cm^{-1} are assigned to the first overtones of CH_3 symmetric and asymmetric stretching modes and their combinations. Those below 5200 cm^{-1} are due to various combination modes. We need the aid of quantum chemical calculations for convincing band assignments [9]. We will discuss about the quantum chemical calculation result of methanol in Chap. 13.

(c) Inorganic functional material-an example of electronic spectrum

Let us show one example of NIR electronic spectra. Figure 2.3a, b depict NIR diffuse-reflectance (DR) spectra in the region of 12000–4000 cm^{-1} and their second-derivative spectra in the region of 10000–5000 cm^{-1} of powders of high reflective green-black (HRGB; $Co_{0.5}Mg_{0.5}Fe_{0.5}Al_{1.5}O_4$) pigments, Co_3O_4, and α-Fe_2O_3, respectively [10]. The HRGB pigment developed at Toda Kogyo Co. (Hiroshima,

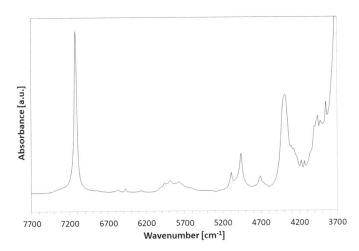

Fig. 2.2 A NIR spectrum in the 7700–3700 cm^{-1} region of low concentration (0.005 M, CCl_4) methanol

Fig. 2.3 **a** NIR DR spectra
of powders of HRGB,
Co_3O_4, and α-Fe_2O_3 in the
12000–4000 cm^{-1} region. **b**
Second derivatives of the
NIR DR spectra of powders
of HRGB, Co_3O_4 and
α-Fe_2O_3 in the
10000–5000 cm^{-1} region.
Reproduced from Ref. [10]
with the permission

(a)

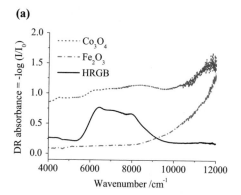

(b)

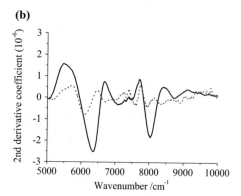

Japan) shows black color, but it absorbs little sunlight. It is noted in the second-derivative spectra that HRGB depicts bands at 6354, 7069, 7590 and 8024 cm^{-1} and that Co_3O_4, which has a similar spinel structure to HRGB, yields those at 6094, 6713, 7569, 7951, and 8320 cm^{-1}. The above bands of Co_3O_4 are ascribed to d-d transitions, $^4A_2 \rightarrow {}^4T_1$, of Co(II) at a tetrahedral cite. A NIR DR spectrum of α-Fe_2O_3 gives a long tail band in the region of 12000-10000 cm^{-1} due to a charge-transfer (CT) transition that has maxima at 17000 and 14000 cm^{-1} [10]. HRGB shows characteristic peaks of Co(II) in spinel structure, but it does not give a tail originating from Fe(III). In this way one can explore the structure of inorganic functional materials using NIR electronic spectra.

2.1.6 Comparison of an NIR Spectrum with an IR Spectrum

Whenever one studies the NIR spectrum of a sample, it is often important to compare the NIR spectrum with the corresponding IR spectrum to interpret the NIR spectrum. Figure 2.4a, b show chemical structure of poly(3-hydroxybutyrate) (PHB) and time-dependent variations in IR spectra and their second derivative spectra in the 3050–2850 cm^{-1} region of a PHB film during the melt-crystallization process at 125 °C,

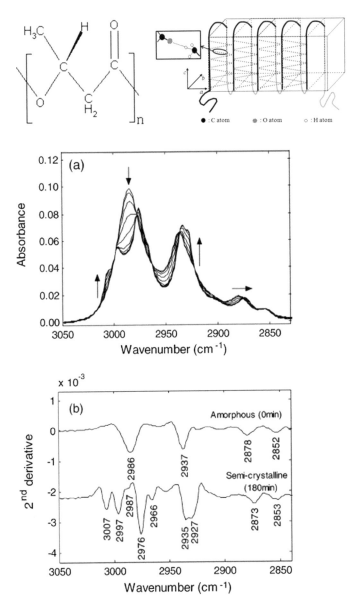

Fig. 2.4 Chemical structure and lamellar structure of poly(3-hydroxybutyrate) (PHB). **a** Time-dependent variations in IR spectra in the 3050–2850 cm^{-1} region of a PHB film during the melt-crystallization process at 125 °C. **b** Second derivative spectra of the spectra shown in **a** for 0 and 180 min. Reproduced from Ref. [11] with the permission

respectively [11]. In Fig. 2.4b the second derivative spectra for 0 and 180 min are shown. PHB is a well-known biodegradable polymer. Sato et al. [12, 13] investigated IR spectra of PHB and found that a crystalline $C = O$ stretching band at 1723 cm^{-1} shows a large downward shift by about 20 cm^{-1} compared with an amorphous $C = O$ stretching band at $\sim$1740 cm^{-1} and that a crystalline CH$_3$ asymmetric stretching band appears at an anomalously high frequency (3009 cm^{-1}) [12, 13]. On the basis of the IR and x-ray crystallography studies, they concluded that the CH$_3$ and $C = O$ groups of PHB form a peculiar C–H...O $= C$ hydrogen bonding.

Time-dependent variations in the NIR spectra in the 6050–5650 cm^{-1} region of a PHB film during the melt-crystallization process at 125 °C are shown in Fig. 2.5a [11]. The second derivatives of the spectra measured at 0 and 180 min are shown in Fig. 2.5b. The second-derivative spectrum obtained at 0 min shows four amorphous bands at around 5954, 5913, 5828, and 5768 cm^{-1}. On the other hand, the spectrum collected at 180 min gives at least seven bands at 5973, 5952, 5917, 5900, 5811, 5757, and 5681 cm^{-1} in the same region. Note that the NIR spectral changes in the 6050–5650 cm^{-1} region and the corresponding IR spectral variations in the 3050–2840 cm^{-1} region show significant similarities. For example, the NIR band with the highest wavenumber at 5973 cm^{-1} and the corresponding IR band at 3007 cm^{-1} show similar temporal variations. The former may be due to the first overtone of the latter [11].

Fig. 2.5 a Time-dependent variations in NIR spectra in the 6050–5650 cm^{-1} region of a PHB film during the melt-crystallization process at 125 °C. **b** Second derivative spectra of the spectra shown in **a** for 0 and 180 min. Reproduced from Ref. [11] with the permission

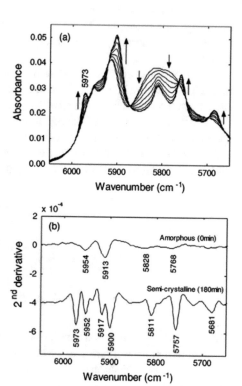

2.2 Principles of NIR Spectroscopy

Before we study the principle of NIR spectroscopy we have to learn the principle of IR spectroscopy because IR spectroscopy deals with fundamentals while NIR spectroscopy treats overtones and combinations which originate from fundamentals [1, 2]. Therefore, learning the fundamentals is the base for understanding NIR spectroscopy.

2.2.1 Principles of IR Spectroscopy

When a molecule is irradiated with IR light, it absorbs the light under some conditions. The energy $h\upsilon$ of the absorbed IR light is equal to an energy difference between a certain energy level of vibration of the molecule (having an energy E_m) and another energy level of vibration of a molecule (having an energy E_n). In the form of an equation,

$$h\upsilon = E_n - E_m \tag{2.1}$$

holds. This equation is known as Bohr frequency condition. In other words, absorption of IR light takes place based on a transition between energy levels of a molecular vibration. Therefore, an IR absorption spectrum is a vibrational spectrum of a molecule.

Note that satisfying Eq. (2.1) does not always mean the occurrence of IR absorption. There are transitions which are allowed by a selection rule (i.e., allowed transition) and those which are not allowed by the same rule (i.e., forbidden transition). In general, transitions with a change in the vibrational quantum number by ± 1 are allowed transitions and other transitions are forbidden transitions under harmonic approximation. This is one of selection rules of IR absorption. Another IR selection rule is a selection rule which is defined by the symmetry of a molecule [1, 2].

$$(\mu_x)_{mn} = \int_{-\infty}^{\infty} \psi_n \mu_x \psi_m \, dQ \tag{2.2}$$

$$\mu_x = (\mu_x)_0 + \left(\frac{\partial \mu_x}{\partial Q}\right)_0 Q + \frac{1}{2}\left(\frac{\partial^2 \mu_x}{\partial Q^2}\right)_0 Q^2 + \cdots\cdots \tag{2.3}$$

$$(\mu_x)_{mn} = (\mu_x)_0 \int \psi_n \psi_m \, dQ + \left(\frac{\partial \mu_x}{\partial Q}\right)_0 \int \psi_n Q \psi_m \, dQ \tag{2.4}$$

The latter selection rule is a rule that IR light is absorbed when the electric dipole moment of a molecule varies as a whole in accordance with a molecular vibration.

The above two selection rules can be introduced by quantum-mechanical considerations. According to quantum mechanics, for a molecule to transit from a certain state m to another state n by absorbing or emitting IR light, it is necessary that the following definite integral or at least one of $(\mu_y)_{mn}$ and $(\mu_z)_{mn}$ which are expressed by a similar equation to (2.2) is not 0, where μ_x denotes an x-component of the electric dipole moment; ψ denotes the eigenfunction of the molecule in its vibrational state; and Q denotes a displacement along a normal coordinate (i.e., a normal vibration expressed as a single coordinate). Now, let us consider only $(\mu_x)_{mn}$. A distribution of electrons in the ground state changes as the coordinate expressing a vibration varies, and thus, the electric dipole moment is a function of the normal coordinate Q. Hence, μ_x can be expanded as follows.

Expressed by a displacement of atoms during the vibration, Q has a small value. This allows to omit Q^2 and the subsequent terms in the equation above. Substituting the terms up to Q of Eq. (2.3) in Eq. (2.2), is obtained. Due to the orthogonality of the eigenfunction, the first term of this equation is 0 except when $m = n$ holds. The first term denotes the magnitude of the permanent dipole of the molecule. For the second term to have a value other than 0, both $(\partial \mu_x / \partial Q)_0 \neq 0$ and $\int \psi_n Q \psi_m d Q \neq 0$ must be satisfied. These two conditions lead to the two selection rules. The nature of the eigenfunction allows the integral to have the value other than 0 only when $n = m \pm 1$ holds. Considering Q^2 and the subsequent terms of Eq. (2.3) as well, we can prove that even when $n = m \pm 1$ fails to hold, $(\mu_x)_{mn}$ has a value, even though small, other than 0. The first selection rule regarding IR absorption is thus proved. The other selection rule, which is based upon the symmetry of a molecule, comes from $(\partial \mu_x / \partial Q)_0 \neq 0$. The relationship $(\partial \mu_x / \partial Q)_0 \neq 0$ indicates that IR absorption takes place only when a certain vibration changes the electric dipole moment. The vibration is **IR active** when $(\partial \mu_x / \partial Q)_0 \neq 0$ holds, but is **IR inactive** when $(\partial \mu_x / \partial Q)_0 = 0$ holds.

Most molecules are in the ground vibrational state at room temperature, and thus, a transition from the state $v'' = 0$ to the state $v'' = 1$ (first excited state) is possible. Absorption corresponding to this transition is called **the fundamental**. Although most bands which are observed in an IR absorption spectrum arise from the fundamental, in some cases, also in the IR spectrum one can observe bands which correspond to transitions from the state $v'' = 0$ to the state $v'' = 2, 3,...$ They are called first, second, **overtones**. Bands due to combinations are also observed in the IR spectra. However, since overtones and combinations are forbidden with harmonic oscillator approximation, overtone and combination bands are very weak even in real molecules. Because of anharmonicity, although the intensities are weak, the forbidden bands appear.

2.2.2 Molecular Vibrations

One must learn molecular vibrations to understand all kinds of vibrational spectroscopy; IR, NIR, FIR/terahertz, and Raman spectroscopy. Vibrations of a polyatomic molecule are, in general, complex, however, according to **harmonic oscillator approximation** (i.e., an approximation on the assumption that the restoring force which restores a displacement of a nucleus from its equilibrium position complies with the Hooke's law; vibrations in harmonic oscillator approximation are called harmonic vibrations), any vibrations of the molecule are expressed as composition of simple vibrations called **normal vibrations**. Normal vibrations are vibrations of nuclei within a molecule, and in the normal vibrations, translational motions and rotational motions of the molecule as a whole are excluded. In each normal vibration, all atoms vibrate with the same frequency (**normal frequency**), and they pass through their equilibrium positions simultaneously. Generally, a molecule with N atoms has $3N–6$ normal vibrations ($3N–5$ normal vibrations if the molecule is a linear molecule). Since normal vibrations are determined by the molecular structure, the atomic weight and the force constant, when these three are known, it is possible to calculate the normal frequencies and the normal modes.

2.2.2.1 A Vibration of a Diatomic Molecule

Let us consider a vibration of a diatomic molecule as the simplest example of molecular vibrations. A diatomic molecule has only one normal mode ($3 \times 2 - 5 = 1$); it is a **stretching vibration** where the molecule stretches and contracts (Fig. 2.6a). One can delineate the stretching vibration using classic mechanics. Assuming that the nuclei are masses, m_1 and m_2, and the chemical bond is the "spring" with spring constant k following the Hooke's law (Fig. 2.6b), the vibration of the molecule can be explained in accordance with classic mechanics. The classic mechanical equation of vibration of a diatomic molecule can be solved by a few methods, but here we use **a Lagrange's equation of motion**, which is equivalent to Newton's equation of motion.

We assume that the masses m_1 and m_2 deviate Δx_1 and Δx_2, respectively, from their equilibrium positions. Then, the potential energy of the system shown in Fig. 2.6b is:

$$V = \frac{1}{2}k(\Delta\chi_2 - \Delta\chi_1)^2 \tag{2.5}$$

Meanwhile, the kinetic energy of the system is:

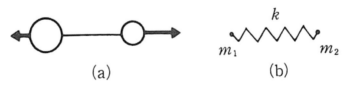

(a)   (b)

Fig. 2.6 a A stretching mode of a diatomic molecule. **b** A model for a diatomic molecule (two masses combined by a spring)

$$T = \frac{1}{2}m_1\dot{\chi}_1^2 + \frac{1}{2}m_2\dot{\chi}_2^2, \quad \text{where} \left(\dot{\chi}_i = \frac{d\chi_i}{dt}\right). \tag{2.6}$$

Now that V and T are known, motions of the system can be determined by solving a Lagrange's equation of motion:

$$\frac{d}{dt}\left(\frac{\partial T}{\partial \dot{\chi}_i}\right) + \frac{\partial V}{\partial \chi_i} = 0 \tag{2.7}$$

Note that Lagrange's equation of motion is more convenient in discriminating the translational motion and the vibrational motion. Before solving the Lagrange's equation of motion, let us introduce new coordinates Q and X

$$Q = \Delta\chi_2 - \Delta\chi_1 \tag{2.8}$$

$$X = \frac{m_1\Delta\chi_1 + m_2\Delta\chi_2}{m_1 + m_2} \tag{2.9}$$

$$\mu = \frac{m_1 m_2}{m_1 + m_2} \quad \text{(where } \mu \text{ is a reduced mass)} \tag{2.10}$$

Now, Q is a coordinate regarding a displacement of a distance between the two masses, while X is a coordinate regarding a displacement of the center of gravity of the system. Using Q and X, the potential energy V and the kinetic energy T are written as:

$$T = \frac{1}{2}\mu\dot{Q}^2 + \frac{1}{2}(m_1 + m_2)\dot{X}^2 \tag{2.11}$$

$$V = \frac{1}{2}kQ^2 \tag{2.12}$$

We substitute V and T in the Lagrange's equation of motion (2.7). First, applying to the coordinate X ($x_i = X$), we obtain

$$\ddot{X} = 0 \tag{2.13}$$

This expresses a free translational motion which is not bounded by the potential energy. On the other hand, from the Lagrange's equation of motion regarding the coordinate Q ($x_i = Q$), we get

$$\mu\frac{d^2Q}{dt^2} + kQ = 0 \tag{2.14}$$

From the differential equation like Eq. (2.14), we can find a solution as the follows:

$$Q = Q_\circ \cos 2\pi \nu t \qquad (2.15)$$

Equation (2.15) implies that the system illustrated in Fig. 2.6b has a simple harmonic motion with the frequency ν and the amplitude Q_0. Substituting Eq. (2.15) in Eq. (2.14),

$$\left(-4\pi^2 \mu \nu^2 + k\right)Q = 0 \qquad (2.16)$$

Finally, we get the frequency of the spring as:

$$\nu = \frac{1}{2\pi}\sqrt{\frac{k}{\mu}} \qquad (2.17)$$

The frequency of the spring corresponds to that of the molecular vibration, and the spring constant parallels to the force constant of the chemical bond, and hence, it can be seen from Eq. (2.17) that the frequency of the molecular vibration is proportional to the square root of the force constant and inversely proportional to the square root of the reduced mass of the atoms. It can be seen from Eq. (2.17) that the stronger a chemical bond is and the smaller the masses of atoms are, the larger the stretching frequency of a molecule is. H_2, which has small masses of atoms and relatively small force constant, gives the highest frequency among the all diatomic molecules (4160 cm^{-1}). The frequency of a vibrational more higher than 4000 cm^{-1} is only this one by H_2. This band is not IR active but Raman active, so that it cannot be observed in an IR spectrum. As a result, all bands due to all fundamentals appear below 4000 cm^{-1} in the IR spectra. This is the reason why 4000 cm^{-1} is the border between IR and NIR regions.

2.2.2.2 Quantum Mechanical Treatment of a Vibration of a Diatomic Molecule

Energy levels of vibrations of diatomic molecules can be described using quantum mechanics. In quantum mechanics, the first step is to write down a Schrödinger's equation, $\hat{H}\Psi = E\Psi$. The second step is to solve the equation to calculate an eigen value and an eigen function. In terms of classic mechanics, the total energy H of a vibration of a diatomic molecules is the sum of a kinetic energy $1/2\mu\,\dot{Q}^2$ (Eq. 2.11) and a potential energy $(1/2)k\,Q^2$ (Eq. 2.12),

$$H = T + V = \frac{1}{2}\left(\mu\dot{Q}^2 + kQ^2\right) \qquad (2.18)$$

Replacing $\dot{Q}$ with an operator $-ih/2\pi \cdot d/d\,Q, \hat{H}$ is calculated as:

$$\widehat{H} = -\frac{h^2}{8\pi^2\mu}\frac{d^2}{dQ^2} + \frac{1}{2}kQ^2 \tag{2.19}$$

Now, we got Hamiltonian. Substituting this in $\widehat{H}\Psi = E\Psi$ and processing the formula, a Schrödinger equation on harmonic oscillator of a diatomic molecule is obtained.

$$\frac{d^2\psi}{dQ^2} + \frac{8\pi^2\mu}{h^2}\left(E - \frac{1}{2}kQ^2\right)\psi = 0 \tag{2.20}$$

It is not easy to solve this differential equation, but one can do it rigorously. As how to solve this equation is described in detail in a number of textbooks, we will explain only results. Formula 2.20 yields a solution only to the following eigen value E_v:

$$E_v = \left(v + \frac{1}{2}\right)h\nu \tag{2.21}$$

where v is a quantum number of a vibration ($v=$ 0, 1, 2,…). It can be seen from Eq. 2.21 that under the harmonic oscillator approximation, the energies take discreet values and their spacings are equal. Figure 2.7 shows the energy levels of vibration of a diatomic molecule. It is noted that the lowest vibrational energy is not 0 but $E_0 = 1/2\ h\nu$. E_0 is called zero point energy. Energies have discrete values; $E_1 = 3/2\ h\nu$, $E_2 = 5/2\ h\nu$, $E_3 = 7/2\ h\nu$,…, and an energy difference between adjacent energy levels is always $h\nu$.

An eigen function to each value of E_v is expressed as:

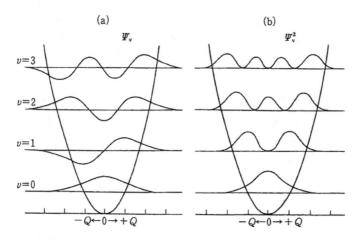

Fig. 2.7 Potential energy curve for a harmonic oscillator and allowed energy levels. **a** wave functions and **b** probability density functions of the harmonic oscillator

$$\psi_\upsilon = N_\upsilon H_\upsilon(\sqrt{\alpha}\,Q)\exp\left(-\frac{\alpha Q^2}{2}\right) \tag{2.22}$$

where N_υ denotes a normalization constant, H_υ is a Hermite polynomial, and $\alpha = \frac{2\pi\sqrt{\mu k}}{h}$

Wave functions to $\upsilon = 0$, 1, and 2 are as follows.

$$\psi_0 = (\alpha/\pi)^{1/4}\exp(-\alpha Q^2/2)$$
$$\psi_1 = (\alpha/\pi)^{1/4}(2\alpha)^{1/2}Q\exp(-\alpha Q^2/2)$$
$$\psi_2 = (\alpha/\pi)^{1/4}\left(1/\sqrt{2}\right)(2\alpha Q^2 - 1)\exp(-\alpha Q^2/2) \tag{2.23}$$

These formulas clearly show that a wave function of harmonic oscillator is an even function when a quantum number is an even number but is an odd function when a quantum number is an odd number. Figure 2.7 shows a potential energy, (a) wave functions, Ψ_υ, (b) probability density function, Ψ_υ^2 and energy eigen values, E_υ, of the harmonic oscillator.

2.2.2.3 Vibrations of Polyatomic Molecules

As examples of vibrations of polyatomic molecules let us consider normal vibrations of carbon oxide and water; these molecules are examples of linear and non-linear triatomic molecules, respectively. CO_2 has $3 \times 3-5=4$ normal vibrations. Figure 2.8 exhibits its four normal modes in CO_2, 1, 2, 3a and 3b. The normal vibrations 1 and 2 are vibrations where two CO bonds stretch and contract in phase (1) and out of phase (2), respectively, called **symmetric** and **anti-symmetric stretching vibrations**. Meanwhile, the vibrations 3a and 3b are both vibrations that the angle of OCO changes and called **bending vibrations**. While the vibrations 3a and 3b are independent of each other, energies required for the vibrations are principally equal

Fig. 2.8 Normal modes in CO_2. (+ and – denote vibrations going upward and downward, respectively, in the direction perpendicular to the paper plane). 1, symmetric stretching vibration. 2, antisymmetric stretching vibration. 3a, 3b, degenerate bending vibrations

1

2

3a

3b

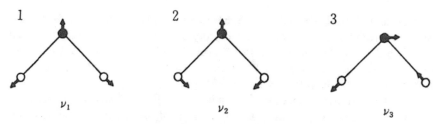

Fig. 2.9 Normal modes of vibration of water. 1: symmetric stretching vibration (v_1). 2: bending vibration (v_2). 3: antisymmetric stretching vibration (v_3)

to each other, only with planes of the vibrations differing 90 degrees from each other. That is, the two vibrations, 3a and 3b, have exactly the same energy. Such vibrations which have principally the same energy are called **degenerate vibrations.**

To know whether the normal vibrations 1, 2, 3a and 3b are IR active or not, we have to examine a change in the electric dipole moment at an equilibrium position $(\partial \mu_x / \partial \ Q)_0$. In the normal vibration 1, the electric dipole moment is always 0. Hence, the normal vibration 1 is IR inactive. Conversely, the electric dipole moment largely changes in the normal vibration 2, and thus, it is IR active. In a similar manner, the normal vibrations 3a and 3b accompany a change in the electric dipole moment, and therefore, are IR active. By the way, with respect to a molecule such as a CO_2 molecule which has the center of symmetry, a general rule holds true that an IR active vibration is a Raman inactive and a Raman active vibration is IR inactive. This rule is called **the mutual exclusion rule**.

Water, being a nonlinear triatomic molecule, has three normal vibrations as shown in Fig. 2.9. Normal vibrations, 1, 2, and 3 (v_1, v_2, and v_3) are named symmetric stretching, bending, and antisymmetric stretching modes. The normal vibrations 1 and 3 have different frequencies from each other, because of different $H_1 \ldots H_2$ interactions between the two vibrations. The three modes are all IR active but their first overtones are inactive. One can understand if overtones and combinations are active or inactive based on group theory. Bands due to their first overtones are very weak in NIR spectra, being almost impossible to be identified. Bands due to water observed in the NIR region are all due to combinations such as $v_1 + v_3$ and $v_2 + v_3$ (Fig. 2.1).

Both in the cases of CO_2 and H_2O molecules, the frequencies of stretching vibrations are larger than that of a bending vibration. This indicates that the stretching vibrations require larger energies than the bending vibration.

2.2.2.4 Group Frequencies

In general, **group frequencies** are useful to consider vibrations of a polyatomic molecule. Group frequencies are vibrations of functional groups such as $C = O$ stretching vibration of a carbonyl group, stretching vibration of an OH group, and symmetric and antisymmetric vibrations of a CH_2 group. The concept of group frequencies hold truth when certain normal vibrations are determined substantially by movements of two or more atoms (atomic group). Group frequencies play prominent roles in analysis of IR and Raman spectra. And even for NIR spectroscopy the idea of group frequency is useful 1–7 For example, NIR spectra show bands due to the overtones and combinations of CH_2 and CH_3 groups.

Next, let us consider vibrations of atomic groups. Figure 2.10 displays six vibrational modes of an AX_2 group (e.g., CH_2, NH_2). Of the six, two vibration modes are stretching vibrations, one being symmetric stretching vibration and the other antisymmetric stretching vibration. The remaining four are bending vibrations, i.e., **scissoring, rocking, wagging**, and **twisting vibrations**. Among the four bending vibrations, scissoring and rocking vibrations are bending vibrations in the plane of CH_2 (**in-plane vibrations**), while wagging and twisting vibrations are vibrations which displace vertically to the plane of CH_2 (**out-of-plane vibrations**).

The idea of group frequencies is beneficial ever for a very complex molecule such as a protein and a polymer. Let us consider normal vibrations of an amide group as an example. Normal vibrations of an amide group have been calculated in detail, taking N-methylacetamide (Fig. 2.11) as a model of the amide group. Considering a methyl group as one atom, N-methylacetamide is a six-atom molecule, and hence, has twelve normal vibrations (3×6-$6 = 12$). Of the twelve, the normal vibrations shown in Fig. 2.11 are amides I, II and III modes which are key vibrations for studying the structure of proteins and nylons. As clearly seen in Fig. 2.11, the amide I has a strong

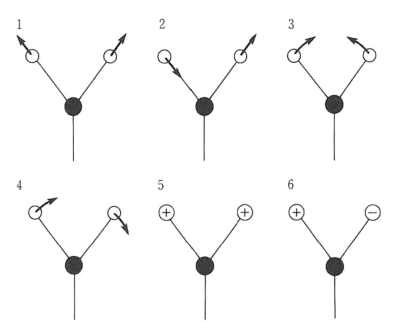

Fig. 2.10 Vibrations of AX_2 group. 1: symmetric stretching vibration. 2: antisymmetric stretching vibration. 3: scissoring vibration. 4: rocking vibration. 5: wagging vibration. 6: twisting vibration

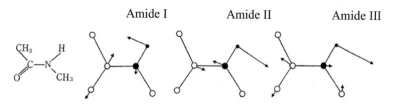

Fig. 2.11 Three normal vibrations of N-methylacetamide which is a model of an amide group

characteristic of $C = O$ stretching vibration. Meanwhile, the amides II and III are coupling modes of C-N stretching vibrations and N-H in-plane bending vibrations. Of the three, the amides I and II appear strongly in IR spectra, and in Raman spectra the amide I and III appear intense. The amide I, II and III bands of proteins are found generally in the regions of 1690-1620 cm^{-1}, 1590-1510 cm^{-1} and 1320–1210 cm^{-1}, respectively. These modes are known to sensitively reflect secondary structures of polyaminoacids, peptides, and proteins. In NIR spectra bands due to the overtones and combinations of amide modes such as the combination of NH stretching mode and Amide II appear.

2.2.2.5 Quantum Mechanical Treatment of Vibrations of Polyatomic Molecules

Now, let us study vibrations of polyatomic molecules using quantum mechanics. A kinetic energy T and a positional energy V are expressed as:
Therefore, a total energy H is:

$$T = \frac{1}{2}\sum_{i=1}^{n}\dot{Q}_i^2 \tag{2.24}$$

$$V = \frac{1}{2}\sum_{i=1}^{n}\lambda_i Q_i^2 \tag{2.25}$$

Therefore, a total energy H is:

$$H = T + V = \frac{1}{2}\sum_{i=1}^{n}\dot{Q}_i^2 + \frac{1}{2}\sum_{i=1}^{n}\lambda_i Q_i^2 \tag{2.26}$$

Replacing $\dot{Q}$ with $-ih/2\pi \cdot d/d\, Q$ again and calculating $\hat{H}$, we can obtain a Schrodinger equation of vibrations of polyatomic molecules.

$$-\frac{h^2}{8\pi^2}\sum_{i=1}^{n}\frac{\partial^2\psi}{\partial Q_i^2} + \frac{1}{2}\sum_{i=1}^{n}\lambda_i Q_i^2\psi = E\psi \tag{2.27}$$

As normal vibrations are independent of each other, the above formula can be separated into n wave equations, respectively, corresponding to the respective normal vibrations, an eigen value E_v is expressed as the sum of eigen values E_i of the respective normal vibrations, and an eigen function ψ_v is given as a product of eigen functions ψ_i representing the respective normal vibrations. Since Formula 2.27 has the same style as Formula 2.20, the eigen value E_i is also the same as Formula 2.21.

$$E_i = \left(v_i + \frac{1}{2} \right) h v_i \qquad (2.28)$$

Therefore, a total of vibrational energies whose frequencies are $v_1, v_2, \ldots, v_n$ is:

$$E_v = E_1 + E_2 + \cdots \cdots E_n = \left(v_1 + \frac{1}{2} \right) h v_1$$
$$+ \left(v_2 + \frac{1}{2} \right) h v_2 + \cdots \cdots + \left(v_n + \frac{1}{2} \right) h v_n \qquad (2.29)$$

The lowest ground state of water can be represented as (0,0,0), and (1, 0, 0), (0, 1, 0), and (0, 0, 1) denote fundamental states where v_1, v_2, and v_3, respectively, have a quantum number of 1. Transitions between the lowest ground state and the fundamental levels are called **fundamentals**. Next, (2, 0, 0), (0, 2, 0), and (0, 0, 2) represent states where v_1, v_2, and v_3 have a quantum number of 2, respectively, and are called **overtone levels**. (3, 0, 0)… are also overtone levels. **Overtones** are transitions between the lowest ground state and these overtone levels. Combination mode levels are levels, such as (1, 0, 1) and (0, 1, 1), where two or more normal vibrations are excited. Transitions between the lowest ground state and the combination mode levels are called **combination modes**.

2.2.3 Anharmonicity

Until now, we have treated molecular vibrations as a harmonic oscillator. However, in reality, the harmonic oscillator model is not a good model for molecular vibrations except for the vicinity of the bottom of a potential energy curve. If the harmonic oscillator model were correct, molecules should never dissociate no matter how large the amplitude is (Fig. 2.7). Therefore, it is necessary to consider a potential energy function $V(Q)$ (Q denotes an inter-nuclear distance) which more accurately expresses vibrations of molecules. In accordance with our instinct, $V(Q)$ must be such a function which rapidly increases when $Q < <0$ but gradually comes close to a dissociation energy, De, where $Q \gg \varrho_e$ (ϱ_e is an equilibrium distance) holds. As a function which satisfies this condition, a Morse's function expressed as below is well-known:

Fig. 2.12 Morse's function

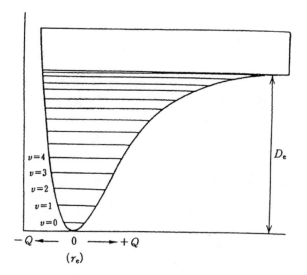

$$V(\varrho) = D_e\big[1 - \exp\{-a(\varrho - \varrho_e)\}\big]^2 \tag{2.30}$$

In Formula 2.30, ϱ is an inter-nuclear distance and a is a constant. This function was proposed by P. M. Morse in 1929. Figure 2.12 deliniates the Morse's function. Assuming that $Q(=\varrho - \varrho_e)$ is always small and expanding $V(\varrho)$ by a Taylor's series into a polynomial with respect to Q in the vicinity of ϱ_e,

$$V(\varrho) = V(\varrho_e) + \left(\frac{\partial V}{\partial \varrho}\right)_{\varrho_e} Q + \frac{1}{2}\left(\frac{\partial^2 V}{\partial \varrho^2}\right)_{\varrho_e} Q^2 + \frac{1}{6}\left(\frac{\partial^3 V}{\partial \varrho^3}\right)_{\varrho_e} Q^3$$

$$+ \frac{1}{24}\left(\frac{\partial^4 V}{\partial \varrho^4}\right)_{\varrho_e} Q^4 + \cdots\cdots \tag{2.31}$$

As the first term on the right-hand side is a constant term, this term is regarded 0. With respect to the second term as well, since V is extremely small to ϱ_e, the second term is also regarded 0. Now, ignoring the fourth and the higher-order terms and applying $(\partial^2 V/\partial Q^2)_{\varrho e} = k$, the following formula holds:

$$V(\varrho) = \frac{1}{2}kQ^2 \tag{2.32}$$

In other words, the Morse's function is equivalent to a function which expresses harmonic oscillator approximation in the region close to the equilibrium inter-nuclear distance ϱ_e (Second derivative on Formula 2.30 provides $k = 2a^2 D_e$).

A potential energy V is generally expressed as:

$$V = k_2 Q^2 + k_3 Q^3 + k_4 Q^4 + \cdots \cdots \tag{2.33}$$

The high-order terms such as Q^3 and Q^4 are called **anharmonic terms**. Calculating an eigen value E'_v considering up to the Q^3-term, we obtain,

$$E'_v = \left(v + \frac{1}{2}\right)h\nu_e - \left(v + \frac{1}{2}\right)^2 h\nu_e \chi_e \tag{2.34}$$

where $\nu_a = a/\pi \sqrt{D_e/2\pi}$. The symbol χ_e is a constant called an **anharmonic constant**. One can estimate the degree of anharmonicity from the value of this constant. Table 2.2 summarizes the values of anharmonic constants for several diatomic molecules. The constant χ_e becomes large for a molecule with a hydrogen atom which has a light mass while it is much smaller for molecules which do not have a hydrogen atom (Table 2.2). Since the anharmonic constant χ_e holds the following relationship with respect to a, D_e, etc., one can calculate the shape of a Morse's function and a dissociation energy of the molecules, etc.

$$\chi_e = \frac{h\nu_e}{4D_e} = \frac{ha}{4\pi \sqrt{2\mu D_e}} \tag{2.35}$$

From Formula 2.34, it is possible to calculate an energy difference, ΔE_v, between energy levels of vibrational quantum numbers v and $v + 1$.

$$\Delta E_{(v \to v+1)} = h\nu_e - 2h\nu_e x_e (v + 1) \tag{2.36}$$

Formula 2.36 indicates that the larger v is, the smaller ΔE_v is (Fig. 2.12) and the larger x_e is, the smaller ΔE_v is. In this formula, a transition $v = 0 \to 1$ is:

Table 2.2 Values of anharmonic constant for several diatomic molecules

Molecule	Anharmonic constant
H_2	0.02685
D_2	0.02055
HF	0.02176
HCl	0.01741
HBr	0.01706
HI	0.01720
N_2	0.006122
O_2	0.007639
Cl_2	0.007081
I_2	0.002857
NO	0.007337

$$\Delta E_{(0\to1)} = h\nu_e - 2h\nu_e x_e = h\nu_e(1 - 2x_e) \tag{2.37}$$

Thus, $\Delta E_\nu = h\nu_e$ does not hold. For the IR, NIR, and Raman spectra, we always use a frequency in a unit of cm^{-1} as $\tilde{\nu} = \nu/c$. The value $\tilde{\nu}_{obs} = \Delta\tilde{E}_{(0\to\nu)}$ (which is an observed value in the unit of cm^{-1}) is obtained as

$$\tilde{\nu}_{obs} = \Delta\tilde{E}_{(0\to\nu)} = \tilde{\nu}_e\nu - \chi_e\tilde{\nu}_e\nu(\nu + 1) = \tilde{\nu}_e\nu[1 - \chi_e(\nu + 1)] \tag{2.38}$$

We will now describe a method of calculating ν_e from $\tilde{\nu}_{obs}$. HCl yields a strong band at 2886 cm^{-1} due to a fundamental ($\nu = 0$ to 1) and a weak band at 5668 cm^{-1} due to a first overtone ($\nu = 0$ to 2). From these observed values one can calculate an absorption wavenumber ν_e and an anharmonic constant χ_e.

With respect to $\nu = 0\to1$ and $\nu = 0\to2$,

$$\Delta\tilde{E}_{\nu(0-1)} = \tilde{\nu}_e(1 - 2x_e)$$
$$\Delta\tilde{E}_{\nu(0-2)} = 2\tilde{\nu}_e(1 - 3x_e) \tag{2.39}$$

Therefore,

$$2886cm^{-1} = \tilde{\nu}_e(1 - 2x_e)$$
$$5668cm^{-1} = 2\tilde{\nu}_e(1 - 3x_e) \tag{2.40}$$

Solving these simultaneous equations, we obtain $\chi_e = 0.0174$ and $\tilde{\nu}_e = 2990\,cm^{-1}$. We must consider $\tilde{\nu}_e$ to discuss the strength of a chemical bond, because considering $\tilde{\nu}_{obs}$ is not enough for this purpose.

2.2.4 Overtones and Combination Modes

It is anharmonicity that permits overtones and combination modes to be observed. Let us consider selection rules of IR spectroscopy once more. This time, we will consider anharmonicity on a dipole moment.

$$\mu_x = (\mu_x)_0 + \left(\frac{\partial\mu_x}{\partial Q}\right)_0 Q + \frac{1}{2}\left(\frac{\partial^2\mu_x}{\partial Q^2}\right)_0 Q^2 + \cdots\cdots \tag{2.41}$$

$$(\mu_x)_{nm} = (\mu_x)_0 \int \psi_n\psi_m dQ + \left(\frac{\partial\mu_x}{\partial Q}\right)_0 \int \psi_n Q\psi_m dQ$$
$$+ \frac{1}{2}\left(\frac{\partial\mu_x^2}{\partial Q^2}\right)_0 \int \psi_n Q^2\psi_m dQ + \cdots\cdots \tag{2.42}$$

The third term of Formula 2.42 has a value other than 0 when $(\partial^2 \mu_x / \partial\, Q^2) \neq 0$ and $\int \psi_n\, Q^2 \psi_m \mathrm{d}\, Q \neq 0$ both hold. The latter integral has a value other than 0 when $\upsilon' = \upsilon$ and $\upsilon \pm 2$. Therefore, even a first overtone is not forbidden any more if we include the term of Q^2. Similarly, second, third… overtones are not forbidden any more as we take higher-order terms into consideration. However, the intensities of these overtones are far weaker than those of fundamentals. The frequencies of first, second, third… overtones are smaller than double, triple, quadruple of the frequencies of fundamentals, respectively, as already described. This is because the differences between the vibrational energy levels become narrower as the quantum number υ increases, as clearly depicted in Fig. 2.12 and indicated by Formula 2.36. Anharmonicity excludes combination modes as well from those forbidden in a similar manner. The intensities of combination bands are also weak.

References

1. H. W. Siesler, Y. Ozaki, S. Kawata, H. M. Heise, *Near-Infrared Spectroscopy* (Wiley-VCH, 2002)
2. Y. Ozaki, W. F. McClure, A. A. Christy, eds., *Near-Infrared Spectroscopy in Food Science and Technology* (Wiley, 2007)
3. Y. Ozaki, Near-infrared spectroscopy-Its versatility in analytical chemistry. Anal. Sci. **28**, 545–563 (2012)
4. D. A. Burns, E. W. Ciurczak eds., *Handbook of Near-Infrared Analysis*, 3rd edn. (Practical Spectroscopy) (CRC Press, 2007)
5. Y. Ozaki, C. W. Huck, K. B. Beć, Near-IR spectroscopy and its applications, in *Molecular and Laser Spectroscopy: Advances and Applications*, edited by V. P. Gupta (2017, Elsevier), pp. 11–38
6. J. Workman, Jr., L. Weyer, *Practical Guide and Spectral Atlas for Interpretive Near-Infrared Spectroscopy*, 2nd edn. (CRC Press, 2012)
7. M.A. Czarnecki, Y. Morisawa, Y. Futami, Y. Ozaki, Advances in molecular structure and interaction studies using near-infrared spectroscopy. Chem. Rev. **115**, 9707 (2015)
8. K.B. Bec, D. Karczmit, M. Kwaśniewicz, Y. Ozaki, M.A. Czarnecki, Overtones of $\nu C \equiv N$ Vibration as a Probe of Structure of Liquid CH_3CN, CD_3CN, and CCl_3CN: Combined Infrared, Near-Infrared, and Raman Spectroscopic Studies with Anharmonic Density Functional Theory Calculations. J. Phys. Chem. A **123**, 4431–4442 (2019)
9. K.B. Bec, Y. Futami, M.J. Wojcik, Y. Ozaki, Spectroscopic and theoretical study in the nearinfrared region of low concentration aliphatic alcohols. Phys. Chem. Chem. Phys. **18**, 13666–13682 (2016)
10. Y. Morisawa, S. Nomura, K. Sanada, Y. Ozaki, Monitoring of a calcination reaction of high reflective green-black (HRGB) pigments by using near-infrared electronic spectroscopy: calcination temperature-dependent crystal structural changes of their components and calibration of the extent of the reaction. Appl. Spectrosc. **66**, 666–676 (2012)
11. Y. Hu, J. Zhang, H. Sato, Y. Futami, I. Noda, Y. Ozaki, C-H···O = C hydrogen bonding and isothermal crystallization kinetics of poly(3-hydroxybutyrate) investigated by near-infrared spectroscopy. Macromolecules **39**, 3841–3847 (2006)
12. H. Sato, M. Nakamura, A. Padermshoke, H. Yamaguchi, H. Terauchi, S. Ekgasit, I. Noda, Y. Ozaki, Thermal behavior and molecular interaction of poly(3-hydrobutyrate-co-hydroxyhexanoate) studied by wide-angle X-ray diffraction. Macromolecules **37**, 3763–3769 (2004)

13. H. Sato, R. Murakami, A. Padermshoke, F. Hirose, K. Senda, I. Noda, Y. Ozaki, Infrared spectroscopy studies of CH···O hydrogen bondings and thermal behavior of biodegradable poly(hydroxyalkanoate). Macromolecules **37**, 7203–7213 (2004)

Chapter 3
Theoretical Models of Light Scattering and Absorption

Kevin D. Dahm and Donald J. Dahm

Abstract When light interacts with a single particle, there are three possible outcomes: absorption, scattering, or transmission. In spectroscopy, one measures the remission from and/or transmission through a macroscopic sample. Such a sample might contain countless locations at which there is a change in refractive index, each of which gives rise to scattered light. This fact poses a challenge in building theoretical models applicable to spectroscopy: even if our theoretical understanding of single interactions is very good, the number of individual interactions is typically too big to make accounting for all of them realistic. This chapter presents an overview of modeling strategies that can be of use in near infrared spectroscopy. Recognizing that no one approach is uniformly applicable, care is taken to call attention to assumptions made in each modeling approach and limitations that are imposed by these assumptions.

Keywords Absorption · Absorbance · Scattering · Remission · Transmission · Diffuse reflectance

This chapter explores the physical behavior of light. When light interacts with a material, it can:

(1) Continue in the direction it was going, (2) be absorbed by the material, or (3) be diverted in a different direction and continue traveling along a new path. It is common, and broadly accurate, to refer to these three possible outcomes as (1) "transmission," (2) "absorption," and (3) "scattering."

This chapter examines several theoretical approaches to understanding and modeling these interactions. No one of these approaches can be considered definitive or universally applicable for spectroscopy. Care is taken to state the assumptions that underlie each approach, and the limitations that result from these assumptions.

K. D. Dahm (✉)
Rowan University College of Engineering, Glassboro, USA
e-mail: dahm@rowan.edu

D. J. Dahm
Retired, North Palm Beach, USA

© Springer Nature Singapore Pte Ltd. 2021
Y. Ozaki et al. (eds.), *Near-Infrared Spectroscopy*,
https://doi.org/10.1007/978-981-15-8648-4_3

3.1 Early Explorations of Absorption, Scattering, and Extinction

The eighteenth-century contributions of Pierre Bouguer were foundational in developing our current understanding of how light interacts with matter. Bouguer studied the phenomenon of light becoming dimmer as it passed through the atmosphere. He discovered a first-order logarithmic relationship between the remaining intensity of the light and the thickness of atmosphere it had penetrated [1]. Using modern terminology, one way to express this is:

$$\frac{I}{I_0} = \exp(-\in t) \tag{3.1}$$

In which I_0 is the intensity of the light at its source, I is the intensity of light that reaches the detector, and t is the thickness of atmosphere. The ε in Eq. 3.1 is a parameter that has different values for different wavelengths, but what specifically does it measure? Bouguer himself did not necessarily attribute the dimming of the light to a particular physical phenomenon. We now use the term "extinction" for the total observed attenuation of a beam, and thus ε quantifies the ability of the atmosphere to "extinguish" a specific wavelength of light. Indeed, ε has been termed the "extinction coefficient." [2]. We now understand that both absorption and scatter contribute to extinction in an experiment like Bouguer's. If one is looking at a distant object (e.g., the sun), any light that was either scattered from its original path or absorbed by the air molecules does not reach one's eyes.

Another foundational body of work was that of Beer in the nineteenth century, which demonstrated that chemical compounds have an ability to absorb light at particular wavelengths, and that the extent of the "absorbance" of the light is proportional to the concentration of the absorber. Mathematically, one can define absorbance as:

$$Absorbance = -log_{10}\frac{I}{I_0} \tag{3.2}$$

I_0 again represents the intensity of incident light, I represents the amount of light that reached the detector, and consequently, I/I_0 is the fraction of the incident light that penetrated the sample and was detected. When this fraction is 1, the "absorbance" is by definition 0. As the fraction of detected light decreases, the "absorbance" as defined in Eq. 3.2 increases, representing the inference that more light has been absorbed. Note that the absorbance is not equal to the "fraction of light that was absorbed"; absorbance can be greater than one and it approaches infinity as I/I_0 approaches zero. The quantity defined in Eq. 3.2 is sometimes called the "decadic absorbance" to emphasize that base 10 logarithms were used, since it is also possible to use natural logarithms.

"Beer's Law" can be expressed as [3]:

$$Absorbance = -log_{10}\frac{I}{I_0} = \kappa ct \qquad (3.3)$$

In which, t represents the thickness of the sample, c represents the concentration of the absorber, and κ is often termed the "absorptivity" or "molar absorptivity" of the material. The absorbance is a dimensionless quantity and the units of κ depend upon the specific units used to express thickness and concentration. Units that are often used are cm for thickness, mol/L for concentration, and L/mol·cm for molar absorptivity.

Note that in Bouguer's experiments, the absorbing "sample" was the atmosphere. A gaseous medium like the atmosphere and a liquid solution are both examples of systems that can plausibly be considered uniform in composition. We now understand that the relationships in Eqs. 3.1 and 3.3 can be derived from continuous mathematical functions. These can only be expected to apply to samples that are uniformly distributed, and therefore reasonably modeled as continuous. Note, too, that Bouguer did not have the ability to vary the concentrations of the absorbers in his experiments. We now understand that the ε in Eq. 3.1 embodies both the concentrations of the gases in the atmosphere (c) and their ability to extinguish light (κ), consistent with Eq. 3.3. However, the parameter κ merits further examination.

Consider an experimental arrangement like that shown in Fig. 3.1, in which a "small area" detector is aligned with the incident beam. The most straightforward case is that of a "clear solution"; one in which scattering can be considered negligible. In such a solution, absorption is the only process that leads to attenuation. However, absorption is a molecular-level phenomenon, and normally molecules do not exist in isolation. Intermolecular interactions such as induced dipoles can impact the ability of a molecule to absorb light. Thus, one cannot simply assume κ has a single, constant value for a compound. For example, a compound in a liquid solution could have a different κ when it is dissolved in water vs. when it is in a non-polar solvent. Further, one must distinguish between dilute solutions and highly concentrated ones. In a dilute solution, one can reasonably assume that each solute molecule is surrounded by and interacting with only solvent molecules. In a concentrated solution, solute molecules interact significantly with each other. Experimentally, it has been observed

Fig. 3.1 Transmission experiment with a small area detector

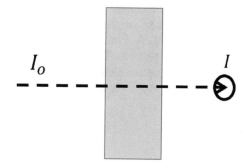

that real solutions can exhibit "departures" from Beer's Law at high concentrations. We understand these departures as symptoms of the fact that in using Beer's Law κ is generally assumed constant, when in reality, there is reason to expect that κ would become dependent upon composition as c increases and solute–solute interactions become more significant.

Now imagine we are still using the experimental arrangement in Fig. 3.1, but the sample produces both scattering and absorption. Like the sunlight in Bouguer's experiments, any light that is *either* absorbed or scattered will fail to reach the detector. Thus, both scattering and absorption contribute to the extinction modeled by κ. Here again, the distinction between dilute vs. concentrated solutions could complicate the use of Beer's Law.

By contrast, if we use an experimental arrangement like the one as shown in Fig. 3.2, the light that is scattered at least once but still exits the sample can be detected. A hemispherical detector can be used to measure as "transmission" (I in Eq. 3.3) all of the light that penetrates the sample, regardless of the specific path. The presence of scatter also gives rise to the phenomenon of "remission," which is light that emerges from the sample's front surface, and which can be quantified by another hemispherical detector. (In practice, the experiment pictured in Fig. 3.2 is most straightforwardly carried out using an integrating sphere.) The classical definition of "absorbance" in Eq. 3.2 does not acknowledge the phenomenon of remission. In most of the models we discuss in Sects. 3.6–3.10, we will instead consider A, R, and T, which we define as the fractions of incident light that were absorbed by, remitted from, and transmitted through the sample, respectively. Absorbance can then be more broadly defined as:

$$Absorbance = -log_{10}(1 - A) \tag{3.4}$$

Note that when R = 0, 1-A is equal to T, which is in turn equal to I/I_0. Thus, this definition is equivalent to Eq. 3.2 for the special case of a non-scattering sample.

In sum, Beer's Law is an equation that has great practical appeal, because it represents a simple linear relationship between a readily measured quantity (I/I_0) and the concentration c of the absorber, which is usually what we are trying to deduce in spectroscopy. However, Beer's Law can only be expected to be a good model in specific circumstances (clear solution, dilute solution, etc.). Furthermore, even when Beer's Law proves to be a good model, one must recognize the limitations of the value of κ. A chemical compound does not have a single "absorptivity" that is uniformly applicable. The value of κ depends upon factors (e.g., solvent, sample thickness, and sample geometry) that are specific to the context of an experiment and should only be considered valid in that specific context.

3.2   The Application of Spectroscopy

Section 3.1 introduced Beer's Law, which is likely to be the first model a person learning about spectroscopy will encounter. It is a simple equation that dates from the nineteenth century and still has practical value. Section 3.1 also introduced some of the reasons why models that are more complex are needed. In spectroscopy, our specific interest is in directing light at a sample, making measurements of the light that penetrates the sample and/or the light that is reflected from the sample, and then using those measurements to make deductions about the composition of the sample. Challenges stem from the fact that the sample is macroscopic in scale. A measurement of light that reaches a detector, such as I in Eqs. 3.1–3.3, is a single number that actually represents the net effect of countless molecular-level absorption and scattering interactions. In the context of a macroscopic sample, the phenomena of absorption and scattering strongly influence each other. As an example, in the experimental arrangement of Fig. 3.2, light that is scattered near the front of the sample penetrates the sample along a longer path than light that continues on the direct path of the incident beam. Thus, the scattered light has more opportunity to be attenuated by absorption than does the light on the original path.

Even the words "scattering," "absorption," and "transmission" can be problematic when applied to a macroscopic sample. Consider for example:

- Light that has been absorbed no longer exists and cannot be detected. When we "measure" absorption, we are actually measuring the amount of light we are able to detect, and assuming that the remainder of the original light was absorbed.
- The experimental arrangement in Fig. 3.1 is incapable of distinguishing between absorption and scatter. What is sometimes loosely called "absorption" in such an experiment is more properly termed "extinction."

Fig. 3.2 Experiment in which transmission and remission are both measured with hemispherical detectors

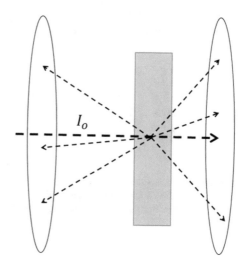

- The experimental arrangement in Fig. 3.2 does not distinguish between the light that truly was "transmitted" through the entire sample, and light that experienced one or more scattering interactions but still penetrated the entire thickness of the sample. The "transmission" that is measured is in reality the latter, or a combination of both.

The remainder of this chapter is broadly divided into two categories. Sections 3.3 through 3.5 discuss light and its microscopic interactions, such as light interacting with a single particle or a single surface. Sections 3.6 through 3.10 discuss strategies for modeling the net amounts of light absorbed by, transmitted through, and remitted from a macroscopic sample.

3.3   The Physics of Light

Light can be understood as electromagnetic radiation propagating through space as a wave. A changing electric field gives rise to a changing magnetic field, and vice versa, and the speed of light is the velocity of the resulting waves. The waves have electrical and magnetic vectors that oscillate, and the maximum extent of the oscillation is called the amplitude. The electric and magnetic oscillations are perpendicular to each other and also perpendicular to the direction of propagation, as illustrated in Fig. 3.3. The intensity of a beam of light is proportional to the square of the amplitude of the wave.

When light waves encounter each other, "interference" occurs and the resulting wave can be modeled as the vector sum of the two original waves. They can interfere:

- Constructively, meaning that the crests overlap each other and the amplitude of the resulting wave is equal to the sum of the amplitudes of the original waves.
- Destructively, meaning that the crest of one wave overlaps the trough of the other wave, and the result is a subtraction of amplitudes.

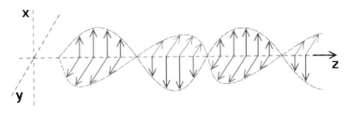

Fig. 3.3 Electromagnetic wave, propagating in the "z" direction, in which the electric field oscillates in the x-direction and the magnetic field oscillates in the y-direction. Adapted from Super-Manu Image:Onde electromagnetique.png, available at https://commons.wikimedia.org/w/index.php?curid=2107870

When light enters a medium other than a vacuum, it slows down. The frequency of the oscillation is maintained, but the wavelength changes. This is quantified by the traveling wave equation:

$$v = f\lambda \tag{3.5}$$

where v represents velocity, f represents frequency, and λ represents wavelength.

When light encounters a molecule, it can either be scattered or absorbed. The origin of the scatter is the charged particles within each atom. These particles vibrate in time with the electric vector of the incident light (electromagnetic radiation). Because electrons are orders of magnitude lighter than the nuclei, the electrons vibrate far more vigorously than do the nuclei and are the main source of scattered light. In "elastic" scatter, the accelerating (vibrating) charges emit radiation that has the same wavelength as the incident light, but it is emitted in all directions. There are multiple charged particles, and therefore multiple sources of scattering, within a single atom or molecule. The emissions from the various charged particles interfere with each other and give rise to a scattering pattern that is dependent upon the relative placement of the vibrating charges within the molecule. (There is also such thing as "inelastic scatter," in which the wavelength of the emitted radiation is not the same as that of the incident radiation. In the field of vibrational spectroscopy, inelastic scattering is primarily encountered in Raman spectroscopy [4]).

Turning to absorption, when light transfers energy to a material, it does so in discrete quantities of energy called "photons." A photon is sometimes envisioned as a "particle" of light, but this is a departure from the wave model of light we are describing here. We picture a beam of light as a wave that contains a certain amount of radiant energy, which can be sub-divided into a certain number of photons. The amount of energy in one photon is given by:

$$E = hf \tag{3.6}$$

Where E is the energy, h is Plank's constant, and f is the frequency. When SI units are used throughout, E is expressed in Joules, f is expressed in hertz, and h = 6.63 × 10^{-34} J·s.

3.4 Reflection and Refraction of Light at a Surface

In this section, we consider the case of a beam of light encountering a surface. When light is traveling through a medium, such as air or glass, it travels in a straight line. The speed of light is not uniform; it varies based upon the refractive index of the medium in which it is traveling. When light encounters a glass window, the surfaces of the glass are locations at which the light makes the transition from one medium to another (air to glass and glass to air) and the refractive index changes.

Fig. 3.4 Light encountering
a surface where the index of
refraction changes

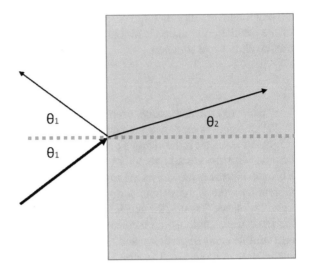

We define θ_1 as the "angle of incidence"; the angle at which the light strikes the surface. As illustrated in Fig. 3.4, a portion of the light is reflected from the surface and a portion continues into the new medium. Figure 3.4 also shows that the angle of reflection, for the reflected light, is identical to the angle of incidence. Light that enters the new medium changes velocity due to the change in refractive index and is diverted from its original path, according to Snell's Law [5]:

$$\frac{\eta_1}{\eta_2} = \frac{\sin \theta_2}{\sin \theta_1} \tag{3.7}$$

In which, η_1 and η_2 are the refractive indices of the two mediums, θ_1 is the angle of incidence, and θ_2 is the "angle of refraction" at which the light continues into the new medium. Note that in Fig. 3.4, θ_1 is pictured as greater than θ_2, which means that $\eta_1 < \eta_2$. A lower refractive index is associated with a higher velocity of the light.

An instructive special case is that of "directed normal illumination," in which the incident light is perpendicular to the surface it is striking, and thus $\theta_1 = 0$. Mathematically, this means that according to Snell's Law, regardless of the specific values of the refractive indices n_1 and n_2 (but assuming they are both finite numbers), then θ_2 must also be 0. Physically, this means that the transmitted light continues along its original path despite the change in refractive index, and the angle of reflection is also 0; the reflected light is still perpendicular to the surface but reverses direction. Here again, however, the distinction between the macroscopic and microscopic scales is important. If the surface is not smooth, then the incident beam may be perpendicular to the surface in a macroscopic sense, but at the microscopic scale, it is actually experiencing a range of angles of incidence, as illustrated in Fig. 3.5. The result is a broadening of both the reflected beam and the transmitted beam. A surface that is

Fig. 3.5 Light interacting with a rough surface

optically very rough and produces highly diffuse reflected light is called a "matte surface."

Note that in Figs. 3.4 and 3.5, the incident light is portrayed as individual rays encountering a surface that extends indefinitely. Thus, an important premise throughout this section was that *the width of the beam is small compared to the size of the object the beam encounters*. Section 3.5 considers the opposite case, in which a beam of light encounters an object whose diameter is small compared to the width of the beam. Thus, the object is surrounded by, or "bathed" in, the incident beam.

3.5 Scatter from a Particle that is Bathed in a Beam

In this section, we consider a beam of light that encounters a particle that is significantly smaller than the width of the beam, such that the particle is immersed in the beam. We will assume that the light is directed, meaning that all of the light is traveling in the same direction prior to encountering the particle. Since the particle is smaller than the beam, a portion of the light is unaffected by the particle. By contrast, in the previous section, the surface was larger than the beam, and all of the light was affected by the surface (either reflected or refracted).

In deciding how best to model the scatter (and absorption) from a particle, a primary consideration is how large is the particle compared to the wavelength of the incident light. Rayleigh scattering is applicable when the particle size is small (~1/10) compared to the wavelength. Scattering (and absorption) happens at the atomic level, but if a particle is very small compared to the wavelength of the incident light, one need not distinguish between the locations of the individual atoms. According to the Rayleigh formula, the intensity of scattered light is proportional to:

$$I \sim I_0 \left(\frac{1 + cos^2\theta}{R^2} \right) \left(\frac{1}{\lambda} \right)^4 \left(\frac{\eta^2 - 1}{\eta^2 + 2} \right)^2 \qquad (3.8)$$

In which:

- R represents the distance from the scattering center. The intensity of the scattered light drops off as the inverse square of the distance.

- θ is the scattering angle. While the intensity of the scattered light is a function of angle, the consequence of the $\cos^2$ functionality is that the scattering pattern is symmetrical, and essentially the same amount of light gets scattered "forward" as "backward," relative to the incident beam.
- λ is the wavelength, which is raised to the −4 power. This means that scatter is essentially zero at very high wavelengths.
- η is the refractive index. It is possible to incorporate absorption into the calculation by expressing the refractive index as a complex number, with the real component representing scatter and the imaginary component representing absorption.

For larger particles, modeling the particle as a single scattering center becomes unrealistic-one must distinguish between scatter from different locations on the particle. Given enough time and computing power, for a specific particle size and shape, it is theoretically possible to solve wave equations for scatter emanating from every point on the particle, sum these, and quantify the intensity of light emanating from the particle in any direction. In practice, the shape that has been studied the most is the sphere. Building a theoretical model of a sphere is simplified by the fact that a sphere presents the same dimensions (e.g., cross-sectional area, depth) to the beam regardless of the orientation of the beam. Mie scattering theory can be applied to spheres of any size, but is especially useful when the wavelength of the light and the size of the particle are of comparable magnitude, or when the particle is larger than the wavelength.

While Mie computations are complex, some outcomes will here be discussed qualitatively. Mie predicts vanishingly small scatter for very small particles and/or very large wavelengths, the latter result being consistent with the $(1/\lambda^4)$ functionality of the Rayleigh equation. The intensity of scatter increases as particle size becomes larger relative to wavelength and reaches a maximum when the circumference of the sphere is equal to the wavelength, as shown in Fig. 3.6. Scattering intensity oscillates as the ratio of circumference to wavelength increases further. Another result predicted by the Mie equations is that scatter from spheres is *not* isotropic. According to Mie scattering theory, the majority of the scattered light continues in a generally "forward" direction, as illustrated in Fig. 3.7. In a limiting case, the largest spheres will approximate the scattering pattern of a planar surface. A crucial point is that Mie's equations were derived specifically for spheres and cannot be applied to other shapes. Bass et al. have noted frequent misuse of the Mie theory, stating "in defiance of logic and history every particle under the sun has been dubbed a 'Mie scatterer,' and Mie scattering has been promoted from a particular theory of limited applicability to the unearned rank of general scattering process… Using Mie theory for particles other than spheres is risky, especially for computing scattering toward the backward direction." [5].

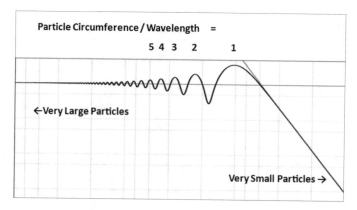

Fig. 3.6 Illustration of how scattering intensity changes with particle size according to Mie theory. The y-axis shows the log of relative scattering intensity, while the x-axis shows inverse log of particle circumference/wavelength. Adapted from a public domain image available at https://com mons.wikimedia.org/wiki/File:Radar_cross_section_of_metal_sphere_from_Mie_theory.svg

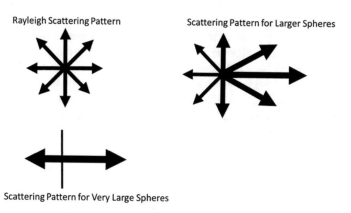

Fig. 3.7 Summary of the scattering patterns expected from different size spheres. The largest spheres will approximate the behavior of a planar surface

3.6   A Modeling Framework for Macroscopic Samples

The models described in Sects. 3.3 through 3.5 describe a variety of possible *single* interactions between light and matter. When a spectroscopic sample is made up of distinct particles, it is typically unrealistic to account for and sum the effects of every interaction with every individual particle. This section presents a framework for building models of particulate samples that has two aspects: the use of "plane parallel layers" and the "two-flux" model.

Figure 3.8 illustrates the notion of plane parallel layers. Each layer is a semi-infinite, rectangular slab. It has a finite thickness d in one direction, which in Fig. 3.8 is also the direction of travel of the incident beam. In the other directions normal to

Fig. 3.8 Sample composed
of plane parallel layers, in
which each layer absorbs
exactly half and transmits
exactly half of the light that
arrives at its front surface

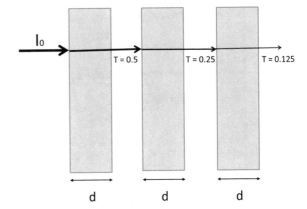

this, it is assumed to be infinite. Thus, each layer has a "front surface" and a "back surface" that extend indefinitely. The incident light arrives at the front surface of the first layer. Some fraction of light will be absorbed by, and some fraction transmitted through, the layer. Thus, some of the light emerges from the back of the first layer and reaches the front surface of the second layer. Envisioning a sample as a series of plane parallel layers in this manner is well established in the literature, with one early example having been published by Stokes [6].

Once we have divided the sample into layers, we seek to quantify the passage of light between the layers. To do this, we first introduce the "absorption coefficient," which will be defined as the fraction of light absorbed (A) by a thin layer of material, divided by the thickness (d) of that thin layer:

$$\mu_a = \frac{A}{d} \tag{3.9}$$

For the case of a <u>non-scattering sample</u>, light is attenuated by absorption in each layer, as illustrated in Fig. 3.8. Note that in Fig. 3.8, each layer absorbs exactly half of the light that reaches its front surface. Thus, the values of T as shown in Fig. 3.8 are discrete: $1, 0.5, 0.25, 0.125$, etc. This phenomenon can be generalized as:

$$T_n = (1 - A)^n = (1 - \mu_a d)^n \tag{3.10}$$

With T_n representing the transmission through n identical layers, and A representing the fraction of light absorbed by each individual layer, which is then related to the sample thickness d through Eq. 3.9.

A real sample has a specific finite thickness, which we will call t. We can imagine subdividing a sample into any number of identical plane parallel layers. As n, the number of layers, gets larger, the thickness d of each individual layer gets smaller (d $= t/n$). By applying the limit as n approaches ∞, we can derive a continuous equation that models the exponential fall-off in intensity with sample thickness:

$$T = \frac{I}{I_0} = exp(-\mu_a t) \tag{3.11}$$

Equation 3.11 is compatible with the Bouguer (3.1) and Beer (3.3) equations that opened this chapter, though the absorption coefficient μ_a is framed differently than either the ε in Eq. 3.1 or the κ in Eq. 3.3. Beer's Law, for example, commonly uses base 10 logarithms (as in Eq. 3.3) rather than natural logarithms. Beer's Law also separates the concentration and the molar absorptivity into two separate parameters, while the μ_a in Eq. 3.11 is a holistic coefficient that is effected by both of these factors.

In the absence of scatter, light only moves in one direction. It is attenuated by absorption, so the intensity of the light (T) is a function of position (d) as quantified in Eq. 3.11, but only one "flux" is needed to describe the process: at any given point, all of the light that has not been absorbed is moving forward along its original path. By contrast, in a scattering sample, light can be moving in literally any direction when it enters or exits a layer.

In a "two-flux" model of a scattering sample, we simply consider the light that enters and exits a particular layer as moving "forward" or "backward." If light strikes the front surface of a layer and exits the back surface, then we say it has been transmitted through the layer and moves "forward" into the next layer. The model does not distinguish between the various angles and paths the light could have followed through the layer. Similarly, if light strikes the front surface of a layer, reverses direction, and is re-emitted from the front surface, it contributes to the "backward" flux that enters the rear surface of the previous layer.

Thus, when one envisions a scatting sample as a series of plane parallel layers and applies a "two-flux" approximation, the resulting modeling framework is as summarized in Fig. 3.9. There are "forward" and "backward" fluxes arriving at each layer, "R" represents the fraction of incident light that emerges "backward" from the first layer and is therefore remitted from the sample as a whole, and "T" represents the fraction of incident light that emerges "forward" from the last layer and is therefore transmitted through the sample as a whole. The next few sections present some of the mathematical outcomes that can be obtained using this modeling framework.

Fig. 3.9 Schematic of a "two-flux" model applied to a scattering sample

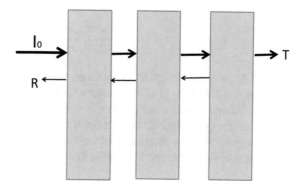

3.7   The Schuster and Kubelka–Munk Equations

The use of a two-flux model for modeling absorption and scattering, as outlined in the previous section, is well established in the literature. More than a century ago, Schuster applied a two-flux approximation in his classic work "Radiation Through a Foggy Atmosphere." [7]. One of Schuster's results described an "infinitely thick" sample, which means a sample that is thick enough that no light penetrates it (T = 0). This is a case that has practical significance in the application of spectroscopy: A sample that is too optically thick for a transmission experiment can still be analyzed by collecting remission data. For an infinitely thick sample, using Schuster's approach, Kortüm obtained [8]:

$$\frac{(1 - R_\infty)^2}{2R_\infty} = \frac{k}{s} \tag{3.12}$$

R_∞ represents fraction of incident light that is remitted from the infinitely thick sample, k represents an absorption coefficient, and s is a scattering coefficient, defined analogously to the absorption coefficient: It is the fraction of light scattered by a layer, divided by the thickness of the layer. (While the symbols k and s were used by Kortüm, μ_a and μ_s are now more commonly used symbols for absorption and scattering coefficients, as introduced in Eqs. 3.9–3.11).

In the derivations, Schuster and Kortüm also made use of the assumption of "isotopic scatter," which means that light is scattered equally in all directions. In the context of a two-flux model, this simply means that half of the scattered light moves "forward" and half "backward." In a two-flux model, the light that is scattered "forward" into the next layer is indistinguishable from light that is directly transmitted into the next layer.

Another well-known result derived using the two-flux approach is the Kubelka–Munk equation [9]. The equation was originally devised in an investigation of paint layers, a very different physical system than Schuster's "foggy atmosphere." Nonetheless, they too used a two-flux model and divided their sample into infinitesimally thin plane parallel layers, so they obtained a functionally identical equation, reported by Kortüm [8] as:

$$\frac{(1 - R_\infty)^2}{2R_\infty} = \frac{K}{S} \tag{3.13}$$

With K and S used to represent the absorption and scattering coefficients. Thus, the Schuster and Kubelka–Munk treatments both lead to the conclusion that remission from an infinitely thick sample depends upon the ratio of the absorption coefficient to the scattering coefficient, but not upon their specific magnitudes. However, this conclusion is subject to the limitations of the assumptions made in their derivations. Schuster articulated the approximation inherent in the "two-flux" model as follows:

The equations… have been deduced under the assumption that the radiation throughout the absorbing mass is uniformly distributed in such a way that it does not depend on the angle between any direction considered and the normal drawn toward the same side. This supposition is obviously incorrect, for it appears that, even if it were to hold at any surface, e.g., the first surface of the layer dx, absorption in that layer would destroy the uniformity owing to the greater absorption which the oblique rays suffer. To some extent, the effect of scattering would act in the sense of partly restoring the equality of distribution…

Kubelka published a later paper with a mathematical treatment that was intended to be more general [10]. Rather than using a two-flux model, he assumed that scatter was isotropic and accounted for the various angles of travel within a layer. He concluded that Eq. 3.13 was an "exact" mathematical solution only in two cases: When the incident light striking the sample was perfectly diffuse, and when the incident light was striking the front surface of the sample at an angle of exactly 60°.

Other authors have gone beyond the "two-flux" approach. Burger et al. used a three-flux approximation in which the three fluxes were each modeled at 120° from each other [11]. Giovanelli published exact solutions for the cases of directed and diffuse illumination encountering a semi-infinite slab [12]. However, even these more general treatments used continuous mathematics, which is itself a limitation when applied to particulate samples.

In the Shuster and Kubelka–Munk approaches, as in the derivations of Beer's Law and Eq. 3.11 for non-scattering samples, the individual plane parallel layers were assumed to be infinitesimal in thickness. Stated in physical rather than mathematical terms, it is assumed that the fractions of light absorbed and scattered by a single inter-action with a single layer are extremely small. In effect, the mathematics used treat the sample as a homogeneous continuum, in which either absorption or scattering can occur at any location. However, as outlined in Sect. 3.4, it is the discontinuities (surfaces) within a sample that are responsible for the phenomenon of scatter, so the assumption of a homogeneous continuum is a limitation in such a case. Kubelka and Munk in their original paper were investigating a system in which the individual particles that made up the sample were small, and therefore it was justifiable to model them mathematically as infinitesimal. Shuster was investigating a system in which the density of particles was extremely low, and it was therefore quite reasonable to say that only a tiny fraction of light would be absorbed or scattered within any single layer. However, what if no such justification exists? Sect. 3.8 addresses this question by outlining an approach that uses discontinuous mathematics.

3.8 Quantifying Absorption, Transmission, and Remission in Plane Parallel Layers

Section 3.6 introduced the concepts of the plane parallel layer and the two-flux model. Here, we outline a two-flux modeling approach that quantifies each of the individual "forward" and "backward" fluxes traveling between individual layers. Throughout this section, we will define A as the fraction of incident light absorbed by a layer, T as the fraction of light that is transmitted through a layer, and R as the fraction of light remitted by a layer. In this approach, we do not assign particular physical phenomena to these three outcomes. Thus, R represents all light that "reversed direction," whether due to reflection or scatter. Similarly, T represents all light that penetrates the layer, whether it did so directly or along a more complex path that included one or more scattering interactions.

Benford derived a set of algebraic equations that can be used to determine A, R, and T for a sample that is composed of multiple layers, assuming that A, R, and T are known for individual layers [13]. Consider, for example, a series composed of two layers, called x and y. A fraction of light A_x is absorbed by the first layer, another fraction R_x is remitted from the first layer and therefore remitted from the sample. The fraction of light that is transmitted through the first layer, T_x, encounters the second layer y, where again fractions will be absorbed, remitted, and transmitted. The fraction $T_x T_y$ is transmitted through both layers and therefore transmitted through the sample. The fraction $T_x R_y$ is transmitted through the first layer and remitted by the second, so it again encounters the first layer, now representing a "backward" flux. The fraction $T_x R_y T_x$ is then transmitted through the first layer and therefore remitted from the sample, while the fraction $T_x R_y R_x$ changes directions again and returns to the front of the second layer. This repetitive remission between the two layers can continue indefinitely. Benford used infinite series to derive the following [13]:

$$T_{x+y} = \frac{T_x T_y}{1 - R_x R_y} \tag{3.14}$$

$$R_{x+y} = R_x + \frac{T_x^2 R_y}{1 - R_x R_y} \tag{3.15}$$

$$A_{x+y} = 1 - T_{x+y} - R_{x+y} \tag{3.16}$$

With T_{x+y}, R_{x+y}, and A_{x+y} representing the transmission, remission, and absorption fractions for the two-layer sample as a whole. Notice that Benford's treatment assumes that the A, R, and T values for a layer are the same whether the light is traveling "forward" or "backward" at the time it encounters the layer.

The Eqs. 3.14–3.16 do not require that layers x and y be identical to each other. However, if a sample is composed of a uniformly distributed material, it can be modeled as a series of layers that are all identical to each other. If x and y are considered identical, the above equations can be used to show:

$$T_{2d} = \frac{T_d^2}{1 - R_d^2} \tag{3.17}$$

$$R_{2d} = R_d(1 + T_{2d}) \tag{3.18}$$

$$A_{2d} = 1 - T_{2d} - R_{2d} \tag{3.19}$$

With T_d, R_d, and A_d representing transmission, remission, and absorption by a sample of thickness d, and T_{2d}, R_{2d}, and A_{2d} representing transmission, remission, and absorption for a sample composed of the same material but with thickness 2d. Inverting these equations allows one to calculate the transmission, remission, and absorption for a sample of the same material with thickness d/2:

$$R_{d/2} = \frac{R_d}{1 + T_d} \tag{3.20}$$

$$T_{d/2} = \left[T_d \left(1 - R_{d/2}^2 \right) \right]^{0.5} \tag{3.21}$$

$$A_{d/2} = 1 - T_{d/2} - R_{d/2} \tag{3.22}$$

Repetitive application of these formulas can be used to calculate A, R, and T for a sample of any thickness from A, R, and T for a sample of the same material with any other thickness. For example, for the special case of no remission (R = 0), the denominators of the right hand sides of Eqs. 3.14 and 3.17 become 1. Repetitive application of these equations can be used to derive Eq. 3.10 which was presented in Sect. 3.6. Another special case is that of a non-absorbing material (A = 0). For this case, repetitive application of the Benford equations can be used to derive:

$$R_n = \frac{nR}{nR + T} \tag{3.23}$$

$$T_n = \frac{t}{nR + T} \tag{3.24}$$

With n representing the number of identical non-absorbing layers in the sample, R and T representing the remission and transmission by a single layer, and R_n and T_n representing the remission from and transmission through the sample as a whole.

For the general case in which both absorption and remission occur, the following expression was derived empirically using Benford's equations [14]:

$$A(R_n, T_n) = \frac{\left((1 - R_n)^2 - T_n^2 \right)}{R_n} = \frac{A}{R}(2 - A - 2R) \tag{3.25}$$

The function $\frac{((1-R_n)^2 - T_n^2)}{R_n}$ has been termed the "absorption-remission function" and $A(R_n, T_n)$ is here introduced as a symbol for that function. $A(R_n, T_n)$ is thus distinct from A, which represents the absorption by a single layer. We recognize the potential for confusion, especially since A or A_{10} is also frequently used in the literature as a symbol for the "absorbance" (defined in Eq. 3.2) which is distinct from either of these. The crucial point is that for a given material with specific A, R, and T values, the absorption-remission function will have a constant value regardless of the number of layers n. This will be illustrated through an example in Sect. 3.10.

It is straightforward to validate the equations presented throughout this section for samples that are literally divided into distinct identical layers. For example, they proved to work well for samples consisting of different numbers of identical plastic sheets [15]. However, how can the concept of a "plane parallel layer" be meaningfully applied to a particulate sample? The next section explores this question.

3.9   The Representative Layer

Previous sections have alluded to the mathematical approach of modeling a sample as a series of identical plane parallel layers. A particulate sample, such as illustrated in Fig. 3.10, clearly is not literally sub-divided into distinct identical "layers" that the incident beam encounters sequentially. Dahm and Dahm proposed that *for spectroscopic purposes*, a particulate sample can be modeled as composed of a series of layers, each one representative of the sample as a whole [16].

In envisioning a "representative" layer, it is instructive to consider the large and small spherical particles included as shown in Fig. 3.10. Let us assume that the large and small spheres that are of identical color have identical composition and differ only in size. The diameter of the large spheres is twice the diameter of the small

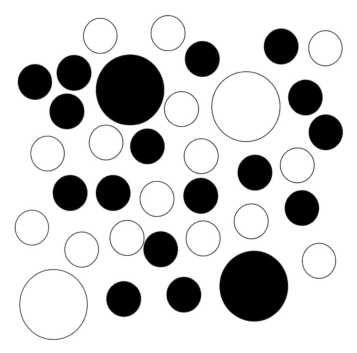

Fig. 3.10 Particulate sample composed of particles with two different compositions, each in two different sizes

spheres, and therefore one large sphere contains the same volume (and mass) of material as eight small spheres. However, the surface area of one large sphere is only four times that of one small sphere. Based upon our discussions of microscopic light interactions (Sects. 3.3–3.5), we regard scattering as predominantly a surface phenomenon, while absorption can occur anywhere within a particle. Thus, the eight smaller particles, taken together, have essentially the same ability to absorb light as the one large particle, but the eight small particles present more surface area to the beam and have more ability to scatter light. The effect has been seen experimentally: Devaux et al., for example, have conducted NIR reflectance experiments on mixtures of two types of particles (wheat and rapeseed) and found that in mixtures containing two different particle sizes, the smaller particles were more prominently represented in the spectra than were the larger particles [17]. Consequently, when considering the different types of particles that make up a sample, we must distinguish between "types" of particles based not only upon composition, but also upon volume and surface area.

The proposed criteria for the "representative layer" are:

- The volume fraction for each particle type is the same in the layer as in the sample as a whole
- The cross-sectional surface areas of different particle types in the layer are in the same proportion as they are in the sample as a whole
- The void fraction in the layer is the same as the void fraction of the sample as a whole
- The layer is only one particle thick. (This means the representative layer does not have a uniform thickness, as different particles have different sizes.)

The first three criteria ensure that the layer is representative of the sample as a whole, in terms of physical properties that are important in determining outcomes for interactions with light. The last criteria is included so that we can assume a *single interaction*. A representative fraction of the light will interact with each of the particles (or voids) present in the layer, but a given ray of light will only interact with a given layer once. This means that we can assume the kinds of "microscopic" models and phenomena described in Sects. 3.3–3.5 would apply to the representative layer.

Mathematical expressions of these criteria can be found in [16] and [18]. Examples of applications of the representative layer theory to real samples can be found in [19] and [20].

3.10 Obtaining Linear Absorbance Data for Scattering Samples

At this point, it is instructive to think back to Beer's Law. The practical appeal of the equation is that it is a simple linear relationship between the concentration of the absorber (which is usually what we are trying to determine) and the absorbance (which is readily measured experimentally). But this usage requires a known value of κ, which in Sect. 3.1 was termed the "molar absorptivity." For non-scattering samples, extinction of the beam is attributed entirely to absorption. It is often reasonable to model κ as a constant that represents the inherent ability of the material in question to absorb light (though limitations to such a model were also noted in Sect. 3.1). It would be convenient if an analogous linear relationship existed for scattering samples.

As introduced in Sect. 3.6, the classical definition of an "absorption coefficient" is the fraction of incident light absorbed by a layer of material, divided by the thickness of that layer. This is a number that can be obtained experimentally for a macroscopic sample of known thickness. However, previous sections have also illustrated that scatter and absorption influence each other, and scatter is dependent upon factors like particle size and shape that are not repeatable from sample to sample. As noted by Burger et al. in a study of pharmaceutical powders, "many parameters, such as particle size distribution, packing density, or homogeneity of the investigated powder mixtures, strongly influence the reflectance spectra [11]. A variation in, for example, the particle size distribution changes the scattering coefficient and leads to different reflectance values even for chemically identical samples." Consequently, one can calculate an "absorption coefficient" from a remission/transmission experiment on a scattering sample, and one might view it as somewhat analogous to the Beer's Law κ, but one cannot generally use it the same way, as one cannot typically assume it is "constant."

Is it possible to separate the effects of absorption and scatter in a scattering sample? Consider a very thin layer such as that as shown in Fig. 3.11. In the example, 99.5% of the incident light is transmitted through the layer, 0.3% is absorbed, and 0.2% is remitted. Consider:

- In general, when light is reflected from the front surface of a layer, it has no opportunity to be absorbed by that layer, and thus the observed absorption has been influenced by the process of remission. But in this example, the remission and absorption are both negligible, so essentially all of the light had the *opportunity* to be either remitted or absorbed.
- In general, when light is diverted from its original path by scatter, it is now penetrating the sample along a different (usually longer) length path, which means it has a different (usually larger) probability of subsequently being absorbed by the sample. But this consideration does not apply in Fig. 3.11, since the layer is very thin and the light *only interacts with it a single time*.

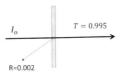

Fig. 3.11 Hypothetical thin layer in which 99.5% of the incident light is transmitted and 0.2% is remitted, with the other 0.3% being absorbed

Thus, in this example, the fraction of light absorbed (0.3%) is determined by the ability of the layer to absorb light and has not been influenced by scatter. Consequently, if one calculated an absorption coefficient from this layer, one could say that the resulting coefficient was a true measurement of the absorbing power of the material for light at that wavelength. This, then, is a more fundamental property of the material than an absorption coefficient obtained from a macroscopic sample, and therefore expected to be more broadly applicable to other samples involving the same material.

It is generally not realistic to fashion an extremely thin sample like the one pictured in Fig. 3.11. However, Benford's equations (Sect. 3.8) give us a way to calculate A, R, and T for a sample of any thickness, given A, R, and T for a sample of any other thickness. Thus, we can mathematically model the behavior of a hypothetical extremely thin sample of a material, using the process as summarized in Table 3.1. In this example, for a sample that has $d = 1$ cm thickness, 78% of the incident light is absorbed, 21% is remitted, and 1% is transmitted. Thus, the calculated absorption coefficient from the macroscopic sample is 0.78 cm^{-1} (the fraction absorbed, 0.78, divided by the thickness, 1 cm).

Using repetitive applications of the Benford equations (specifically 3.20–3.22), Table 3.1 shows the calculated A, R, and T for successively thinner layers of the same material. Also shown is the absorption coefficient calculated from A and d for each layer. Notice that as the layers get thinner, the absorption coefficient converges. The converged value of ~ 2.98 cm^{-1} is very different than that obtained from the original macroscopic sample and illustrates the significance of mathematically separating the effects of absorption and scatter. Table 3.1 also shows the absorption-remission

Table 3.1 Application of Benford's equations to a hypothetical sample 1 cm thick that remits 21% of incident light and transmits 1%

Thickness (cm)	A	R	T	μ_a (cm^{-1})	$-log_{10}(1 - A)$	$A(R_n, T_n)$
1	0.78	0.21	0.01	0.780	0.65758	2.971
½	0.6943	0.2079	0.0978	1.389	0.51465	2.971
¼	0.5035	0.1894	0.3071	2.014	0.30409	2.971
1/8	0.3068	0.1449	0.5483	2.454	0.15914	2.971
1/16	0.1692	0.0936	0.7372	2.707	0.08050	2.971
1/32	0.0888	0.0539	0.8574	2.840	0.04037	2.971
1/64	0.0454	0.0290	0.9256	2.908	0.02020	2.971
1/128	0.0230	0.0151	0.9619	2.943	0.01010	2.971
1/256	0.0116	0.0077	0.9808	2.960	0.00505	2.971
1/512	5.80E-03	3.88E-03	0.9903	2.968	0.00253	2.971
1/1024	2.90E-03	1.95E-03	0.9951	2.973	0.00126	2.971
1/2048	1.45E-03	9.76E-04	0.9976	2.975	0.00063	2.971
1/4096	7.27E-04	4.89E-04	0.9988	2.976	0.00032	2.971
1/8192	3.63E-04	2.44E-04	0.9994	2.977	0.00016	2.971

function $A(R_n, T_n)$, which was introduced in Eq. 3.25 of Sect. 3.8). The value of this function is identical regardless of the assumed thickness of the sample.

Table 3.1 also shows the separation of the effects of absorption and scatter through the "absorbance," $-log_{10}(1 - A)$, as defined in Eq. 3.4. According to Beer's Law, absorbance should be linear with sample thickness. Thus, if the data in Table 3.1 were following Beer's Law, the absorbance on each row would be one-half the value of the row above it. This is clearly not the case at the top of the table, which represents the actual sample. However, if one views the results for thicknesses of ~ 1/32 cm and below, one sees that absorbance is indeed linear with thickness. Thus, if we produce layers that are thin enough that the effects of absorption and scatter have been successfully isolated and separated, Beer's Law is a good model, even for scattering samples. Applying the fact that Beer's Law is linear with thickness, we can use the absorbance that was determined for a thin layer (e.g., 1/128 cm or 1/256 cm) to compute an expected absorbance for the actual sample thickness of 1 cm:

$$\text{Absorbance} = (128)(0.0101) = 1.29 \qquad (3.26)$$

$$\text{Absorbance} = (256)(0.00505) = 1.29 \qquad (3.27)$$

Notice the values obtained using these two thin layers are essentially identical to each other but very different from the measured absorbance from the original sample, 0.658. The value of 1.29 can be termed the "scatter-corrected absorbance." It can be interpreted as the absorbance that *would* be observed from a 1 cm thick sample of a hypothetical material that had the same absorbing power as the real sample, but with a complete absence of scatter. It was noted in Sect. 3.1 that the absorbance of a material is not repeatable from one sample to another because absorbance is not solely a measure of the absorbing power of the sample material; it is also influenced by sample size, geometry, etc. The "scatter-corrected absorbance" has the potential to address these limitations and provide a more genuine metric for the absorbing power of a material.

Mathematically, one can continue halving the thickness of a layer indefinitely, until the obtained value of the scatter-corrected absorbance and/or the absorption coefficient becomes constant, to as many significant figures as desired. Recall that one of the assumptions underlying Benford's equations was that the sample was uniformly distributed; the approach presented throughout this section relies upon this assumption. Even if this assumption is reasonable for the sample, one must take care to consider whether the hypothetical "thin layers" are physically meaningful. Suppose the original sample as illustrated in Table 3.1, which was 1 cm thick, was made up of particles that were approximately 1 mm in diameter. Table 3.1 includes a row in which d = 1/1024 cm, or approximately 1 mm. (The calculated A and R values for this particular row are similar to those as shown in Fig. 3.9). Thus, one could plausibly consider a layer of thickness d = 1/1024 cm to be approximately "one particle thick" and representative of the sample as a whole. Mathematically, the scatter-corrected absorbance and the absorption coefficient do not change significantly if the process of halving the thickness is continued beyond this point. This, then, is an example in which the effects of absorption and scatter have been successfully separated, and the calculated absorption coefficient truly represents the ability of the material to absorb light.

By contrast, suppose the A, R, and T data in Table 3.1 were obtained from a sample composed of particles that are 0.25 cm, or 250 mm, in diameter. The calculated absorption coefficient for a layer of thickness 0.25 cm is significantly different from the converged value of ~ 2.98 cm^{-1}. Subdividing a particle creates new surfaces, and scattering is a surface phenomenon. (The significance of surface area to volume ratio in absorption and scattering was introduced in Sect. 3.9.) Consequently, one cannot plausibly expect a layer with a thickness of 0.01 or 0.001 cm to be "representative" of the original sample. In this scenario, one might regard all rows of Table 3.1 with $d < 0.25$ cm as mathematical constructs that have no physical significance. Notice that according to Table 3.1, $R = 0.189$ when $d = 0.25$ cm. A plausible interpretation of this result is that even if the layer is only one particle thick, almost 19% of the incident light is remitted by a single interaction, and this 19% has no opportunity to be absorbed. In such a case, separating the effects of absorption and scatter is likely not a realistic goal.

References

1. W. E. Knowles Middleton, *Bouguer, Lambert, and The Theory of Horizontal Visibility*. Isis 51, no. 2 (1960): 145–49. https://www.jstor.org/stable/226845.
2. W.W. Wendlandt, H.G. Hecht, *Reflectance Spectroscopy* (John Wiley and Sons, New York, 1966)
3. J.M. Chalmers, P.R. Griffiths, *Handbook of Vibrational Spectroscopy* (John Wiley & Sons Ltd., Chichester, UK, 2002)
4. C.V. Raman, A new radiation. Indian J. Phys. **2**, 387–398 (1928)
5. M. Bass, E. W. Van Stryland, D. R. Williams, W. L. Wolfe, *Handbook of Optics*, 2nd ed., vol. 1, McGraw-Hill, Inc., New York (1995
6. G.G. Stokes, On the intensity of the light reflected from or transmitted through a pile of plates. Proceedings of the Royal Society of London **11**, 545 (1862)
7. A. Schuster, Radiation Through a Foggy Atmosphere. Astrophysical J.**21** , 1 (1905)
8. G. Kortüm, *Reflectance Spectroscopy* (Springer, Berlin, 1969)
9. P. Kubelka, F. Munk, Ein Beitrag zur Optik der Farbanstriche. Z.Tech. Phys. (Leipzig), **12** (1931)
10. P. Kubelka, New Contributions to the Optics of Intensely Light-Scattering Materials. Part I J. Optical Soc. America **38**, 5 (1948)
11. T. Burger, J. Kuhn, R. Caps, J. Fricke, Quantitative Determination of the Scattering and Absorption Coefficients from Diffuse Reflectance and Transmittance Measurements: Application to Pharmaceutical Powders. Appl. Spectrosc. **51**(3), 309–322 (1997)
12. R.G. Giovanelli, Reflection by Semi-infinite Diffusers. Optica Acta: International Journal of Optics **2**(4), 153–162 (1955)
13. F. Benford, Radiation in a Diffusing Medium. J. Optical Soc. America **36**, 524–554 (1946)
14. D. J. Dahm, K.D. Dahm, Bridging the continuum-discontinuum gap in the theory of diffuse reflectance. J. Near Infrared Spectroscopy **7**, 47-53 (1999)

15. D.J. Dahm, K.D. Dahm, K.H. Norris, Test of the representative layer theory of diffuse reflectance. J. Near Infrared Spectrosc. **8**, 171–181 (2000)
16. D.J. Dahm, K.D. Dahm, Representative layer theory for diffuse reflectance. Appl. Spectrosc. **53**, 647–654 (1999)
17. M. Dexaux, N. Nathier-Dufour, P. Robert, D. Bertrand, Effects of particle size on the near-infrared reflectance spectra of Wheat and Rape Seed meal mixtures. Appl. Spectrosc. **49**(1), 84–91 (1995)
18. D.J. Dahm, K.D. Dahm, *Interpreting Diffuse Reflectance and Transmittance* (NIR Publications, Chichester, UK, 2007)
19. B. G. Yust, D. K. Sardar, and A. Tsin, "A Comparison of Methods for Determining Optical Properties of Thin Samples." Proceedings of SPIE--the International Society for Optical Engineering, 7562 (2010): 75620C.
20. A. Gobrecht, R. Bendoula, J.M. Roger, V. Bellon-Maurel, Combining linear polarization spectroscopy and the Representative Layer Theory to measure the Beer-Lambert law absorbance of highly scattering materials. Anal. Chim Acta **853**, 486–494 (2015)

Part II
Spectral Analysis and Data Treatments

Chapter 4
Spectral Analysis in the NIR Spectroscopy

Yukihiro Ozaki, Shigeaki Morita, and Yusuke Morisawa

Abstract This chapter is concerned with the introduction to spectral analysis in the NIR spectroscopy. It consists of two major parts, conventional spectral analysis and spectra pretreatments. In the former, various conventional spectral analysis methods such as group frequency analysis, derivative spectra, difference spectra, spectral analysis based on perturbation, comparison of a NIR spectrum with the corresponding IR spectrum, and isotope exchange experiments are explained. In the latter part smoothing, derivative methods, multiplicative scatter correction (MSC), standard normal variate (SNV), centering methods, and normalization are described.

Keywords Spectral analysis · Chemometrics · Group frequency · Derivative · Difference spectra · Spectral pretreatment · Baseline correction noise

4.1 Introduction to Spectral Analysis in the NIR Region

As described partly in Chaps. 1 and 2 there are various kinds and various types of NIR spectra [1–7]. First of all, NIR spectra can be divided into electronic spectra and vibrational spectra. However, in this chapter, we treat only vibrational spectra of solids and liquids. Compared with IR spectroscopy diversity of the types of NIR spectra is quite large because NIR spectroscopy is concerned with so many kinds of materials from pure samples such as pure liquids, solutions, and crystals to bulk materials including raw materials, industrial products, and natural products. To look at the diversity of NIR spectra let us compare the spectrum of methanol (0.005 M, in CCl_4; see Fig. 2.3) with that of flour (Fig. 4.1). The former is rather simple

Y. Ozaki (✉)
School of Science and Technology, Kwansei Gakuin University, Sanda, Japan
e-mail: yukiz89016@gmail.com

Toyota Physical and Chemical Research Institute, Nagakute, Japan

S. Morita
Faculty of Engineering, Osaka Electro-Communication University, Neyagawa, Japan

Y. Morisawa
School of Science and Engineering, Kindai University, Higashi-Osaka, Japan

© Springer Nature Singapore Pte Ltd. 2021
Y. Ozaki et al. (eds.), *Near-Infrared Spectroscopy*,
https://doi.org/10.1007/978-981-15-8648-4_4

Fig. 4.1 A NIR spectrum of flour. Measured by A. Ikehata

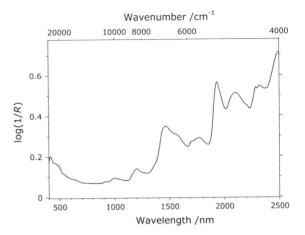

although several bands are overlapped in the 4500–4000 cm^{-1} region. It has very little baseline change. On the other hand, the spectrum of flour consists mainly of the spectra of water, starch, and proteins. Bands are broad, and its baseline increases with the increase in the wavelength. Like this NIR spectra show significant diversity depending on samples and conditions. Therefore, spectral analysis methods are also diverse in NIR spectroscopy [1–3]. In other words, one must select the best spectral analysis method for a target. The spectral analysis methods must change with samples, sample conditions, measurement methods, and the purpose of analysis. The purpose of analysis is also wide spread varies from quantitative analysis, qualitative analysis, and sample identification to studies of molecular structure and chemical reaction. Therefore, when one selects a spectral analysis method, one must consider the purpose of analysis. For many purposes, chemometrics is very useful but for some purposes it is almost meaningful. In this chapter, general introduction to the spectral analysis methods and spectral pretreatments for NIR spectra are outlined. The detailed explanations of representative spectral analysis methods such as chemometrics (Chap. 7), two-dimensional correlation spectroscopy (2D-COS, Chap. 6), and quantum chemical calculations (Chap. 5) will be given in each chapter and session.

As in the cases of IR and Raman spectroscopy, band assignments are always the base for spectral analysis of NIR spectroscopy. However, the assignments are generally not straightforward since a number of bands originating from overtones and combinations overlap each other. In some cases, bands arising from combinations including combinations of overtones appear; for such cases, it is very difficult to make accurate band assignment. In NIR spectroscopy detailed band assignments are often not necessary, but even in such cases, one should know from which functional group a band arises.

Spectral analysis methods in NIR spectroscopy can be divided into conventional spectral analysis method, chemometrics [3], quantum chemical calculation [5, 8], and 2D-COS [1]. The conventional spectral analysis methods are, more or less, common among NIR, IR, Raman, and Terahertz/far-IR(FIR) spectroscopy. One must know that they yield also a base for chemometrics, quantum chemical calculation, and 2D-COS. Chemometrics has most often been employed to extract rich quantitative and qualitative information from NIR spectra (Chap. 7). A major part of chemometrics is multivariate data analysis such as principal component analysis/regression (PCA/PCR) and partial least squares regression (PLSR), however, self-modeling curve resolution (SMCR), which is used to predict pure component spectra and pure component concentration profiles from a set of NIR spectra, is also becoming more and more significant (Chap. 7). Using quantum chemical calculations such as density function theory (DFT) calculations, one can calculate the intensities and frequencies of overtones and combination bands (Chap. 5). Quantum chemical calculation is still not always popular in NIR spectroscopy but it has already been applied not only to simple compounds but also to rather complicated molecules such as long-chain fatty acids, nucleic acid bases, and rosemaric acid [8]. 2D-COS is not a general method but it is often useful to unravel complicated NIR spectra (Chap. 6). In addition, neural network, AI, and machine learning have been started to be used to analyze NIR spectra. They are very promising methods for the spectral analysis in NIR spectroscopy (Chap. 7). However, one should know that in the early 1990 s neural network has already been tried to be applied to NIR spectra [9].

4.2   Conventional Spectral Analysis Method

Various kinds of conventional spectral analysis methods are used in NIR spectroscopy [1, 2, 6, 7]. They are summarized as follows:

(1) *Spectral analysis based on group frequencies*

This is a traditional method established in IR and Raman spectroscopy. Spectral analysis based on group frequencies built for the fundamentals is modified for overtones and combinations. Each functional group such as OH and CH groups shows characteristic bands in particular regions. One can find tables for group frequencies in the NIR region in a few NIR textbooks [1, 6].

(2) *Calculation of derivative spectra*

Derivative methods have long been popular in various spectroscopies [1–3, 6]. They are useful for resolution enhancement as well as baseline correction. Figure 4.2A, B shows a good example demonstrating the usefulness of the second derivative [10]. In Fig. 4.2A, NIR spectra in the 7500–5500 cm^{-1} region are shown for water-methanol mixtures with a methanol concentration of 0–100 wt% at increments of 5 wt% at 25 °C. Figure 4.2B, a gives an enlargement of the 6000–5700 cm^{-1} region of the NIR spectra as shown in Fig. 4.2a. In the 6000–5700 cm^{-1} region, many bands due to the overtones and combination of the CH stretching modes of CH_3 group of methanol are expected to appear. Figure 4.2B, b displays the second derivative of the spectra in Fig. 4.2B, a. Note that a broad feature in the 6000–5750 cm^{-1} region can be divided

Fig. 4.2 **A** NIR spectra in the 7500–5500 cm^{-1} region of water-methanol mixtures with a methanol concentration of 0–100 wt% at increments of 5 wt% at 25 °C. **B a** An enlargement of the 6000–5700 cm^{-1} region of the NIR spectra shown in **A** and **B, b** the second derivative of the spectra in **a**. Reproduced from Ref. [10] with permission

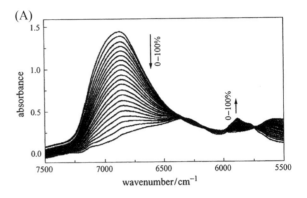

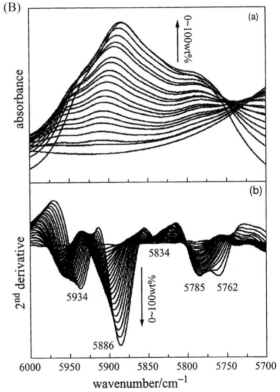

into many bands which show clear concentration-dependent variations. Derivative methods will be explained in more detail in Chap. 4.3.2.

(3) *Calculation of difference spectra*

Calculation of difference spectra is also useful to unravel overlapping bands and to find out a weak feature hidden by a strong band [1, 2, 5, 7]. The difference spectrum between a spectrum of sample *a* and that of sample *b* can be calculated by subtracting the spectrum of sample *a* from that of sample *b*. The calculation of difference spectra is effective to analyze perturbation-dependent NIR spectra such as temperature-dependent, concentration-dependent, and pH-dependent spectra. To calculate accurate difference spectra one must obtain spectra with very high wavelength accuracy. Let us show very simple but important example of the calculation of difference spectra. Figure 4.3a displays NIR spectra of water collected over a temperature range of 5–85 °C [11]. From this figure, it is clear that the intensity at 7050 cm^{-1} increases while that at 6844 cm^{-1} decreases but it is not clear whether there is a band shift or not in the 7300–6200 cm^{-1} region. The calculation of the difference spectra clearly answers this question. Figure 4.3b displays the difference spectra of water obtained by subtracting the spectrum at 5 °C as a reference spectrum from other spectra in Fig. 4.3a [11]. It can be seen from Fig. 4.3b that the broad water feature consists of two bands at 7089 and 6718 cm^{-1} and that there is no significant band shift.

Generally speaking, difference spectra method is a reliable method but even so care must be taken. Using spectra of a model system, let us explain a problem of difference spectra. Figure 4.4a shows spectral changes of a system consisting of two components having Gaussian type bands with different peak wavenumbers (7300 and

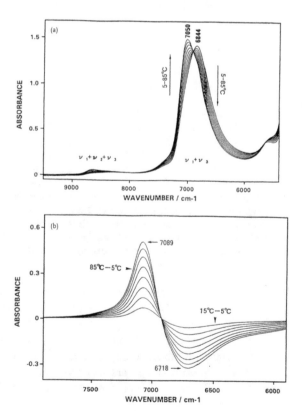

Fig. 4.3 a NIR spectra of water measured in a temperature range of 5–85 °C. **b** The difference spectra of water obtained by subtracting the spectrum at 5 °C as a reference spectrum from other spectra in **a**. Reproduced from Ref. [11] with permission

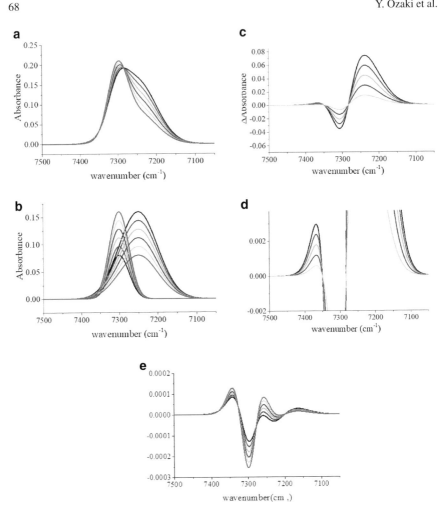

Fig. 4.4 a Spectral changes of a system consisting of two components having Gaussian type bands with different peak wavenumbers (7300 and 7250 cm^{-1}), different intensities (10→5 and 10→20), and different band widths (59 and 118 cm^{-1}). **b** Variations of each band. **c** The difference spectra calculated using the first spectrum (pink color) as a reference. **d** An enlargement of **c**. **e** The second derivative of the spectra shown in **a**. Prepared by Y. Morisawa

7250 cm^{-1}), different intensities (10→5 and 10→20), and different band widths(59 and 118 cm^{-1}). Figure 4.4b depicts variations of each band and Fig. 4.4c displays the difference spectra calculated using the first spectrum (pink color) as a reference. Note that peak positions (7309, 7239 cm^{-1}; Fig. 4.4c) are shifted in the difference spectra compared with positions in the original spectra. Figure 4.4d exhibits an enlargement of Fig. 4.4c. A ghost peak appears near 7380 cm^{-1} because the widths of these two bands are significantly different from each other. Therefore, a special care must be taken for feeble peaks.

(4) *Spectra-structure correlations*

The NIR spectrum of a compound can be compared with those of similar compounds to make band assignments. For instance, NIR measurements of a series of alcohols or that of fatty acids allow one to make assignments of bands due to OH, CH_2, and CH_3 groups. Figure 4.5 compares NIR spectra in the 7500–4000 cm^{-1} region of saturated (stearic acid, arachidic acid, and palmitic acid) and unsaturated (oleic acid, linolenic acid, and linoleic acid) long-chain fatty acids (0.05 M in CCl_4) [12]. By comparison, one can easily discriminate saturated and unsaturated long-chain fatty acids; particularly see the 5800–5670 cm^{-1} region. A common band near 6908 cm^{-1} can be assigned to the first overtone of OH stretching mode. The unsaturated fatty acids yield characteristic bands near 4663 and 4590 cm^{-1}. Grabska et al. have assigned the 4663 cm^{-1} band to the combination of C=C stretching and CH stretching modes by quantum chemical calculation [12].

(5) *Spectral analysis based on perturbation*

NIR measurements of perturbation-dependent spectral variations, such as temperature-dependent, concentration-dependent, and pH-dependent spectra variations, often give valuable information about the band assignments [1, 2, 7]. As a good example, temperature-dependent NIR spectra changes of octanoic acid in the pure liquid over a temperature range of 15–90 °C are shown in Fig. 4.6 [13]. Octanoic acid in the pure liquid forms a cyclic dimer with hydrogen bonds at room temperature but with the temperature increase the dimer dissociates gradually and a free OH

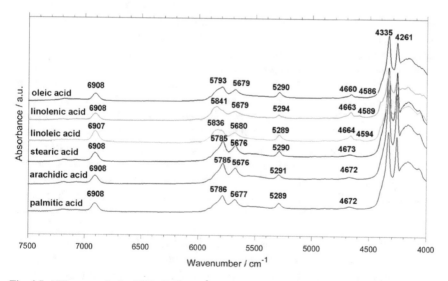

Fig. 4.5 NIR spectra in the 7500–4000 cm^{-1} region of saturated (stearic acid, arachidic acid, and palmitic acid) and unsaturated (oleic acid, linolenic acid, and linoleic acid) long-chain fatty acids (0.05 M in CCl_4). Reproduced from Ref. [12] with permission

Fig. 4.6 Temperature-
dependent NIR spectra
changes of liquid octanoic
acid collected in a
temperature range of
15–90 °C. Reproduced from
Ref. [13] with permission

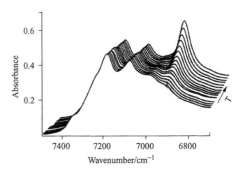

group emerges. It is noted that the intensity of a band at 6920 cm^{-1} increases as a function of temperature, while those of other bands in the 7300–7000 cm^{-1} region are almost temperature independent. Thus, the band at 6920 cm^{-1} is assigned to the first overtone of the OH stretching mode of the monomeric species of octanoic acid, and the rest is due to combinations of CH vibrations [13].

The spectra of Figs. 4.2 and 4.3 are also good examples of perturbation-dependent spectra changes.

(6) *Comparison of an NIR spectrum with the corresponding IR spectrum*

The significance of comparison of an NIR spectrum with the corresponding IR spectrum was pointed out in Chap. 2.1.6. Here, let us compare NIR spectra of poly(3-hydroxybutyrate) (PHB) with the corresponding IR spectra again but this time in a different region. Figure 4.7a shows time-dependent changes in the NIR spectra in the 5200–5060 cm^{-1} region of a PHB film during the melt-crystallization process at 125 °C [14]. The corresponding IR spectra in the region of 1780–1670 cm^{-1} are depicted in Fig. 4.7b [14]. Of note is that a band at 5127 cm^{-1} in the NIR spectra gradually increases during the crystallization process, while a broad feature centered at 5160 cm^{-1} decreases with time, suggesting that the former band is assigned to the crystalline band and the latter band to the amorphous one. In the corresponding IR spectra bands at 1722 and 1743 cm^{-1} are ascribed to the crystalline and amorphous C=O bands, respectively. Of note is that the NIR spectra and the IR spectra show very clear correspondence. The comparison of the NIR spectra with the IR spectra led Hu et al. [14] to ascribe the band at 5127 cm^{-1} to the second overtone of the C=O stretching mode of the C–H...O=C hydrogen bonding in the crystalline state and the broad feature near 5160 cm^{-1} to the corresponding band due to the amorphous state.

Figure 4.8a, b gives rise to another example of comparison between a NIR spectrum and an IR spectrum. They are the IR spectrum in the 3800–3000 cm^{-1} region and the NIR spectrum in the 7600–6000 cm^{-1} region of diluted methanol in CCl$_4$. A peak at 3630 cm^{-1} and that at 7090 cm^{-1} are due to a fundamental and a first overtone of a stretching mode of free OH group of methanol while broad features in the regions of 3500–3200 and 6900–6100 cm^{-1} arise from a fundamental and a first overtone of a stretching mode of hydrogen-bonded OH groups of methanol dimer, trimer, and oligomers. Thus, there is the correspondence between the NIR spectrum

Fig. 4.7 **a** Time-dependent
NIR spectra changes in the
5200–5060 cm^{-1} region of a
PHB film during the
melt-crystallization process
at 125 °C. **b** The
corresponding IR spectra
variations in the
1780–1670 cm^{-1} region.
Reproduced from Ref. [14]
with permission

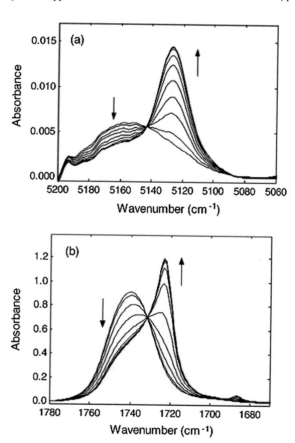

and the IR spectrum, but the relative intensity of bands between the free OH bands
and the hydrogen-bonded OH bands is largely changed between them. Accordingly,
special care must be taken in comparison with the relative intensity between a NIR
spectrum and an IR spectrum.

(7) *Spectral interpretation by polarization measurement*

Polarization measurement, which is popular in IR and Raman spectroscopy, is not
often used in NIR spectroscopy, but it is useful for the determination of the molecular-
orientation of solid-oriented compounds such as uniaxially stretched polymers. For
more details, see Ref. [14].

(8) *Isotope exchange experiments*

The use of an isotope shift, which is associated with isotopic substitution, is a traditional method for band assignment in IR and Raman spectroscopy. It is also useful
for NIR spectra analysis, particularly, a deuteration shift. The isotope shift provides
convinced assignment of bands in a number of cases. Force constants can be assumed

Fig. 4.8 **a** An IR spectrum
in the 3800–3000 cm^{-1}
region and **b** a NIR spectrum
of the 7600–6000 cm^{-1}
region of diluted methanol in
CCl$_4$

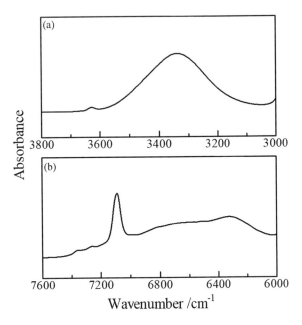

not to vary due to isotopic substitution, and hence, isotope shifts involve only with mass effects. Taking a diatomic molecule as an example, one can calculate the magnitude of an isotope shift. Frequency of a stretching vibration of the diatomic molecule is given by Eq. (4.1)

$$\nu = \frac{1}{2\pi}\sqrt{\frac{k}{\mu}} \tag{4.1}$$

if ν' is the frequency for replacing an atom having a mass m_1 with an isotope having a mass m_1', the following relation holds:

$$\frac{\nu}{\nu'} = \sqrt{\frac{\mu'}{\mu}} \tag{4.2}$$

where $\mu' = m_1' m_2/(m_1' + m_2)$. As Eq. (4.2) reveals, the larger the difference between m_1 and m_1', the larger the isotope shift is. Since $\nu/\nu' = 1.36$ if H is replaced with D, a C–H stretching vibration of saturated hydrocarbon, which is located in the vicinity of 2900 cm^{-1}, shifts close to 2100 cm^{-1}. Figure 4.9 shows calculated three vibrational modes (Amide I′, II′, and III′) of deuterated N-methylacetamide and the corresponding modes of the nondeuterium-substituted one (Amide I, II, and III modes) are displayed in Fig. 2.11 [16]. Band shifts induced by the deuterium substitution are rather large for the Amide II and III modes since NH bending vibrations contribute to these two modes, but the Amide I mode, being principally a C=O

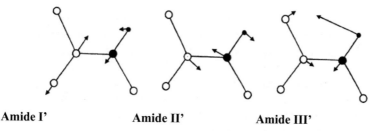

Amide I'   **Amide II'**   **Amide III'**

Fig. 4.9 Amide I', Amide II', and Amide III' modes of deuterated *N*-methyacetamide. Reproduced from Ref. [16] with permission

stretching vibration, yields a very small isotope shift [15]. It is noted with respect to isotope shifts of polyatomic molecules that vibrational modes vary more or less with isotopic substitution, which can be evidently understood from the comparison of Fig. 4.9 and Fig. 2.11. In studies of IR, Raman, and NIR spectra, [15] *N*-substitution, [13] *C*-substitution, and the like are often used in addition to deuterium substitution. Although an isotope shift is small when such a heavy atom is replaced, a variation in a vibrational mode associated with the isotopic substitution is also small.

4.3   Pretreatment Methods in NIR Spectroscopy

NIR spectra often encounter the problems of unwanted spectral variations and baseline shifts [1–3]. They come from the following sources.

1. Various kinds of noises such as those from a detector, an amplifier, and an AD converter.
2. Light scattering from cloudy liquids or solid samples.
3. Changes in temperature, density, and particle size of samples.
4. Poor reproducibility of NIR spectra caused by, for example, path length variations.
5. The use of optical fiber cable may cause baseline shifts.

   The above interferences may become obstacles for conventional spectral analysis and 2D-COS. More importantly, they may easily violate the assumptions on which chemometrics equations are based [3]. For instance, the simple linear relationship stated by Beer's law does not hold any more, and the additivity of individual spectral responses is not guaranteed. Accordingly, data pretreatment is often necessary [1–3]. (Chap. 7) Whenever one attempts to improve SN ratio or to correct baseline fluctuations, one should explore the cause of poor SN ratio and that of baseline changes. Otherwise, one cannot find proper pretreatment methods. One interesting example of studies of baseline changes was reported by Geladi et al. [17] They modeled the reflectance spectra of milk by optical effects and chemical light absorption effects. The former induces variations in the direction of the light, and the latter is concerned with light absorption. In some cases, the former brings about more prominent variations to spectra than the latter. The response of the spectral data to the physical effects is significant baseline variations. On the basis of this study, Geladi et al. [17] proposed multiplicative scattering collection (MSC) as a preprocessing tool to correct the light scattering problems in the NIR spectra.

This section explains four kinds of data pretreatment methods, noise reduction methods, baseline correction methods, resolution enhancement methods, and centering and normalization methods [1–3].

4.3.1 Noise Reduction Methods

In NIR spectroscopy, several kinds of noise are caused by a variety of interfering physical and/or chemical process [1–3]. The most general noise is high-frequency noise associated with the instrument's detector and electronic circuits. There are other forms of noise as well; for example, low-frequency noise and localized noise. Low-frequency noise is induced, for instance, by instrument drift during the scanning measurements. The reduction of the low-frequency noise may be more difficult because it often resembles the real information in the data.

Most standard method to improve SN ratio in spectra is accumulation-average processing that requires to increase the accumulation number and calculate an average. This reduces the effects of high-frequency noise significantly, but technically it is not a "pretreatment" but a normal, integrated part of collecting spectra. If the noise reduction by the accumulation average is still insufficient, one can employ smoothing to remove high-frequency noise. The most commonly used smoothing methods are moving-average method and Savitzky–Golay method [1, 2].

The moving-average method is the simplest type of smoothing [1–3]. In this method, the reading A_i' (A is, for instance, absorbance) at each variable $i = 1, 2, ---k$ is replaced by a weighted average of itself and its nearest neighbors. From $i-n$ to $i+n$:

$$A_i = \sum_{k=-n}^{n} w_k A_{i+k} \qquad (4.3)$$

w_k, defining the smoothing, is called the convolution weights.

The Savitzky–Golay method originated from the idea that in the vicinity of a measurement point a spectrum can be fitted by low-degree polynomials [18]. Practically, w_k is determined by fitting the spectrum with low-degree polynomials using least squares regression. Savitzky and Golay calculated w_k for the different orders of polynomials and N ($N = 2n+1$) [18]. One can find these calculated convolution weights in a numeral table. For example, when N is equal to 5, smoothed values can be obtained by substituting $w_k = -3/35, 12/35, 17/35, 12/35, -3/35$ ($k = -2, -1$, 0, 1, 2) into Eq. (4.3). It is noted that if one tries to increase the effect of smoothing by increasing the number of the point of w_k, a band shape would be distorted. This distortion may lead to the decrease in spectral resolution and band intensity.

There are other methods for the noise reduction such as wavelets, eigenvector reconstruction, and artificial neural networks (ANN) [2, 3, 19, 20].

4.3.2 Baseline Correction Methods

As described above in NIR spectra baselines vary for various reasons [1–3]. An observed NIR spectrum, $A(\lambda)$, can be represented as follows;

$$A(\lambda) = \alpha A_0(\lambda) + \beta + e(\lambda) \tag{4.4}$$

Here, $A_0(\lambda)$, α, β, and $e(\lambda)$ are a real spectrum, a multiplicative scatter factor (amplification factor), an additive scatter factor (offset deviation), and noise, respectively. There are several methods to eliminate or reduce the effects of α and β. We explain three of them.

Derivative methods

Derivative methods are utilized in NIR spectra for both resolution enhancement and baseline correction [1–3]. (Chap. 7) A derivative spectrum is an expression of derivative values, $d^n A/d\lambda^n$ ($n = 1, 2, \ldots$), of a spectrum $A(\lambda)$ as a function of λ. The second derivative, $d^2 A/d\lambda^2$, is most often encountered. The superimposed peaks in an original spectrum turn out as clearly separated downward peaks in a second-derivative spectrum. Another important property of second-derivative method is the removal of the additive and multiplicative baseline changes in an original spectrum. Figure 4.10a displays NIR spectra of 16 kinds of linear low-density polyethylene (LLDPE) and one kind of high-density polyethylene (HDPE), and Fig. 4.10b shows the second derivative obtained with the Savitzky–Golay method of the spectra as shown in Fig. 4.10a [21]. It can be seen from Fig. 4.10b that the second derivative is powerful in removing additive and multiplicative baseline variations of the spectra, and at the same time, it enables to detect a number of bands clearly. A drawback in the derivative methods is that the SN ratio deteriorates every time a spectrum is differentiated.

Let us explain derivative methods using equations. If the band shape is a Gaussian shape as below;

$$A(x) = \alpha \exp\left\{-\left(\frac{x - x_0}{w}\right)^2\right\}, \tag{4.5}$$

Fig. 4.10 **a** NIR spectra of
16 kinds of linear
low-density polyethylene
(LLDPE) and one kind of
high-density polyethylene
(HDPE). **b** The second
derivative was calculated
with the Savitzky–Golay
method of the spectra in **a**.
c NIR spectra shown in
a after the MSC treatment.
Reproduced from Ref. [21]
with permission

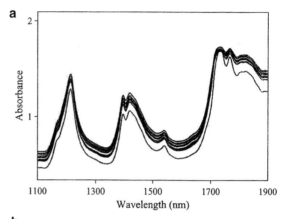

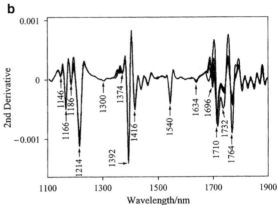

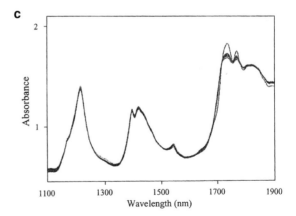

Analytical solutions of the first and second derivatives of the band are

$$\frac{dA}{dx} = -2\frac{x - x_0}{w^2}A = -2\frac{\alpha(x - x_0)}{w^2}\exp\left\{-\left(\frac{x - x_0}{w}\right)^2\right\} \qquad (4.6)$$

$$\frac{d^2A}{dx^2} = -2\frac{A}{w^2}\left(1 - 2\frac{(x - x_0)^2}{w^2}\right) = -2\frac{\alpha}{w^2}\left(1 - 2\frac{(x - x_0)^2}{w^2}\right)\exp\left\{-\left(\frac{x - x_0}{w}\right)^2\right\}$$
$$(4.7)$$

respectively, where α is a peak height of the band, and w is proportional to a full width at the half maximum (FWHM) of the band ($\omega = $ FWHM/$\{$Ln (4)$\}^{0.5}$). As shown in Fig. 4.11, at the peak position ($x = x_0$), the first derivative coefficient is 0, and the second-derivative coefficient is $-2\alpha/\omega^2$. The positions and intensities of the maximum and minimum of the first derivative spectra are $x_0 \pm \omega\sqrt{2}$ and $\mp\sqrt{2}\alpha\omega e^{0.5}$, respectively. As can be seen in these formulas, the maximum and minimum of the first- and second-derivative coefficients are proportional to an area of the band, if the width of the band is not changed.

Fig. 4.11 Original spectrum, its first and second derivatives. Prepared by Y. Morisawa

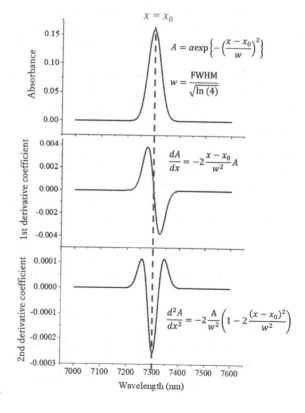

Since, in general, spectrum data take discrete values, and the calculation of derivatives with various orders is performed by algebraic differences between data taken at closely spaced wavelengths. Transformation to first- and second-derivatives is then

$$dA_i = A_{i+k} - A_{i-k}$$

$$d^2A_i = d(A_{i+k} - A_{i-k})$$
$$= A_{i+2k} - 2A_i + A_{i-2k} \tag{4.8}$$

Multiplicative scatter correction (MSC)

MSC is an effective method for correcting vertical variations of the baseline (additive baseline variation) and inclination of the baseline (multiplicative baseline variation) [17]. The idea of MSC originates from the fact that light scattering has the wavelength dependence different from that of chemically based light absorbance. Thus, we can use data from a number of wavelengths to distinguish between light absorption and light scattering.

MSC corrects spectra according to a simple linear univariate fit to a standard spectrum. α and β are estimated by least squares regression using the standard spectrum. As the standard spectrum, a spectrum of a particular sample or an average spectrum is used. Figure 4.10c shows the NIR spectra as shown in Fig. 4.10a after the MSC treatment [21]. The spectra demonstrate the potential of MSC in correcting offset and amplification in the NIR spectra. Generally, MSC improves essentially the linearity in NIR spectra. While it is generally an very useful technique, care must be exercised since the use of MSC may generate unwanted artifacts.

Standard normal variate (SNV)

Standard normal variate (SNV) is also a powerful method for correcting vertical baseline drift of a set of spectra [22]. For each spectrum $A(\lambda)$, SNV is calculated as $A_{SNV}(\lambda) = (A(\lambda) - \dot{A})/\sigma$, where $\dot{A}$ and σ are mean and standard deviation of the intensities in the spectrum, respectively. Therefore, mean and standard deviation of the intensities for each spectrum after SNV are standardized as 0 and 1, respectively.

4.3.3 Resolution Enhancement Methods

Resolution enhancement methods are very important to unravel overlapping bands and elucidating the existence of obscured bands [2, 3]. In NIR spectroscopy derivative methods, difference spectra, mean centering, and Fourier self-deconvolution are employed as resolution enhancement methods. Note that PCA loadings plots are often effective for resolution enhancement [21]. 2D correlation spectroscopy can also make resolution enhancement, but since it is not a pretreatment method, it will be outlined in Chap. 6. Mean centering will be discussed in centering and normalization section.

Derivative methods

We already explained the usefulness of derivative method in resolution enhancement as shown in Fig. 4.10a, b. Here, a problem in the derivative methods which we often encounter is pointed out. Figure 4.4e shows the second derivative of the spectra as shown in Fig. 4.4a. It can be seen from Fig. 4.4e that the 7250 cm^{-1} peak is much weaker than the 7300 cm^{-1} peak. In the second-derivative spectra, a broad band is often underestimated, and thus, care must be taken for the second derivative of a broad band.

4.3.4 Centering and Normalization Methods

Centering and normalization are often effective in chemometrics analysis of NIR data [1–3]. (Chap. 7) Mean centering is simply an adjustment to a data set to reposition the centroid of the data to the origin of the coordinate system [2, 3]. Normalization is an adjustment to a data set that equalizes the magnitude of each spectrum. [2, 3]

Centering methods

Mean centering is a method where from every element of the jth spectrum (row) the column mean is subtracted:

$$X_{jcent} = X_j - \left(\frac{1}{n} \sum_{j=1}^{n} X_{ij} \right) \tag{4.9}$$

X_j and X_{ij} are an element of the jth spectrum and that of a data matrix X, respectively. After this step, all means are zero and variances are spread around zero. Each mean centering spectrum can be regarded as a difference spectrum between the individual spectrum and an averaged spectrum. Mean centering is often powerful in resolution enhancement. Figure 4.12a displays NIR spectra in the region of 6000–5500 cm^{-1} of nylon 12 collected in a temperature range from 30 to 150 °C, and Fig. 4.12b exhibits their mean-centered spectra [23]. The mean-centered spectra show that the intensity of a band at 5770 cm^{-1} arising from the first overtone of CH$_2$ stretching mode varies markedly with temperature. Mean centering is used also as a pretreatment for constructing 2D correlation spectra (Chap. 6).

Normalization

Two popular normalization procedures have been known in common practice [2, 3]. Most normalization methods employ vectors normalized to constant Euclidean norm. That is,

Fig. 4.12 a NIR spectra in
the 6000–5500 cm^{-1} region
of nylon 12 measured over a
temperature range from 30 to
150 °C. **b** Mean-centered
spectra of the spectra in **a**.
Reproduced from Ref. [23]
with permission

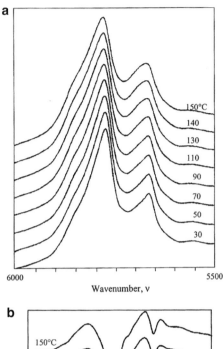

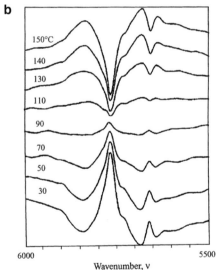

$$x_{j,\,norm} = x_j / ||\mathbf{x}|| \qquad (4.10)$$

where ||**x**|| is the Euclidean norm of the spectral vector **x**. This normalization trans-
forms the spectral points on a unit hypersphere, and all data are approximately in
the same scaling. This normalization has a good property that the similarity between
two spectral vectors may be estimated by the scalar product of these two vectors.
However, the normalization induces the geometric configuration of the data points,

either the clustering structure or the spreading directions, substantially different from the original one, which may result in a misleading in the understanding of the data in exploratory data analysis. In addition, the variation in spreading directions has a significant effect on principal component analysis (PCA)-related analysis. Therefore, one must be careful enough in using normalization in situations where exploratory data analysis and PCA-related procedures, such as PCA, partial least squares (PLS), and so on, are concerned.

Another normalization procedure is so-called mean normalization, where all points of the jth spectrum are divided by its mean value

$$X_{j\mathrm{norm}} = X_j / \left(\frac{1}{m} \sum_{i=1}^{m} X_{ij} \right) \tag{4.11}$$

where m is a total number of spectral points. After mean normalization, all the spectra have the same area. Essentially, mean normalization is equivalent to normalize the spectral vectors to constant 1-norm, that is, the sum of spectral values (always positive) equals to a constant. This means that the geometry of mean normalization is to transform the spectral points to be contained in a convex set, and the dimensionality of the spectral space is thus decreased by 1. This transformation is very useful in self-modeling curve resolution (SMCR).

References

1. H. W. Siesler, Y. Ozaki, S. Kawata, H. M. Heise, Eds., *Near-Infrared Spectroscopy, Principles, Instruments, Applications* (Wiley-VCH, 2002)
2. Y. Ozaki, W. F. McClure, A. A. Christy, eds., *Near-Infrared Spectroscopy in Food Science and Technology* (Wiley-Interscience, 2007)
3. H. Martens, M. Martens, *Multivariate Analysis of Quality; An Introduction* (John Wiley and Sons, 2001)
4. D. A. Burns, E. W. Ciurczak eds., *Handbook of Near-Infrared Analysis*, 3rd edn. (Practical Spectroscopy) (CRC Press, 2007)
5. Y. Ozaki, C. W. Huck, K. B. Beć, Near-IR spectroscopy and its applications, in *Molecular and Laser Spectroscopy: Advances and Applications,* edited by V. P. Gupta (Elsevier, 2017), p. 11
6. J. Workman, Jr., L. Weyer, *Practical Guide and Spectral Atlas for Interpretive Near-Infrared Spectroscopy*, 2nd edn. (CRC Press, 2012)
7. M.A. Czarnecki, Y. Morisawa, Y. Futami, Y. Ozaki, Advances in molecular structure and interaction studies using near-infrared spectroscopy. Chem. Rev. **115**, 9707–9744 (2015)
8. K.B. Beć, J. Grabska, C. W. Huck, Y. Ozaki, Quantum mechanical simulation of near-infrared spectra: Applications in physical and analytical chemistry, in *Molecular Spectroscopy; A Quantum Chemistry Approach*, edited by Y. Ozaki, M. J. Wojcik, J. Popp, Wiley-VCH, vol. 1, pp. 353–388 (2019)
9. P.J. Gemperline, J.R. Long, V.G. Gregoriou, Nonlinear multivariate calibration using principal components regression and artificial neural networks. Anal. Chem. **63**, 2313–2323 (1991)
10. Y. Katsumoto, D. Adachi, H. Sato, Y. Ozaki, Useless of a curve fitting method in the analysis of overlapping overtones and combinations of CH stretching modes. J. NIR Spectrosc. **10**, 85–91 (2002)

11. H. Maeda, Y. Ozaki, M. Tanaka, N. Hayashi, T. Kojima, Near infrared spectroscopy and chemo-metrics studies of temperature-dependent spectral variations of water: relationship between spectral changes and hydrogen bonds. J. NIR Spectrosc. **3**, 191–201 (1995)

12. J. Grabska, K.B. Beć, M. Ishigaki, C.W. Huck, Y. Ozaki, NIR spectra simulations by anharmonic DFT-saturated and unsaturated long-chain fatty acids. J. Phys. Chem. B **122**, 6931–6944 (2018)

13. M.A. Czarnecki, Y. Liu, Y. Ozaki, M. Suzuki, M. Iwahashi, Potential of Fourier transform near-infrared spectroscopy in studies of the dissociation of Fatty acids in the liquid phase. Appl. Spectrosc. **47**, 2162–2168 (1993)

14. Y. Hu, J. Zhang, H. Sato, Y. Futami, I. Noda, Y. Ozaki, C-H...O=C hydrogen bonding and isothermal crystallization kinetics of poly(3-hydroxybutyrate) investigated by near-infrared spectroscopy. Macromol. **39**, 3841–3847 (2006)

15. L. Bokobza: Origin of near-infrared absorption bands, in Ref. [1], p. 11–42

16. Y. Sugawara, A.Y. Hirakawa, M. Tsuboi, In-plane force constants of the peptide group: Least-squares adjustment starting from ab initio values of N-methylacetamide. J. Mol. Spectrosc. **108**, 206–214 (1984)

17. P. Geladi, D. MacDougall, H. Martens, Linearization and scatter-correction for near-infrared reflectance spectra of meat. Appl. Spectrosc. **39**, 491–500 (1985)

18. A. Savitzky, M.J.E. Golay, Smoothing and differentiation of data by simplified least squares procedures. Anal. Chem. **36**, 1627–1639 (1964)

19. F.T. Chau, T.M. Shih, J.B. Gao, C.K. Chan, Application of the fast wavelet transform method to compress ultraviolet-visible spectra. Appl. Spectrosc. **50**, 339–348 (1996)

20. C.L. Stok, D.L. Veltkamp, B.R. Kowalski, Detecting and identifying spectral anomalies using wavelet processing. Appl. Spectrosc. **52**, 1348–1352 (1998)

21. M. Shimoyama, T. Ninomiya, K. Sano, Y. Ozaki, H. Higashiyama, M. Watari, M. Tomo, Near infrared spectroscopy and chemometrics analysis of linear low-density polyethylene. J. NIR Spectrosc. **6**, 317–324 (1998)

22. R. Barnes, M. Dhanoa, S.J. Lester, Standard normal variate transformation and de-trending of near-infrared diffuse reflectance spectra. Appl. Spectrosc. **43**, 772–777 (1989)

23. Y. Ozaki, Y. Liu, I. Noda, Two-dimensional near-infrared correlation spectroscopy study of premelting behavior of nylon 12. Macromol. **30**, 2391–2399 (1997)

Chapter 5
Introduction to Quantum Vibrational Spectroscopy

Krzysztof B. Beć, Justyna Grabska, and Thomas S. Hofer

Abstract In this chapter, the quantum mechanical basis for computational studies of near-infrared spectra (NIR) is discussed. Since this topic is rarely covered in detail in the literature, the necessary prerequisites are provided as well, which include (i) the coordinate frame for the description of molecular vibrations, (ii) methods for the determination of the vibrational potential, (iii) the principles of the harmonic approximation, and (iv) its role as the foundation for methods taking anharmonic effects into account. The details of various anharmonic approaches in quantum vibrational spectroscopy are discussed, including methods based on the vibrational self-consistent field (VSCF) approach, vibrational perturbation theory (VPT) as well as one- and multidimensional grid-based methods. The merits and pitfalls of these approaches are critically assessed from the perspective of applications in NIR spectroscopy. Selected examples from recent literature are included to demonstrate how these methods can be applied to solve practical problems in spectroscopy. The aim of this chapter is to provide a comprehensive presentation of the topic aimed at a spectroscopic audience, while remaining accessible and focused on the key details. Although primarily intended for readers interested in NIR spectroscopy, the essential information provided in this chapter represents a fundamental perspective on quantum vibrational absorption spectroscopy and is useful for a more general readership as well.

Keywords Quantum vibrational spectroscopy · Harmonic approximation · Overtone and combination transitions · Anharmonic methods

K. B. Beć · J. Grabska
CCB-Center for Chemistry and Biomedicine, Institute of Analytical Chemistry and Radiochemistry, Leopold-Franzens University, Innrain 80/82, 6020 Innsbruck, Austria
e-mail: Krzysztof.Bec@uibk.ac.at

J. Grabska
e-mail: Justyna.Grabska@uibk.ac.at

T. S. Hofer (✉)
CCB-Center for Chemistry and Biomedicine, Institute of General, Inorganic and Theoretical Chemistry, Leopold-Franzens University, Innrain 80/82, 6020 Innsbruck, Austria
e-mail: t.hofer@uibk.ac.at

© Springer Nature Singapore Pte Ltd. 2021
Y. Ozaki et al. (eds.), *Near-Infrared Spectroscopy*,
https://doi.org/10.1007/978-981-15-8648-4_5

5.1 Introduction

The aim of this chapter is to present the essential information required to obtain a fundamental understanding of quantum vibrational absorption spectroscopy, in particular near-infrared (NIR) spectroscopy. The discussion highlights the critical aspects to provide an in-depth and accessible overview aimed at a spectroscopic audience. The necessary basics include the commonly used coordinate frame for the description of molecular vibrations, an overview of the role of the vibrational potential and methods for its determination, as well as the critical factor in all applications of quantum chemistry being the computational complexity of a given approach. Considerable attention is focused toward the harmonic approximation, the fundamental framework underlying most applications of theoretical vibrational spectroscopy. The harmonic approximation is in general not sufficiently accurate for the needs of NIR spectroscopy. However, it is an essential foundation for advanced anharmonic treatments. The majority of these methods are either built on the basis of a harmonic Hamiltonian (VPT2), adopt a harmonic Hessian as the reference state (VSCF) or use the harmonic analysis (i.e., harmonic normal modes) in the process of probing the true vibrational potential (grid-based methods). The details of various anharmonic approaches are discussed, and their specific merits and shortcomings examined from the point of view of applications in NIR spectroscopy. This outline is based on several examples selected from recent literature.

5.2 Normal Modes of Vibration

Commonly, literature introducing the principles of vibrational spectroscopy mainly employs the example of a simple diatomic molecule. However, this kind of two-body system is limited to a single mode of vibration resulting from a one-dimensional potential and is not suitable for a complete presentation of the main concepts in vibrational analysis [1–4]. The total number of degrees of freedom (DOF) in a chemical system is $3N$, where N is the number of atoms. Translational and rotational motion can only be defined in an external coordinate system; thus, the translational and rotational DOF are invariant in the molecule's frame of reference. This sets them apart from the internal DOF (vibrational DOF; vibrational modes). The number of vibrational DOF equals to $3N - N_{inv}$. N_{inv} is generally partitioned into three translational and three rotational DOF (along the x, y, z directions); however, no change in the potential energy is associated to the rotation over the main rotational axis of linear molecules (including diatomic ones). Additionally, N_{inv} of periodic systems only considers uniform translational DOF (x, y, z) of the entire lattice. This effectively leaves $3N - 6$ modes for nonlinear molecules, $3N - 5$ modes for linear molecules, and $3N - 3$ modes for periodic systems. Note that these different DOFs need to be separated, e.g., no translation of the molecule's center of mass may occur along the vibrational mode. The concept of normal modes in computational chemistry has its origin in the formalism of the harmonic approximation and will be outlined in detail in Sect. 5.4 of this chapter.

In polyatomic systems, symmetric and antisymmetric modes occur due to symmetry factors [1, 2]. In addition, deformation modes appear as well; they involve change of valence and dihedral angles between the atoms in the system. As a rule, force constants associated with stretching modes are typically higher, and thus, the wave numbers of these vibrations are higher compared to associated deformation modes. For instance, CO_2 is a linear molecule and thus has $3N - 5 = 4$ modes of vibration (Fig. 5.1). Among them are the two stretching modes, being symmetric and antisymmetric. The CO_2 symmetric stretch (v_1) is IR inactive because there is no change in the dipole moment of the molecule along the associated vibrational coordinate. In contrast, the antisymmetric stretching vibration (v_3) generates a significant net change of the dipole moment giving rise to a strong IR band observed at ca. 2345 cm^{-1} in gas phase. The two deformation modes of CO_2 involve the bending of the OCO angle in the molecule. These two modes differ only from the point of view of an external coordinate system; the vibrations occur along perpendicular planes. However, from the molecule's point of view, they are indistinguishable, and their energies (and thus wavenumbers) are degenerate giving rise only to a single IR absorption band v_2 located at 667 cm^{-1} in gas phase. Therefore, despite possessing four vibrational degrees of freedom, only two fundamental bands of CO_2 are observed in the respective IR spectrum. However, IR spectra of gaseous molecules are further complicated because of rotational–vibrational coupling. Water serves as an archetypical nonlinear molecule; it has $3N - 6 = 3$ vibrational degrees of freedom, with only a single deformation mode v_2 (Fig. 5.1).

Since the center of mass of the vibrating molecule may not change its position in space, the atomic displacements associated to these normal modes often involve displacements of all atoms in the molecule. These are not necessarily large amplitude motions, however. Water serves a good example, as large amplitude motions of the light-weighted hydrogen atoms are accompanied by a low-amplitude motion of the heavy oxygen atom; this is reflected in an exaggerated way in Fig. 5.1. These complex

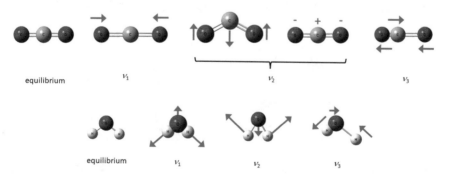

Fig. 5.1 Equilibrium geometries and normal modes of a carbon dioxide (top) and a water (bottom) molecule, respectively

atom displacements defined in normal coordinate system are difficult to interpret in larger systems; normal coordinates are also specific to a given molecular symmetry and not transferrable to other ones. Nevertheless, similarities exist between vibrations of molecules constituting similar functional groups and having comparable structures. A more useful description of vibrational motion is achieved by defining an alternative coordinate system based on the structural parameters of the systems such as bond lengths, valence, and dihedral angles. The commonly accepted standard is the internal coordinate system proposed by Pulay et al. [5]; often referred to as natural coordinate system. The deformation vibrations of functional groups most commonly found in organic molecules (methyl, $-CH_3$; and methylene sp^3; $>CH_2$) defined in natural coordinates are presented as an example in Table 5.1 and Fig. 5.2. The definitions of the other vibrations can be found in Ref. [5]. The transformation of

Table 5.1 Recommended internal coordinate system at the example of methyl and methylene (sp^3) groups. The complete definition for other types of functional groups can be found in Ref. [5]

Bond stretchings	Individual coordinates rather than combinations; possible exceptions: methyl and methylene groups where symmetrized combinations of the CH stretchings may be used
Methyl deformation	Sym. def. $= \alpha_1 + \alpha_2 + \alpha_3 - \beta_1 - \beta_2 - \beta_3$
	Asym. def. $= 2\alpha_1 - \alpha_2 - \alpha_3$
	Asym. def.$' = \alpha_2 - \alpha_3$
	Rocking $= 2\beta_1 - \beta_2 - \beta_3$
	Rocking$' = \beta_2 - \beta_3$
Methylene (sp^3) deformation	CH_2 scissoring $= 5\alpha + \gamma$
	CXY scissoring $= \alpha + 5\gamma$
	CH_2 rocking $= \beta_1 - \beta_2 + \beta_3 - \beta_4$
	CH_2 wagging $= \beta_1 + \beta_2 - \beta_3 - \beta_4$
	CH_2 twisting $= \beta_1 - \beta_2 - \beta_3 + \beta_4$

Sym.—symmetric; asym.—antisymmetric; def.—deformation
Adopted with permission from Pulay et al. [5]. Copyright (1979) American Chemical Society

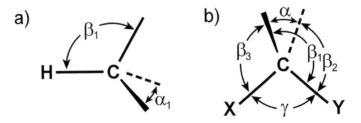

Fig. 5.2 Definition of internal coordinates in: **a** a methyl ($-CH_3$) group; **b** a methylene sp^3 ($>CH_2$) group

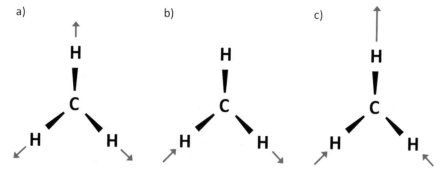

Fig. 5.3 Methyl stretching vibrations; **a** symmetric mode; **b, c** two kinds of antisymmetric modes

molecule-specific normal coordinates into natural coordinates enables a straightforward comparison between the vibrational properties of different molecules, providing considerable benefits when analyzing IR spectra.

The concept of Pulay's natural coordinate system suggest not to group stretching vibrations, but rather treat them as individual bonds. The allowed exceptions include methyl and methylene group, for which symmetrized combinations of C–H stretching vibrations may be used. The number of stretching vibrations specific to these functional groups is ruled by the number of involved DOF. Considering an archetypical system with a methyl group, X–CH$_3$ ($N = 5$), the number of vibrations is $3N - 6 = 9$. This is partitioned into five deformation vibrations (symmetric, two kinds of antisymmetric, and two kinds of rocking vibrations; Table 5.1) and four stretching vibrations. One stretching mode involves the X–C(H$_3$) bond, leaving three possible stretching vibrations of the CH$_3$ moiety itself; one symmetric and two kinds of antisymmetric stretching modes (Fig. 5.3). A methylene group features just two degrees of freedom due to stretching vibrations, being symmetric and antisymmetric.

5.3 The Underlying Phenomena

5.3.1 The Potential Energy of a Molecular Oscillator

From the point of view of quantum vibrational spectroscopy, the primary problem focuses on the determination of the potential energy function along the spatial coordinate describing the molecular oscillator, or in other words, the motion of the nuclei (Fig. 5.4) [6]. The potential is the key property that dictates the quantum states (i.e., the vibrational wavefunctions) of a molecular oscillator. Following the fundamental approximation of quantum chemistry, the Born–Oppenheimer approximation, the motion of nuclei can be treated separate from the motion of the electrons in the majority of cases. Consequently, in vibrational problems, the electronic structure is reduced to the source of an external potential energy. Therefore, prior to any step

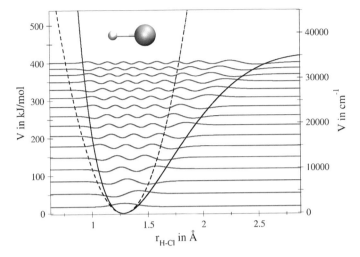

Fig. 5.4 Interatomic potential (solid black) and the associated harmonic approximation (dashed black) of the HCl molecule in the gas phase at CCSD(T)/aug-cc-pVQZ level and the respective vibrational wavefunctions (red) obtained via the Numerov approach. The associated energy differences between the individual states correspond to the wavenumbers measured via vibrational spectroscopy

made into quantum vibrational spectroscopy, the electronic structure of the system under consideration needs to be determined [6]. This can be accomplished with a large array of different approaches, aimed at providing various approximations to balance the accuracy of results and the computational demand.

5.3.2 Quantum Chemical Methods for the Determination of the Electronic Structure of Molecular Systems

An approach for the determination of the electronic structure of a quantum system (Fig. 5.5) may follow two principal ways of categorization; conceptual and practical. From the conceptual point of view, these methods differ by how the energy of the system is described. The major approaches will be presented in Sects. 5.3.2.1, 5.3.2.2, 5.3.2.3, 5.3.2.4, 5.3.2.5, and 5.3.2.6. More exhaustive information can be found in topic-oriented textbooks [6]. The practical categorization mostly concerns their computational complexity, with the cost versus accuracy factor of the available methods being a key consideration, which translates into their respective applicability to certain problems.

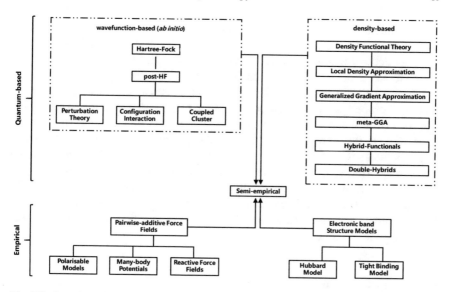

Fig. 5.5 Overview of computational chemistry methods commonly used for the determination of electronic structure in molecular systems and crystals

5.3.2.1 Hartree–Fock Theory

Any quantum-based treatment assumes that a wavefunction is the primary entity describing the state and therefore all observables such as the energy of a quantum many-body system. Hartree–Fock (HF) theory provides the fundamental and most straightforward approach in this category. This method is based on solving an approximate time-dependent Schrödinger equation (HF equation), which describes the state of a quantum mechanical system and the associated energy. All methods that are in practical use are based on the Born–Oppenheimer approximation, which limits the role of the nuclei to the source of an external potential. The interaction between the electrons involves the exchange and the correlation potential. Within the HF formalism, only the former is treated appropriately. The exchange energy results from the indistinguishability of electrons and is reflected in the HF procedure by antisymmetric properties of the wavefunction. This step is accomplished by deriving a single Slater determinant, an antisymmetrized product of one-electron wavefunctions (i.e., orbitals), to approximate the wavefunction of an N-body quantum system. In other words, the HF formalism assumes that the problem of interactions between the many electrons in the molecule is separable into a set of electron–electron problems, coupled through an averaged effective potential that describes the interaction with all other electrons in the system.

A solution to the HF equation is found by invoking the variational principle, in which a set of N-coupled equations for the N spin orbitals is derived, yielding the Hartree–Fock wavefunction and energy of the system. The HF framework belongs to the family of self-consistent field (SCF) methods, as self-consistency is a criterion that needs to be fulfilled within the iterative procedure of solving the Hartree equations. The most far-reaching approximation assumed in the HF approach is neglecting the Coulomb correlation, which is often described as the mean-field charge distribution approximation of the electron correlation, since the HF method effectively averages the electron–electron interactions. This causes an inherent inability of the method to properly describe London dispersion. In order to step beyond a mean-field approximation of independent particles, so-called post-HF methods have been developed. The HF theory has a critical historical importance, being the first developed quantum theory with practical implementation. Nowadays, pure HF calculations are rarely used. However, the method is still widely adopted for calculations of the initial wavefunction of a quantum system, thus representing the preliminary step for calculations at higher levels of theory. On the other hand, a hierarchy of increasingly accurate methods based on the HF results exist, in which more than one Slater determinant is employed.

5.3.2.2   Post-HF Approaches

Different populations of atomic orbitals by electrons or electron configurations in a quantum system are possible. When any given electron changes its configuration, which can be described as an excitation into another orbital, the distribution of the other electrons in the molecule adjusts to minimize the total energy of the system. Thus, the motion of electrons is not independent but correlated, which lowers the total energy of the system. However, in the HF approach, any given electron only interacts with the average potential of all the other electrons in the system. To amend this shortcoming, post-HF methods aiming at a more accurate treatment of electron correlation effects were introduced. This may be accomplished in different ways, but unequivocally increases the computational complexity of the method by orders of magnitude.

In the simplest case, the electron correlation energy can be treated as a perturbation of the electronic state described in the HF formalism. As long as electron correlation has a relatively small contribution to the total energy, it can be expressed via a perturbing Hamiltonian corresponding to a correction added to the HF Hamiltonian. Since the unperturbed HF state is known, the perturbative correction is solvable using approximate methods, e.g., via an asymptotic series. The practical formulation of this approach is based on Møller–Plesset theory (MP) of a given order k. Zeroth-order wavefunction corresponds to an unperturbed HF state, and the first-order perturbation correction (MP1) to the HF energy can be shown to be equal to zero, which implies that only second- (MP2) and higher-order MP expressions are practically meaningful. Among those, MP2 bears the highest practical usefulness and finds broad applications. In most cases, higher-order perturbations (such as MP3, MP4, and MP5) do not improve the accuracy by an acceptable margin and display a huge computational demand. The MP2 method has become particularly widely applied since it is the most efficient approach to take electron correlation effects into account. However, there are limitations in the applicability of MP theory, which led to the development of more advanced approaches.

Unlike the HF formalism, configuration interaction (CI) theory utilizes multiple Slater determinants to construct configuration state functions (CSF), which are then linearly combined to describe the wavefunction of the quantum system. The first term in the expansion of the CI wavefunction is equivalent to the HF ground-state wavefunction, while the higher terms capture the effects of the correlated motion of the electrons. In the CI formalism, the wavefunction is a combination of the HF reference states plus all possible excited states. This is reflected by mixing the ground CSFs and the excited CSFs. If all possibilities of orbital occupations are included (full-CI, FCI), an exact solution to the electron correlation problem can be achieved. Unfortunately, the number of excited configurations is enormously large, and in practice, the number of CI terms representing the electronic excitations needs to be truncated. The abbreviations for truncated CI variants reflect the excitation levels treated; 'S' for single excitations, 'D' for double, 'T' for triple, 'Q' for quadruple. This leads to CI single and double excitations (CISD), CI single, double, and triple excitations (CISDT), etc. From the point of view of quantum theory, CI is the most complete approach to describe the electronic structure of molecular systems. However, this corresponds only to FCI, that is, the case in which all orbital occupations possible for the quantum system are treated. The FCI method is useful for validation and benchmarking purposes of lower-level quantum methods, where its extensive computational cost remains manageable. In practical terms, unless FCI conditions are achieved, the application of truncated variants is often linked to considerable inaccuracies. Truncated CI methods capture a rapidly decreasing amount of the 'exact' correlation energy with an increase of the system size, which limits their usefulness in treating larger molecules and heavy atoms. Multiconfigurational self-consistent field (MCSCF) is an analogous approach that additionally applies a similar CI-like concept also to derive the one-electron functions that are subsequently used to construct CSFs.

Coupled-cluster theory (CC) expands the molecular orbitals obtained at HF level using an exponential cluster operator (acting as the excitation operator) and constructs a multi-electron wavefunction that includes electron correlation. The CC formalism may be considered as an alternative to CI, which produces an equivalent combination of one-electron functions to yield the multi-electron wavefunction. However, unlike linear combination assumed in the latter, the exponential expansion used in the former grants its size-extensivity resulting in an improved limiting behavior of the CC correlation energy upon truncation to a given excitation level (e.g., CCSD). Similar practical limitations as those found in CI apply here as well. The number of treated electronic excitations needs to be limited in order to make the method applicable in terms of the associated computational demand. This leads to variants, abbreviated analogous to CI variants, e.g., CCSD, CCSDT, etc. Unfortunately, in practical use, the CCSD level yields moderately correct results considering its cost, while the more accurate CCSDT proves to be too expensive for most applications. For this reason, CCSD(T) was introduced as a variant approximating the triple excitations via perturbation theory. Note, however, that when truncated at the same level,

CC approaches still capture a higher fraction of the correlation energy than their CI counterparts do, albeit at a higher computational cost. The CCSD(T) variant is highly valued for its high accuracy achieved at a relatively acceptable computational cost, and is often considered as the "golden standard of quantum chemistry."

5.3.2.3 Density-Functional Theory

Inclusion of electron correlation in wavefunction-based methods leads to a steep increase in their computational complexity. This gave an impulse for the development of a fundamentally different concept known as density-functional theory (DFT) [7, 8]. It is based on the Hohenburg–Kohn theorems postulating that the state of a many-electron system can be described based on a unique functional (i.e., a function acting on another function), which in this case is the spatially dependent electron density function. The benefit of this formalism is a reduction of the dimensionality of the problem from that of a multidimensional ($3N$) N-electron wavefunction to a three-dimensional electron density function. The practical implementation of DFT became possible due to the formulation of the Kohn–Sham equations, which enabled the reduction of an intractable problem of interacting electrons in a static external potential to a tractable problem of non-interacting electrons in a local effective potential, i.e., the Kohn–Sham potential. The latter is constituted by the external potential plus electron exchange and correlation effects expressed via the associated exchange–correlation functional E_{xc}. Unfortunately, the exact E_{xc} is unknown except for the limiting case of a free electron gas, which became known as the local-density approximation (LDA; E_{xc}^{LDA}). While this formalism is applicable in case of metals and simple ionic solids, the LDA approximation fails to deliver satisfactory results for more complex systems. The meaningful development of DFT within the regime of chemistry started with the introduction of the generalized-gradient approximation (GGA) level, which was followed by more advanced approximations, such as meta-GGA functionals. A significant progress in the underlying theory was marked with the introduction of hybrid Kohn–Sham theory and the resulting hybrid formulation for E_{xc}. A hybrid E_{xc} is constructed as a linear combination of GGA and/or LDA (explicit) functionals and a HF 'exact' exchange functional (implicit functional). This inclusion of an ab initio electron exchange term in hybrid functionals greatly improved the accuracy and applicability of DFT. Popular hybrid functionals include the B3LYP, PBE0, HSE, and M06 functionals. Further advancement was achieved with the development of double-hybrid functionals. These approaches represent a natural progression from hybrid functionals, as in addition to the exchange term, ab initio correlation is included as well. The correlation is calculated similar to post-HF methods, e.g., MP2 correlation is employed in B2PLYP, mPW2PLYP, PBE0DH, or PBEQIDH double-hybrids. In addition to a much improved treatment of electron correlation, double-hybrid functionals also enable a better implementation of HF exchange; however, they are significantly more expensive than single-hybrid functionals.

The DFT concept is a rigorous re-interpretation of the quantum many-body problem. It offers a significant improvement in the affordability of calculations. Unfortunately, its practical implementation needs to include approximated electron exchange and correlation. The formulation of DFT limits the ability to improve its quality systematically. Instead, different functionals have been parametrized (i.e., calibrated) toward better accuracy when applied to certain systems. Inherent limitations of DFT, e.g., poor description of long-range (dispersive or non-covalent) interactions were recently mitigated by introduction of empirical corrections of dispersion, e.g., the series of Grimme's dispersion models (GD). Despite some shortcomings, DFT offers highly favorable cost versus accuracy level that made it particularly widely used in spectroscopic studies.

5.3.2.4   Semi-empirical Concept

Semi-empirical quantum chemistry methods are derived by insertion of pre-determined parameters into quantum mechanical calculation schemes. The most straightforward semi-empirical treatment replaces the relatively most time-consuming calculation procedures in the HF ansatz, i.e., two-electron integrals are omitted and their values are provided as empirical parameters to produce the expected results. These parameters are most often obtained from higher-level quantum mechanical calculations performed for small-scale models, and then used universally. Semi-empirical methods are significantly more affordable than their corresponding quantum mechanical frameworks, and thus suitable for the treatment of large molecules. Conceptually, in some cases semi-empirical schemes are relatively more complete, as empirical parameters may better describe some phenomena (e.g., electron correlation effects) than the ab initio approach with necessary approximations. Accordingly, as long as the considered system fits the conditions of the parametrization, semi-empirical calculations may yield more accurate results than when treated with a pure HF formalism. However, semi-empirical calculations are prone to produce erroneous results if they are applied outside of their area of parametrization. Therefore, they need to be used with care. Semi-empirical schemes based on a wavefunction ansatz include the Austin Model 1 (AM1), the parametric model family of methods (e.g., PM3, PM5) that implement the neglect of differential diatomic overlap (NDDO) principle (all two-electron integrals involving two-center charge distributions are neglected) as well as a number of additional approximations and corrections, depending on the particular method. A similar concept may also be applied to density-based methods. For example, density-functional-based tight-binding (DFTB) inserts pre-calculated parameters into the DFT calculation scheme, in which a minimal basis and only nearest-neighbor interactions are employed. The resulting deficiency in the description of long-range interactions is corrected with empirical dispersion (analogous to those developed for DFT functionals). The resulting approach yields reasonably accurate results at a fraction of the cost of DFT calculations. Although primarily popular decades ago, when the technology barrier prevented wider use of higher-level quantum methods, semi-empirical methods remain continuously evolving with newer variants developed, e.g., PM6, PM7, or new concepts introduced such as self-consistent change density-functional tight binding (SCC DFTB).

5.3.2.5 Molecular Mechanics

An alternative to QM-based approaches is the description of interatomic potentials in an entirely empirical way. These methods are typically referred to as molecular mechanics (MM) or force fields (FF) [9–11]. In this approach, the potential energy is calculated as a function of the nuclear coordinates using empirical (i.e., pre-parametrized) interaction potentials. Accordingly, MM uses classical mechanics to describe the forces acting between the atoms in a molecule. In the most fundamental approach, the interatomic potential energy is described as a sum of non-covalent pairwise interactions resulting from electrostatic (Coulomb) and van-der-Waals (e.g., Lennard-Jones) contributions, while covalent contributions such as bond and valence angle interactions are often represented via harmonic potentials centered on preoptimized equilibria. These pair-wise additive approaches comprise the simplest possible description of the systems and are typically applied in the regime of (bio)organic chemistry (e.g., peptide/protein systems, nucleic acids, organic polymer materials) as well as for the treatment of simple solid-state systems such as oxide materials. In order to improve the accuracy of these approaches over the pair-wise additive character, a variety of improved MM methods have been developed. One of the simplest approaches to improve the pair-wise additive character is the inclusion of explicit coupling terms for bonded interactions with the Urey–Bradly angular term and the Axilrod–Teller three-body potential being typical examples. More advanced frameworks comprise the inclusion of polarization effects, which can for example be achieved using charge-on-spring/shell models, explicit polarization approaches as well as charge equilibration schemes. While these approaches are essentially linked to the Coulombic character of the interaction, many-body potentials such as the Finnis–Sinclair and embedded-atom models (EAMs) attempt to improve the description of the non-Coulombic contributions with typical applications being in the area of metals, alloys, and semiconductors. A comparably challenging yet highly intriguing development enjoying increased success in recent years is the formulation of dissociative/reactive force field approaches, capable of adequately describing the formation and cleavage of chemical bonds along the calculation.

The approximate nature of the interatomic forces described this way implies that force fields need to be heavily parametrized to yield an accurate description of the potential energy surface of a molecular system. The practical concept of MM is based on the assumption that a force field parametrized on the basis of a small-scale model, for which more accurate QM methods may be used, is reasonably well transferrable to larger systems. The parametrization may be also based on experimental data, if available. This fundamentally different approach has a significant consequence in the terms of accuracy versus complexity factor. Consequently, MM is applicable to extensively complex molecular systems counting up to millions of atoms.

Therefore, MM is the only method of computational chemistry presently capable of treating multiscale chemical systems. Examples include large biological systems, solvated systems involving a large solvent volume, as well as composite materials. The unmatched affordability of MM makes it useful for molecular dynamics simulations. It is possible, e.g. to obtain vibrational spectra of the molecular models treated by molecular dynamics by calculating the dipole moment autocorrelation function. From the point of view of NIR spectroscopy, however, the MM potentials are too approximate to yield useful results. Briefly mentioned here should be hybrid quantum mechanics/molecular mechanics (QM/MM) approaches, in which only the chemically most relevant part of the molecular system is treated quantum mechanically while MM potentials are considered as sufficiently accurate to model all remaining interactions. These QM/MM schemes enable a more accurate treatment of the potential in key molecular fragments important from the point of view of a particular study.

5.3.2.6 The Fundamental Dilemma in Computational Chemistry; Cost Versus Accuracy Factor

With few exceptions, in computational chemistry, a higher accuracy can only be achieved with a significant increase in the demand for resources, understood mostly as calculation time or/and memory requirements. The nominal complexity of a method is limited to the number of electrons/atoms in the systems and scales distinctly different among the methods presented here. From the point of view of practical applications in spectroscopy, this should be a fundamental consideration as the application of higher levels of theory to the molecular system of interest may become prohibitively expensive. In the most straightforward case, the computational complexity of MM simulations is proportional to the square of the number of treated atomic centers N, $O(N^2)$, whereas advanced implementations are capable of reducing the scaling to $O(N \log N)$. The simplest ab initio HF method formally scales as $O(N^4)$. However, those schemes are widely regarded as not being sufficiently accurate for spectroscopic applications. The significant improvement in accuracy of post-HF approaches comes at a steep increase in their complexity, e.g., starting from $O(N^5)$ for MP2, $O(N^7)$ for CCSD(T), and $O(N^8)$ scaling for CCSDT. The CI formalism elevates this trend further, with CISD $O(N^6)$, CISDTQ $O(N^{10})$, while FCI is known to scale factorial with respect to the system size. In addition, post-HF methods require a larger number of functions describing the distribution of each electron (i.e., basis sets of one-electron functions) to provide accurate results. This gives an answer to the question that may arise at some point, about the root cause for numerous approximations that have been introduced to quantum theory in practical implementations. Such consideration explains the impact that DFT has in the field of practical applications, as it scales as $O(N^3)$, proportionally to the spatial dimensionality of the electron density function. Calculations performed with popular hybrid functionals such as B3LYP nominally scale as $O(N^4)$ but their practical effectiveness is enhanced by a decisively more rapid basis set convergence typical for DFT in comparison with wavefunction-based approaches. In combination with further computational techniques such as efficient pre-screening approaches, the use of sparse matrix algebra routines, as well as convergence accelerators aimed at keeping the number of iterations small, DFT offers a remarkably favorable level of efficiency, an advantage which is well-reflected by the popularity of its use in spectroscopic studies.

5.4 Harmonic Frequency Evaluation

5.4.1 Molecular Geometry Optimization Toward the Energy Minimum

Geometry optimization, or energy minimization, is the procedure of determination of the atomic (nuclear) coordinates of a molecule, which result in the lowest total potential energy of the system. A molecule's potential energy $V(Q)$ is a many-parameter function of its atomic coordinates, represented as the vector $Q = \{q_1, q_2, ..., q_{3N-N_{inv}}\}$. In principle, geometry optimization is a purely mathematical optimization problem of finding Q that minimizes $V(Q)$. In other words, it is a search for atomic coordinates of the molecule that minimize its potential energy. For a stationary point on the potential energy surface (PES), the energy gradient (the derivative of the energy with respect to all atomic coordinates, $\partial V/\partial q_i$) is zero. A generic implementation of the geometry optimization procedure is an iterative process of adjusting Q by following the gradient toward zero. Note, the definition of the atomic coordinates is not implicitly imposed. These may be, e.g., Cartesian coordinates, or internal coordinates describing bond lengths, bond angles, and dihedral angles. The quantum theory model that provides $V(Q)$ is also not imposed from the point of view of the optimization problem. As it will be demonstrated in the next section, geometry optimization performed in order to bring the system to its local minimum on the potential energy surface is a mandatory step prior to the execution of a harmonic frequency analysis.

5.4.2 Harmonic Approximation

Quantum chemical approaches to vibrational motion are in many points analogous to the problem of electronic structure. Accordingly, the theory of the vibrational structure is based on the time-independent vibrational (nuclear) Schrödinger equation [2]. The Born–Oppenheimer approximation still applies, but in this case, the electronic structure is reduced to the role of the source of an external potential upon which the motion of nuclei depends. The vibrational Hamiltonian of a polyatomic oscillator can be expressed as (Eq. 5.1)

$$H = -\frac{1}{2}\sum_i \frac{1}{m_i}\frac{\partial^2}{\partial q_i^2} + \frac{1}{2}\sum_i m_i\omega_{0i}^2 q_i^2 + \sum_{i\leq j\leq k} k_{ijk}q_iq_jq_k + \sum_{i\leq j\leq k\leq l} k_{ijkl}q_iq_jq_kq_l + \cdots$$

$$(5.1)$$

where m_i is the reduced mass of the i-th normal mode and ω_{0i} the corresponding harmonic frequency given as

$$\omega_{0i} = \sqrt{\frac{k_i}{m_i}}$$

$$(5.2)$$

with k_i being the harmonic force constant. The third and higher terms in the expansion describe anharmonic contributions to the vibrational Hamiltonian via the associated cubic and quartic force constants, k_{ijk} and k_{ijkl}, respectively. Commonly, anharmonic contributions diminish consecutively toward higher terms, with the third (cubic) and fourth (quartic) terms capturing the majority of the total anharmonicity.

As it will be demonstrated further, taking into account anharmonic contributions staggeringly increases the complexity of the vibrational problem. However, a universal rule in physics states that the harmonic motion is a generic feature for sufficiently low-amplitude vibrations. This applies reasonably well for a number of molecular vibrations as reflected by relatively low contributions from the anharmonic terms in Eq. 5.1. Based on this premise, an approach called harmonic approximation is constructed. Within this approximation, no coupling between modes is permitted, which implies that all k_{ijk}, k_{ijkl}, and higher-order constants are set to zero. In other words, the normal vibrations of harmonic oscillator are entirely independent. Therefore, in Eq. 5.1, all terms beyond the second one are ignored, in many cases with an acceptable loss of accuracy. Next, the potential in the vicinity of the equilibrium is approximated as a Taylor series (Eq. 5.3)

$$V(Q) = V_0(Q) + \Delta Q^T \cdot g(Q) + \frac{1}{2}\Delta Q^T \mathbf{H}\Delta Q + \cdots \qquad (5.3)$$

with the higher terms in the expansion being neglected. At a stationary point on the PES, i.e. minima and transition states, the gradient $g(Q)$, and hence the second term in Eq. 5.3, is equal to zero as well. This results in a quadratic function as the approximation of the potential, corresponding to a harmonic potential. In practical applications, the mass-weighted second-derivative matrix of the potential, or mass-weighted Hessian $\mathbf{H}$ is introduced, which elements are given as:

$$H_{i,j}^{mw} = \frac{1}{\sqrt{m_im_j}}\frac{\partial^2 V(Q)}{\partial q_i\partial q_j} \qquad (5.4)$$

Diagonalization of the mass-weighted Hessian yields a matrix with $3N - N_{inv}$ columns consisting of orthonormal eigenvectors that describe the vibrational motion of the system within the harmonic approximation, the so-called mass-weighted normal modes. The $3N - N_{inv}$ diagonal elements of the eigenvalue matrix $\mathbf{h}$ are proportional to the square frequency of the associated normal mode.

$$\mathbf{h} = \mathbf{U}^T \mathbf{H} \mathbf{U} \tag{5.5}$$

The example of how the harmonic approximation simplifies the true behavior of a vibrating molecule is demonstrated for the case of a water molecule, H1OH2 (Fig. 5.6), considering a two-dimensional example limited to the two OH stretching vibrations. The corresponding two-dimensional potential energy surface $V(r_{OH1}, r_{OH2})$ is described by the interatomic distances r_{OH1} and r_{OH2} (the corresponding coordinates are depicted in Fig. 5.6a as black lines). In this example, the true potential was determined with high accuracy using the CCSD(T)/aug-cc-pVTZ level of theory employing a tight grid spacing.

As outlined above, the problem of the harmonic oscillator is only solvable at a stationary point of the molecule's PES. In the present example, this means that prior to the evaluation of the Hessian, r_{OH1} and r_{OH2} need to be optimized to identify the

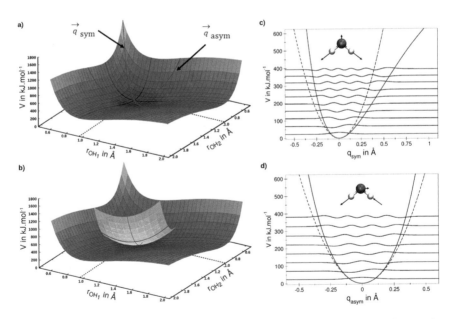

Fig. 5.6 Harmonic analysis at the example of the stretching vibrations of water v_1 (symmetric: $\mathbf{q}_{sym}$) and v_3 (antisymmetric, $\mathbf{q}_{sym}$); **a** the true nature of normal modes on the potential energy surface (red line: $\mathbf{q}_{sym}$; blue line: $\mathbf{q}_{asym}$); **b** the nature of the harmonic approximation applied to these modes; **c** harmonic and anharmonic Morse-like potential curve of $\mathbf{q}_{sym}$; the spacing between subsequent energy levels is increasing; **d** harmonic and quartic anharmonic potential curve of $\mathbf{q}_{asym}$; in contrast to $\mathbf{q}_{sym}$, the spacing between the levels demonstrates a increase upon higher excitation

minimum on the $V(r_{OH1}, r_{OH2})$ surface. To obtain harmonic modes, $V(r_{OH1}, r_{OH2})$ needs to be approximated via the potential of a 2D harmonic oscillator V^{harm}. The key approximation in this case is that the harmonic potential is additive.

$$V^{harm} = V(r_{OH1}) + V(r_{OH2}) \qquad (5.6)$$

Equation 5.6 requires that the potential does not depend simultaneously on r_{OH1} and r_{OH2}, i.e., there is no coupling potential. This implies that the vibrational wavefunction is a product of 1D wavefunctions, and the respective energy eigenvalues are additive (same as the potential case shown above), as described by Eqs. 5.7 and 5.8.

$$|\Psi(r_{OH1}, r_{OH2})\rangle = |\Psi(r_{OH1})\rangle \cdot |\Psi(r_{OH1})\rangle \qquad (5.7)$$

$$E(r_{OH1}, r_{OH2},) = E(r_{OH1}) + E(r_{OH2}) \qquad (5.8)$$

In this case, the Hessian would be diagonal. To match the latter criterion, a reorientation of the coordinate frame is required, which corresponds mathematically to the diagonalization of the mass-weighted Hessian described in Eq. 5.5. The frequency of the harmonic vibration are obtained from the square root of the diagonal entries in $\mathbf{h}$, while the columns in the matrix $\mathbf{U}$ (i.e., the eigenvectors) provide the new coordinates highlighted in red and blue in Fig. 5.6. The data in the matrix $\mathbf{U}$ lead to Eqs. 5.9 and 5.10 describing how to recombine r_{OH1} and r_{OH2} to obtain the harmonic normal modes, $\mathbf{q}_1$ and $\mathbf{q}_3$.

$$q_1 = q_{sym} = \frac{1}{\sqrt{2}} r_{OH1} + \frac{1}{\sqrt{2}} r_{OH2} \qquad (5.9)$$

$$q_3 = q_{asym} = \frac{1}{\sqrt{2}} r_{OH1} - \frac{1}{\sqrt{2}} r_{OH2} \qquad (5.10)$$

This means for $\mathbf{q}_1$ that if r_{OH1} increases, so does r_{OH2}. In contrast, for $\mathbf{q}_3$, if r_{OH1} is elongated, r_{OH2} is shortened (and vice versa). Therefore, $\mathbf{q}_1$ and $\mathbf{q}_3$ refer to symmetric and antisymmetric stretching normal modes, respectively. The representation of the potential in the harmonic approximation now corresponds to the paraboloid depicted in yellow in Fig. 5.6b. Every position on this paraboloid (any point on the harmonic potential surface) is given as the addition of the points lying on the main axes of the re-oriented coordinate frame, i.e., the red and blue line shown at the surface of the paraboloid. This surface dictates the stretching vibrations of the water molecule; symmetric (v_1) and antisymmetric (v_3). Note that the additive character of the harmonic potential directly implies its paraboloid shape in a geometrical sense.

In the following the principles of the harmonic approximation, a fundamental simplification that has found extensive use in spectroscopy, are summarized. A complex shape of the true vibrational potential is replaced by the corresponding harmonic potential; as described in Eq. 5.3, this step is determined by the diagonalization of the Hessian evaluated at the respective energy minimum. In the process, a paraboloid approximating the shape of the true potential is derived. This process may be interpreted as the rotation of the coordinate system until Eq. 5.3 is fulfilled.

A positive-definite Hessian (all-positive eigenvalues) corresponds to a positive curvature of the potential along all directions from the reference point. On the other hand, in case a negative curvature is present along a specific direction, an imaginary frequency is obtained in the solution of the harmonic approximation, which is for instance employed to evaluate the properties of a transition state or/and reaction coordinates. Hence, the analysis of the Hessian at the stationary point (at which $g(Q)$ is equal to zero, i.e., no slope of the potential) enables the identification of the local minima (positive curvatures), local maxima (negative curvatures), and transition states (mixed occurrence of positive and negative curvature).

Since the harmonic potential depends on the Hessian, the efficiency of its determination by means of electronic structure theory is critical. The methods for which an analytical solution to the Hessian is available (e.g., HF, DFT, MP2, CIS) are far more efficient as the basis for a harmonic analysis than those for which the Hessian can only be calculated numerically (e.g., CC). Regardless, the harmonic approximation leads to a dramatic simplification of the vibrational problem in terms of complexity. The diagonalization of the Hessian yields the full vibrational solution: harmonic frequencies and the associated normal modes. For small to intermediate-sized molecules, this is a computationally inexpensive step (although it may become a bottleneck in studies of large systems using FF approaches), which made the harmonic approximation particularly important for early advances in vibrational spectroscopy. However, it is an extensive approximation of the real molecular oscillator. Firstly, the shape of the potential is fixed as a quadratic function. This is well-reflected in Fig. 5.6, as seen in three-dimensional space (Fig. 5.6a, b) as well as in one-dimensional projections respective to each of the modes (Fig. 5.6c, d). In this example, the true potential along the symmetric stretching mode of H_2O is asymmetric with respect to the equilibrium position (i.e., anharmonic) and resembles a Morse-like curve (Fig. 5.6c). This type of anharmonic potential is well-known, as it is often discussed in case of diatomic molecular oscillators (Fig. 5.4). Unlike the harmonic solution, the true vibrational levels are not equidistant. Morse-like anharmonicity (high contribution from the cubic terms in Eq. 5.1) leads to a subsequent reduction of the energy gaps between consecutive levels. However, the potential of the antisymmetric stretching mode of H_2O, although symmetric in shape, also deviates from the harmonic potential (Fig. 5.6d). This is due to significant contribution in the quartic terms in Eq. 5.1. The quartic anharmonicity leads to widened distances between consecutive vibrational levels.

Nevertheless, with some exceptions, e.g., of X–H stretching modes, in many cases the deviation between the harmonic approximation and the true molecular oscillator is relatively moderate. Consequently, in case of fundamental transitions, harmonic frequencies corresponding to those vibrations remain overestimated but not dramatically. This effect can be mitigated by an empirical correction, applied a posteriori in the form of a scaling of harmonic frequencies. Hence, the calculation of harmonic normal modes provides an effective route to enable approximate computational IR and Raman spectroscopy. However, in the majority of cases, this approach is too error prone to provide a reasonable prediction of overtones. In addition, the fundamental point of the harmonic approximation, the additive nature of the harmonic potential, does not take the coupling between individual modes into account, as reflected by the assumed zero cross-derivatives, or anharmonic force constants in Eq. 5.1. This fact leads to a critical limitation of the harmonic approximation, being its inability to describe combination transitions, rendering it inapplicable to NIR spectroscopy.

5.5 Beyond the Harmonic Approximation

5.5.1 Anharmonic Approaches Formulated on the Basis of the Harmonic Approximation

For the reasons explained above, the harmonic approximation is unsuitable for the calculation of NIR transitions. The inclusion of anharmonic effects to vibrational structure theory may be treated in an analogous way as electron correlation is included into the theory of the electronic structure. Accordingly, vibrational self-consistent field (VSCF) is the most straightforward anharmonic approach and an analogy to HF theory. The VSCF method is based on the concept that for each vibrational state k of the oscillator, the wavefunction Ψ is separable into a product of single-mode (harmonic) wavefunctions ϕ_i^k (Eq. 5.11), or a Hartree product.

$$\Psi_k(q_1, \ldots, q_n) = \prod_i^n \phi_i^k(q_i) \tag{5.11}$$

Through this, the multidimensional vibrational Schrodinger equation for the molecular oscillator in mass-weighted coordinates $q_1, \ldots, q_n$ (Eq. 5.11) is given as:

$$\left[-\frac{1}{2} \sum_{i=1}^n \frac{\partial^2}{\partial q_i^2} + V(q_1, \ldots, q_n) \right] \Psi_n(q_1, \ldots, q_n) = E_n \Psi_n(q_1, \ldots, q_n) \tag{5.12}$$

which leads to a set of one-dimensional (single-mode) equations (Eq. 5.13)

$$\left[-\frac{1}{2} \frac{\partial^2}{\partial q_i^2} + \bar{V}_i^{(n)}(q_i) \right] \Psi_i^{(n)}(q_i) = \varepsilon_i^{(n)} \Psi_i^{(n)}(q_i) \tag{5.13}$$

To fulfill the condition of separability, an effective potential $\bar{V}_i^{(n)}$ has to be introduced, through which the modes are coupled in form of a mean-field. It implies that there is no explicit mode–mode correlation, which is the major simplification in the VSCF concept. In other words, the potential for each normal mode is averaged over all other normal modes. Interestingly, the accuracy of the basic VSCF method increases relatively with the system size, as the average treatment of mode couplings applies better to extensively multidimensional (i.e., multimodal) systems.

To reduce the complexity of the problem further, a truncated pair-wise representation of the potential may be applied (Eq. 5.14)

$$V(q_1, \ldots, q_n) = \sum_{i=1}^{n} V_i^{\text{diag}}(q_i) + \sum_i \sum_{j>1} W_{ij}^{\text{coup}}(q_i, q_j) \qquad (5.14)$$

This way, the potential is approximated by a sum of single-mode potentials and interactions W_{ij}^{coup} between pairs of normal modes. Pair-wise potentials neglect contributions from any higher-order couplings (triplets, quartets, etc.).

Since the treatment of mode coupling in the basic VSCF scheme is approximated, no explicit correlations between modes is considered. As long as the coupling is relatively small, its impact may be evaluated more accurately through the addition of a correction by means of second-order perturbation theory. This leads to the VSCF-PT2 approach sometimes also called correlation-corrected VSCF, CC-VSCF. In this variant, the correction to the energy E_k^{corr} results from a potential V_k^{pert} defined as a small perturbation to the effective potential. Accordingly, the VSCF-PT2 ansatz leads to a perturbed VSCF Hamiltonian (Eq. 5.15)

$$H = H^{\text{SCF},(n)} + \Delta V(q_1, \ldots, q_n) \qquad (5.15)$$

and the associated correlation-corrected energy (Eq. 5.16)

$$E_n^{\text{VSCF-PT2}} = E_n^{\text{VSCF}} + \sum_{m \neq n} \frac{\left| \prod_{i=1}^{n} \Psi_i^{(n)}(q_i) | \Delta V | \prod_{i=1}^{n} \Psi_i^{(m)}(q_i) \right|^2}{E_n^{(0)} - E_m^{(0)}} \qquad (5.16)$$

denotes for n-th state coupling with all other m-states of the oscillator.

Energy corrections obtained through higher-order levels of perturbation theory return no meaningful improvements. The VSCF-PT2 method yields more accurate vibrational energies, however, at a sizeable increase in its computational complexity. Moreover, it is prone to behave erroneously in the case of nearly degenerated states (i.e., with similar energies; $E_n^{(0)} - E_m^{(0)} \approx 0$); thus, it is not applicable to strongly coupled modes.

A more advanced concept of including explicit mode correlations into the VSCF wavefunction has been formulated in the form of vibrational configuration interaction (VCI) theory. Per analogiam to the HF scheme, the VSCF solution yields a number of unoccupied virtual 'excited' modals. In a CI-like approach, the VSCF modals can be linearly combined to yield a correlated vibrational wavefunction. In an alternative approach, instead of a linear one, an exponential expansion using a cluster operator is proposed, leading to the vibrational coupled-cluster (VCC) scheme. VCI/VCC wavefunctions provide very good approximations to the exact vibrational wavefunction, and at the level of theory are not limited to any particular systems (such as those with weakly coupled modes). These approaches are capable of yielding very accurate results; however, they are extremely costly in their application to multimodal systems, and thus not suitable for spectroscopic studies of even moderate sized chemical systems.

On the other hand, vibrational perturbation theory (VPTn) adopts the Møller–Plesset formalism of n-th order (e.g., second-order perturbation leading to VPT2) to re-introduce the anharmonic terms in Eq. 5.1 as a perturbation to the (harmonic) vibrational Hamiltonian. The VPT ansatz separates the anharmonic contributions in the vibrational Hamiltonian H (Eq. 5.17) into a set of individual terms (Eqs. 5.18–5.20).

$$H = H^{(0)} + H^{(1)} + H^{(2)} \tag{5.17}$$

$$H^{(0)} = \frac{1}{2} \sum \omega_i \left(p_i^2 + q_i^2\right) \tag{5.18}$$

$$H^{(1)} = \frac{1}{6} \sum \phi_{ijk} q_i q_j q_k \tag{5.19}$$

$$H^{(2)} = \frac{1}{24} \sum \phi_{ijkl} q_i q_j q_k q_l + \sum_{\tau=x,y,z} B_e^\tau \zeta_{ij}^\tau \zeta_{kl}^\tau \left(\frac{\omega_j \omega_l}{\omega_i \omega_k}\right) q_i p_j q_k p_l \tag{5.20}$$

with $H^{(0)}$ being the harmonic Hamiltonian. The first-order Hamiltonian $H^{(1)}$ includes the cubic anharmonic terms, while the second-order Hamiltonian $H^{(2)}$ the quartic terms.

Unlike in the VSCF-PT2 scheme, in which a perturbative correction is added to the VSCF Hamiltonian, the VPT2 ansatz operates on a harmonic Hamiltonian and a perturbative treatment is inserted at the lower level of the vibrational structure theory. Compared with the VSCF approach, VPT2 calculations typically require a lower number of potential evaluations to achieve a comparable accuracy. Hence, in practical implementations, the VPT2 approach may be more efficient. However, in its original formulation, this method is highly unreliable in treating nearly degenerated modes. The number of degeneracies rapidly increases for larger molecules, which makes VPT2 unsuitable for the description of such systems. With aim of providing a universal methodology, the 'deperturbed' VPT2 (DVPT2) ansatz was formulated, in which the terms describing nearly degenerated states are removed entirely from the calculation. Thus, the DVPT2 energies have a more approximate character, but are not likely to be affected by large errors. Further development of this concept led to its generalized variant GVPT2, in which the removed terms are re-evaluated using a variational approach. In principle, the GVPT2 method is applicable to any system regardless of its size, while maintaining a favorable cost versus accuracy ratio.

5.5.2 Grid-Based Approaches

Implementations of VSCF or VPT2 theory are constructed for an efficient treatment of moderately anharmonic modes. The relatively low number of energy evaluations, and the approximate probing of the vibrational potential yield high efficiency of these approaches. However, the amount of anharmonicity they effectively capture is limited. In order to predict the vibrational energy eigenstates with high accuracy, solving the time-independent Schrödinger as given in Eq. 5.21 for a one-dimensional problem yields a nearly exact solution of the vibrational problem for an accurately evaluated potential.

$$
\frac{\partial^2 \Psi(q)}{\partial q^2} = \left\{ \frac{2m}{\hbar^2} \cdot \left(V(q) - E \right) \right\} \Psi(q) = f(q) \cdot \Psi(q) \tag{5.21}
$$

Here, Ψ denotes the vibrational wavefunction along the respective normal coordinate q, while m and $\hbar$ are the reduced mass of the vibrational mode and the reduced Planck constant, respectively. Typically, the potential $V(q)$ is provided on an equispaced grid with step length Δq, and E denotes the associated energy eigenvalue. The solution to Eq. 5.21 can be obtained by means of grid-based approaches such as discrete variable representation (DVR) and Numerov's method. The latter is based on a Taylor series of $\Psi(q)$ expanded around the point q of the normal coordinate with $\Psi^{(n)}$ representing the n-th derivative of the wavefunction with respect to Δq:

$$
\Psi(q + \Delta q) = \Psi + \frac{1}{1!} \|\Delta q\| \Psi^{(1)} + \frac{1}{2!} \|\Delta q\|^2 \Psi^{(2)} + \frac{1}{3!} \|\Delta q\|^3 \Psi^{(3)} + \frac{1}{4!} \|\Delta q\|^4 \Psi^{(4)} + \cdots \tag{5.22}
$$

Summation of the Taylor expansion in forward and backward direction (i.e., $\pm \Delta q$) leads to the cancellation of all odd, and the doubling of all even entries. Next, higher-order derivatives (i.e., $\Psi^{(n)}$ with $n = 4, 6, 8, \ldots$) are expressed via their associated finite differences employing the appropriate number of grid points $V(\pm m \cdot \Delta q)$ to achieve the desired accuracy. In the simplest case, the time-independent Schrödinger equation can be expressed via a three-point expression employing the two neighboring grid points $\pm 1 \cdot \Delta q$ of any given point on the equispaced grid (labeled as $\Psi_{-1}, \Psi_0, \Psi_{+1}$ for convenience)

$$
-\frac{\hbar^2}{2m} \cdot \frac{\Psi_{-1} - 2\Psi_0 + \Psi_{+1}}{\|\Delta q\|^2} + \frac{V_{-1}\Psi_{-1} + 10V_0\Psi_0 + V_{+1}\Psi_{+1}}{12}
$$

$$
\approx E \cdot \frac{\Psi_{-1} + 10\Psi_0 + \Psi_{+1}}{12} \tag{5.23}
$$

Initial implementations of Numerov's approach employ an iterative process based on an initial guess in the energy eigenvalue E and are sometimes referred to as *shooting methods*. However, modern approaches assure Dirichlet boundary conditions (i.e., the wavefunction outside the considered interval is zero) which enables the implementation of Numerov's approach in the form of a matrix eigenvalue problem. Accordingly, using the matrices $\mathbf{A}$ and $\mathbf{B}$ as well as the diagonal matrix $\mathbf{V}$, containing $V(q)$ as elements, the solution can be written as

$$\left(-\frac{\hbar^2}{2m}\mathbf{A} + \mathbf{BV}\right)\Psi \approx \mathbf{B}E\Psi \tag{5.24}$$

Rearrangement of Eq. 5.24 leads to the matrix representation of the time-independent Schrödinger equation $\mathbf{H}\Psi = E\Psi$, with

$$\mathbf{H} = -\frac{\hbar^2}{2m}\mathbf{B}^{-1}\mathbf{A} + \mathbf{V} \tag{5.25}$$

Eigen decomposition of $\mathbf{H}$ simultaneously yields all energy eigenvalues along the diagonal of the energy matrix $\mathbf{E}$, and the associated eigenvectors are collected in Ψ. As the key advantage, these grid-based approaches do not require any assumption or pre-defined building blocks (e.g., basis sets) to formulate the wavefunction Ψ. These approaches are not limited to questions in vibrational spectroscopy and similar methods have also been employed in the description of quantum tunnelling and the electronic structure of atoms and small molecular systems.

The method can be extended to arbitrary orders in the numerical derivatives by truncating the Taylor series of higher degree (e.g., $\mathbf{A}$ and $\mathbf{B}$ matrices with seven diagonal entries would require a Taylor series of eight degree). To predict IR intensities, the transition moment integral μ_{mn}, consisting of the respective wavefunctions of the two involved states Ψ_m and Ψ_n, as well as the transition moment operator $\hat{\mu}(q)$, has to be calculated.

$$\mu_{mn} = \int_{-\infty}^{\infty} \Psi_m \hat{\mu} \Psi_n d\tau \tag{5.26}$$

In case of infrared spectroscopy, $\hat{\mu}(q)$ equals the dipole moment μ as a function of the molecule's normal coordinates q, making infrared measurements especially sensitive on polar function. The transition dipole moment is then employed to calculate the associated oscillator strength f_{mn}

$$f_{mn} = \frac{4\pi m_e}{3e^2\hbar}\|\mu_{mn}\|^2 v_{mn} \tag{5.27}$$

where m_e denotes the electron mass, e the elementary charge, and v_{mn} the transition energy between the two states m and n. Typically, the oscillator strength is normalized to that of the fundamental mode.

Within the 1D formalism, grid-based methods are capable of inherently taking arbitrary anharmonicities into account. However, the main benefit lies in the generalization of the grid-based methods to higher dimensions, which enables the inclusion of intermode coupling contributions in addition to the adequate treatment of anharmonic effects. Especially when combined with sparse matrix algebra routines and advanced interpolation techniques to reduce the associated computational effort, grid-based methods are capable of delivering a highly accurate description of complex quantum mechanical systems.

5.6 Applications of Anharmonic Approaches in NIR Spectroscopy

Applications of VSCF theory in investigations of mid-infrared (MIR) spectra are reasonably popular in literature [12], yet relatively few examples aimed at the NIR region can be found. Although even the basic VSCF approach is capable of predicting up to third-order overtones and combination bands, it seems that an improved description of mode correlations (e.g., by means of PT2-VSCF, or VCI) is often necessary to yield a qualitatively correct prediction of NIR modes [13]. These approaches frequently prove to be prohibitively expensive for treating larger molecules, although examples exist of successful applications of the PT2-VSCF approach to molecules counting ca. 15 atoms (e.g., malic acid), when certain approximations are assumed (e.g., the application of a quartic force field, QFF) [13]. The anharmonic frameworks featuring a robust treatment of mode correlations (i.e., VSCF-VCI, VCC) are far more expensive. The applicability of these methods may improve in the future, however.

The primary advantage of the DVPT2-GVPT2 approach is efficiency and applicability to molecules that are in the center of attention of applied NIR spectroscopy. Additionally, the GVPT2 framework fully mitigates the typical shortcoming of perturbation theory being prone to produce meaningless description of tightly coupled modes that becomes increasingly probable upon an increase of the system size. Therefore, this framework finds a remarkably widening application area in solving spectroscopic problems. A good example is a recent investigation of the NIR spectroscopic properties of melamine [14]. This compound is of key interest to analytical NIR spectroscopy in the context of food quality control. However, as in many other cases, the NIR spectrum of melamine remained shallowly understood before. Spectra calculations by means of the DVPT2 and GVPT2 methods performed at B3LYP-GD3BJ/SNST level were able to accurately reconstruct all essential NIR absorption bands of melamine (Fig. 5.7). This yielded detailed and unambiguous band assignments for the compound, enhancing the ability to interpret the essential features of the multivariate models used for analyzing melamine content. It is noteworthy that at the same time, an interesting comparison was made. The present implementation of GVPT2 includes the ability to predict three quanta transitions (i.e., second overtones, ternary combination bands). The appearance of such bands in the experimental NIR spectrum of melamine could be directly assessed. As demonstrated in Fig. 5.7, this improves the interpretability of minor bands. However, the

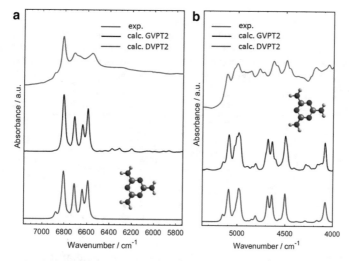

Fig. 5.7 Experimental diffuse reflectance NIR spectrum of polycrystalline melamine compared to the calculated spectra (DVPT2 and GVPT2 at B3LYP-GD3BJ/SNST level) in the regions **a** 7150–5750 cm^{-1}, and **b** 5400–4000 cm^{-1}. Reproduced from Ref. [14] under Creative Commons Attribution 4.0 International (CC BY 4.0)

spectrum of melamine in 7500–4000 cm^{-1} region is for the most part decided by two quanta transitions (i.e., first overtones and binary combinations).

Grid-based methods offer a nearly exact solution of the vibrational problem well exemplified in the case of the simplest molecular oscillator, a diatomic molecule such as HCl in gas phase. Figure 5.4 shows the associated interaction energy obtained using the accurate yet comparably expensive CCSD(T) method (i.e., coupled cluster with single, double, and perturbative triples) in conjunction with a large one-electron basis (augmented correlation consistent polarization valence quadruple zeta basis set, aug-cc-pVQZ). This bonding potential has been scanned in tight intervals of 0.005 Å in the region near the equilibrium distance $r_{eq} = 1.278$ Å. The Morse-like character of the bond showing a steep increase in the potential at low distances is clearly visible, which is lost when the harmonic approximation is applied to the region near r_{eq} (dashed line in Fig. 5.4). Solving the vibrational Schrödinger equation (in this case using Numerov's approach) yields the vibrational wavefunctions and the associated energy eigenvalues. The respective differences between the eigenvalue of a particular excited state and the ground state correspond to the frequency measured in vibrational spectroscopy (see Table 5.2). It can be seen from the experimental values that the spacing between the energy levels is decreasing, which is adequately described when taking anharmonicity into account. The harmonic approximation on the other hand is known to perform poorly for the fundamental excitation in many cases and is effectively useless when aiming at the associated overtone vibrations. In case of a diatomic molecule, the inclusion of rotational coupling is comparably simple and can be realized by taking the changes in the moment of inertia of the molecule upon bond stretching into account. As can be seen from Table 5.2, this contribution referred to

Table 5.2 Vibrational wave numbers of the fundamental and four lowest overtones of HCl(g) in cm^{-1} obtained at CCSD(T)/aug-cc-pVQZ level via the harmonic approximation and the Numerov treatment (grid spacing 0.005 Å) with and without the rotational Watson potential, respectively. It can be seen that an explicit inclusion of anharmonic effects to the vibrational excitations is vital to obtain reliable estimates for higher excitations. Rotational coupling on the other hand only plays a minor role in this example

Transition	Harmonic	Numerov	Numerov–Watson	Experimental[a]
$0 \to 1$	2990.2	2885.4	2885.5	2885.9
$0 \to 2$	5980.3	5667.6	5667.7	5668.0
$0 \to 3$	8970.5	8347.1	8347.3	8347.0
$0 \to 4$	11960.6	10924.1	10924.4	10923.1
$0 \to 5$	14950.8	13398.9	13399.2	13396.5

[a]Ref. [4], p. 193

as Watson potential has only a minor influence on the vibrational wave numbers in the case of HCl(g).

Grid-based approaches yield highly accurate solutions of the vibrational problem, but presently their applicability to larger molecules is limited because of their excessive cost. However, for studies of such systems, they remain effective in selective treatments of a particular mode of interest (one-dimensional grid). For example, in several cases, these methods have been used for an accurate prediction of the OH stretching overtone band [15, 16]. This strong band is highly sensitive to the chemical environment and is an important spectral feature frequently investigated by NIR spectroscopy (refer to the Chapter *NIR spectroscopy in physical chemistry*). Therefore, accurate calculations of the frequency and intensity is essential, e.g., for obtaining detailed insights into solvent effects [15]. On the other hand, grid-based methods may be used to improve theoretical NIR spectra obtained with different methods. Although in principle, the VPT2 approach is applicable to Morse-like potentials, in some cases, it provides unreliable results. For instance, the $2v(OH)$ peak delivers a relatively easily accessible information on the conformational state of hydroxyl bearing molecules. In the case of cyclohexanol, it consists of two components due to the two major conformers. The wavenumber difference Δv between these components was found to be 27 cm^{-1} in the experimental spectrum. A recent study reported a strongly underestimated VPT2 frequency for the major conformer, resulting in the splitting of the predicted peak ($\Delta v^{VPT2} = 260$ cm^{-1}). However, the application of a grid-based approach yielded a much more reliable value of 30 cm^{-1} [16].

Further, grid-based methods are applicable universally, including low-lying torsional modes that are typically challenging for generalized methods (e.g., VSCF, VPT2). The highest potential for future advances is associated with multidimensional grid-based approaches covering full vibrational configuration of the system. Presently, state-of-the-art enables treatment of triatomic linear molecules (e.g., CO_2, BeH_2, HCN), which requires a four-dimensional grid [17]. In such case, the entirety of mode coupling is explicitly included and the predicted frequencies deviate by less than 1% from experimental data [17]. Feasible implementation of higher-dimensional grid-based approaches should enable nearly exact prediction of NIR spectra of more complex molecules, which will form an essential progress in our understanding of NIR spectroscopy.

5.7   Summary and Future Prospects

Practical applications of the methods of quantum chemistry in NIR spectroscopy have mostly been limited by their computational cost. In the recent decade, a remarkable rise in practical applications of theoretical calculations in NIR spectroscopy was observed. This resulted from the development of quantum-based approaches and their implementation, as well as from the progress in technology resulting in a continuous increase in computational capacities. This allows for the anticipation of further advances in the forthcoming years, and a twofold development can presently be witnessed in this field. Firstly, studies of NIR spectra of increasingly complex systems are becoming feasible. This opens new opportunities, as the complexity of NIR spectra tends to scale steeply with the system size, and their interpretability by conventional spectroscopic methods is limited. Secondly, highly accurate grid-based methods are capable of yielding nearly exact results. At the moment, computational complexity of grid-based methods limits their applicability to few-atom systems. Nevertheless, they form an essential aid at the moment, as the established 'universal' anharmonic frameworks (e.g., VSCF, VPT2) have primarily been formulated with efficiency in mind. This could only be achieved through various approximations affecting their robustness and accuracy. Grid-based methods demonstrate their usefulness in directly correcting VPT2 results for a few selected modes of interest; this even applies for seemingly manageable modes such as OH stretching vibrations. As the primary factors limiting the applicability of computational chemistry in NIR spectroscopy are consistently challenged, a general conclusion may be drawn that in the next decade a markedly rapid expansion of NIR studies utilizing methods of quantum chemistry will be observed.

References

1. D.C. Harris, M.D. Bertolucci, Symmetry and spectroscopy. *An Introduction to Vibrational and Electronic Spectroscopy* (Dover Publications, INC., New York, 1980)
2. E.B. Wilson, J.C. Decius, P.C. Cross, *Molecular Vibrations: The Theory of Infrared and Raman Vibrational Spectra* (Dover Publications, INC., New York, 1980)
3. K. Nakamoto, *Infrared and Raman Spectra of Inorganic and Coordination Compounds*, 6th edn. (Wiley, Hoboken, New Jersey, 2009)
4. F.S. Levin, *An Introduction to Quantum Theory* (Cambridge University Press, 2002)
5. P. Pulay, G. Fogarasi, F. Pong, J.E. Boggs, Systematic ab initio gradient calculation of molecular geometries, force constants, and dipole moment derivatives. J. Am. Chem. Soc. **101**, 2550–2560 (1979)

6. T. Helgaker, P. Jorgensen, J. Olsen, *Molecular Electronic—Structure Theory* (Wiley, New York, 2000)
7. W. Koch, M.C. Holthausen, *A Chemist's Guide to Density Functional Theory*, 2nd edn. (Wiley, VCH Verlag GmbH, 2001)
8. D.S. Sholl, J.A. Steckel, *Density Functional Theory: A Practical Introduction*, 1st edn (Wiley, 2009)
9. A.R. Leach, *Molecular Modelling*, 2nd edn. (Prentice Hall, Essex, 2001)
10. A. Stone, *The Theory of Intermolecular Forces*, 2nd edn. (Oxford University Press, Oxford, 2016)
11. F. Jensen, *Introduction to Computational Chemistry*, 3rd edn. (Wiley, Chichester, 2017)
12. T.K. Toy, R.B. Gerber, Vibrational self-consistent field calculations for spectroscopy of biological molecules: new algorithmic developments and applications. Phys. Chem. Chem. Phys. **15**, 9468–9492 (2013)
13. M. Schmutzler, O.M.D. Lutz, C.W. Huck, Analytical pathway based on non-destructive NIRS for quality control of apples, in *Infrared Spectroscopy: Theory, Developments and Applications*, ed. by D. Cozzoliono (Nova Science Publisher, New York, USA, 2013)
14. J. Grabska, K.B. Beć, C.G. Kirchler, Y. Ozaki, C.W. Huck, Distinct difference in sensitivity of NIR vs. IR bands of melamine to inter-molecular interactions with impact on analytical spectroscopy explained by anharmonic quantum mechanical study. Molecules **24**, 1402 (2019)
15. M.J. Schuler, T.S. Hofer, C.W. Huck, Assessing the predictability of anharmonic vibrational modes at the example of hydroxyl groups—ad hoc construction of localised modes and the influence of structural solute–solvent motifs. Phys. Chem. Chem. Phys. **19**, 11990–12001 (2017)
16. K.B. Beć, J. Grabska, M.A. Czarnecki, Spectra-structure correlations in NIR region: spectroscopic and anharmonic DFT study of n-hexanol, cyclohexanol and phenol. Spectrochim. Acta A **197**, 176–184 (2018)
17. U. Kuenzer, T.S. Hofer, A four-dimensional Numerov approach and its application to the vibrational eigenstates of linear triatomic molecules—the interplay between anharmonicity and inter-mode coupling. Chem. Phys. **520**, 88–99 (2019)

Chapter 6
Two-Dimensional Correlation Spectroscopy

Mirosław A. Czarnecki and Shigeaki Morita

Abstract Two-dimensional (2D) correlation spectroscopy is a well-established method for analysis of perturbation-induced spectral changes in various kinds of data. Due to selective correlation of the peaks and resolution enhancement, it provides useful information on the dynamics of spectral changes and enables more reliable band assignments. The generalized 2D correlation approach permits to apply various kinds of perturbations and makes possible for correlation between data obtained from different experiments (hetero-correlation). At the beginning of this chapter are shown the basic principles of 2D correlation spectroscopy together with the rules for interpretation of the synchronous and asynchronous spectra. Next, we report new developments in this method like sample–sample correlation spectroscopy and perturbation–correlation moving-window 2D correlation spectroscopy. Finally, are shown selected examples of successful applications of 2D correlation spectroscopy for study of interactions and molecular structure.

Keywords 2D correlation spectroscopy · Sample–sample correlation spectroscopy · Perturbation-correlation moving-window 2D correlation spectroscopy · Applications of 2D correlation NIR spectroscopy

6.1 Introduction

Two-dimensional (2D) correlation spectroscopy was introduced to vibrational spectroscopy by Isao Noda in 1986 [1]. In its original form, this method was dedicated to small-amplitude periodic perturbations, and hence, the area of its applications was limited—mainly to polymer studies [2, 3]. A breakthrough has started in 1993 after development of the generalized two-dimensional correlation spectroscopy (2DCOS)

M. A. Czarnecki (✉)
Faculty of Chemistry, University of Wrocław, F. Joliot-Curie 14, 50-383 Wrocław, Poland
e-mail: miroslaw.czarnecki@chem.uni.wroc.pl

S. Morita
Faculty of Engineering, Osaka Electro-Communication University, Neyagawa, Japan
e-mail: smorita@isc.osakac.ac.jp

[4]. This new approach allows to accept any kind of perturbations and permits the hetero-correlations between the data obtained from different methods, such as MIR-NIR, MIR-Raman, UV-MIR and so on. In this way, the generalized 2D correlation analysis has begun a powerful and versatile tool for analysis of spectral data from various experiments [5]. This approach appears to be particularly useful in near-infrared (NIR) region since NIR spectra are very complex due to overlap of numerous overtones and combination bands [6, 7]. In addition, 2DCOS spectroscopy solves the problem of multicollinearity by independent correlation of variables.

The idea behind 2DCOS is very simple and is displayed in Fig. 6.1. The sample is subjected to external perturbation, which generates the specific changes at a molecular level. These changes can be monitored by any kind of electromagnetic radiation, including NIR. As a result, the measurements provide a series of perturbation-dependent spectra. From the single spectra $y(v, t)$, a perturbation-ordered data matrix is assembled:

Fig. 6.1 General scheme of 2D correlation spectroscopy

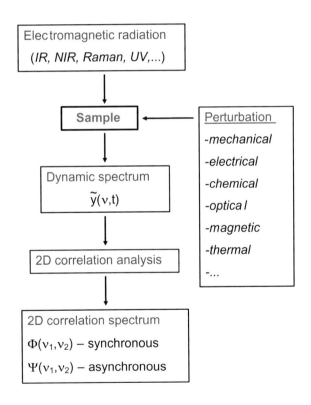

$$\begin{bmatrix} y(\nu_1, t_1) & y(\nu_2, t_1) & \cdots & y(\nu_m, t_1) \\ y(\nu_1, t_2) & y(\nu_2, t_2) & \cdots & y(\nu_m, t_2) \\ y(\nu_1, t_3) & y(\nu_2, t_3) & \cdots & y(\nu_m, t_3) \\ \cdots & \cdots\cdots & & \cdots \\ y(\nu_1, t_n) & y(\nu_2, t_n) & \cdots & y(\nu_m, t_n) \end{bmatrix} \tag{6.1}$$

where ν means wavenumber (or the other units), t is the value of perturbation, n is the number of spectra and m is the number of data points in the spectrum. Usually, this matrix is row-oriented; in the other case, one has to transpose the data. Prior to 2D correlation analysis, the perturbation-ordered data matrix is converted into the dynamic spectrum ($\tilde{y}$) by subtraction of the reference spectrum ($\hat{y}$):

$$\tilde{y}(\nu, t) = \begin{cases} y(\nu, t) - \hat{y}(\nu) & \text{for } t_{\min} \leq t \leq t_{\max} \\ 0 & \text{otherwise} \end{cases} \tag{6.2}$$

In principle, one can select an arbitrary reference spectrum, but usually, a perturbation-average spectrum is used as a reference:

$$\hat{y}(\nu) = \frac{1}{n} \cdot \sum_{i=1}^{n} y(\nu, t_i) \tag{6.3}$$

The proper selection of reference spectrum appreciably simplifies synchronous and asynchronous contour plots, since the peaks are developed only at the positions where intensity changes occur. This means that if the applied perturbation does not induce the spectral changes at given position, this peak does not appear in the correlation spectrum. 2D correlation analysis yields synchronous (Φ) and asynchronous (Ψ) spectra, which are a product of two or three matrices:

$$\Phi(\nu_i, \nu_j) = \frac{1}{n-1} \tilde{y}(\nu_i, t)^T \cdot \tilde{y}(\nu_j, t) \tag{6.4}$$

$$\Psi(\nu_i, \nu_j) = \frac{1}{n-1} \tilde{y}(\nu_i, t)^T \cdot M \cdot \tilde{y}(\nu_j, t) \tag{6.5}$$

where M is Hilbert–Noda transformation matrix [8]:

$$M_{i,j} = \begin{cases} 0 & \text{if } i = j \\ \frac{1}{\pi \cdot (j-i)} & \text{otherwise} \end{cases} \tag{6.6}$$

The specific information obtained from 2DCOS primarily depends on the sample properties, kind of external perturbation and the electromagnetic probe. The perturbation should stimulate the sample and generate specific variations of physicochemical properties at a molecular level. These variations may result from changes of time, sample composition, temperature, pressure, pH and so on. Each kind of perturbation yields unique information about studied system. It is also important to apply an appropriate probe to successfully monitor the perturbation-induced changes in the studied system.

As mentioned before, the generalized 2DCOS permits for hetero-correlation of two unlike types of data; however, both data sets have to be recorded under the same perturbation values. If $\tilde{y}$ and $\tilde{u}$ are the dynamic spectra from two different experiments, then the synchronous and asynchronous hetero-correlation spectra are expressed:

$$\Phi\left(v_i, \mu_j\right) = \frac{1}{n-1}\, \tilde{y}(v_i, t)^T \cdot \tilde{u}\left(\mu_j, t\right) \tag{6.7}$$

$$\Psi\left(v_i, \mu_j\right) = \frac{1}{n-1}\, \tilde{y}(v_i, t)^T \cdot M \cdot \tilde{u}\left(\mu_j, t\right) \tag{6.8}$$

The properties of the synchronous and asynchronous spectra were explained by using the simulated spectra (Fig. 6.2). A data series consist of 11 spectra and each spectrum includes five peaks. The arrows point the direction of intensity changes. Figure 6.3 shows the corresponding synchronous and asynchronous spectra. As can be seen, the synchronous spectrum includes both the diagonal and cross-peaks. The diagonal peaks are always positive and represent the overall extent of intensity changes at individual wavenumbers. The cross-peaks are positive or negative and yield information on similarities of spectral changes at two different wavenumbers (v_1, v_2). The synchronous cross-peaks are positive if the spectral changes at v_1 and v_2 are in the same direction (both increasing or both decreasing) (Fig. 6.3a). The negative sign means the opposite. Such positive synchronous cross-correlation suggests that the changes at v_1 and v_2 originate from the same molecular fragment or two different fragments strongly interacting. In contrast, the asynchronous spectrum includes only the cross-peaks and yields information on differences of spectral

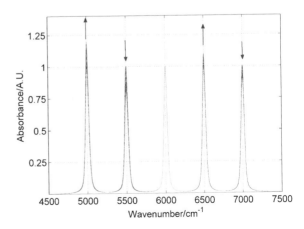

Fig. 6.2 A series of 11 simulated spectra. Each spectrum includes five peaks approximated by a product of Gauss and Lorentz function. The initial intensities were 1 and the final were 1.2 ($5000\ cm^{-1}$), 0.8 ($5500\ cm^{-1}$), 1 ($6000\ cm^{-1}$), 1.1 ($6500\ cm^{-1}$) and 0.9 ($7000\ cm^{-1}$). The arrows show direction of the changes

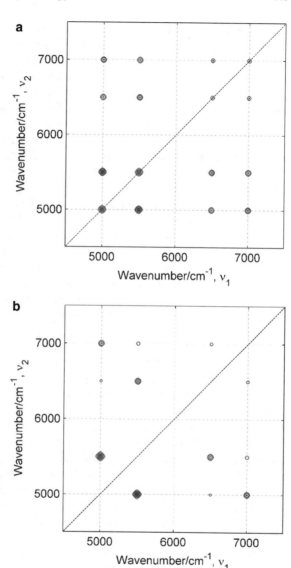

Fig. 6.3 Synchronous and asynchronous spectra obtained from the spectra shown in Fig. 6.2

changes at ν_1 and ν_2 (Fig. 6.3b). The presence of the asynchronous peak at (ν_1, ν_2) evidences that the spectral changes at ν_1 and ν_2 occur at different rate or are shifted in-phase. This way, one can differentiate the spectral responses from various components of the sample. This is an important feature of 2DCOS, which allows for the resolution enhancement. Irrespective of the separation, the peaks are resolved in the asynchronous spectrum as long as their spectral responses follow different pattern. To easy interpretation of 2D asynchronous contour plots, one can multiply the asynchronous intensity at (ν_1, ν_2) by the sign of the synchronous intensity at the same coordinate. Hence, the presence of the positive asynchronous peak at (ν_1, ν_2) means that the spectral changes at ν_1 occur earlier/faster than those at ν_2. The negative sign of the asynchronous peak means the opposite behavior. Selective correlation of the peaks in 2DCOS spectra allows for establishing of the origin of the peaks and easy the band assignment. Particularly useful are hetero-correlation spectra, which show the selective correlation between known and unknown spectral features.

The rules for interpretation of 2DCOS spectra are straightforward [4, 5]; however, correct interpretation of the real-world data is not always easy. Firstly, an application of the external perturbation is often accompanied by side effects. To obtain a 'net' information on the effect of interest, at first, one has to remove these side effects. The procedure, which removes these side effects depends on their specific nature, but in many cases, the normalization of the spectra significantly improves the quality of 2D contour plots [9]. Secondly, 2DCOS spectra, particularly the asynchronous ones, are very sensitive to noise, baseline fluctuation and other distortions. Besides, interpretation of 2DCOS spectra is complicated by band position and/or width variations. All these effects may generate artifacts in the synchronous and asynchronous spectra. Therefore, the systematic studies were undertaken to recognize and eliminate (where possible) these effects from 2D correlation spectra [10–13].

Sometimes, the spectral changes of interest are obscured by the noise, baseline fluctuation or the other effects. The proper pretreatment of the experimental spectra may significantly improve the quantity and quantity of the information obtained from 2DCOS [7, 9, 10]. An extensive baseline fluctuation generates long streaks observed in the synchronous and asynchronous contour plots. In many cases, a simple offset of the spectra at selected reference point can significantly reduce this effect [7]. Sometimes are necessary more advanced corrections by using polynomial functions. In an extreme case, one can use the second derivative spectra, instead of the original data, for the analysis [12]. The high level of noise will produce a lot of artifacts, especially in the asynchronous spectrum. The most popular methods of smoothing are based on Savitzky–Golay algorithm. The more advanced methods employ Fourier or wavelet filtering, or principal component analysis (PCA). Also, normalization of the spectra is often used as a pretreatment method. This way, one can eliminate the effects of varying concentration, temperature, pressure or sample thickness on the 2DCOS spectra [9].

2D correlation spectroscopy offers a significant simplification of the complex NIR spectra. However, the most important advantage of using 2DCOS in NIR region is the ability of resolving of highly overlapped peaks. Besides, selective correlation between MIR (or Raman) and NIR spectra allows for reliable band assignment in NIR region and obtain information on the molecular structure and interactions [6, 7].

6.2 New Developments in Two-Dimensional Correlation Spectroscopy

6.2.1 Sample–Sample Correlation Spectroscopy

For the last two decades or so, several new ideas regarding 2DCOS have been proposed such as sample–sample (SS), moving-window two-dimensional (MW2D) and perturbation-correlation moving-window two-dimensional (PCMW2D). Here, SS, MW2D and PCMW2D methods will be outlined. The first idea of sample–sample correlation, i.e., opposite to conventional variable–variable correlation, was proposed by Zimba [14], and this idea was subsequently refined by Šašić et al. [15, 16]. As given in Eq. (4), the conventional synchronous 2D correlation spectrum $\Phi(v_i, v_j)$ is calculated as a covariance matrix of $y(v, t)$. The synchronous sample–sample correlation (Φ_{SS}) is given as a covariance matrix of transposed $y(v, t)$ matrix, and the asynchronous sample–sample correlation (Ψ_{SS}) is calculated as:

$$\Phi_{SS}(t_k, t_l) = \frac{1}{m-1} \tilde{y}(v, t_k) \cdot \tilde{y}(v, t_l)^T \qquad (6.9)$$

$$\Psi_{SS}(t_k, t_l) = \frac{1}{m-1} \tilde{y}(v, t_k) \cdot M \cdot \tilde{y}(v, t_l)^T \qquad (6.10)$$

Figure 6.4 shows temperature-dependent diffuse reflectance NIR spectra of microcrystalline cellulose (MCC) and their second derivative spectra [17]. Figure 6.5 represents synchronous sample–sample 2D correlation spectrum constructed from the second derivative spectra shown in Fig. 6.4b. In the case of conventional 2D correlation spectra (not shown), correlation maps are spread between two spectral variable

Fig. 6.4 Temperature-dependent diffuse reflectance NIR spectra of microcrystalline cellulose (MCC) (**a**) and their second derivative spectra (**b**)

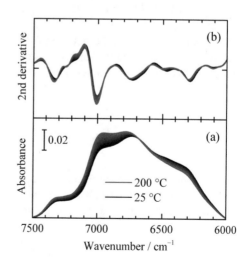

Fig. 6.5 Synchronous
sample–sample 2D
correlation map constructed
from the second derivative
spectra shown in Fig. 6.4

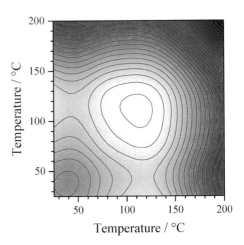

axes, e.g., wavenumber–wavenumber axes. In contrast, in the case of sample–sample
correlation, 2D correlation maps are spread between two sample variable axes, e.g.,
temperature–temperature axes, as shown in Fig. 6.5. Therefore, some informative
sample points are visually identified in the 2D correlation maps by this method.

6.2.2 Perturbation-Correlation Moving-Window Two-Dimensional (PCMW2D) Correlation Spectroscopy

Thomas and Richardson proposed the first idea of MW2D correlation spectroscopy
[18]. For a set of obtained spectra $y(v, t)$, jth window of submatrix consisting of $2w$
$+ 1$ spectra $y_j(v, t_J)$ is considered, where j and J are the index of window and that
of a spectrum within the window, respectively. The MW2D correlation spectrum is
obtained by incrementally sliding the window position along the perturbation variable
direction from $j = 1 + w$ to $n-w$, where n is the number of spectra in $y(v, t)$, and
calculating

$$\Omega_{A,j}(v, t_j) = \frac{1}{2w} \sum_{J=j-w}^{j+w} \tilde{y}_j^2(v, t_J) \tag{6.11}$$

This is an auto-correlation spectrum or variance spectrum calculated using the $2w$
$+ 1$ spectra in the window. Morita et al. [19] reported that the MW2D correlation
intensities are proportional to a squared perturbation derivative, i.e.,

$$\Omega_A(v, t) \sim \left[\frac{\partial y(v, t)}{\partial t}\right]^2 \tag{6.12}$$

Another type of moving-window technique, PCMW2D correlation spectroscopy, was proposed by Morita et al. [20] In this case, both synchronous and asynchronous correlation spectra were calculated as

$$\Pi_{\Phi,j} = \frac{1}{2w} \sum_{J=j-w}^{j+w} \tilde{y}(v, t_J) \cdot \tilde{t}_J \qquad (6.13)$$

$$\Pi_{\Psi,j} = \frac{1}{2w} \sum_{J=j-w}^{j+w} \tilde{y}(v, t_J) \cdot \sum_{K=j-w}^{j+w} M_{JK} \cdot \tilde{t}_K \qquad (6.14)$$

where $\tilde{t}$ and M are dynamic perturbation and Hilbert–Noda transformation matrix, respectively. As similar to MW2D correlation spectroscopy, following relations were found by Morita et al.

$$\Pi_{\Phi}(v, t) \sim \left[\frac{\partial y(v, t)}{\partial t} \right]_v \qquad (6.15)$$

$$\Pi_{\Psi}(v, t) \sim - \left[\frac{\partial^2 y(v, t)}{\partial t^2} \right]_v \qquad (6.16)$$

i.e., synchronous and asynchronous PCMW2D correlation intensities are proportional to a perturbation derivative and the opposite sign of a perturbation second derivative [20]. In the case of linear perturbation, therefore, synchronous and asynchronous PCMW2D correlation intensities are proportional to a gradient and a curvature of the spectral intensity variations along the perturbation direction, respectively [20].

Figure 6.6 shows synchronous PCMW2D correlation map constructed from the temperature-dependent NIR spectra of MCC shown in Fig. 6.4a. Positive and negative correlation intensities in the map represent increase and decrease of the spectral intensities along the temperature direction, respectively. A slice spectrum at 90 °C is also plotted in the figure. A positive correlation peak located at 6961 cm^{-1} is reported to be intermediate hydrogen bonds in MCC [17].

6.3 Applications of Two-Dimensional Correlation NIR Spectroscopy

The simplicity in obtaining of 2D correlation spectra resulted in a fast development of this approach. In a short time, a number of successful applications of 2DCOS to various fields of chemistry were reported. In 1996, Noda et al. published the first application of 2DCOS in NIR region (2DCOS-NIR) to study self-association of oleyl alcohol in the liquid phase [21]. Due to the resolution enhancement, a number

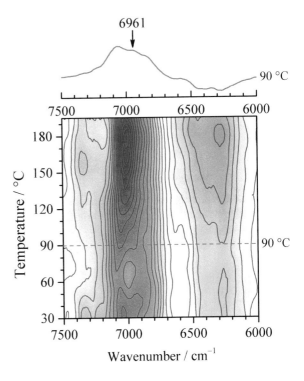

Fig. 6.6 Synchronous PCMW2D correlation map constructed from the temperature-dependent NIR spectra shown in Fig. 6.4. Positive and negative correlation intensities are colored by red and blue, respectively

of new peaks, not seen in the raw spectra, were resolved. The selective correlation of the peaks allowed proposing the molecular mechanism of the temperature-induced changes in hydrogen bonding for neat oleyl alcohol. Introduction of 2DCOS approach to NIR region inspired a number of interesting studies on various kinds of samples. The first applications were devoted to studies of hydrogen bonding in pure liquids like alcohols, water, NMA or fatty acids. A next important step was an application of 2DCOS-NIR to more complex systems including water solutions of proteins. These studies were focused mainly on examination of the secondary structure of proteins as a function of temperature, concentration or pH. An important area of employing of 2DCOS-NIR is polymer studies, with particular attention paid on the exploration of the structural response of the hydrogen bonds and hydrocarbon chains during heating, as well as stress-induced molecular chain deformation. Numerous works have been devoted to 2DCOS-NIR examination of the structure and interactions in binary mixtures of aqueous solutions of organic solvents like alcohols, NMA diols, aminoaclohols and diamines. Most of these studies have been recently reviewed [5–7]; hence, we focus our attention on new works.

Kwaśniewicz and Czarnecki explored the spectra–structure correlations in MIR and NIR regions [22, 23]. The authors applied 2DCOS and chemometric methods for analysis of ATR-IR, Raman and NIR spectra of eight n-alkanes and seven 1-chloroalkanes in the liquid phase [22]. In both cases, the chain length variation was used as a perturbation. As can be seen (Fig. 6.7), the overtones of the CH stretching

Fig. 6.7 Asynchronous 2D
correlation spectrum
constructed from NIR
spectra of n-alkanes. In red
and blue are drawn the
positive and negative peaks,
respectively. (Reprinted from
Ref. [22] with permission
from Elsevier)

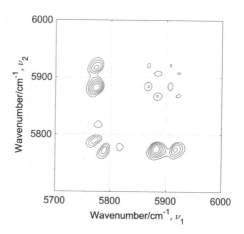

bands (near 5800 and 5900 cm^{-1}) show characteristic pairs of the peaks close to
the diagonal. Owing to resolution enhancement in the asynchronous spectrum, for
the first time, the contributions from the terminal and midchain methylene groups
were observed in the spectra of liquid n-alkanes and 1-chloroalkanes. The same
authors examined the effect of chain length on MIR and NIR spectra of aliphatic
1-alcohols from methanol to 1-decanol [23]. A negative correlation between the first
overtone of the hydrogen-bonded OH (near 6300 cm^{-1}) and the second overtone of
the methylene group (near 8225 cm^{-1}) (Fig. 6.8) reveals that the intensity changes
for these two groups are in the opposite direction. Hence, an application of 2DCOS
approach provides direct evidence that the degree of self-association of liquid 1-
alcohols decreases with the chain length increase. It is worth to mention that the

Fig. 6.8 Synchronous
2DCOS-NIR spectra of
1-alcohols from 1-butanol to
1-decanol. In red and blue
are drawn the positive and
negative peaks, respectively

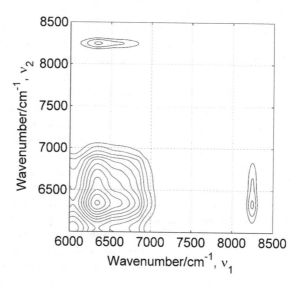

asynchronous contour plots made possible to identify the peaks from the terminal CH_2 next to OH and the peaks from the midchain CH_2.

A lot of efforts were undertaken for examination of microheterogenity in binary mixtures [24–27]. 2DCOS-NIR studies on propyl alcohols/water mixtures reveal the separation at a molecular level and the presence of homoclusters of water and alcohol existing in equilibrium with the mixed clusters (heteroclusters) [24]. The presence of these clusters is responsible for macroscopic structure of the mixtures and leads to anomalous physicochemical properties. In the water-poor region, the molecules of alcohols are in the same environment as those in the pure liquid alcohols, while the molecules of water are dispersed in the organic phase. When the water content increases, the molecules of water form clusters interacting with the OH groups of the alcohols. These results clearly show that the degree of microheterogeneity in alcohol/water mixtures is closely related to the extent of self-association of the alcohol.

Interestingly, similar conclusion was obtained from 2DCOS-NIR and chemometric studies of binary mixtures of methanol with short-chain aliphatic alcohols [25]. The degree of deviation from the ideal mixture is correlated with the chain length and the order of the alcohol. For most of the mixtures, the largest deviation from the ideality appears at equimolar mixture. The heteroclusters were observed in the whole range of mole fractions, while the homoclusters occur above a certain concentration limit. It is interestingly to note that the homoclusters of both components are similar as those observed in neat liquids.

In spite of similar structure and properties of methanol and its deuterated derivative, CH_3OH/CD_3OH mixture also deviates from the ideal mixture [26]. The extent of this deviation is much smaller as compared with the mixtures of unlike alcohols [25], and it results mainly from the difference between the CH_3 and CD_3 groups. It is of note that the contribution to heterogeneity from the OH groups is relatively small. The CH_3OH/CD_3OH mixture is composed of the homoclusters of both alcohols and the mixed clusters. The homoclusters in the mixture are similar to those present in neat alcohols. The highest population of the heteroclusters and the largest deviation from the ideal mixture appears at equimolar mixture.

2DCOS-NIR and chemometric study on microheterogeneity in binary mixtures of aliphatic and aromatic hydrocarbons has shown that even relatively weak interactions like π-π or differences in molecular shapes may give rise to deviation from the ideality [27]. The extent of these deviations is small for aromatic/aromatic or aliphatic/aliphatic mixtures and increases for aromatic/aliphatic mixtures. The shape of molecules has a significant effect on the extent of deviation from the ideal mixture. If both components of the mixture have similar shapes (linear or cyclic), the molecules with the same probability form the homo- and heteroclusters, otherwise, increases the tendency for formation of the homoclusters. Since the homoclusters of both components resemble those in neat liquids, one can conclude that deviation from the ideal mixture is due to presence of the heteroclusters. Interesting information provides 2D correlation moving-window spectrum. Figure 6.9 displays the composition-dependent moving-window spectrum of n-hexane/benzene mixture. It is of note that the spectral changes from the aromatic and aliphatic parts are clearly

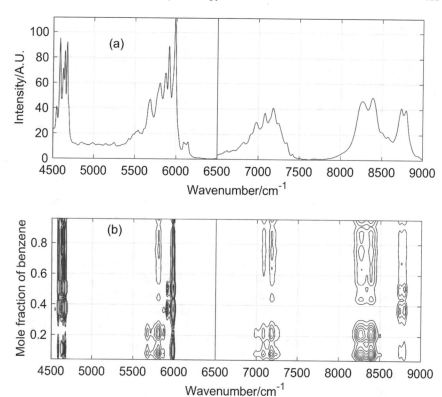

Fig. 6.9 A composition-average (**a**) and moving-window (**b**) spectrum for n-hexane/benzene mixture. Intensities in the 6500–9000 cm^{-1} range were enlarged to appear in this scale (Reprinted from Ref. [27] with permission from Elsevier)

separated. The maximum of spectral changes for n-hexane appears at mole fraction of benzene smaller than 0.3, while the largest changes for benzene are observed for mole fraction from 0.3 to 0.5. The differences observed in the mowing-window spectra are nicely confirmed in the asynchronous spectra (Fig. 6.10). The spectrum develop the peaks between the overtones from the aliphatic and aromatic components of the mixture.

Shinzawa and Mizukado examined hydrogen/deuterium (H/D) exchange of gelatinized starch by using 2DCOS-NIR [28]. The time-dependent spectra reveal a series of subtle changes, which were resolved in the asynchronous contour plots. As shown, during the isotopic substitution, the exchange rate becomes different depending on solvent accessibility of various parts of the molecule. This way, it is possible to explore the local structure and dynamics of the sample. Zhou et al. have studied interactions in C_2H_5OH/CH_3CN binary mixture by using NIR and MIR spectroscopy [29]. The data were converted to the excess absorption spectra and then analyzed by 2DCOS. The resolution enhancement in the excess and 2DCOS spectra permitted to identify a series of species including dimers, trimers and multimers of ethanol as

Fig. 6.10 Asynchronous 2D correlation spectrum constructed from NIR spectra of n-hexane/benzene mixture. In red and blue are drawn the positive and negative peaks, respectively

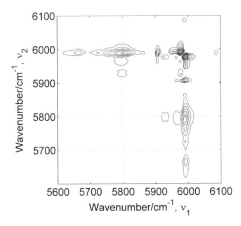

well as C_2H_5OH-CH_3CN complex. As shown, the dissociation of ethanol multimers is correlated with an increase in the concentration of acetonitrile. At mole fraction of $X_{CH3CN} = 0.7$, all multimers of ethanol are dissociated. Chang et al. applied 2DCOS-NIR to investigate combination bands of water perturbed by the presence of four different inorganic acids including: HCl, H_2SO_4, H_3PO_4 and HNO_3 [30]. Analysis of the concentration-dependent 2DCOS contour plots evidenced that each of these acids has a different effect on NIR spectra of water.

Due to the resolution enhancement and selective correlation of various peaks, 2DCOS spectroscopy is a powerful tool for analysis of complex NIR spectra. The proper data pretreatment can substantially reduce the noise or baseline fluctuations and provide more reliable results. Sometimes, it is necessary to perform the normalization of the experimental data before application of 2D correlation analysis. Since publication of the principles of the generalized 2D correlation spectroscopy by Isao Noda in 1993, numerous modifications of this method were reported. These new developments extend the usefulness of the generalized 2DCOS and opens new possibilities of the spectral analysis. Among them, the most popular is the moving-window analysis, which provides the information on the dynamic changes in very simple and straightforward form. Similarly like chemometrics, 2DCOS prefers large data sets, especially for examination of complex processes. Nowadays, 2D correlation analysis is a routine tool for the spectral analysis, and its codes are included in the spectroscopic software.

References

1. I. Noda, Two-dimensional infrared (2D-IR) spectroscopy of synthetic and biopolymers. Bull. Am. Phys. Soc. **31**, 520 (1986)
2. I. Noda, Two-Dimensional Infrared (2D IR) spectroscopy: theory and applications. Appl. Spectrosc. **44**, 550–561 (1990)
3. I. Noda, Two-dimensional infrared spectroscopy. J. Am. Chem. Soc. **111**, 8116–8118 (1989)

4. I. Noda, Generalized two-dimensional correlation method applicable to infrared, raman, and other types of spectroscopy. Appl. Spectrosc. **47**, 1329–1336 (1993)
5. I. Noda, Y. Ozaki, *Two Dimensional Correlation Spectroscopy Applications in Vibrational and Optical Spectroscopy* (Chichester, UK, Wiley, 2004)
6. M.A. Czarnecki, Y. Morisawa, Y. Futami, Y. Ozaki, Advances in molecular structure and interaction studies using near-infrared spectroscopy. Chem. Rev. **115**, 9707–9744 (2015)
7. M.A. Czarnecki, Two-dimensional correlation analysis of hydrogen-bonded systems: basic molecules. Appl. Spectrosc. Rev. **46**, 67–103 (2011)
8. I. Noda, Determination of two-dimensional correlation spectra using the Hilbert transform. Appl. Spectrosc. **54**, 994–999 (2000)
9. M.A. Czarnecki, Two-dimensional correlation spectroscopy: effect of normalization of the dynamic spectra. Appl. Spectrosc. **53**, 1392–1397 (1999)
10. A. Gericke, S.J. Gadaleta, J.W. Brauner, R. Mendelsohn, Characterization of biological samples by two-dimensional infrared spectroscopy: simulation of frequency, bandwidth, and intensity changes. Biospectroscopy **2**, 341–351 (1996)
11. P.J. Tandler, P.B. Harrington, H. Richardson, Effects of static spectrum removal and noise on 2D-correlation spectra of kinetic data. Anal. Chim. Acta **368**, 45–57 (1998)
12. M.A. Czarnecki, Interpretation of two-dimensional correlation spectra: science or art? Appl. Spectrosc. **52**, 1583–1590 (1998)
13. M.A. Czarnecki, Two-dimensional correlation spectroscopy: effect of band position, width, and intensity changes on correlation intensities. Appl. Spectrosc. **54**, 986–993 (2000)
14. C. Zimba, *Presented at the Second International Symposium on Advanced Infrared Spectroscopy (AIRS II)* (Durham, NC, 1996)
15. S. Šašić, A. Muszynski, Y. Ozaki, A new possibility of the generalized two-dimensional correlation spectroscopy. 1. Sample—Sample Correlation Spectroscopy. J. Phys. Chem. A **104**, 6380 (2000)
16. S. Šašić, A. Muszynski, Y. Ozaki, A new possibility of the generalized two-dimensional correlation spectroscopy. 2. Sample—Sample and Wavenumber—Wavenumber Correlations of Temperature-Dependent Near-Infrared Spectra of Oleic Acid in the Pure Liquid State. J. Phys. Chem. A **104**, 6380–6387 (2000)
17. A. Watanabe, S. Morita, Y. Ozaki, Temperature-Dependent structural changes in hydrogen bonds in microcrystalline cellulose studied by infrared and near-infrared spectroscopy with perturbation-correlation moving-window two-dimensional correlation analysis. Appl. Spectrosc. **60**, 611–618 (2006)
18. M. Thomas, H.H. Richardson, Two-dimensional FT-IR correlation analysis of the phase transitions in a liquid crystal, 4′-n-octyl-4-cyanobiphenyl (8CB). Vib. Spectrosc. **24**, 137–146 (2000)
19. S. Morita, H. Shinzawa, R. Tsenkova, I. Noda, Y. Ozaki, Computational simulations and a practical application of moving-window two-dimensional correlation spectroscopy. J. Mol. Struct. **799**, 111–120 (2006)
20. S. Morita, H. Shinzawa, I. Noda, Y. Ozaki, Perturbation-correlation moving-window two-dimensional correlation spectroscopy. Appl. Spectrosc. **60**, 398–406 (2006)
21. I. Noda, Y. Liu, Y. Ozaki, M.A. Czarnecki, Two-dimensional Fourier transform near-infrared correlation spectroscopy studies of temperature-dependent spectral variations of oleyl alcohol. J. Phys. Chem. **99**, 3068–3073 (1995)
22. M. Kwaśniewicz, M.A. Czarnecki, MIR and NIR group spectra of n-alkanes and 1-chloroalkanes. Spectrochim. Acta A **143**, 165–171 (2015)
23. M. Kwaśniewicz, M.A. Czarnecki, The Effect of chain length on mid-infrared and near-infrared spectra of aliphatic 1-alcohols. Appl. Spectrosc. **72**, 288–296 (2018)
24. P. Tomza, M.A. Czarnecki, Microheterogeneity in binary mixtures of propyl alcohols with water: NIR spectroscopic, two-dimensional correlation and multivariate curve resolution study. J. Mol. Liq. **209**, 115–120 (2015)
25. W. Wrzeszcz, P. Tomza, M. Kwaśniewicz, S. Mazurek, R. Szostak, M.A. Czarnecki, Microheterogeneity in binary mixtures of methanol with aliphatic alcohols: ATR-IR/NIR spectroscopic, chemometrics and DFT studies. RSC Adv. **6**, 37195–37202 (2016)

26. W. Wrzeszcz, S. Mazurek, R. Szostak, P. Tomza, M.A. Czarnecki, Microheterogeneity in CH3OH/CD3OH mixture. Spectrochim. Acta A **188**, 349–354 (2018)

27. P. Tomza, W. Wrzeszcz, M.A. Czarnecki, Tracking small heterogeneity in binary mixtures of aliphatic and aromatic hydrocarbons: NIR spectroscopic, 2DCOS and MCR-ALS studies. J. Mol. Liq. **276**, 947–953 (2019)

28. H. Shinzawa, J. Mizukado, Hydrogen/deuterium (H/D) exchange of gelatinized starch studied by two-dimensional (2D) near-infrared (NIR) correlation spectroscopy. Spectrochim. Acta A **197**, 138–141 (2018)

29. Y. Zhou, Y. Zheng, H. Sun, G. Deng, Z. Yu, Hydrogen bonding interactions in ethanol and acetonitrile binary system: a near and mid-infrared spectroscopic study. J. Mol. Struct. **1069**, 251–257 (2014)

30. K. Chang, Y.M. Jung, H. Chung, Two-dimensional correlation analysis to study variation of near-infrared water absorption bands in the presence of inorganic acids. J. Mol. Struct. **1069**, 122 (2014)

Chapter 7
NIR Data Exploration and Regression by Chemometrics—A Primer

Klavs Martin Sørensen, Frans van den Berg, and Søren Balling Engelsen

Abstract This chapter is a primer on the use of multivariate data analysis—or chemometrics—to near-infrared spectra. The extraordinary synergy between near-infrared spectroscopy and the data analysis methods called chemometrics has led to a green analytical revolution in practically all areas of life sciences and related industries for quality control and process monitoring. The near-infrared spectroscopy method is nondestructive, rapid and environmentally friendly. However, the most unique advantage of near-infrared spectroscopy is that it can measure samples remotely and unbiased, as is, i.e., solids and liquids without interfering with the sample or sample preparation. The success of near-infrared spectroscopy would not have been possible without the chemometric data processing. This chapter gives an overview, including tricks of the trade, of the most common chemometric techniques for analysis of near-infrared spectral ensembles illustrated by downloadable data examples.

Keywords Chemometrics · Spectroscopy · Quality control · Process analytical technology · Second green analytical revolution

K. M. Sørensen · F. van den Berg · S. B. Engelsen (✉)
Department of Food Science, Faculty of Science, University of Copenhagen, Frederiksberg, Denmark
e-mail: se@food.ku.dk

K. M. Sørensen
e-mail: kms@food.ku.dk

F. van den Berg
e-mail: fb@food.ku.dk

© Springer Nature Singapore Pte Ltd. 2021
Y. Ozaki et al. (eds.), *Near-Infrared Spectroscopy*,
https://doi.org/10.1007/978-981-15-8648-4_7

7.1 Introduction

> All models are wrong, but some are useful
> —George Edward Pelham Box, British statistician

The revolutionary progression of near-infrared (NIR) spectroscopy has evolved hand-in-hand with the development of the personal computer, which is essential for the comprehensive data analysis of NIR spectra. If the PC had not been developed, NIR spectroscopy as a widespread analytical discipline would probably not exist today. As evident from previous chapters, NIR spectra contain no baseline-separated peaks that can be integrated and quantified, but rather deeply convoluted and strongly overlapped spectral features. Retrieving information from such signals is a demanding numerical exercise. This can however be managed well by the computer, and together the NIR spectrometer and the PC have revolutionized quality control in practically all areas of primary food and feed production in the form of ultra-rapid, noninvasive, remote and chemical-free analysis (Fig. 7.1).

The remarkable potentials of NIR spectroscopy (NIRS) were discovered and demonstrated by pioneers such as Karl Norris, Phil Williams and Harald Martens, and multiple books, chapters and reviews have been written on the multivariate data

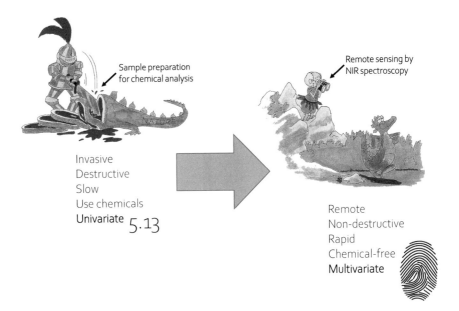

Fig. 7.1 Advantages of using NIR spectroscopy for analysis. Multivariate analysis of spectroscopic data provides a change from the traditional univariate, chemical measurement, where systems can be observed nondestructively and provide a much broader and holistic description—a complete fingerprint Adapted from Engelsen [1]

analysis of NIR spectra [2–4]. While many of these are indeed excellent, many newcomers are understandably overwhelmed by the amount of equations and abundance of methodologies, guidelines and recommendations. To assist beginners in multivariate analysis of NIR data, this chapter instead takes the form of a pragmatic step-by-step tutorial for multivariate analysis of ensembles of NIR spectra obtained on similar classes of samples.

In classical empirical research, a model requires that the number of variables must be less than or equal to the number of observations. In spectroscopy, this would correspond to the number of intensities measured at individual wavelengths and number of samples. If these samples are measured by NIR spectroscopy, such as in a conventional quality control analysis setup, at least 1000 spectral variables are recorded. A typical dataset of 100 samples will thus have the dimensions 100 × 1000, incompatible with traditional empirical models, but effectively dealt with using chemometrics, which can handle collinear data structures with many more variables than samples.

The trick in chemometrics is the reduction of the complex dataset into a limited set of latent (or principal) variables, which in turn can be used to model (unsupervised) and visualize class belonging, identify outliers, suggest trends, etc. If response variables measured by another reference method are available as well, the latent structures can be used for supervised regression modeling and prediction. In general, chemometrics assumes additivity of underlying components (in spectroscopic terms called Lambert–Beer law) and bilinear relations between the spectra ($\mathbf{X}$) and response variables ($\mathbf{y}$). Within this framework of the "straight-line tyranny," we generally decompose the spectral datasets as follows:

$$\mathbf{X} = \mathbf{A} \cdot \mathbf{B}^{\mathrm{T}} + \mathbf{E} \tag{7.1}$$

where $\mathbf{X}$ ($n \times m$) is a set of n sample spectra $\mathbf{x}$ of length m:

$$\mathbf{X} = \begin{bmatrix} x_{1,1} & \cdots & x_{1,m} \\ \vdots & \ddots & \vdots \\ x_{n,1} & \cdots & x_{n,m} \end{bmatrix} \tag{7.2}$$

and $\mathbf{A}$ ($n \times f$) and $\mathbf{B}$ ($m \times f$) contain f latent variables (see Fig. 7.2).

$\mathbf{A}$, with typically $f \ll n$ and m, then contains the contributions (or pseudo-concentrations) of the hidden latent phenomenon modeled by $\mathbf{B}$. $\mathbf{E}$ contains the residuals or unexplained information (in a least squares sense). Appearing some 40 years ago on the scientific scene [5], chemometrics has established itself as an indisputable effective and valuable multivariate data analysis toolset, extensively used in spectroscopic applications to extract information from data that would otherwise remain hidden for classical univariate methods. It exploits the multivariate advantage, while at the same time facilitating noise reduction and allowing for outlier removal.

Before going into deeper facets of chemometrics, we will briefly introduce the datasets that will be used throughout this chapter to illustrate the different algorithms.

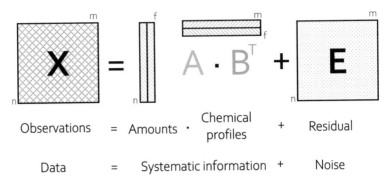

Fig. 7.2 Factor analysis of an ensample of NIR spectra. The measured spectra X, with n observations on m wavelengths (m variables), are factored into f latent variables, consisting of concentration scores A and chemical profile loadings B. The number of latent variables, f, is chosen so that the factored components $A \cdot B^T$ only contain the "relevant information" or systematic part of the data, and the remaining noise is summarized into the residual matrix E. Superscript ^{T} means the transposed matrix

7.1.1 Dataset 1: Degree of Esterification in Pectins

Dataset 1 consists of NIR spectra recorded on powder samples, which stem from different pectin extractions, including a pure polygalacturonan polymer (Fig. 7.3). The samples are expected to be of slightly different purities and slightly different particle sizes [6]. For demonstration purposes, the set is divided into two: A containing only seven samples with very extreme degrees of esterification (%DE; range 0–93%) and B including the remaining 25 samples with a much more narrow span (60.1–68.8%DE).

Dispersive NIR data were collected using a NIRSystems Inc. (model 6500) spectrophotometer. The instrument uses a split detector system with a silicon (Si) element between 400 and 1100 nm and a lead sulfide detector from 1100 to 2500 nm. The angle of incident light was 180°, and reflectance was measured at 45°. The NIR/VIS reflection spectra were recorded using a rotating sample cup with a quartz window, and spectral data were converted to $\log(1/R)$ units. In this chapter, we use only the NIR spectral range 1100–2500 nm (collecting every 2 nm interval), giving as dimensions 7 samples × 700 wavelengths (set A) and 25 samples × 700 wavelengths (set B). The dataset is available for download at http://food.ku.dk/foodomics.

7.1.2 Dataset 2: Glucose, Fructose and Sucrose Powder Mixture Design

This is a spectral ensemble in which we have mixed sucrose and its two monomers fructose and glucose in a three-component powder mixture design. Each compound

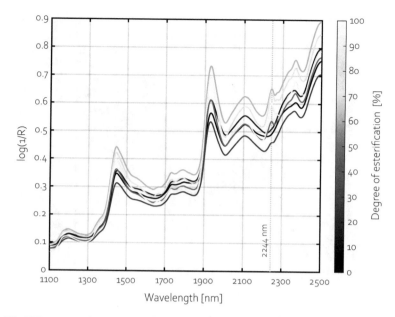

Fig. 7.3 NIR spectra of Dataset 1. The spectra—a representative subset—are colored according to the degree of esterification. Most significant is the peak at 2244 nm originating from the ester groups

is varied at 21 levels (5% w/w steps), and when mixed in a three-component mixture design, this results in $n = 231$ powder blends, all measured by NIR spectroscopy (Fig. 7.4). No attempts were made to homogenize the particle size of the three powders. Data were recorded on a spectrometer identical to the one used in Dataset 1, resulting in a matrix **X** of dimensions 231 samples × 700 wavelengths. The dataset is available for download at http://food.ku.dk/foodomics.

7.1.3   Dataset 3: Authenticity of Gum Arabic

A total of 260 *gum arabic* samples were selected from a large industrial collection provided by the Toms Group A/S (Ballerup, Denmark). The set is composed of the two Acacia species: *Acacia senegal* (L.) ($n = 19$), which is considered to be the best in quality, due to its low quantity of tannins, and comprises the majority of global trade, and *Acacia seyal* ($n = 7$), which produces a lower grade of *gum arabic*. The two grades differ in price by a factor of ten, and trained specialist can distinguish the good from the bad form when evaluating the raw material called "tears." For practical purposes, they are traded in the form of freeze-dried powder and a rapid authentication method is required. For each of the 26 selected *gum arabic* samples,

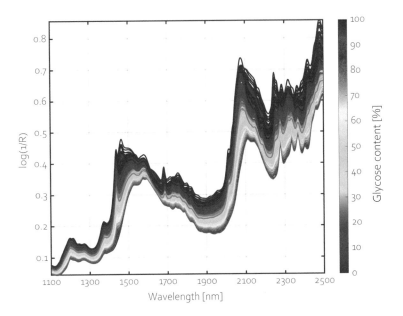

Fig. 7.4 NIR spectra of Dataset 2. The spectra are colored according to the glucose contents, one of the three factors in the experimental design

10 subsamples were prepared by randomly taking approximately 15 g of the non-uniform *gum arabic* tears and ground into a fine powder with a coffee grinder mill [7].

NIR spectra of the $n = 260$ samples (Fig. 7.5) were recorded in random order using a QFA Flex Fourier transform spectrometer (Q-Interline A/S, Roskilde, Denmark) equipped with a reflectance kinetic powder sampler that rotates a vial with the *gum arabic* powder over the instrument window. Spectra were recorded in the range from 1100 to 2500 nm using an InGa detector with a 16 cm^{-1} resolution and 512 scans averaged. The spectra were converted to $\log(1/R)$ units using a PFTE filled vial as a background reference, measured every hour of the experiment using the same instrumental conditions. The result is a NIRS dataset of the dimensions 260 samples × 700 wavelengths. The dataset is available for download at http://food.ku.dk/foo domics.

7.1.4 Dataset 4: Single-Seed NIR Spectra

A subset of NIR spectra of single wheat kernels ($n = 264$) were taken from a larger experiment spanning four different field trials in Denmark and Germany [8]. Single-kernel NIR transmittance spectra (Fig. 7.6) were collected on an Infratec 1255 Food and Feed Analyzer (Tecator AB, Höganäs, Sweden). Each kernel was placed in a

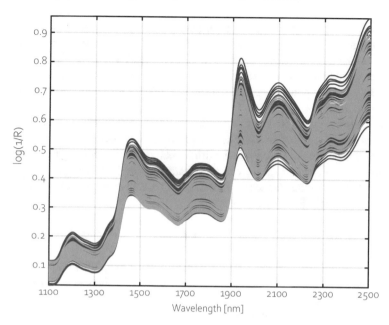

Fig. 7.5 NIR spectra of Dataset 3. The 260 spectra of gum arabic samples—190 spectra of Acacia senegal (red lines) and 70 spectra of Acacia seyal (blue lines)

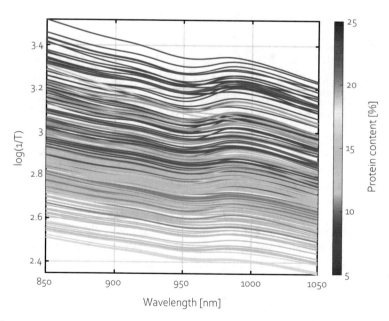

Fig. 7.6 NIR spectra of Dataset 4. The single-seed spectra have been colored according to the seed protein content

single-seed sample cassette, and transmittance spectra were recorded in the range 850–1050 nm in 2 nm steps. A tungsten lamp (50 W) and a diffraction grating were used to create monochromatic light. The light passed through the kernel, reaching the silicon detector. The time required for scanning (single scans) 23 single kernels in the cassette was about 90s. The dataset is available for download at http://food.ku.dk/foodomics.

The dataset is assembled from 7 specimens of 5 kernel varieties. Each of these 35 seeds has been measured in two alternative orientations in the carrousel—with the germ facing either up or down—each with four alternative positions (front, left, back and right); the different measurements are made by four times stopping the carrousel and manually changing the same 35 kernels. The dataset composition can thus be written as (in total, 280 NIR samples):

$$\text{varieties (5)} \times \text{individuals (7)} \times \text{positions (4)} \times \text{orientations (2)} \qquad (7.3)$$

Unfortunately, 6 kernel samples were measured using a faulty carousel well and we thus have only 264 samples. After recording the spectra of the intact wheat kernels, each one was crushed, and the single kernel protein content was determined directly by a modified Kjeldahl method [9].

7.2 Spectral Inspection and Pre-processing

Before any spectral exploration, regression or classification, it is of fundamental importance to visualize and understand the quantitative aspects of the recorded spectra

NIR spectra consist of a complex superimposition of linear concentration effects and several different nonlinear contributions such as intermolecular interactions, light scattering by particles, surfaces and phase transitions. In theory, these are highly complex phenomena, but in practice they can be removed by simple *pre-processing* techniques.

A necessary and implicit pre-processing step for spectral data, automatically applied in most spectrometer software, is the linearization of chemical concentration in the measured spectra to absorbance through application of the Lambert–Beer law on the amount of light transmitted through the sample:

$$A_\lambda = -\log_{10}(T_\lambda) = \varepsilon_\lambda \cdot l \cdot c \qquad (7.4)$$

where A_λ is the absorbance at wavelength λ, T_λ is ratio of light transmitted through the sample (a.k.a. transmittance) at wavelength λ, ε_λ is the wavelength-dependent molar absorptivity for the chemical constituent of interest, l is the effective path length of the light through the sample matrix and c is the concentration of the constituent of interest. It is further assumed that the absorption from multiple analytes is additive, which is a fundamental conjecture in applying chemometrics to spectral ensembles. For a sample with several chemical species each defined by a concentration c_s and a molecular absorptivity $\varepsilon_{\lambda s}$:

$$A_\lambda = A_1 + A_2 + \cdots + A_n = l(\varepsilon_{\lambda 1} c_1 + \varepsilon_{\lambda 2} c_2 + \cdots + \varepsilon_{\lambda s} c_s) \qquad (7.5)$$

In order to apply the Lambert–Beer law in Eq. 7.4, it is necessary to include a *blank* or empty sample in the experiment, providing a background signal which is used as a reference to all other measurements [10]. Thus, Lambert–Beer can be reformulated as:

$$A = -\log_{10}(T) = -\log_{10}\left(\frac{I}{I_0}\right) \qquad (7.6)$$

where I is the light observed as passed through the sample and I_0 is the background or blank sample.

However, most NIR applications are made in diffuse reflectance mode and Lambert–Beer is only valid for pure transmittance systems with no optical artifacts. For reflectance measurements, the reflectance R is defined—in analogy to Lambert–Beer law for transmittance—as:

$$R \cong -\log_{10}\left(\frac{I_R}{I_{R0}}\right) \qquad (7.7)$$

where, as previously, I_R is the incident light of the sample (the reflected light) and I_{R0} the light emitted by the spectrometer using the "perfect reflector" such as Spectralon.

When working with NIRS data, one of the most essential provisions for a successful application of chemometrics is the pre-processing of the spectral data. Data modification by pre-processing is introduced in order to augment the linear relationship between the apparent absorbance or reflectance and the concentration of the analytes. In other words, the purpose of pre-processing is to eliminate artifacts and nonlinearities from the spectral data before the actual modeling phase. A great number of techniques have been proposed, addressing many distinct influences from physical, chemical or mechanical sources [11]. The idea is that the spectra, before modeling, should contain only additive chemical information that follow the Lambert–Beer law.

In reflectance mode, the NIR electromagnetic radiation that is reflected by the samples will be influenced by the true absorption plus some "apparent absorption," which is primarily due to scattering of light by small particles, bubbles, surface roughness, droplets, crystalline defects, micro-organelles, cells, fibers, density fluctuations, etc. For NIR, two scattering phenomena are relevant: *Rayleigh scattering,* which is strongly wavelength-dependent ($\sim \lambda^{-4}$) and occurs, when the particles are much smaller in diameter than the wavelength of the electromagnetic radiation ($< \lambda/10$), and *Lorentz–Mie scattering* which is predominant when the particle sizes are larger than the wavelength.

There exists a forest of pre-processing techniques for NIR spectra to alleviate errors introduced by scattering. They can be roughly divided into two groups: scatter correction methods and spectral derivatives. In this chapter, we will only briefly introduce the multiplicative scatter correction (MSC) method [12, 13] and the second-derivative spectra. For almost any practical application in which the need is to analyze less than a few hundred NIR spectra, these two pre-processing methods are fit for purpose. Selection of more advanced spectral pre-processing methods to optimize the quantitative results is generally not advisable unless many more sample spectra and expert domain knowledge are present.

Before considering pre-processing, it is always worthwhile to do a simple visual inspection of the data. The aim should be to inspect the spectral variations and to observe if there exist faulty measurements or if parts of the spectral region are noisy/saturated and no sample information is to be expected. In both cases, such spectral data (or spectral ranges) should be removed prior to the application of pre-processing and chemometrics since they might deteriorate the result. When applied correctly, NIR spectroscopy is a very robust measurement technique and faulty measurements thus rarely occur. Typically, it will be the results of the occasional poor sample presentation. Noisy variables, on the other hand, are a frequent phenomenon in NIR spectroscopy, when the sample is absorbing too strong, which often occurs in the long-wavelength NIR region, where the molar extinction coefficients (molecular absorptivity) are high.

In addition, some spectrometers introduce artifacts in the spectra that might be hard to spot without a visual inspection of the data. This includes cutting or truncating absorbance values above a certain threshold (typically 3 absorbance units and above) or detector overload resulting in strange reporting values. Such phenomena are typically seen in samples containing high amounts of water, where the OH information can be very distorted, unless great care is taken. These artifacts are poison for most pre-processing techniques and must be addressed before any other handling of the data.

The spectral region covered by NIR spectrometers is a pragmatic compromise between optical materials (e.g., quartz), light sources (e.g., halogen bulb) and detector

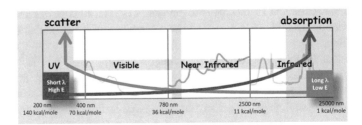

Fig. 7.7 The NIR scatter-absorption valley in the electromagnetic spectrum. Toward longer wavelengths, the absorption becomes prohibitive for transmission, and toward shorter wavelengths particle and molecular scatter becomes prohibitive for efficient measurements of chemistry

Fig. 7.8 Removal of bad (noisy) region. Including bad variables in any subsequent analysis will not serve any purpose and in worst cases lead to deterioration of the multivariate models.
(© Newlin & Engelsen)

principles (e.g., InGaAs). The whole NIR region is situated in a sweet spot in the electromagnetic spectrum, where both the molecular absorption and the scatter intensity are relatively low ([1] Fig. 7.7). However, different applications may require information from different subregions in the NIR spectral region and if for example the long-wavelength region is too strongly absorbing (noisy with no or too few photons reaching the detector) it should be removed prior to further analysis (Fig. 7.8).

Depending on the distribution of particle and microstructure sizes and density, scatter is perhaps the strongest effect that needs to be removed from the NIR signal. The scattering effect can be both frequency-dependent (proportional to λ^{-4} for Rayleigh scattering) and dependent on particle size and/or shape (Mie–Lorentz scattering). An example of scattering of data recorded on crystalline sugar powders is shown in Fig. 7.9. As the granular size changes, the spectra show offsets for the longer wavelengths, resulting in large deviations in the apparent absorption. Note that the samples have exactly the same chemistry—only the crystal size is different. If extreme variations, like the ones observed in Fig. 7.9, are observed in an NIR spectral ensemble, the pre-processing is likely to fail. But if the scatter variations are of a lesser magnitude, the pre-processing will often be able to eliminate the apparent differences in absorption.

7.2.1 Multiplicative Scatter Correction (MSC)

The multiplicative scatter correction (MSC) method was introduced by Martens and coworkers [12, 13], and together with the standard normal variate (SNV) method [14] it is the most widely applied NIR pre-processing technique. MSC removes frequency-linear imperfections, both additive and multiplicative, by fitting a first-order function between the recorded spectra $\mathbf{x_{org}}$ and a reference spectrum $\mathbf{x_{ref}}$ of the form:

$$\mathbf{x_{org}} = b_0 + b_1 \cdot \mathbf{x_{ref}} + \mathbf{e} \tag{7.8}$$

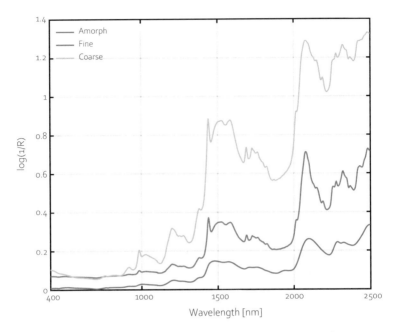

Fig. 7.9 NIR spectra of sucrose. The same chemistry (sucrose) measured as large sucrose crystals (coarse), downsized crystalline sucrose (fine) and amorphous sucrose (amorph). The graph shows clearly that the different crystal sizes have a big influence on the spectral shape

where b_0 provides the additive part and b_1 the frequency-dependent multiplicative part. In most applications, the reference spectrum $\mathbf{x_{ref}}$ is the mean spectra of all samples in the data ensemble. A scatter-corrected spectrum $\mathbf{x_{corr}}$ is thus calculated from $\mathbf{x_{org}}$ as:

$$\mathbf{x_{corr}} = \frac{\mathbf{x_{org}} - b_0}{b_1} \tag{7.9}$$

The interpretation of the scalar parameters is illustrated in Fig. 7.10. It is important to note that the MSC is recommended to be calculated only from those regions in the spectra, which do not contain the analyte-relevant variation (i.e., excluding the "curly regions" in Fig. 7.10). Thus, use the regions that are only influenced by the scatter to calculate the transform and then apply the transform to the entire spectrum. Unfortunately, this important feature has been forgotten in most, if not all, commercial software where MSC is calculated and applied to the full spectrum.

The result of the MSC procedure is thus a set of scatter-corrected spectra plus two new variables for each spectrum: b_0 and b_1. It is not a common practice to include the two new variables in the subsequent data analysis, mainly because it is not well supported by commercial software. However, there is a risk of *"throwing the baby out with the bathwater"* as the two new variables may contain important

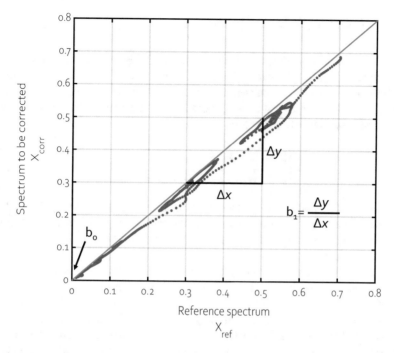

Fig. 7.10 Multiplicative scatter correction. A spectrum $\mathbf{x}_{corr}$ to be corrected is plotted against a reference spectra $\mathbf{x}_{ref}$. The linear relationship between the two spectra is the offset b_0 and the slope, $b_1 \sim \Delta x/\Delta y$

information about the scatter and thus about the physics of the samples. Discarding the coefficients thus eliminates information from the subsequent analysis.

The MSC method has been expanded into the extended multiplicative scatter correction (EMSC) method [15, 16] by introducing a second-order polynomial fitted to the reference spectrum, fitting of a baseline or, optionally, fitting of reference spectra of known analyte to target specific wavelength regions of interest. The EMSC method can in limited cases lead to slightly improved pre-processing, but will not be discussed further here. Finally, it should be mentioned here that the MSC method has the previously mentioned sibling SNV transformation, which has wide spread use and will yield very similar results for most practical applications [11]. SNV is performed by reducing the spectra with its own mean value and normalizing it to unit variation—similar to the autoscaling procedure for variables, but across the sample direction rather than the variable direction. It has the advantage, like the derivative pre-processing, that it can be applied to individual samples. This is in contrast to the MSC that needs a dataset-common reference, typically the mean spectrum.

7.2.2 Spectral (Second) Derivatives

The classical method to eliminate spectral offsets (additive effects) and slopes (multiplicative effects) is to calculate derivatives. The first-order derivative is calculated as the difference between two subsequent spectral variables; the second-order derivative is then calculated by calculating the difference between two successive points of the first-order derivative spectra:

$$x'_m = x_m - x_{m-1}$$
$$x''_m = x'_m - x'_{m-1} = x_m - 2x_m + x_{m-2} \tag{7.10}$$

The second derivate of a spectrum will have two implications: (a) The additive effect will be eliminated by the first derivative, and (b) the multiplicative effect will be eliminated by second derivative. This is illustrated in Fig. 7.11 for a double peak of two individual analytes.

The derivative approach to pre-processing has the advantage that it can be calculated independent for each sample—no other information is needed. The disadvantage of second-derivative spectra is that the spectral appearance is changed and that the peaks in the raw spectra are now turning downward. It is therefore a common practice to multiply the second derivative with −1 for visual inspection. More problematic is however the numerical calculation of the derivative on real-world, imperfect data with a significant level of noise. This noise, perhaps even causing discontinuous signal transitions, would cause considerable noise inflation in the smooth strongly overlapped spectral features of NIR when derivatized.

One way to avoid noise inflation is to use Savitzky–Golay [17] derivatization. In this method, a polynomial is fitted symmetrically around w neighboring points of data for each data point in the spectra. This produces a smoothed version of the spectra, which makes the subsequent derivation much less prone to noise artifacts. The width of the moving smoothing window and the order of the polynomial fitted

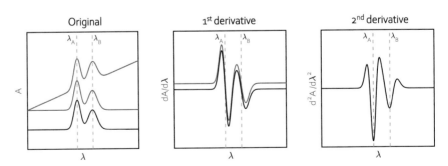

Fig. 7.11 Effect of calculating spectral derivatives. For the first derivative, the blue and black signals have become identical (the constant offset has been removed). The multiplicative effect in the red signal is seen as a constant offset in the first derivative. As a second derivative, the three signals become identical, and all spectral artifacts have been eliminated

in this window must be decided prior to analysis. It can be shown that the highest derivative that can be calculated is equal to the polynomial order incurred with the smoothing.

7.2.3 Application of Pre-processing to NIR Spectra

There is no simple answer to the question "which pre-processing method (and what configuration) suits a given data type." It will all depend on the data, data collection and purpose of the analysis. However, the primary target of pre-processing is clear: a linearization of the spectra to match the Lambert–Beer law.

To demonstrate this, Dataset 1 is used. Beforehand, it is known that the primary information about the degree of esterification (%DE) is located at 2244 nm. In Fig. 7.12, the effect of pre-processing can be observed via colors and by correlating the reflectance at 2244 nm to the known degree of esterification. In the raw spectra (A), the colors and the sequence of spectra relating to the %DE appear random and the correlation between the reflectance at 2244 nm and %DE is weak ($R^2 = 0.67$, B).

In Fig. 7.12b, the spectra have been pre-processed by MSC, and immediately we observe that the sequence of the spectra is now sorted according to %DE and the correlation between the reflectance of 2244 nm and the %DE now is improved considerably ($R^2 = 0.95$). Similarly, in Fig. 7.12c in which the spectra have been pre-processed by second derivatives, we observe that the correlation between the

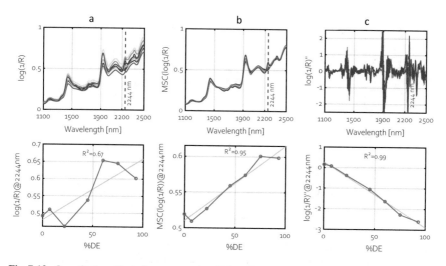

Fig. 7.12 Quantitative effect of pre-processing. Selected spectra from Dataset 1 are shown as **a** raw, **b** MSC processed spectra and **c** spectral derivatives. The lower figures show the correlation between the reflectance at 2244 nm and the degree of esterification

reflectance at 2244 nm and the degree of esterification %DE is nearly perfect ($R^2 = 0.99$) albeit in opposite direction (second-derivative points peak downward).

When the pre-processing has been decided and when regression is the target, it is often useful (but not so often supported by software packages) to plot a covarygram. This is simply a plot of the correlation of the response value (%DE in Dataset 1) for each spectral variable.

Figure 7.13 shows such a covarygram made on the MSC transformed spectra. It immediately visualized that one dominant spectral variable, namely 2244 nm, has a correlation coefficient of 1.0 to the reference variable %DE. The figure also shows that large parts of the NIR spectrum contain data that are uncorrelated to the %DE and therefore can (in principle) be excluded in a multivariate regression model. This will be further discussed in the variable selection subchapter.

As computers are getting faster and chemometric software packages more and more complete, most if not all, relevant pre-processing methods will be present in the software packages. It is sometimes even a possibility to test many alternative pre-processing methods in an automated way and optimize the desired classification or regression performance.

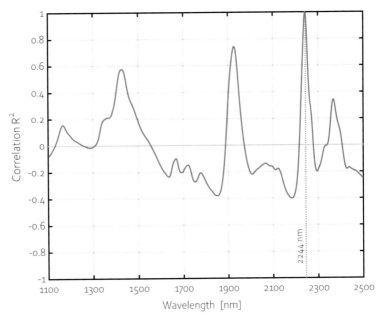

Fig. 7.13 Correlation (R) between all NIR variables and the response function (covarygram). The covarygram shows the correlations between each spectral variable (wavelength) and the response function, the %DE of Dataset 1

7.2.4 *Outro*

The pre-processing methods mentioned here, and sometimes a larger selection of additional methods, are always included in chemometric software. However, great care must be taken in the selection of pre-processing methods and especially when they are used to optimize quantitative results. No matter how elaborate the portfolio of methods, the NIR spectroscopists time is typically much better spent in getting familiar with the spectral data and target variables (plotting spectra with intelligent use of colors, generating ratio plots, covarygrams, etc.) than by spending time investigating more complex pre-processing procedures. It has been estimated that the maximum regression improvement of any pre-processed model when compared to the global model is approximately 25% in RMSE. This is hardly what makes the difference in multivariate feasibility studies, and it is thus recommendable to select pre-processing in order to achieve parsimonious, interpretable models [11].

7.3 Unscrambling Spectral Mixtures by Self-Modeling Multivariate Curve Resolution (MCR)

Ideally, we want to resolve complex mixture spectra into contributions of pure analyte spectra (S) weighted by their concentrations (C):

$$X = C \cdot S^T + E$$

If several analytes with varying concentrations are present in a mixture, the pure spectra and the associated relative concentrations can be estimated under certain conditions. The method used is called self-modeling curve resolution [18] or just multivariate curve resolution (MCR) [19].

The MCR model attempts to approximate the variation in the data, $\mathbf{X}$, with a bilinear model of two factor matrices. MCR fits f components simultaneously into a set of concentration profiles $\mathbf{C}$ ($n \times f$) and pure spectral profiles $\mathbf{S}$ ($m \times f$):

$$\mathbf{X} = \mathbf{C} \cdot \mathbf{S}^{\mathbf{T}} + \mathbf{E} \tag{7.11}$$

under the least squares constraint:

$$\min_{\mathbf{C},\mathbf{S}} \sum_{n,m} \left\| \mathbf{X}_{n,m} - \sum_{f=1}^{F} \mathbf{C}_{n,f} \mathbf{S}_{m,f}^{\mathbf{T}} \right\| \tag{7.12}$$

The matrix $\mathbf{E}$ contains the spectral variation that could not be explained by the model (e.g., noise and unsystematic structure/interferences).

Equation 7.12 can be solved using alternating least squares (ALS) [20] in which both concentration profiles ($\mathbf{C}$) and pure spectral profiles ($\mathbf{S}$) are optimized simultaneously in an iterative manner. The value of f must be determined before starting the algorithm, but very often the correct choice is not known for real systems. Several procedures have been suggested to solve this issue. Most of them are based on the principle that there are as many components as linearly independent elements (e.g., chemical constituents) in the $\mathbf{X}$ matrix (practical rank of the matrix). A very useful method is to get the eigenvectors and eigenvalues of the $\mathbf{X}$ matrix by performing a singular value decomposition (SVD) on the cross product $\mathbf{X}^T \cdot \mathbf{X}$. The chemical rank can be expressed as the number of eigenvalues higher than eigenvalues associated with the noise level. Also, the shape of the eigenvector (or length m corresponding to the spectral length) can be useful to estimate the correct number of absorbing components. When the number of components f has been decided, the ALS goes as follows.

It is straightforward to estimate $\mathbf{C}$ if you already know $\mathbf{S}$. It will be equivalent to estimating the concentrations when you know the pure spectra:

$$\mathbf{C} = \mathbf{X} \cdot \mathbf{S} \cdot \left(\mathbf{S}^T \cdot \mathbf{S}\right)^{-1} \tag{7.13}$$

The ALS solution needs to be initialized with a random or a sensible first estimate. This can be found if there is prior knowledge on the system, e.g., pure spectra of some of the components. Accordingly, MCR-ALS is often initiated by guessing the pure spectra $\mathbf{S}$ and then calculates an estimate of the concentrations $\mathbf{C}$. This estimate of $\mathbf{C}$ can now be used to improve the estimate of the pure spectra $\mathbf{S}$:

$$\mathbf{S}^T = \left(\mathbf{C}^T \cdot \mathbf{C}\right)^{-1} \cdot \mathbf{C}^T \cdot \mathbf{X}^T \tag{7.14}$$

By alternating between Eqs. 7.13 and 7.14 until convergence, at least a local solution to the problem Eq. 7.11 can be obtained. The ALS optimization has converged, when the model improvement between consecutive iterations is below a certain threshold value (typically less than a tolerance of 10^{-12}).

The strength of MCR-ALS is its capacity to resolve the pure underlying spectra and obtain their relative concentrations. However, the challenge with MCR-ALS is its dependence of the initial guess of $\mathbf{S}$ or $\mathbf{C}$, its slow convergence and its indeterminacy in the solution [21]. Ambiguities in general render the MCR models more inconsistent and dependent of initial guesses of the spectral profiles. In many cases, the solution of ALS-MCR will reach a local minimum and not the global minimum. Imposing constraints to the MCR solution can help in decreasing the risk of local minima and "false" solutions. Common constraints employed in MCR are nonnegativity, unimodality (i.e. peak has only a single highest value), closure (e.g., all components add up to 100%), equality (e.g., two components are equal in concentration) and selectivity (e.g., some variables carry only information about one analyte) [21]. In most works concerning spectroscopy, a nonnegativity constraint in the spectral mode (**S**) and in the concentration mode (**C**) is employed to guide the algorithm, i.e., using the knowledge that the NIR spectra only have positive absorbances and that the concentrations can only be positive. Detailed description of other constraints, limitations and other aspects of MCR is discussed in the literature [19].

Even in the absence of error, three indeterminacies exist for the MCR solution [22]: (i) Permutation indeterminacy—there is no defined order of the components in **C** and **S** and no sequential calculation of the components. This is a minor bookkeeping problem, which should be solved when, for example, repeating the model in cross-validation scenarios (see validation section 7.6). (ii) Intensity indeterminacy—two identical spectra, but scaled differently, will provide the same model fit since the concentrations will be adjusted accordingly. This provides two different solutions. The problem is easily solved by normalizing the spectral profiles to have the norm 1 or by constraining the concentrations to add up to 1 (closure constraint). (iii) Rotational indeterminacy—similarly to the intensity indeterminacy, a rotation of the concentration profiles and consequently of the spectral profiles can in some cases reproduce the original data with the same fit quality. The closure constraint will not help solving this problem, but nonnegativity constraint on both concentrations and spectra will reduce the solution space considerably, often sufficiently to finding the correct solution.

Even with constraints, MCR does not always provide a unique solution, and the result will sometimes depend on the initial guess of **C** or **S**; therefore, only specialized chemometric software packages include the MCR. However, by repeating the MCR model with many different, random initial guesses and subsequently analyzing the solution space, it is possible to find a unique global solution.

The number of components in the MCR model can be validated by inspecting the explained variation as a function of the number of MCR components (f) in the same way as the number of components in other chemometric algorithms is validated (see validation section 7.6). However, due to the ambiguity in the solution space the MCR results should always be validated by a priori knowledge about the chemical system being investigated. Spectral pre-processing that is focused on "cleaning" the spectra from scatter and artifacts will normally be an advantage, while other operations such as centering and autoscaling will "destroy" the pure spectral information sought and thus the MCR model.

7.3.1 Application of MCR to NIR Spectra

MCR is in general best suited for relative simple and well-behaving systems. It has nevertheless been applied to numerous NIRS applications such as whey powder [23], protein denaturation [24], edible oils [25], porcine fat tissue [26], process analytical technology [27] and many more studies.

Figure 7.14 shows the result of a 3-component MCR model fitted to the Dataset 2. Figure 7.14a shows the concentration matrix **C** from the MCR model plotted in the

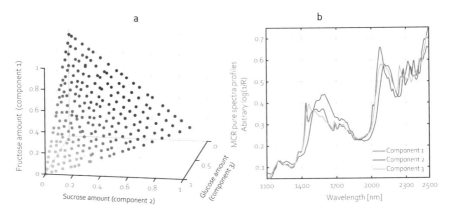

Fig. 7.14 Multivariate curve resolution applied to Dataset 2. The data have been MSC pre-processed, and a 3-component solution calculated. **a** The concentrations **C** for the three sugars closely following the design. The concentrations are colored according to the mixture content of the three pure sugars (red: sucrose; blue: fructose and green: glucose). **b** The pure MCR spectral profiles **S** for the three identified sugars

three dimensions. This plot almost perfectly reproduces the triangular experimental design. Even better, the MCR pure spectra profiles **S** (Fig. 7.14b) are exactly similar to pure sugar spectra. Since the pure spectra (**S**) can be regarded as estimates of the real spectra, so are **C** the relative estimates of the concentrations.

This "textbook" application shows how MCR in ideal cases can resolve a mixture spectrum into its pure single components and their relative concentrations which just need to be scaled using a single known sample or standard in order to obtain the real concentrations.

7.3.2 *Outro*

The main advantage offered by MCR is the spectral/chemically meaningful solution. The concentration values are nonnegative (often by virtue of the applied constraints), and the spectral profiles can be related back to the spectra of specific compounds.

In NIRS, many applications involve nonlinear and nonadditive effects (e.g., hydrogen bonding effects), but for some well-behaving systems MCR may provide a unique, interpretable solution, plus the recovery of pure component spectra. However, due to the labile nature of the MCR models, it is not implemented in most commercial chemometric software, which typically relies on rapid convergence and robust solutions requiring little user interaction. For NIRS, the most used software for performing MCR models is the academic implementations such as the *PLS Toolbox* (Eigenvector Research, Manson, WA, USA, http://www.eigenvector.com) and the MCR-ALS toolbox [28] both implemented in MATLAB (MathWorks, Natick, MA, USA, www.mathworks.com).

7.4 Spectral Exploration by Principal Component Analysis (PCA)

In practice, we can resolve the NIR spectral ensembles into a low number of orthogonal latent variables by Principal Component Analysis:

$$X = T \cdot P^T + E$$

Principal component analysis (PCA) is the workhorse of chemometrics. The method is now more than 100 years old [29–31], is used in many research disciplines for different purposes and is therefore unfortunately known under many different names. Spectroscopic data are characterized by high colinearity, i.e., that two neighboring wavelengths are positively correlated. PCA is tailored to handle this type of data, and it is in the analysis of spectroscopic data that PCA really shows its worth. However, before we go to the analysis of NIR spectroscopic data, we will briefly outline the principle of PCA. There is a striking similarity between the PCA and the MCR models, as they both attempt to approximate the variation in the data with a bilinear model. The difference between the two models lies in how the system of equations is solved. For PCA, an algorithm is used that successively finds orthogonal components in a multivariate dataset **X**. This principle is, in contrast to the MCR algorithms, extremely efficient and robust, but the solution has the interpretative disadvantage that the extracted components are forced to be orthogonal while spectra in a mixture are not.

7.4.1 The PCA Method

A key concept in chemometric analysis is the reduction of variance in data into a lower-dimensional space of *principal components* or *latent variables*. In PCA, the multivariate dataset is decomposed into orthogonal components, whose linear combinations approximate the original dataset in a least squares sense.

If we have an experiment of n observations (samples) of m independent variables (wavelengths), a line describing the maximum observed variance in the variable space can be defined as the least squares solution of minimizing each of the orthogonal projected distances l_n from the nth sample point onto to the principal line:

$$\min \left\| \sum_n l_n^2 \right\| \tag{7.15}$$

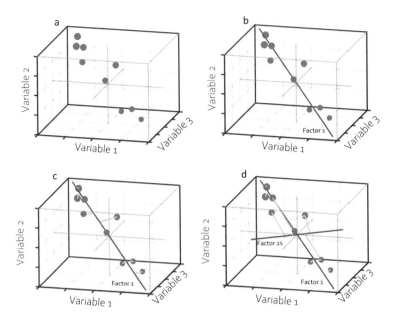

Fig. 7.15 Principle of principal component analysis. For an artificial dataset of nine samples with only three wavelengths (**a**), the first principal component (**b**) is found that spans the most of the sample variation and which minimizes the sample residuals (**c**) represented as the orthogonal projections to the line. The second principal component is orthogonal to the first principal component and spans the most residual variance left by the first component (**d**)

is illustrated in Fig. 7.15, for a toy system with 8 samples and 3 variables. The principal line shown in Fig. 7.15b corresponds to the direction in the data that spans the most variance, and all sample points can now be defined or "fixed" by their orthogonal distance to (or *projection on*) the principal line. This principal line is called the *principal component* or *loading* and the orthogonal distances from the sample points to the line for the *scores*. We see that this principal line does not represent completely the systematic variance structure of the measured data as none of the observations lies exactly on the line. A second component can be found, orthogonal to the first principal component, which describes as much of the remaining variance in the samples (Fig. 7.15d). This is the second principal component, and each sample point will have a related score, which is again the orthogonal distance from the sample to this principal component line. In a three-dimensional system, it is only possible to extract three components, but for more realistic systems (e.g., a NIR data ensemble) the process of extracting subsequent principal components can continue. If the samples are projected onto the principal component, and this projection is subtracted from the original set of data, a new principal component can be determined on the remaining variance (the *deflated* **X** matrix). In fact, this process can be repeated until there is no more systematic variance left to explain.

If the data of n samples and m variables are represented as a matrix **X**, of size n x m, the PCA is defined as:

$$\mathbf{X} = \mathbf{T} \cdot \mathbf{P}^{\mathbf{T}} + \mathbf{E} \qquad (7.16)$$

with the least squares solution:

$$\mathbf{min} \left\| \sum_{n,m} \left(\mathbf{X}_{n,m} - \sum_{f=1}^{F} \mathbf{T}_{n,f} \mathbf{P}_{m,f}^{\mathbf{T}} \right)^2 \right\| \qquad (7.17)$$

where the sample scores $\mathbf{T}$ are the projection of variance onto the variable loadings $\mathbf{P}$ using f components.

The values in $\mathbf{T}$ (the *scores*) are the projections of the samples on the principal directions defined by $\mathbf{P}$ (the *loadings*). It is possible to view the PCA process as a breakdown of information from the raw data matrix ($\mathbf{X}$) in which PCA creates two new, smaller data matrices: one containing information about the samples (scores, $\mathbf{T}$) and one containing information about the variables (loadings, $\mathbf{P}$). The splitting of the information is done in such a way that the two parts, $\mathbf{T}$ and $\mathbf{P}$, explain as much variation in the original data matrix ($\mathbf{X}$). The PCA algorithm finds the weights (loadings) so that this happens. No other weights will be able to describe more of the systematic variation in the given dataset. In fact, an additional matrix is created, namely the model *residual* $\mathbf{E}$ equal in size to $\mathbf{X}$, a remainder of the data that is not explained by the two-component model. Residuals are a core concept of chemometric analysis as real-life data always tend to be imperfect. The principal components represent the systematic or *explained* variance in the data—the remainder, measurement error, biological variance, etc., are kept out as the residual.

In Eq. 7.17, f indicates the number of principal components calculated in the model. Not surprisingly, the described accumulated variance of the components will be ever increasing as more and more components are determined for a system. The maximum number of components to be found, before no more systematic variance can be modeled, is governed by the chemical or practical rank of the data. The mathematical rank of $\mathbf{X}$ determines the maximum number of principal components that could be determined, and is equal to the maximum number of independent linear combinations that can be made from the matrix (chemical rank $f \ll \min(n,m)$ = mathematical rank). Data originating from real-world experiments will naturally have imprecisions, originating from measurement errors, sampling methodology, biological variations, etc. These imprecisions will be independent of the experiment and can thus be seen as unsystematic variation or noise.

A very important premise of conducting PCA is centering of the data. It is normally not very interesting to model the absolute level of the data, but rather to model the variance of the "data cloud" around the center of gravity. In the process called *mean centering,* the mean value of each variable column in $\mathbf{X}$ is subtracted from the variable itself:

$$\hat{\mathbf{x}}_m = \mathbf{x}_m - \bar{\mathbf{x}}_m \qquad (7.18)$$

where $\mathbf{x}_m$ is the *m*th column of $\mathbf{X}$. The correct form for PCA is then written:

$$\mathbf{X} = \overline{\mathbf{X}} + \mathbf{T} \cdot \mathbf{P}^{\mathbf{T}} + \mathbf{E} \qquad (7.19)$$

Mean centering is often considered together with its analog in correlation analysis, namely *autoscaling*. In spectral data, each wavelength is expressed in absorbance (or reflectance) units and, thus, is approximately equal in scale and variance, which in turn can be weighed equal in the analysis. It is thus sensible to only apply mean centering in spectroscopic analysis. However, if the variables were composed from different measurement types, with different units, they will be weighed unequally in the analysis. *Autoscaling* seeks to rectify this by scaling each variable to have unit variance [11], where the *m*th column of $\mathbf{X}$ is mean centered and normalized by the standard deviation of the *m*th column:

$$\hat{\mathbf{x}}_m = \frac{\mathbf{x}_m - \overline{\mathbf{x}}_m}{\text{std}(\mathbf{x}_m)} \qquad (7.20)$$

Due to the orthogonality constraint imposed in the model, PCA has a simple and unambiguous solution that can be calculated rapidly using, e.g., singular value decomposition (SVD). PCA is thus present in all commercial chemometric software packages due to its extraordinary robust data reduction and data summarizing capabilities.

7.4.2 Explained Variance

It is very useful to be able to quantify how much of the information in the data that a given component—or the residual—is describing. In this respect, each set of *t*'s and *p*'s is called a component; thus, $\mathbf{T}$ is of size *n* by *f* and $\mathbf{P}$ is of size *m* by *f*. The graphical representation of a PCA with *f* components is identical to Fig. 7.2. The total variance of $\mathbf{X}$ is explained by the sum of the individual components (plus the residual):

$$\mathbf{X} = \mathbf{T} \cdot \mathbf{P}^{\mathbf{T}} + \mathbf{E} = \mathbf{t}_1 \cdot \mathbf{p}_1^{\mathbf{T}} + \mathbf{t}_2 \cdot \mathbf{p}_2^{\mathbf{T}} + \cdots + \mathbf{t}_f \cdot \mathbf{p}_f^{\mathbf{T}} + \mathbf{E} \qquad (7.21)$$

where $\mathbf{t}_f$ is the *f*th column vector of $\mathbf{T}$ and $\mathbf{p}_f$ is the *f*th column vector of $\mathbf{P}$.

The *sum of squares* of $\mathbf{X}$ (size $n \times m$) is defined as the summation of the squared of each value of $\mathbf{X}$:

$$\text{SSQ}(\mathbf{X}) = \sum_n \sum_m \mathbf{X}_{n,m}^2 \qquad (7.22)$$

Similarly, the sum of squares can be calculated for an individual component *f*:

$$\mathrm{SSQ}(\mathbf{t}_f \cdot \mathbf{p}_f) = \sum_n \sum_m (\mathbf{t}_f \cdot \mathbf{p}_f)^2_{n,m} \qquad (7.23)$$

When combined, Eqs. 7.22 and 7.23 yield the percentage of explained variance for a given component f as:

$$\mathrm{Variance}(f) = \frac{\mathrm{SSQ}(\mathbf{t}_f \cdot \mathbf{p}_f)}{\mathrm{SSQ}(\mathbf{X})} \cdot 100\% \qquad (7.24)$$

It is customary, when reporting PCA results, to state how much variance the individual components explain. The explained variance by the PCs is often indicated on the PCA score plot axes, where, e.g., "PC1 (50%)" means that PC1 explains fifty percent of the total systematic variance in the dataset.

7.4.3 Application of PCA to NIR Spectra

The application of PCA to spectroscopic data is best illustrated by an example. Figures 7.16 and 7.17 demonstrate PCA applied to the designed Dataset 2, which

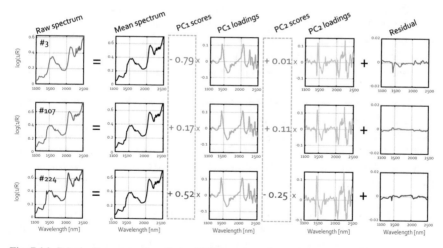

Fig. 7.16 Principal component analysis of NIR spectra selected from Dataset 2. The plot shows how PCA decomposes Dataset 2, visualized for 3 selected samples #43, #107 and #224. The first column shows the input spectra, and the second column shows the mean spectrum (black) which is equal for all three samples. The third column shows the first loading (green) which is also equal for all three samples, but the amount (score) of this loading is different for the three samples. The fourth column shows the second loading (blue) with the corresponding scores, and the last column shows the residuals, i.e., what is left when the first two principal components have been extracted to the three sample spectra. The residuals are different for the three samples, but note the low magnitude of these compared to the loadings

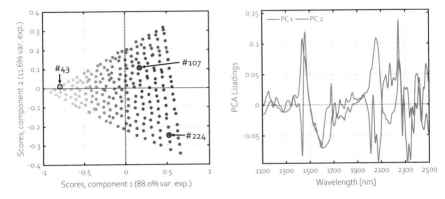

Fig. 7.17 PCA scores and loadings' plot of Dataset 2. Left: Scatter plot of scores for PC1 versus PC2. The scores are mixture colored according to the mixture content of the three sugars (red: sucrose; blue: fructose and green: glucose). Right: The loadings of PC1 (blue) and PC2 (red)

includes three chemical components (pure sugars) in a mixture design. In this analysis, the data was first corrected for light scattering using the MSC method. The first step in PCA modeling is to mean center the spectroscopic data. This is done to focus on the variations between the individual samples rather than the general signal level.

In this example, it is only necessary to inspect the first two principal components based on the number of chemical variation sources in the samples. Three chemical components in a mixture design (summing to 100% by definition) ideally give rise to two independent sources of variation. For more complex systems, the optimal number of components in a PCA model can be determined mathematically as described in the validation subchapter (7.6).

In Fig. 7.16, the principle in PCA is illustrated for three selected samples, but note that the PCA model is calculated for all 231 samples. Column A to the left in Fig. 7.16 shows the raw spectra for three samples: #43 (blue), #107 (red) and #224 (purple) coming directly from the spectrometer. Column two shows the average spectrum that is subtracted from each sample spectrum corresponding to the *mean centering* of the data. The average spectrum is the same for all samples and therefore shown in the same color (black).

The first loading vector (green—third column) is the spectral structure that best describes the variation in the centered data (Fig. 7.16). No other underlying structure can explain more of the variation in data than this one. The first loading is common to all the samples, and what makes the samples different is the amount (or "concentration") of this structure in their spectrum. This amount is called the *score* value of the sample. Sample #43 has, e.g., the *score* value −0.79 for the first loading, and the other 230 samples in the dataset have other scores. The *loading vector* multiplied by −0.79 is the best possible description of sample #43 using one principal component, when this loading vector is determined to also describe (in the least square sense) all other samples.

The second loading (cyan) is the structure that describes the second largest amount of variation in the dataset where this second loading vector also has to have the constrained of being orthogonal to the first loading. Again, the difference in the samples is evident only from the *score* value, which is −0.01 for sample #43.

The part of the variation in the dataset that is not described by the first two principal components is shown in the residuals in the right most column of Fig. 7.16. The residuals are specific for each sample and may for example be used to detect aberrant patterns in single measurements. Note the y-axis of the residuals: The numerical values fluctuate within ±0.002. These values can be directly comparable to the variation in the mean-centered spectral data (Fig. 7.16), which varies between 0.1 and 0.7.

By comparing the size of the residuals with the variation of the mean-centered data, the explained variance can be calculated for each principal component. In this case, the first component (PC1) explains 88.0% of the initial total variation, the second component (PC2) explains 11.6% of the remaining variation, and overall the two components thus explain 99.6% of the variation in the dataset.

Plotting all 231 score values for the first principal component against the corresponding values for the second component yields a *score scatter plot* (Fig. 7.17) in which each point represents a NIR spectrum with originally 700 variables. In the given case, sample #43 can be seen in the coordinate system with the coordinates (−0.79; 0.01), sample #107 at (0.17; 0.11), sample #224 at (0.52; -0.25) and so on for the remaining samples.

As shown by the example, PCA is a good tool for exploratory data analysis of highly colinear data as often seen in spectroscopy. As a result, one can observe the behavior and characteristics of single samples and study, which wavelength ranges are important for the similarity or difference between samples. PCA can be perceived as a "reverse" Lambert–Beer model: The model estimates latent spectra (loadings) and determines the (pseudo)concentrations of these in the samples (scores) from the measured spectra. For spectroscopists, the disadvantage is the tricky interpretation of the loadings, which are not pure analyte spectra representative of the underlying chemistry. The main take-home message of this PCA application is that samples which are close to each other in composition are also close in the score plot; i.e., biological replicates in your dataset, e.g., should be found close to each other. For Dataset 2, we observe the experimental design and it is characteristic for PCA that the score plot of this 3-component mixture is completely described by two components (the chemical rank is 2 because of closure where the three concentrations add up to 100%). This is in contrast to the MCR model (Fig. 7.14) which models one component for each chemical component.

7.4.4   PCA for Outlier Detection

The ability of the PCA to reveal the behavior and characteristics of single samples as part of the complete sample set makes it a powerful tool in the detection of *outliers*.

Outliers can be defined as samples that have a different variable pattern compared to other samples in the dataset. When many variables have been measured (spectral data), it can often be difficult to find such patterns via direct inspection, i.e., plotting, of measurement data. Using PCA, it is possible to find deviating spectra (outliers) using relevant graphical images that lead directly back to measurement data. In PCA, the pattern or relationship between all variables is analyzed and handled through the calculated loadings ($\mathbf{P}$). The basic assumption in PCA is that all samples can be described with the same set of loadings. Samples for which this does not apply will have a variable pattern that differs from all other samples in the dataset.

In PCA, there are two distance measures that differentiate how much a sample differs from the rest of the dataset: the size of the residual and Hotelling's $\mathbf{T}^2$. The residual of a sample can be calculated directly from the residual $\mathbf{E}$ matrix. For a given sample, the square sum of all the elements of the corresponding row in $\mathbf{E}$ is calculated (see, e.g., Fig. 7.16). A sample with higher residual variance will have a pattern or variation in the original data that is not similar to the remaining samples. The second most important distance measure is based on the *score* values ($\mathbf{T}$). The distance to the center of a sample in the score space can be calculated using Hotelling's $\mathbf{T}^2$, which considers the covariance in the data. Combining the two distance measures in a scatter plot, i.e., the residual variance and Hotelling's $\mathbf{T}^2$ provide the most important diagnostic plots in PCA.

It is of fundamental scientific importance to be able to efficiently identify outliers as they may represent new discoveries with completely new functionalities or, as a contrast, identification of samples that ruins the models. In chemometric modeling, outliers are undesirable because they are included in the estimation of model parameters. Thus, the PCA model must be recalculated, when one or more samples are characterized as outliers and discarded. It is thus an iterative process to characterize and eliminate outliers. This is easily done in modern chemometric software where the sample is marked in a residual variance versus Hotelling's $\mathbf{T}^2$ plot, and then the model is recalculated without the selected sample.

While the residual variance versus Hotelling's $\mathbf{T}^2$ plot is very efficient in identifying obvious outliers, it is important to underline that there is no general method for outlier recognition and removal. This is because, among other things, Hotelling's $\mathbf{T}^2$ "outliers" may be desirable as extreme but valid specimens that span the model.

7.4.5   PCA for Data Quality Control

Due to its capability to model-free convey the samples inter-variability, PCA is a very effective tool for quality control of an experimental dataset. Not only can PCA be used to detect outliers as described above, but it also provides information on how samples are related to each other in a quantitative series (such as in Fig. 7.17), in a time series or in discrete groupings, which by PCA can all be scrutinized concerning the smallest detail. Browsing through the PCA plots of a newly recorded dataset can usually reveal more information about the data, than is otherwise possible from the inspection of the obtained data alone, and that in a very short time by using interactive graphical displays with easily interpretable symbols and colors according to the *metadata* of the dataset.

The term metadata is used for any information associated with the data that is not a part of the PCA itself. Typically, such data are categorical and do only exist as discrete levels, like the name of the person who measured a specific sample, or if the measurement is first, second or third of a set of replicate measurements. Metadata can also be numerical, but not relevant to include in the PCA itself—e.g., "seconds elapsed since a reference measurement was made on the equipment," "the content of an analyte in the sample" or "the relative humidity in the laboratory on that day". The most important source of metadata is the experimental design parameters such as harvest year, variety and field. Metadata are very important to gain insight into a dataset and can be exploited in the PCA, in the validation of multivariate models, in PLS-DA models and in ASCA models. Some chemometric programs allow for coloring the samples according to a quantitative response parameter, which is going to be target for regression analysis. As an example, see Fig. 7.17 where the score plot is colored according to the sugar contents. This allows for a quick, graphical investigation on how much of the total variance is related to the quantitative response parameter, and how systematic it is distributed over the sample set.

Using PCA makes it easy to evaluate the validity of a dataset simply by observing the location of the replicates in a score scatter plot. An experiment can contain two types of sample replicates: experimental replicates (i.e., "mixing" or "chemical") and measurement or analytical replicates. Concerning experimental replicates, the samples are to be considered experimentally alike, but have different origins (for instance, the same type of beer, but brewed on three different days). When each of the experimental replicates is measured several times, they become measurement or analytical replicates. This is illustrated in Fig. 7.18, where a score plot for two components resulting from a PCA displays three experimental replicates which each has been measured three times (in a random order).

Based on the location of the colored groups, an inspection reveals that the samples originating from the red group are significantly different (distant), than the green and blue groups, which are very similar (close). In addition, the green and blue measurement collections appear more similar (closer), than the red group, which spreads out more indicating a higher inter-group variance. The next step would be to inspect the loadings of the two components to investigate why the difference in red and green/blue is so significant, or to look back into the experimental logbook to see if there is anything known about red group that can hint at this separation.

It should be noted here that PCA on real spectral data always is able to find and illuminate such replicate variances and groupings, no matter how small. It is thus important to compare the replicate (intra-group) variance to the sample (inter-group) variance. When conducting this exercise, it is important to have always the explained variance of the investigated PCs in mind as they carry the information of the magnitude of the explained variance (i.e., importance) in the two directions.

PCA can be advantageous in analyzing performance of an analytical technique or sample preparation over time. One such diagnostic feature is the *pool sample*,

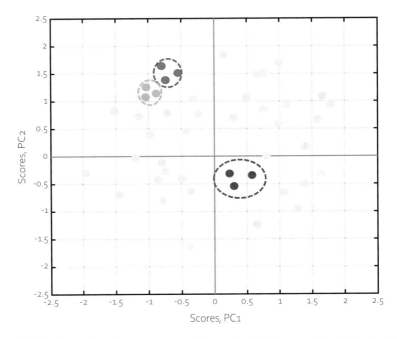

Fig. 7.18 Experimental and measurement replicates in a PCA score plot. In the PCA plot, nine measurements are highlighted. The highlighted samples are three measurement or analytical replicates (same color) of three process samples or experimental replicates (different colors)

where a fragment of all samples is pooled into a single pooled sample by adding the same amount of each of the individual analytes to the pool. The pool sample should be found approximately at the origin of all score plots for all components—as it represents the "chemical average"—especially if the molecular species do not interact. To check that a measurement campaign is progressing without any changes in the analytical instrument, a pool sample measurement can be conducted at regular intervals (e.g., every 25th sample). When during data analysis all the pool samples are found to be at origin of the score space, the analytical system can be trusted.

Another diagnostic is the simple coloring of samples according to measurement preparation or acquisition time. If, in any score plot, a trend can be seen following such a coloring, one should be highly suspicious on how much the instrument itself has influenced the obtained results. If measurements are acquired online, for instance using an NIR probe inserted into a process where data are acquired at set time intervals to study the process, the plotting of scores vs. acquisition or process time represents a unique tool to monitor the dynamics of a process.

7.4.6 Outro

PCA is implemented in all chemometric software packages, and it cannot be emphasized enough that performing a PCA on a recorded spectral dataset is a prerequisite to understand the variability in the data. This also includes the effect of different spectral pre-processing methods. Usually, the more focused and systematic the score plot is, the better and higher the amount of explained variance from the first PCs (so-called parsimony), the better. It should be noted here that PCA score plots do not change by validation, but the explained variance of the principal components does.

7.5 Calibration by Partial Least Squares (PLS) Regression

> PLS is one of the strongest regression methods invented.
> It works where Multiple Linear Regression fails!

The task of multivariate calibration is to find a predictive model that relates the NIR instrumental response space to the analyte concentration space. Here, the purpose of calibration is to model analyte concentrations, $\mathbf{y}$, as linear combinations of absorption spectra $\mathbf{X}$. Next, analyte concentrations in future samples can be predicted based on the absorption spectra only. Where PCA represents an untargeted and unsupervised data exploration, partial least squares (PLS) regression [32] is the targeted and supervised method par excellence.

7.5.1 Regression with Principal Components

When data matrix $\mathbf{X}$ is decomposed with a PCA, it is represented in a model space, represented by the principal components describing the systematic variance. If a relationship can be found between this model space of the data and an independent or reference variable, the independent variable can be explained in terms of the observed dependent variance and hence a regression can be made. This process is called *calibration*.

A classical regression extension of PCA is known as *principal component regression* (PCR) [33]. Having projected a data matrix $\mathbf{X}$ into a PCA model defined by loadings $\mathbf{P}$, resulting in scores $\mathbf{T}$, a regression toward a dependent $\mathbf{y}$ can be made via the *regression vector* $\mathbf{b}$ in the model space:

$$\begin{aligned}
\mathbf{X} &= \mathbf{T} \cdot \mathbf{P}^{\mathrm{T}} + \mathbf{E} \\
\mathbf{y} &= \mathbf{T} \cdot \mathbf{b}^{\mathrm{T}} + \mathbf{q}
\end{aligned} \qquad (7.25)$$

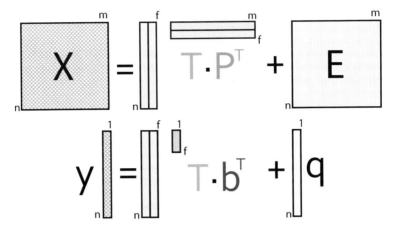

Fig. 7.19 Principal component regression (PCR) generalized to a **y** variable. The score space from a PCA is projected onto the **y** variables using a regression vector **b**. **b** does not use the full space of **X**; the residuals **E** will be the noise, not used in calibration

The advantage of regression onto the component-based variance model space is that **T** does not have to describe the full rank of **X**, and measurement noise in the data can thus be removed from the calibration.

However, PCR has the shortcoming that it is a two-step method in which the scores (**T**) are first calculated from a data table **X** (e.g., the NIR spectra), focusing on explaining **X** variance only, and then a regression model is made toward the dependent variable (**y**), e.g., a quality (see Fig. 7.19). This is equivalent to going into a supermarket (**X**), buying items in different departments such as fruits and vegetables, meats, desserts and wines, and only after the goods are paid for you know what menu (**y**) you want to make for dinner. Obviously, once you select the information in **T** (the goods in the supermarket) without thinking about what it should be used for, then the calibration model that relates **T** to **y** may be unnecessarily complicated. There could, for example, be large interferences (irrelevant peaks) to which the target signal in comparison is much smaller. These interferences will contain most of the variation expressed in the principal components calculated from the data. Hence, regression on the model space will not produce an optimal calibration model since the regression would describe the interferences rather than the sought analyte.

7.5.2 Partial Least Squares Regression

The problem is solved in *partial least squares* (PLS) *regression*, which as the name indicates only partially performs regression onto the variance model space, i.e., only on the part that is relevant to the regression [32, 34]. In PLS regression, **y** is explicitly included in each step of the algorithm to find the relevant **T** from **X**. This is equivalent to going into the supermarket with the menu in hand and therefore having the opportunity to buy exactly the items you need. Thus, it is not necessary to ensure that all systematic information in **X** is represented in **T**, but it is enough to extract the information relevant to (or correlating with) **y**. This principle, which makes the model easier to interpret and understand, is illustrated in Fig. 7.20.

In PLS, an intermediate step is introduced for each component, where the direction of the largest covariance between **X** and **y** is used as a weight vector **w** to "steer" the regression into the direction of most systematic variance in the dependent variable as a function of **y** [35]. The scores **t** calculated in a PLS are the projection of **y** onto **w**, after which the loadings **p** are determined by regression of **t** on **X**. As previously described for PCA, the components are calculated successively, and after each component, the explained variance of both **X** and **y** is subtracted from the initial values, and the next component can be determined. Ultimately, the regression vector **b** can be determined by the regression of **y** onto **T** [36].

In practice, NIR spectroscopy is used primarily as a rapid noninvasive prediction method using PLS regression to a reference method. The text book example is the development of a NIR prediction model for the protein content in wheat samples made by Phil Williams in 1975 for the Canadian Grain Commission [2]. This technological jump saved more than 50 tons of chemicals annually used for the Kjeldahl protein determination [9]. The NIR prediction method is a two-step procedure. A *calibration*

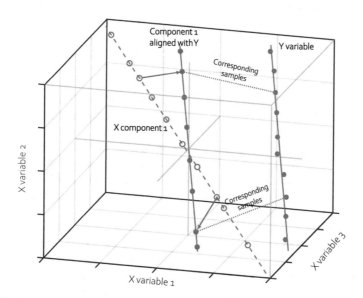

Fig. 7.20 Concept of PLS regression for calibration. This figure illustrates in analogy to the principle of PCA (Fig. 7.15) how the first PCA (stipulated blue line) which describes the main variation in the raw data matrix (X) is twisted toward describing most of the variance in the response vector (y), shown by the orange line. The resulting "new" PLS component (solid blue line) is aligned with the y variance

step in which a PLS regression model is established using a set of samples measured twice. One measurement by NIR spectroscopy to collect **X** and one measurement by a reference method to collect **y**. And next a *prediction* step uses NIR spectroscopy and the PLS calibration model to predict the value of the response variable for unknown samples. The benefit of this approach is the ultra-rapid and sustainable measurement of billions of samples worldwide, but as mentioned in the previous chapter, the accuracy and reliability of the prediction method will depend on the validation of the calibration model.

Two examples of PLS calibrations will be shown after discussion on model validation and error reporting (Section 7.6).

7.5.3 Partial Least Squares Regression—Discriminant Analysis (PLS-DA)

In NIRS analysis, it is common to have a priori knowledge about the spectral data, typically from a controlled experimental design (called metadata in the PCA section). Some metadata are binary or only have discrete levels such as male or female, organic or conventional, active or placebo, authentic or not, variety 1 or 2, and breed 1 or 2. Such labels can be used actively in regression modeling by introducing a so-called dummy variable that contains the a priori knowledge in the form of a response variable **y** vector made of dummy values (typically 0 and 1) distinguishing between the two different classes. The final assignment of a class to a prediction is done on a threshold of the predicted dummy y. For instance, if the predicted value is above 0.5, it is assigned to class 1, and if below 0.5, to class 0.

Accordingly, the PLS-DA is a classification method where the dummy variable is predicted in the best possible way using the information found in the spectral data [37]. This is closely related to a normal PLS prediction model, where a continuous parameter (e.g., protein level) is predicted from a NIR spectrum, but the main difference is that PLS-DA solves a classification task. In PLS-DA, the classes are described in the dummy parameter in the best possible way, providing the best obtainable prediction from a linear combination of the wavelengths, which are weighted via the regression coefficients in **b** according to their importance in the prediction model of the class parameter.

Where a normal PLS model is optimized according to the prediction error (e.g., the root mean square error of prediction: RMSEP), the PLS-DA should be optimized based on classification parameters (e.g., rate or percentage of correct and misclassified samples). PLS-DA is prone to yield overfitted results, and therefore a thorough validation step (see validation section 7.6) is needed.

> When you ask for discrimination, you will get it!
> —Lars Nørgaard, Danish chemometrician

While PCA results often can be presented without considering validation, PLS models must always be validated before presenting scores and loadings and prediction errors (see section on validation). For PLS-DA models, the validation becomes even more crucial since spurious correlations can often lead to excellent, but false, classifications. Moreover PLS-DA score plots should be used with great care (read: not be used) since it can be demonstrated that score plots from a PLS-DA model often can show clear groupings even when random data is assigned to two classes. Similarly, discriminative PLS-DA score plot can be found when sound real data are arbitrarily divided into two classes [38]. Regardless of validation or not, the scores and loading plots would be similar and these plots can thus not be used to access the classification performance of a PLS-DA model.

7.5.4 Outro

In many practical applications, multiple response variables are available and for this purpose there is a variant of PLS called PLS2, which can be used as alternative. It could, for example, be that one would like to predict protein, fat and carbohydrate content of a cereal product. With the help of PLS2, these three different models can be made at once and thus used to directly understand how the three different quality parameters interact. However, if performance is the single objective, then it is highly likely that you will get better performance results by just applying PLS separately to each of the three response variables.

For many PLS applications, the target is to minimize the prediction error. It should be as low as possible, yet maintaining its predictive power. It is important to note that PLS is correlation/covariance-based and will not be able to distinguish between direct correlations (causal) and indirect correlations. A sound and healthy PLS regression model may very well rely on indirect (biological) correlations in the sample set—also called *the cage of covariance* [39].

PLS is implemented in all chemometric software and is probably the strongest regression tool ever developed. Accordingly, good reasons (typically called nonlinearities) are needed for not choosing PLS in multivariate regression. Other alternatives are principal component regression, random forest, neural networks and machine learning, which will be briefly discussed later in this chapter.

7.6 Validation of Multivariate Models

The purpose of validation of multivariate NIRS models is to provide an unbiased evaluation of the model performance.

When selecting the validation method, you should act as the advocate of the devil!

A key concept in multivariate data analysis is *model validation*. Generally, validation is about the models' applicability (extrapolation) to new samples. Unfortunately, it is not enough to use the model fit alone as a validation criterion, since addition of components will nearly always lead to an improved fit in the least squares sense for a finite dataset. This means that we cannot use the calibration diagnostics to see if the model is good or bad. A key question is instead "how will it perform for other and new data?". In order to try to answer this, it is necessary to find a way to validate the number of components used in the multivariate model. If too few are used, the model is said to be underfitted, and if too many, the model is said to be overfitted. The way to estimate the correct number of components in a calibration model is to use a test set, which is a set of sample spectra and related response variables unknown to the model. The model can be developed on the calibration set, and the goodness can be evaluated by the test set. This is called test set validation. Sometimes, when the total dataset is not large enough to be split into both a calibration and test set, there exists another option, which is called cross-validation. In this section, we will summarize the most common types of validation employed in multivariate data analysis.

7.6.1 Model Performance Metrics

Correlation is a key statistic used to gauge regression model performance. Pearson's correlation coefficient, or just R, of known $\mathbf{y}$ and the associated predicted $\hat{\mathbf{y}}$ is:

$$R = \frac{\text{cov}(\mathbf{y}, \hat{\mathbf{y}})}{\text{std}(\mathbf{y}) \cdot \text{std}(\hat{\mathbf{y}})} \tag{7.26}$$

Sometimes expressed as *coefficient of determination*, R^2 provides a measure for the *relationship* between the predicted outcome and the reference. $R^2 = 1$ corresponds to a perfect relation, and $R^2 = 0$ corresponds to no relationship at all. Often, it is stated as a stand-alone indicator for regression accuracy. However, as shown in Fig. 7.21, it is a dangerous assumption to equate a high correlation to a high model quality. In most practical circumstances, a $R = 0.8$ will be considered high, but as it is seen in Fig. 7.21, it hardly reflects a metric that resembles quality.

Clearly, the correlation does indicate the *accuracy* of the given prediction. A prediction may be near perfectly related to the associated reference values (high R),

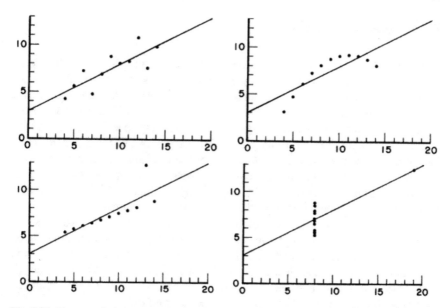

Fig. 7.21 Four correlation scenarios between two variables (for one sample the value of variable 1 is plotted on the x-axis and value of variable 2 is plotted on the y-axis; it could be the PLS predicted values by NIRS versus the response values measured by a reference method) all having the correlation R = 0.816. **a** A linear model with some uncertainty, **b** a nonlinear model, **c** a perfect linear model with one outlier and **d** a nonsense two-group model. Modified from Anscombe [40]

but still with high numerical deviations (high error). Chemometric applications tend to state the root mean square error (RMSE) [41] as a measurement for prediction error of n measured **y**'s and predicted **ŷ**'s:

$$\text{RMSE} = \sqrt{\frac{1}{n}\sum_{i=1}^{n}(y_i - \hat{y}_i)^2} \qquad (7.27)$$

The structure of the RMSE calculation is like that of the standard deviation and has the same rules for interpretation. That is, for given regression, one can expect approximately 68% of the predicted values to lie within ±1 RMSE and approximately 95% to lie within ±2 RMSE.

The optimal model is one that has a high correlation, and a low prediction error, with as few components in the model as possible, and has been validated on a set of independent samples. The act of validation is an absolute necessity for producing reliable results, and it can be argued that any error statistic is worthless, unless it has been validated against an independent set of data. Only then does the produced quality estimate reflect what can be expected from the model "in the real world."

7.6.2   Model Validation

In order to develop reliable error measurements for a regression, the model must undergo a validation, where it is tested how it predicts unknown or new data. In this respect, new or unknown refers to data points which were not included in the dataset used while calibrating the model. That does not exclude them from originating from the same experiment. However, a very crucial rule needs to be enforced for any split of data into a calibration and validation set: The validation data must be completely independent from the calibration data. In practice, this means that the same physical sample measured cannot be present in both datasets, nor as separate measurements and neither as measurement replicates. A good way of thinking of this is that the data conceptually could come from two different measurement campaigns.

As described below, one does not always have the luxury of an isolated dataset used for validation purposes. In those cases, the method of cross-validation can be applied to get validation statistics. But, however way the data are validated, it is important to stress that the chosen validation scheme is determined by the structure of the experiment so as to ensure independent validation data, and hence should be a consideration made very early in any calibration workflow.

A term frequently used in multivariate analysis is the concept of *overfitting*. A model is said to be overfitted, when it loses the ability to optimally predict new or unknown samples. This is particularly relevant to regression models, where the inclusion of too many components will cause the regression to overfit the data and essentially making it useless for any practical purpose. When the model overfits the data, the additional components will start to structure, not just the systematic variation in the data, but also the sample specific noise.

7.6.3   Cross-Validation

In order to avoid overfitting, a model can be validated in one of two ways, namely by applying it on a test set or by *cross-validation*. Cross-validation (CV) is the process of sequentially removing one or more samples, makes a calibration on the remaining samples and uses that to predict the values of the ones removed [42, 43]. The process is illustrated in Fig. 7.22.

Cross-validation can be used to find an optimal number of components, when the RMSE of cross-validated predicted values (named RMSECV) is calculated against the associated reference (original $\mathbf{y}$). When RMSECV is plotted against the number of components included in the model, it will typically reveal a local minimum that yields the most accurate model. A characteristic RMSECV development, along with its corresponding root mean square error of calibration (RMSEC), is shown in Fig. 7.23. As may be expected, the error occurring from the unvalidated calibration (blue line)

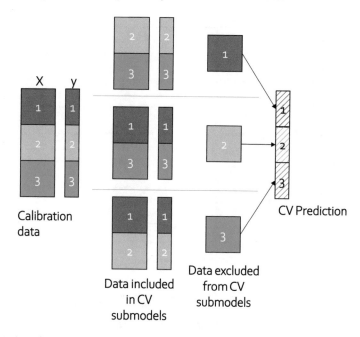

X   y

Calibration data

Data included in CV submodels

Data excluded from CV submodels

CV Prediction

Fig. 7.22 Schematic overview of the cross-validation process. A dataset is split into blocks—here three, which each is removed once from the dataset in turn. As each block is taken out, a model can be developed on the two remaining blocks. The new sub-model can be used to predict the values of the excluded block (CV prediction). After excluding all blocks in turn, a complete y vector of CV predictions has been produced

keeps falling, as more and more components are included in the model. The cross-validated model (red line) instead shows a characteristic low point after four components. From there on, the prediction error starts to increase, showing that the model is overfitting the data. As a rule of thumb, one must select as few components in a model as possible in order to eliminate the possibility of overfitting. The developing model in Fig. 7.23 show clearly that the 3-component model is optimal. Including further components will cause the RMSECV to increase and, hence, overfit. However, other PLS models sometimes display an insignificant improvement in RMSECV when going from a 3-component system to a 4-component model. It can then be argued that the proper conservative choice is to use the more parsimonious 3-component model. Despite being offered as an option in several software packages, automatic selection of the number of components should only be considered as a guidance. An automatic system can never replace prior knowledge of the samples, the measurement system or reference values.

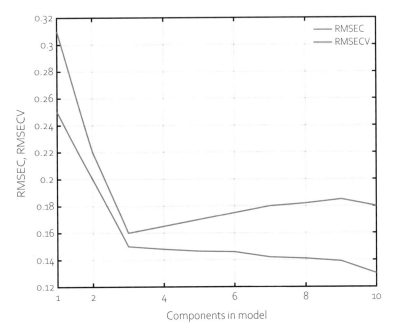

Fig. 7.23 Model fit as a function of number of components. A typical development of the root mean square error of calibration (RMSEC) and root mean square error of cross-validation as a function of the number of PLS components. The RMSECV line has a local minimum at component #3, which is optimal for the model

7.6.4   Cross-Validation Systems

The strength of cross-validation lies in testing the prediction on samples *independent* from the calibration. It requires that the cross-validation segments are constructed in proper correspondence with the experiment—especially if the experiment includes replicate measurements. If 20 samples are measured with five replicates, a regression on the 100 actual sample measurements should be cross-validated in a way so that all five replicates of each sample are removed as a group (20 cross-validation segments), and not by removing each 20th object as a group (5 cross-validation segments). Only the former setup will give independent evaluation of each sample, which is not the case in the latter case, where 4 versions of the same sample are still included per CV segment, which then no longer can be called independent anymore. The former case will test the model's ability to predict unknown samples, where the latter will validate the model against the measurement and tolerances, as this is the changing factor between the cross-validation segments.

Several systems specifying the segmentation of the samples in cross-validation setups are used in the literature and software. These include *venetian blinds*, where the CV segments are grouped in sets 1-2-3-1-2-3-1-2-3-…, *contiguous subsets* where the CV segments are grouped in sets 1-1-1-2-2-2-3-3-3-…, or combinations thereof, as 1-1-2-2-3-3, etc. This nomenclature refers to how the data are structured in the matrices and must be accompanied by other information (number of splits, number of segments, etc.). Metadata can often be used to split the data into sensible nested segments (experimental design factors) such as e.g. animals, varieties, vintages etc. Full cross-validation (or leave-one-out/LOO cross-validation) is often referenced in the literature, but should be used with caution, especially in the case of data with analytical replicates. Leaving out every single sample does rarely provide independent sampling, and the full cross-validation should only be applied in cases with very few samples [44].

7.6.5 Bootstrapping

An alternative to full cross-validation, or when no prior knowledge of the data structure is available, is to divide the dataset into a number of random blocks of each typically 10–20% of the data, called *random subsets*. In case of replicates, all replicates of the same sample must still be kept out at the same time. Repeating the random sampling validation, a high number of times for a dataset, each time with new randomization, gives a robust error estimate by averaging the CV predicted *y* over the repeated CV runs.

7.6.6 Test Set Validation

When enough samples are available, or when the experiment design permits, a very efficient way of validating a model is to do *test set validation*. Ideally, the experimental data can be split into independent *calibration* and *test* parts, each representative of the population of observations in the experiment. As described earlier, replicate measurements of the same physical sample cannot be present in both sets at the same time.

Test set validation is straightforward. A model is calculated for the data in the calibration dataset, which is then applied to the test set. The error of the predicted test set values and the associated reference values determined is referred to as root mean square error of prediction (RMSEP).

7.6.7 Application of PLS to NIR Spectra

In a first example, the application of PLS to the mixture design in Dataset 2 is demonstrated, using the NIR spectra as **X** and the glucose content as the response variable, **y**. The results of this model are shown in Fig. 7.24.

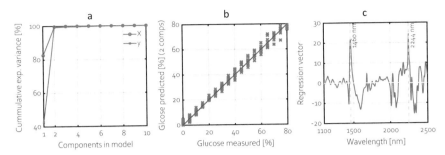

Fig. 7.24 Prediction of glucose content in Dataset 2. The **X** matrix has been MSC pre-processed prior to analysis. **a** shows the cumulative explained variance for **X** and **y**, respectively, and **b** shows the "actual vs. predicted" plot for two components. Frame **c** shows the resulting regression vector

In analogy to PCA, the number of PLS components is calculated using cross-validation. In Fig. 7.24a, it is observed as expected that the first two components explain nearly all **X** and **y** variances (glucose). Indeed, a total of 2 components is optimal for the model and this results in a model performance of $R^2 = 0.99$ and a RMSEC of 2.26% glucose. For the 2 components, the PLS model describes 99.2% of the **y** variance. Similarly, at two components, the explained **X** variance is 99.6%. Inspecting the regression vector (Fig. 7.24c), two positive peaks are identified at 1460 and 2240 nm which accordingly has high importance for the PLS model.

In a second more realistic example, the application of PLS is demonstrated to the prediction of single-seed protein content from NIR transmission measurements in the region from 850 to 1050 nm (Dataset 4). The advantage of this dataset is that it is large ($n = 264$) and has an experimental structure that makes it interesting in studying the effect of different cross-validation schemes.

As is evident from a casual inspection of the raw data (Fig. 7.6), the spectra seem to exhibit, what appears to be, if not just scatter, than a highly varying degree of transmission intensity (path length). Therefore, the NIR transmission spectra need to be pre-processed before any calibration to the underlying chemistry can be performed.

As a first step, a suitable cross-validation scheme should be decided upon. In this experiment, the spectral data originate from 5 different varieties of grain and it will thus be appropriate to use a "leave one variety out at a time" cross-validation scheme with the purpose of selecting a suitable pre-processing method and an optimal number of components. This scheme will sequentially exclude blocks with 20% of the dataset (or 52 spectra).

As observed from Fig. 7.25, the choice of pre-processing has a large impact on the performance of the model. The worst performance is observed for no pre-processing (blue line). It seems that the derivative methods are performing better than multiplicative scatter correction (MSC) alone, and the best performer is a Savitzky–Golay (SG) filter of second order with a width of 7 spectral variables (corresponding to 14 nm). The best performing model, and that quite significantly, is a combination of a second derivative and a subsequent MSC [45]. Combining pre-processing methods can indeed produce more accurate models, as is seen here. The example here is a

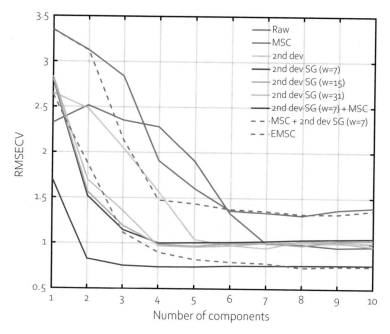

Fig. 7.25 Development of PLS models and comparison of pre-processing methods (Dataset 4). Prediction of the protein content in single wheat seeds from NIR transmission spectra. Testing and comparing different pre-processing methods. The graph shows cross-validated prediction errors from different types of pre-processing method. SG indicates second-order Savitzky–Golay derivatives with the indicated window length. The models are validated "leave one variety out at a time"

testimony to the fact that a thorough inspection of different pre-processing methods and reasonable combinations thereof for a given dataset should always be considered.

Having selected a suitable pre-processing method, and by inspecting the curve in Fig. 7.25, it appears that 4 components may be a reasonable choice. The decrease of prediction error is negligible including further components, and hence, the model will yield a RMSECV of 0.74% protein, as shown in Fig. 7.26.

7.6.8 Application of PLS-DA to NIR Spectra

For demonstration, a PLS-DA classification model is developed on Dataset 3. In this set, several *gum arabic* samples have been measured, and they are known to belong to one of two classes—Acacia seyal or Acacia senegal. In order to define the class of each of the samples, a dummy **y** vector is constructed that has the same number of elements as samples in the **X** data. In this vector, all samples of Acacia seyal are set to "1" and all samples of Acacia senegal are set to "0". When performing

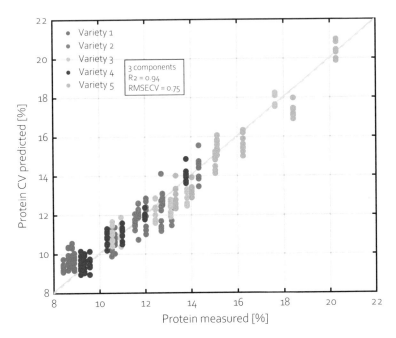

Fig. 7.26 Predicted versus measured plot for the PLS model (Dataset 4). This figure shows the performance of the PLS model for prediction of the protein content in single wheat seeds from NIR transmission spectra. The spectra have been pre-processed using second derivative + MSC and validated using "leave one variety out at a time." The sample points are colored according to the variety

a PLS prediction model on these data, a pseudo-probability will be produced for a measurement to belong to the Acacia senegal variety.

When performing the cross-validation of the model, one physical sample will be left out by removing all of its 10 analytical replicates at a time. An independent sampling for the validation is thus achieved.

The resulting PLS-DA model is shown in Fig. 7.27. Inspecting the cross-validated prediction error (red line) in Fig. 7.27a, a local minimum is found at 8 components. This is a very high number for a system with only 26 physical samples, and by closer inspection, it is decided that 3 components are a more suitable trade-off between error and complexity because the gain from including component 4 and onward is negligible (does not change the number of misclassifications). Given a 3-component solution, the cross-validation predicted dummy **y** is shown in Fig. 7.27b. There is a clear separation between the two classes of samples with only three misclassifications using a threshold of 0.5 (stipulated red line). A so-called *confusion table* that groups the counts of classifications of samples in terms of classification modes can now be developed (see Table 7.1).

The table can be summarized into the *classification accuracy*, determined to be 0.996, calculated as:

$$\text{ACCURACY} = \frac{TP + TN}{N_T + N_F} \qquad (7.28)$$

In fact, the classification accuracy can be calculated for each component added to the model. Such a plot, shown in Fig. 7.27c, is a valuable tool to determine number of components needed. And in this example, three components seem as a reasonable choice: The increase in accuracy by including a fourth component is negligible.

7.6.9   Outro

Selection of a validation method for multivariate models is of crucial importance to the performance of the final prediction model. The chemometric software may have many different validation schemes implemented, and it might at first seem difficult to choose the correct one. However, the choice may be simplified by following a set of simple rules:

1. In general, you want to perturb your data as much as possible when applying cross-validation [43]. Use the highest relevant nested level such as batch, variety,

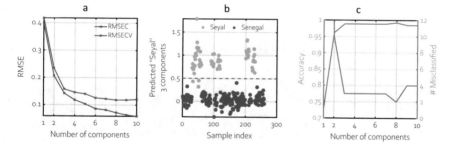

Fig. 7.27 PLS-DA prediction of the "Acacia seyal" class belonging of Dataset 3. **a** shows the prediction error of the dummy y variable. **b** shows the predicted value of the two groups colored according to class (Acacia seyal is red, and Acacia senegal is blue). The red line indicates the classification threshold. **c** shows the prediction accuracy as a function of components

Table 7.1 Confusion table for the Acacia seyal PLS-DA prediction	$n = 260$	Actual Acacia seyal $n = 70$	Actual Acacia senegal $n = 190$
	Predicted Acacia seyal	True positives (TP) 67	False positives (FP) 0
	Predicted Acacia senegal	False negatives (FN) 3	True negatives (TN) 190

location or year as segments. The segments should be representative for what you would like the model to be able to predict.

2. Do not use full cross-validation (leave out one sample at a time) unless you have very few samples [44]!
3. Replicates must always end up in the same cross-validation segments!
4. If you have several experimental design factors, try to use the different factors as segments in the cross-validation, i.e., batch, year, variety, location, etc.
5. If different models are approximately equally good, be conservative and select the model that is most parsimonious (i.e., uses fewest latent factors).
6. If different cross-validation methods suggest different numbers of latent factors, try to repeat a method, where samples are randomly split into (relatively few) segments.

The application of other more advanced validation methods like double cross-validation, permutation and Monte Carlo testing often adds complementary insight of the model performance (Westerhuis et al. 2008). However, in most cases following the rules above will be adequate to validate the multivariate models. While validation allows for the assignment explained of variance in the PCA models, it is primary when it comes to regression/prediction models that validation becomes crucial for assigning measures of accuracy to the models in terms of bias, variance, confidence intervals, prediction errors, etc., and to the determination of number of components.

7.7 Variable Selection in Regression

Too much data—too little information!
—Harald Martens, Norwegian chemometrician

The spectral range of NIR instruments is typically determined by hardware components such as optical materials, light sources and detectors. The spectral region for a given instrument may thus not be optimal for your application. The multivariate advantage has been amply demonstrated for PCA and PLS applications to NIRS data, but how much multivariate is enough and how much is too much?

In principle, two or a few, covarying neighboring variables should suffice to provide the multivariate advantage. Despite the high redundancy in NIR spectra, the multivariate methods can often be improved by variable selection. The primary reason for the improvements is the reduced number of interferences in the reduced set of variables, but also because of the fact that the data structure in the NIR region has different behaviors across the NIR region, in particular in the shortwave (SW) NIR region, the first overtone region and the combination tone region. Feeding all this variation to PCA or PLS may deteriorate the performance. It can be an advantage to get rid of the irrelevant spectral regions and spectral regions which contain mostly

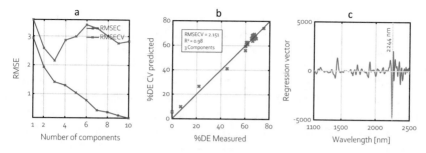

Fig. 7.28 PLS regression to the pectin %DE from the combined Dataset 1. The data have been pre-processed using second-derivative Savitzky–Golay (window width of 9 variables). The model has been cross-validated (leave-one-out, chosen here for simplicity). **a** A local RMSECV minimum is found at 3 components with an error of 2.15%DE. **b** Predicted versus measured plot for a 3-component model with an R^2 of 0.98. **c** The resulting regression vector

noise (e.g., too high absorbance regions). However, when performing spectral region reduction and/or variable selection the strategy must be carefully considered.

> If you keep only the data you think are relevant, you will confirm what you already "know" is important and this will reduce your chances of innovation
> —Frank Westad, Norwegian chemometrician

First, it must be decided for what reason variable selection is performed. Is it in order to obtain parsimonious models with simple interpretation or is it exclusively to increase model performance? In any case, combining a supervised model, such as PLS-DA, with a variable selection method gives a high risk for overfitting and thus creates the need for rigorous validation.

Many strategies exist for variable selection in NIRS regression methods. They come in two flavors: one focusing on finding variables that are good at prediction of the response variable, and one that is focused on uncertainty estimates on the coefficients of the regression vector. In the following, a few pragmatic methods that have found their way into NIRS will be described and compared with the results provided in a straightforward PLS application (Dataset 1). The "baseline" model for this dataset is shown in Fig. 7.28.

7.7.1 Regression Coefficients

The PLS regression coefficients (**b**) represent a measure of association between each variable and the response. If an acceptable global PLS model is obtained, a normal procedure is to inspect the model parameters, for example this regression coefficient.

High absolute regression coefficients are considered important, and small regression coefficients are considered less important and could potentially be eliminated. Eliminating variables that have a small absolute value for the regression coefficients may thus lead to an improvement of the model [46]. This simple idea is the basis for variable selection by so-called jackknifing [47, 48]. For each cross-validation segment, a new regression vector is calculated, and it will thus be possible to estimate the standard deviation of the PLS regression vector. When the distribution of the regression coefficients includes zero, they can be discarded, and a new model calculated. The procedure can be implemented iteratively by recalculating the model and eliminating more variables. It differs from other variable selection methods by not searching directly for variables that are good at predictions or **y**, but instead by eliminating variables that possibly have a regression coefficient close to zero and thus do not contribute (significantly) to the prediction. No matter what value such variables have for a new sample, they will be multiplied by zero and thus not contribute in the prediction. To remove variables, this way makes the risk of overfitting smaller compared to selection methods that are focused on finding variables that are good at predicting the response variable.

It is a prerequisite for this variable selection method to work, that a decent model has already been developed using all spectral variables. As shown in Fig. 7.28c, the majority of the regression coefficients are close to zero despite the fact that the raw spectral data varying significantly in these regions. This indicates that the PLS model has down-weighted these variables.

Since spectroscopic data are normally smooth, so is the regression vector expected to appear smooth. It is thus (normally) not possible that the measurements at, e.g., 2450 nm are positively correlated to the response variable and the measurements at 2452 nm are negatively correlated. Models with noisy regression coefficients indicate that the spectral region should be removed. By these very pragmatic and basic rules, it is possible to reduce the spectral data to a region of interest, which improve the predictive PLS model. This method further has the practical advantage that PLS regression coefficients are readily available as standard output and plots from PLSR software.

7.7.2 Variable Importance in Projection

A variant in using the regression coefficients for variable selection is the variable importance in projection (VIP) estimate, originally proposed by Wold et al. [49]. The VIP score is an estimate of the importance of each variable for the projection of **y** onto **X**. It is found by accumulating the importance of each variable from the PLS loading weights for each component. The average VIP score for all variables is equal to 1, and hence, typically a "larger-than-one" selection rule is applied for variable selection [50]. The VIP score is normally used as an assistance in manual variable selection and can be a valuable tool, when used together with prior knowledge about the measurements. As a selectivity ratio, the VIP number can be used to exclude

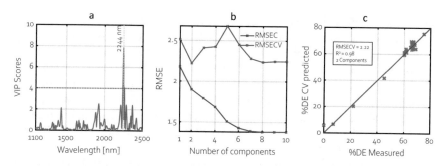

Fig. 7.29 VIP score variable selection of the PLS model to the pectin %DE in the combined Dataset 1. **a** The model VIP scores, where the red dotted line at y = 4 (selected in this instance as half the height of the highest VIP score) indicates the inclusion threshold. **b** The model statistics: RMSECV of 2.22%DE with a two-component model is the optimal choice. **c** The resulting actual versus predicted plot with an R2 of 0.98

variables with insignificant regression coefficients in order to improve the predictive power of the model. However, as they are based on regression against **y**, degrees of freedom are lost in the process and the produced models should be validated using a test set to secure that the model is not overfitted.

Figure 7.29 shows the application of VIP score variable selection to Dataset 1. The performance when imposing the variable selection (VIP scores > 4 which result in 6 variables retained), the model has a similar RMSECV of 2.22%DE to the global PLS model (Fig. 7.28) but uses one component less. This is a typical result of variable selection; i.e., a few variables are selected (good for interpretation), the regression model is deteriorated a bit (not good for scrutiny of performance but perhaps good for robustness), and model is using fewer components (good for interpretation and avoidance of interferences).

Variable selection based on VIP scores is becoming rather common and is available in most commercial software.

7.7.3 Forward Stepwise Selection

The primary target of most variable selection methods is to improve the regression, and this concept is employed in the most direct and brute way in the forward stepwise selection (FSS) procedure. In this method, all single independent variables are tested in finding the one, which provides the best regression model toward the dependent variable. All these single-variable models are test set validated, and the variable with the lowest RMSEP (on the independent test set) is chosen. In a second iteration, one new variable, that improves the PLS model the best, is included to complement the first selected variable. The variable that, in combination with the first chosen variable, gives the lowest RMSEP is selected. Subsequently, additional variables are included one-by-one by their capability to improve the previous model. This procedure is continued as long as the RMSEP (on the independent test set) decreases by the introduction of a new variable. The result of this procedure is a set of variables that represent the best combination in the spectral region, when optimizing for model performance. However, it should be noted that the variable set found may not be the absolute best one when compared to more sophisticated variable selection methods due to the buildup nature of FSS. Like the VIP method, the FSS is based on regression against **y** and it is thus important to use a test set, when evaluating the selection of new variables. An evaluation procedure based on cross-validation only will often lead to severe overfitting.

Figure 7.30 shows the application of FSS variable selection to Dataset 1. The performance when adding variables up to 10 provides a model performance with RMSECV of 1.32%DE which is markedly better than the global PLS model (Fig. 7.28) using only 2 components. The 10 variables that are picked up by the algorithm facilitate "simple" interpretation and in this case make good sense for modeling degree of esterification in pectins. The improvement in performance over the global PLS model should give serious concern to the danger of overfitting, and the model will need a real test set to be confirmed.

FSS variable selection is considered "quick and dirty" and is rarely implemented in commercial software, but it is easy to program.

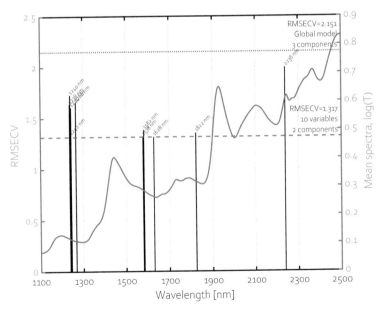

Fig. 7.30 Forward stepwise selection on the PLS model to the pectin %DE in the combined Dataset 1. The plot shows FSS variable selection using 2 components. The height of the individual bars indicates the resulting RMSECV of including that wavelength variable, and the number above each bar indicates the wavelength for that particular variable. The selection stopped after 10 variables/wavelengths, resulting in a RMSECV of 1.32%DE and a R^2 of 0.96

7.7.4   *Recursively Weighted PLS (rPLS)*

Recursively weighted PLS (rPLS) [51] combines the selection of variables by regression coefficients with an automatic iterative variable selection procedure. rPLS iteratively eliminates variables by using the regression coefficients to magnify important variables and thus down-weight less important variables. rPLS is based on a process of repeated PLS models where the current regression coefficients are used as cumulative weights on $\mathbf{X}$:

$$\mathbf{X}_i = \mathbf{X}_{i-1} \cdot \mathrm{diag}(\mathbf{b}_{i-1}) \qquad (7.29)$$

where $\mathbf{X}_i$ is the weighted $\mathbf{X}$ and $\mathbf{b}_i$ is the regression coefficient for iteration i. Using this method, a reduced subset of variables is identified for regression by recursively reweighing of the independent variables ($\mathbf{X}$) with the estimated regression coefficients ($\mathbf{b}$). The algorithm is started with a standard PLS model between $\mathbf{X}_1$ (equal to $\mathbf{X}$) and $\mathbf{y}$, giving $\mathbf{b}_1$. The re-weighting is recursively repeated until no further change in the regression coefficients occurs. The result is a regression vector $\mathbf{b}_{\mathrm{end}}$ that contains only ones and zeros. This binary result is a direct output from the rPLS algorithm; i.e., no rescaling of the final regression vector is performed. The rPLS model has the advantage that it, under normal conditions, will converge to a limited number of variables, normally including colinear neighbor variables, which is very useful for interpretation. This is not the case in more complicated situations.

The method will ultimately and normally converge to a solution that has the same number of variables as the number of principal components included in the regression. rPLS has the advantage in comparison with other iterative variable selection methods that no meta-parameters are required (i.e., interval sizes or number of components), at the "optimum" a relative low number of variables will be included in the model, and after recursive convergence very few variables are retained in the end model. In the latter case, the model performance is slightly worse than the optimal, but the interpretability may be significantly improved.

Figure 7.31 shows the application of rPLS variable selection to Dataset 1 using only 2 components. The optimal performance is reached for iteration #7 and gives a performance of RMSECV of 1.77%DE and a R^2 of 0.97 which is markedly better than the global PLS model (Fig. 7.28) using only 2 components and 13 variables (centered around 1460 nm and 2244 nm). Then, the method is allowed to converge, and it reaches 3 variables (1460 and 2244 nm) and a performance of 1.78%DE. The model results thus give excellent interpretation and demonstrate that the multivariate advantage may be gained by just adding a few or a single covariate neighbor variable. Again, the improvement in performance should give serious concern to the danger of overfitting and the model will need a real test set to be confirmed. The model shown in this section was determined in MATLAB using the open-source rPLS algorithm available at http://www.models.life.ku.dk/algorithms.

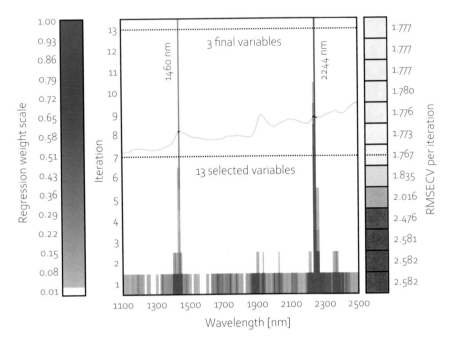

Fig. 7.31 Application of the rPLS algorithm to Dataset 1. The plot shows the output of the rPLS algorithm, converged at 13 iterations for 2 components. The center frame shows the cumulative regression weights per iteration (colored according to the weight shown in the left frame). The lowest RMSECV = 1.77%DE (right frame) was found at iteration #7, as indicated by the horizontal dotted line. The mean spectra are superimposed for reference, with the selected wavelength highlighted

rPLS is a relatively new variable selection model, but it is included in the popular *PLS Toolbox* (Eigenvector Research, Manson, WA, USA, http://www.eigenvector. com) for MATLAB (MathWorks, Natick, MA, USA, http://www.mathworks.com).

7.7.5 Interval PLS (iPLS)

iPLS is an extension to PLS that creates local models on intervals from the full spectrum focusing on important spectral regions, without including interferences and noise from other regions [52]. This very pragmatic algorithm divides the NIR spectrum into intervals for which individual PLS models are made. The performance (RMSECV) of these interval models is then compared with the global, full spectrum model. This allows an immediate localization of those spectral regions that are correlated with the response **y**. This simple exercise is able to provide an excellent overview via the so-called iPLS plot that immediately visualizes which *interval* performs better than the global model and that with *how many* PLS component(s).

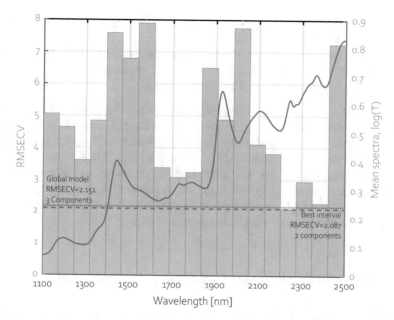

Fig. 7.32 Application of the rPLS algorithm to Dataset 1. The plot shows an iPLS model, where the spectral region has been divided into 19 segments. The height of the individual bars indicates the resulting RMSECV when including only that region. The number above each bar indicates the number of PLS components included in the subregion model, selected as the first occurring local minimum from cross-validation. The small interval around 2244 nm gives a RMSECV = 2.09%DE, using only two components. The blue stipulated line gives the global PLS performance of 2.15% using 3 components

Figure 7.32 shows the iPLS plot when applying this variable selection strategy to the Dataset 1. While the **y** vector containing the response variable remains invariant, the **X** data matrix is split into 19 intervals of equal width. As the figure shows, a single region around the 2244 nm performs dramatically better than the others: RMSECV = 2.09%DE and R^2 of 0.89 using one PLS component less.

iPLS is implemented in several chemometric software packages, including the *PLS Toolbox* (Eigenvector Research, Manson, WA, USA, http://www.eigenvector. com) for MATLAB (MathWorks, Natick, MA, USA, www.mathworks.com). The model shown in this section was determined in MATLAB using the open-source *iPLS Toolbox* available at http://www.models.life.ku.dk/algorithms.

7.7.6 Outro

Variable selection is important to consider, when one is challenged by complex multivariate NIR data. It serves three primary purposes:

1. To improve the performance of multivariate regression models (RMSEP/RMSECV)
2. To simplify multivariate regression models by excluding interferences, local nonlinearities and noisy variables (fewer latent variables)
3. To improve interpretability of multivariate models.

It is normally not a good idea to compare and scrutinize variable selection methods for better performance. The danger of overfitting is too high, and the applicability of the final models will often become too limited and sensitive. The pragmatic compromise is often to use a variation of iPLS in which a spectral region can be selected with the relevant signals and without deteriorating noise or interferences present. Remember the instrumental spectral range was not decided for a specific application! However, when seeking for causality and interpretation, variable selection methods may be a strong tool to combine with a priori knowledge.

7.8 ANOVA Simultaneous Component Analysis (ASCA)

While Variable Selection can be considered as a horizontal elimination of interferences, ASCA can be considered as a vertical elimination of interferences (partition of variances)

As mentioned previously, the most valuable meta-parameters in any spectral recording sets are the experimental design factors. It is a good practice to use this knowledge, and many NIRS studies provide multivariate datasets with an underlying experimental design.

Biological systems exhibit sources of variation due to a large number of factors, such as variety, soil and climate. Realizing this, led Fisher [53] (broadly recognized as the father of modern statistics), to develop experimental designs suited for estimation and handling of the variation based on these factors. The paired t-test and analysis of variance (ANOVA) are examples of models used to analyze data from designed experiments. The backbone of these methods is to estimate variance related to the design factors, including both systematic factors such as treatment, but also a nuisance factor like subject, and hence in turn be able to remove dominating but un-interesting variation. In this way, the variation of interest, like treatment, is emphasized, which in turn increases the chance of finding something interesting. For a wide range of applications, the dominating variation in data is often trivial, while the interesting—and new—variation sources can be minor in comparison, leaving it covered or unresolved if not handled through a proper designed experimental structure and in turn elucidated via a mathematical extraction of the relevant design effects.

An interesting multivariate tool for exploiting the experimental design information, combining the power of ANOVA to separate variance sources with the advantages of simultaneous component analysis (SCA) to modeling of the individual separate effect matrices, is called ANOVA simultaneous component analysis (ASCA)
[54]. It utilizes advantages of ANOVA in terms of both partitioning the sources of
variance and using PCA for multivariate interpretation. In order to exemplify how
this method works, imagine a simple experiment, where n samples are treated by
process A and process B in a randomized crossover fashion. At the end of each
process, a quality is measured. If this response is univariate, the difference between
processes A and B is naturally tested by a t-test. The power of a t-test is that each
sample serves as its own control and the variation is hence split into what origins
from the individual samples and what origins from the processes. More formally, the
model can be written as:

$$x_i = a(\text{process}_i) + \beta(\text{sample}_i) + e_i \qquad (7.30)$$

where α has k levels (the number of different processes) and β has j levels (the number
of samples). e_i is the error term, which often is assumed normally distributed and
may have an arbitrary number of levels. In a testing scenario, the aim is to compare
the effect, i.e., differences between different levels of α, with the magnitude of the
(random) error.

If the response is multivariate (e.g., NIR spectrum), this paradigm simply scales to
the multivariate case. Take the setup from above, but let the response be multivariate
($\mathbf{X}$); the model of $\mathbf{X}$ can now be formalized as follows:

$$\mathbf{X} = \mathbf{X}(\text{process}) + \mathbf{X}(\text{sample}) + \mathbf{E} \qquad (7.31)$$

For a full crossover, the dimension of the $\mathbf{X}$'s is k_j by m (m is the number of
variables). $\mathbf{X}(\text{process})$ describes the information related to the different processes
and has k levels, while $\mathbf{X}(\text{sample})$ describes between sample variations (j levels).
Equation 7.31 hence is merely a concatenation of Eq. 7.30 m times, one for each
variable. $\mathbf{E}$ represents the non-design-related information—it is often systematic but
just not related to the experimental design. The right-hand side of Eq. 7.31 can be
combined or analyzed individually by, for example, PCA (the working principle
in ASCA). Assume that $\mathbf{X}$ is made of NIR spectra, then a PCA on $\mathbf{X}_{\text{process}}$ will
point toward spectral patterns that discriminate between the processes, and likewise
a PCA on $\mathbf{X}_{\text{sample}}$ will reflect where the largest variation due to sample differences
(e.g., variety, soil, climate, etc.) is distributed. If the aim is to investigate the process-
related patterns by taking the error spread into account, a score plot made from
projecting $\mathbf{X}_{\text{process}} + \mathbf{E}$ onto loadings from a PCA on only $\mathbf{X}_{\text{process}}$ will reflect the
process differences in relation to the non-design-related variation in the data. If
the aim is to test for differences, multivariate classification models can be built on
relevant parts of Eq. 7.31. For example, a PLS-DA on $\mathbf{X}_{\text{proces}} + \mathbf{E}$ for classification of
processes would point toward how strong the process-related variation is compared to

the random variation. If formalized as a PLS-DA problem, this is known as multi-level PLS-DA [55].

A limitation of the application of ASCA is that it is most optimally applied to balanced design structures. On the other hand, the statistical significance of the strength of each design factor and their interactions can be evaluated using a permutation test [56], thus providing a link to classical statistics. Needless to say that appropriate pre-processing and centering of the X matrix will be essential to the subsequent ASCA.

7.8.1 Application of ASCA to NIR Spectra

Since the ASCA description above can seem a little theoretical, it will be illustrative to demonstrate the potential of ASCA from an example. Here, it is used to partition (split) the variance due to different sampling geometries in Dataset 4. The ASCA model thus reflects the experimental factors shown in Eq. 7.32 and Fig. 7.33. Only three factors are included in the model, and the "individual" factor is left out.

It will make no sense to include "individual" as a design factor in the ASCA model, as the different individuals are not the same for the different varieties. They are physically different kernels and can thus not be compared across varieties. This leaves us to focus on the partitioned effects of the three design factors:

$$\mathbf{X} = \mathbf{X}_{\text{variety}} + \mathbf{X}_{\text{position}} + \mathbf{X}_{\text{orientation}} + \mathbf{E} \tag{7.32}$$

The ASCA model will be computed on Dataset 4 with the spectra $\mathbf{X}$, pre-processed by a 2nd order, derivative Savitzky-Golay filter of window width 7, followed by a MSC, and finally a mean centering. The effect is evaluated with 1000 permutations. For simplicity, interactions between the three factors are not included. The full rank of each factor is also calculated.

The primary output of the ASCA model is the effect table (Table 7.2). From the table, it is clear that largest effect is found in the residual (81.1%). In the current case, this is primarily caused by the individual seed variation, which we choose to disregard here. The second highest effect stems from *variety*, followed by the two sample presentation parameters of *position* and *orientation*. Based on the permutation test, we can determine the statistical significance of each of the factors. The *orientation* factor is found not to be significant due to its high *p*-value. However, it seems that *variety* and *position* both are highly important and are contributing significantly to the variation in the dataset.

It can be concluded that the *position* is an important experimental factor and can be directly compared to the more important *variety* effect. As previously mentioned, the orientation factor describes the orientation of the single seed in the instrument— which is thus a factor that should be controlled in the future use of the instrument or in, e.g., a high throughput scenario with single-seed sorting!

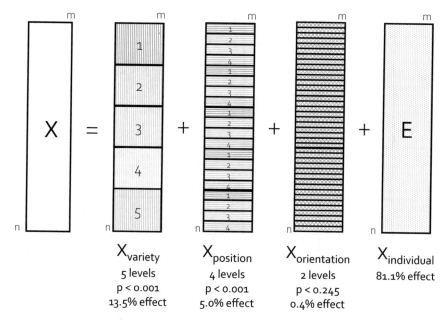

Fig. 7.33 Structure of the nested design in Dataset 4. The three factors are nested, so that each of the 5 varieties contains 4 positions, which in turn each contains 2 orientations. The design is thus balanced—all combinations exist at all levels. As the effect of the mean centering has been partitioned out, the only remaining experimental variance is the effect of the individual kernels. The individuals cannot be nested a factor, as they are biological individual specimens and would break the design balance. See the text for discussion of effect and significance for the design levels. The X shown here has been pre-processed, so that spectral artifacts (scatter) have been removed, in addition to mean centering

Table 7.2 Effect table of the ASCA model

Factor	Principal components (DF)	Effect (%)	Significance (*p*-value)
Variety	4	13.5	0.001
Position	3	5.0	0.001
Orientation	1	0.4	0.246
Residual	>2	81.1	

Further analysis of the ASCA solution includes examination of each of the factor sub-models. Since the *orientation* factor was found to be insignificant, focus is directed toward the *variety* and *position*, shown in Fig. 7.34. Each factor can be analyzed using PCA and described in terms of their scores and loadings.

As for the *variety* (Fig. 7.34a and c), they seem to vary by the significant peak at ~1020 nm. This band relates to the second overtone of the N-H stretching vibrations corresponding to a separation into different protein levels, meaning that each variety will have a different "species"-dependent protein content, and further investigation

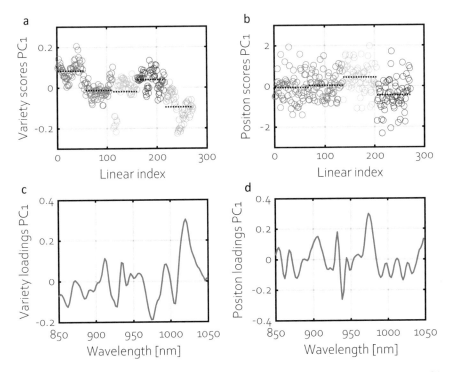

Fig. 7.34 Two sub-models from the ASCA model. **a** and **c** show the scores and loadings for PC1 of the variety model and colored by each of the five varieties. The black lines indicate the score averages of the individual varieties. **b** and **d** show the scores and loadings for the PC1 of the position sub-model and colored by each of the 4 positions. The black lines indicate the score averages of the individual positions

will in fact reveal that PC1 of the variety sub-model shows a high correlation to the protein content. The picture for the position sub-model (Fig. 7.34b and d) is on the other hand a bit more unclear and should be investigated further. Comparing the score averages, it seems that the two first positions (left and right) are identical, whereas the two following (front and back) are of much higher levels, in opposite directions. It shows that rotating the kernel in the sample compartment has an influence on the measurements, but not as much as the morphology of the variety has.

ASCA can be performed using the *PLS Toolbox* (Eigenvector Research, Manson, WA, USA, http://www.eigenvector.com) or academic freeware such as the ASCA package written in MATLAB by Morten Arendt Rasmussen found at https://bitbuc ket.org/modelscat/asca/.

7.8.2 Outro

The idea of using ASCA to partition the variances of the experimental design parameters has great potential and has not been fully exploited in NIRS literature. ASCA brings a valuable link between the multivariate data analysis and statistics, as it is able to provide significance testing of the different design factor effects and interactions.

7.9 Process Analytical Technology, Machine Learning and Other NIRS Trends

> Collecting large quantities of extremely low-quality data will not be the recipe for success!
> —Tom Fearn, British chemometrician

Due to its unique capabilities and complex, holistic spectra, NIR spectroscopy has served as the perfect playground for the development of multivariate data analysis and chemometrics, and there is no sign for this to stop in the near future. The majority of spectroscopic sensors used in process analytical technology (PAT) are based on NIR technology [57, 58]. This may be called "the second green analytical revolution of NIRS analysis." The first was introduced by Williams and Norris in 1975 when replacing sulfuric acid demanding Kjeldahl analysis with clean NIRS analysis [59]. NIRS analysis in the PAT context has perhaps a much larger potential to change the way that we produce sustainably and the way that we optimize processes for a new circular and green economy (see Fig. 7.35). Moreover, portable NIRS sensors are omnipresent in agriculture, we are beginning to see NIRS sensors on drones, NIRS hyperspectral imaging and NIRS sensors on mobile phones are emerging. This will drastically increase the amount of NIR data collected from practically all aspects of life.

This technology revolution has created a strong quest for new and more efficient data analytical tools. Artificial intelligence, machine learning and deep learning are increasingly exploited to facilitate efficient information extraction from the enormous data collections, and applications are starting to emerge for the spectroscopic disciplines [60]. These methods can deal with nonlinear effects (abundant in NIR spectroscopy), but are generally less interpretable (black box) for the scientist.

The developments of machine learning methods are primarily made in the NIR imaging field, where neural networks have proven quite efficient in decoding the complexity of hyperspectral images in their original multidimensional form [61], a natural extension to the more traditional approach of applying PCA on deconvoluted images.

The true application of "deep" neural processing, where the neural network is fed raw sensor data and trained to form self-organizing feature detectors, is an obvious

RECIPE   MONITORED   MONITORED
BASED   RECIPE BASED   & CONTROLLED

© Newlin & Engelsen

Fig. 7.35 Use of NIR spectroscopic monitoring in PAT context. (**a**) When new processes are scaled up to industrial scale, they are often uncontrollable, use excess heat, substrate and stirring, use too much reaction time and may lead to occasional scrap. (**b**) When the NIRS sensors are mounted for online monitoring, the process engineers obtain knowledge and may get ideas for controlling the reaction better. (**c**) When NIRS is used for active feedback control, the process can continue smoothly with minimal energy and substrate use and with faster total reaction times which may increase production capacity

tool for identification of hidden phenomena in data streams. This opens possibilities involving difficult (nonlinear) classification tasks—especially process-based time series data, where NIR spectra can be recorded very frequently in multiple process streams using distributed sensor systems. The spectral data can be analyzed for emerging patterns [62], for a holistic view over all process streams. Such detector systems may be able to produce early warnings for process failures by the use of long short-term memory (LSTM) neural nodes on NIR data directly, much earlier than current PAT tools allow for.

These methods are also often marked as the "magic tool" in connection with NIR sensors of lower quality, but this is not recommendable in praxis. The rule of thumb of data quality also applies to machine learning. The quality of the generated predictions will be just as good as the quality of the modeled data, but not better.

Another related trend is the combination of NIR data with signals from other analytical platforms and metadata for fusion [63] or 2D correlation spectroscopy [64]. This can sometimes be useful to add and co-model complementary information to strengthen multivariate models and their interpretation.

Last but not least, there exist some more academic trends trying to create the calibration-free NIR spectrometer using factor analysis [65], and trying to diagnose, when NIRS calibration models rely on indirect correlations with the aim of understanding the boundaries for the validity of the covariance structures [66]. Indirect NIR calibrations are becoming more widespread, and they can be problematic in terms of accuracy and robustness of the calibration models, since they rely on biological covariance structures, which may not remain constant over time or other (changing) external factors.

References

1. S.B. Engelsen, Near infrared spectroscopy—a unique window of opportunities. NIR News **27**(5), 14 (2016)
2. P.C. Williams, K.H. Norris, *Near Infrared Technology in the Agricultural and Food Industries* (American Association of Cereal Chemists, Inc., St. Paul, Mn, 1987)
3. B.G. Osborne, T. Fearn, P.H. Hindle, *Practical NIR Spectroscopy with Applications in Food and Beverage Analysis* (Longman Scientific & Technical, Harlow, Essex, UK, 1986)
4. R. DiFoggio, Guidelines for applying chemometrics to spectra: feasibility and error propagation. Appl. Spectrosc. **54**(3), 94A (2000)
5. P. Geladi, K. Esbensen, The start and early history of chemometrics. 1. Selected interviews. J. Chemometrics **4** (5), 337 (1990)
6. S.B. Engelsen, E. Mikkelsen, L. Munck, New approaches to rapid spectroscopic evaluation of properties in pectic polymers. Progr. Colloid Polym. Sci. **108**, 166 (1998)
7. Y. Dong, K.M. Sørensen, S. He, S.B. Engelsen, Gum Arabic authentication and mixture quantification by near infrared spectroscopy. Food Control **78** (Supplement C), 144 (2017)
8. E. Tønning, L. Nørgaard, S.B. Engelsen, L. Pedersen, K.H. Esbensen, Protein heterogeneity in wheat lots using single-seed NIT—A Theory of Sampling (TOS) breakdown of all sampling and analytical errors. Chemometr. Intell. Lab. Syst. **84**(1–2), 142 (2006)
9. J. Kjeldahl, A new method for the determination of nitrogen in organic bodies. Anal. Chem. **22**, 366 (1883)
10. H.W. Siesler, Y. Ozaki, S. Kawata, H.M. Heise, *Near-Infrared Spectroscopy: Principles, Instruments* (Wiley-VCH, Applications, 2008)
11. A. Rinnan, F. van den Berg, S.B. Engelsen, Review of the most common pre-processing techniques for near-infrared spectra. TRAC-trends Anal Chem **28**(10), 1201 (2009)
12. P. Geladi, D. McDougall, H. Martens, Linearization and scatter-correction for near-infrared reflectance spectra of meat. Appl. Spectrosc. **39**(3), 491 (1985)
13. H. Martens, S.A. Jensen, P. Geladi, N-4000 Stavanger, Norway, p 205 (1983)
14. R.J. Barnes, M.S. Dhanoa, S.J. Lister, Standard normal variate transformation and de-trending of near-infrared diffuse reflectance spectra. Appl. Spectrosc. **43**(5), 772 (1989)
15. H. Martens, E. Stark, Extended multiplicative signal correction and spectral interference subtraction: New preprocessing methods for near infrared spectroscopy. J. Pharm. Biomed. Anal. **9**(8), 625 (1991)
16. H. Martens, J.P. Nielsen, S.B. Engelsen, Light scattering and light absorbance separated by extended multiplicative signal correction. Application to near-infrared transmission analysis of powder mixtures. Anal. Chem. **75** (3), 394 (2003)
17. A. Savitzky, M.J.E. Golay, Smoothing and differentiation of data by simplified least squares procedures. Anal. Chem. **36**, 1627 (1964)
18. W.H. Lawton, E.A. Sylvestre, Self modeling curve resolution. Technometrics **13**(3), 617 (1971)
19. A. de Juan, J. Jaumot, R. Tauler, Multivariate curve resolution (MCR). Solving the mixture analysis problem. Anal. Methods 6 (14), 4964 (2014)
20. J. de Leeuw, F.W. Young, Y. Takane, Additive structure in qualitative data: An alternating least squares method with optimal scaling features. Psychometrika **41**(4), 471 (1976)
21. A. de Juan, R. Tauler, Multivariate curve resolution (MCR) from 2000: Progress in concepts and applications. Crit. Rev. Anal. Chem. **36**(3–4), 163 (2006)

22. T. Fearn, Multivariate Curve Resolution. NIR News **22**(1), 18 (2011)
23. L. Nørgaard, M. Hahn, L.B. Knudsen, I.A. Farhat, S.B. Engelsen, Multivariate near-infrared and Raman spectroscopic quantifications of the crystallinity of lactose in whey permeate powder. Int. Dairy J. **15**(12), 1261 (2005)
24. S. Navea, A. de Juan, R. Tauler, Modeling temperature-dependent protein structural transitions by combined near-IR and mid-IR spectroscopies and multivariate curve resolution. Anal. Chem. **75**(20), 5592 (2003)
25. K. Wojcicki, I. Khmelinskii, M. Sikorski, E. Sikorska, Near and mid infrared spectroscopy and multivariate data analysis in studies of oxidation of edible oils. Food Chem. **187**, 416 (2015)
26. K.M. Sørensen, S.B. Engelsen, The spatial composition of porcine adipose tissue investigated by multivariate curve resolution of near infrared spectra: Relationships between fat, the degree of unsaturation and water. J. Near Infrared Spectrosc. **25**(1), 45 (2017)
27. T.R.M. De Beer, P. Vercruysse, A. Burggraeve, T. Quinten, J. Ouyang, X. Zhang, C. Vervaet, J.P. Remon, W.R.G. Baeyens, In-line and real-time process monitoring of a freeze drying process using Raman and NIR spectroscopy as complementary Process Analytical Technology (PAT) tools. J. Pharm. Sci. **98**(9), 3430 (2009)
28. J. Jaumot, A. de Juan, R. Tauler, MCR-ALS GUI 2.0: New features and applications. Chemometr. Intell. Lab. Syst. **140**, 1–12 (2014)
29. K. Pearson, On lines and planes of closest fit to systems of points in space. Phil. Mag. **2**, 559 (1901)
30. H. Hotelling, Analysis of a complex of statistical variables into principal components. J. Educ. Psychol. 24, 417 (1933)
31. S. Wold, K. Esbensen, P. Geladi, Principal component analysis. Chemometr. Intell. Lab. Syst. **2**(1–3), 37 (1987)
32. S. Wold, H. Martens, H. Wold, The multivariate calibration-problem in chemistry solved by the PLS method. Lect. Notes Math. **973**, 286 (1983)
33. H. Hotelling, The relations of the newer multivariate statistical-methods to factor-analysis. Br. J. Stat. Psychol. **10**(2), 69 (1957)
34. H. Martens, S.A. Jensen, in *Progress in Cereal Chemistry and Technology* ed. by J. Holas, J. Kratochvil, vol. 5a (Elsevier, Amsterdam, 1983)
35. A. Smilde, R. Bro, P. Geladi, *Multi-Way Analysis with Applications in the Chemical Sciences* (John Wiley & Sons, Ltd, 2005)
36. H. Martens, T. Karstang, T. Næs, Improved selectivity in spectroscopy by multivariate calibration. J. Chemom. **1**(4), 201 (1987)
37. L. Ståhle, S. Wold, Partial least squares analysis with cross-validation for the two-class problem: A Monte Carlo study. J Chemometrics 1 185 (1987)
38. J.A. Westerhuis, H.C.J. Hoefsloot, S. Smit, D.J. Vis, A.K. Smilde, E.J.J. van Velzen, J.P.M. van Duijnhoven, F.A. van Dorsten, Assessment of PLSDA cross validation. Metabolomics **4**(1), 81 (2008)
39. D.T. Berhe, C.E. Eskildsen, R. Lametsch, M.S. Hviid, F. van den Berg, S.B. Engelsen, Prediction of total fatty acid parameters and individual fatty acids in pork backfat using Raman spectroscopy and chemometrics: Understanding the cage of covariance between highly correlated fat parameters. Meat Sci. **111**, 18 (2016)
40. F.J. Anscombe, Graphs in statistical-analysis. Am. Stat. **27**(1), 17 (1973)
41. T. Næs, T. Isaksson, SEP or RMSEP, which is best? NIR News **2**(4), 16 (1991)
42. I.N. Wakeling, J.J. Morris, A test of significance for partial least squares regression. J. Chemom. **7**(4), 291 (1993)
43. S. Wold, Cross-validatory estimation of the number of components in factor and principal components models. Technometrics **20**(4), 397 (1978)
44. H. Martens, P. Dardenne, Validation and verification of regression in small data sets. Chemometr. Intell. Lab. Syst. **44**(1–2), 99 (1998)
45. D.K. Pedersen, H. Martens, J.P. Nielsen, S.B. Engelsen, Near-infrared absorption and scattering separated by extended inverted signal correction (EISC): Analysis of near-infrared transmittance spectra of single wheat seeds. Appl. Spectrosc. **56**(9), 1206 (2002)

46. T. Mehmood, K.H. Liland, L. Snipen, S. Saebo, A review of variable selection methods in partial least squares regression. Chemometr. Intell. Lab. Syst. **118**, 62 (2012)
47. B. Efron, The Jackknife, the Bootstrap and Other Resampling Plans. Soc. Ind. Appl. Math. Philadelphia, Pennsylvania (1982)
48. H. Martens, M. Martens, Modified Jack-knife estimation of parameter uncertainty in bilinear modelling by Partial Least Squares Regression (PLSR). Food Qual. Prefer. **11**(1–2), 5 (2000)
49. S. Wold, E. Johansson, E. Cocchi, ESCOM, Leiden, Holland (1993) p. 523
50. I.G. Chong, C.H. Jun, Performance of some variable selection methods when multicollinearity is present. Chemometr. Intell. Lab. Syst. **78**(1–2), 103 (2005)
51. Å. Rinnan, M. Andersson, C. Ridder, S.B. Engelsen, Recursive weighted partial least squares (rPLS): An efficient variable selection method using PLS. J. Chemom. **28**(5), 439 (2014)
52. L. Nørgaard, A. Saudland, J. Wagner, J.P. Nielsen, L. Munck, S.B. Engelsen, Interval partial least squares regression (iPLS): A comparative chemometric study with an example from the near infrared spectroscopy. Appl. Spectrosc. **54**(3), 413 (2000)
53. R.A. Fisher, The correlation between relatives on the supposition of Mendelian inheritance. Philos Trans R Soc Edinburgh **52**, 399 (1918)
54. A.K. Smilde, J.J. Jansen, H.C.J. Hoefsloot, R. Lamers, J. van der Greef, M.E. Timmerman, ANOVA-simultaneous component analysis (ASCA): a new tool for analyzing designed metabolomics data. Bioinformatics **21**(13), 3043 (2005)
55. J.A. Westerhuis, E.J.J. van Velzen, H.C.J. Hoefsloot, A.K. Smilde, Multivariate paired data analysis: Multilevel PLSDA versus OPLSDA. Metabolomics **6**(1), 119 (2010)
56. G. Zwanenburg, H.C.J. Hoefsloot, J.A. Westerhuis, J.J. Jansen, A.K. Smilde, ANOVA-principal component analysis and ANOVA-simultaneous component analysis: A comparison. J. Chemom. **25**(10), 561 (2011)
57. A.L. Pomerantsev, O.Y. Rodionova, Process analytical technology: A critical view of the chemometricians. J. Chemom. **26**(6), 299 (2012)
58. E. Skibsted, S.B. Engelsen, in *Encyclopedia of Spectroscopy and Spectrometry (Second Edition)* (Academic Press, Oxford, 2010)
59. P.C. Williams, Application of near-infrared reflectance spectroscopy to analysis of cereal-grains and oilseeds. Cereal Chem. **52**(4), 561 (1975)
60. G. Huang, G.B. Huang, S.J. Song, K.Y. You, Trends in extreme learning machines: A review. Neural Netw. **61**, 32 (2015)
61. S. Mahesh, A. Manickavasagan, D.S. Jayas, J. Paliwal, N.D.G. White, Feasibility of near-infrared hyperspectral imaging to differentiate Canadian wheat classes. Biosyst. Eng. **101**(1), 50 (2008)
62. A.P. Teixeira, R. Oliveira, P.M. Alves, M.J.T. Carrondo, Advances in on-line monitoring and control of mammalian cell cultures: Supporting the PAT initiative. Biotechnol. Adv. **27**(6), 726 (2009)
63. E. Borras, J. Ferre, R. Boque, M. Mestres, L. Acena, O. Busto, Data fusion methodologies for food and beverage authentication and quality assessment—A review. Anal. Chim. Acta **891**, 1 (2015)
64. I. Noda, Generalized 2-dimensional correlation method applicable to infrared, Raman and other types of spectroscopy. Appl. Spectrosc. **47**(9), 1329 (1993)
65. E. Alm, R. Bro, S.B. Engelsen, B. Karlberg, R.J.O. Torgrip, Vibrational overtone combination spectroscopy (VOCSY)—A new way of using IR and NIR data. Anal. Bioanal. Chem. **388**(1), 179 (2007)
66. C.E. Eskildsen, M.A. Rasmussen, S.B. Engelsen, L.B. Larsen, N.A. Poulsen, T. Skov, Quantification of individual fatty acids in bovine milk by infrared spectroscopy and chemometrics: Understanding predictions of highly collinear reference variables. J. Dairy Sci. **97**(12), 7940 (2014)

Part III
Instrumentation

Chapter 8
New Trend in Instrumentation of NIR Spectroscopy—Miniaturization

Christian W. Huck

Abstract The emergence of handheld spectrometers in the past decade marked a significant turning point in the evolution of the practical applications of near-infrared (NIR) spectroscopy. Miniaturized sensors enabled a new and previously unattainable spectrum of applications of NIR spectroscopy. Nonetheless, several issues connected with the use of miniaturized spectrometers have become apparent. In contrast to a matured design of a FT-NIR benchtop spectrometer, the handheld devices are much less uniform and incorporate various novel technologies. These compact technologies result in different performance of miniaturized spectrometers, with narrower spectral regions or lower resolution over which they operate. For this reason, current research focus is on thorough systematic evaluation of the applicability limits and analytical performance of these devices in a variety of applications. This chapter aims to present a comprehensive information on the principles of the technology and application potential of miniaturized NIR spectrometers.

Keywords NIR spectroscopy · Portable · Handheld · Miniaturized · Sensor

8.1 General Introduction

In principle, from the point-of-view of instrumentation, near-infrared (NIR) spectroscopy is an optical absorption spectroscopy. Therefore, NIR spectrometers share the general design scheme with those used in more popular types of spectroscopy such as visible (Vis) or infrared (IR) spectroscopy. This similarity goes further, and often instrument dedicated to operate in Vis or IR regions is able to measure at least some part of NIR spectra. This is primarily limited by the emission spectrum of the source and sensitivity of the detector. The performance of these key elements is not curtailed sharply with any arbitrary wavelength boundaries, but rather steadily diminish at the extended ranges. For this reason, ultraviolet–visible (UV–Vis) spectrometers

C. W. Huck (✉)
CCB-Center for Chemistry and Biomedicine, Institute of Analytical Chemistry and Radiochemistry, Leopold-Franzens University, Innrain 80/82, 6020 Innsbruck, Austria
e-mail: Christian.W.Huck@uibk.ac.at

© Springer Nature Singapore Pte Ltd. 2021
Y. Ozaki et al. (eds.), *Near-Infrared Spectroscopy*,
https://doi.org/10.1007/978-981-15-8648-4_8

are often capable of extending measurements to short-wavelength NIR region (SW-NIR). There is no clear definition of SW-NIR wavelength boundaries in the literature, and it mostly depends on the limitations of specific instrument; however, SW-NIR region is quite important from the point-of-view of portable spectroscopy, as it will be explained later on. On the other hand, most IR spectrometers can extend its operation to the long-wavelength NIR as the most commonly used IR detector (deuterated triglycine sulfate detector with a cesium iodide window, DTGS/CsI) offers sufficient sensitivity up to 6400 cm^{-1}. Nonetheless, it requires a dedicated NIR spectrometer equipped with a proper light source (e.g., tungsten halogen lamp) and detector (e.g., high-performing indium gallium arsenide, InGaAs) to measure good-quality spectrum in the entire NIR region. In this regard, laboratory-scale (benchtop) NIR spectrometers are nowadays highly matured. In contrast, it is still a challenge to design a miniaturized sensor that would offer similar capability. The properties of light sources, detectors, and wavelength selection elements are of critical importance for the design of miniaturized devices. It may be stated that the available technology governs the level of miniaturization, performance, and affordability of portable NIR spectrometers.

Further, the properties of molecular excitation in NIR region are meaningful for the requirements issued to the instrumentation. Unlike IR bands, NIR spectra rather feature broad absorption structures resulting from numerous overlapping contributions (combination and overtone transitions). This makes the optical resolution of an NIR spectrometer relatively less critical. Instead, with lower resolution, a better optical gain is achieved, which results in a greater signal-to-noise, and/or faster scanning operation. Such high throughput capacity and rapid analysis are often the critical factors of instrumental nature, which stand behind the wide adoption of NIR spectroscopy in practical applications. IR spectrometers require transparent optical elements to be made of alkali halides (e.g., KBr). In sharp contrast, glass optics is transparent in NIR region. This makes NIR spectrometers easier to adopt to operate in humid conditions, a fact of great importance for on-site analysis, process monitoring, and for ease engineering of portable spectrometers.

8.1.1 Basic Technology Design of NIR Spectrometers

The design blocks of a generic NIR spectrometer constitute of a radiation source, wavelength selector or interferometer and detector, interfaced by optics. Two general classes of spectrometers may be differentiated: wavelength-dispersive and Fourier transform (FT). In the former, the wavelength selector only passes selected, narrow wavelength windows that can reach detector at a time (Fig. 8.1). Note that the conventional dispersive devices are obsolete; however, miniaturization has introduced concepts similar to them, which will be discussed in detail later on.

Benchtop spectrometers have nowadays almost entirely adopted FT principle. Instead of a classical wavelength filter, an interferometer enables a simultaneous incidence of all wavelength on the detector (Fig. 8.2). The spectrum is obtained through

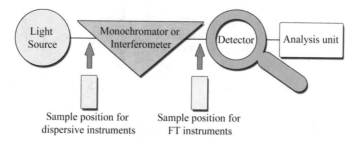

Fig. 8.1 Schematic spectrometer assembly

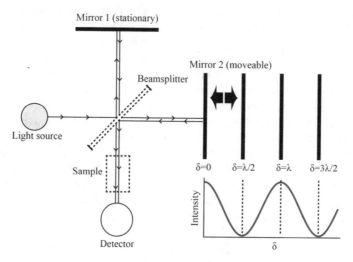

Fig. 8.2 Schematic illustration of a Michelson interferometer

Fourier transform. Noteworthy, Michelson interferometer is the most often employed in laboratory-scale spectrometers. The core element there is formed by a fixed and a moving mirror, onto which the polychromatic beam is simultaneously directed by a beam splitter [1]. One of the beams is used to probe the sample, and afterward both are recombined. The path difference between the beams introduced over the time of the scan leads to periodically alternating interferences (phase differences), from which a spectrum can be reconstructed (Fig. 8.3) [2].

The Fourier transform principle has a meaningful impact on the performance as short scan times are possible and high-optical throughput improves the quality of spectra [3, 4]. Further, optimization and adjustment of optical throughout vs. resolution enable maintaining an excellent signal-to-noise ratio. However, very high precision and stability of operation over time are critically important for the proper function of Michelson interferometer. Nowadays, the precision of motion of the interferometer's elements is maintained pneumatically, with hovering on a layer of air and/or inert gas. However, such features are not suitable for implementation in

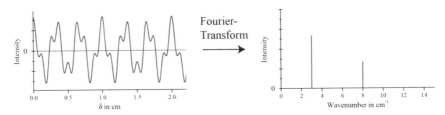

Fig. 8.3 Schematic illustration of applying the Fourier transform

compact spectrometers; firstly, because of the miniaturization, secondly, because of the requirement for ruggedness of mobile, portable devices, and their intended use as on-site sensors. Here again, modern technology could offer alternative solutions that can exclude use of moving parts in miniaturized devices.

Dispersive devices necessarily require repetitive external calibrations in order to prevent the drift and maintain the control of the wavelength/wavenumber. In contrast, in interferometer-based devices, the control over the wavelength axis can be easily and continuously maintained by the interference of a reference laser (usually a He–Ne laser). A highly accurate wavenumber calibration is obtained through correlation of the laser´s wavelength with the interferogram zero-crossing sections [5].

The choice of the detector depends on the investigated wavelength region. There exist two types of detectors, photon detectors (i.e., photodetectors), and thermal detectors. Because of the ability to operate over a broad NIR region, the first class almost exclusively dominates in scientific-grade benchtop spectrometers. However, many types of detectors require stable temperature to operate, while some also need to be actively cooled to deliver useful S/N. This is obviously much more difficult to achieve in miniaturized format. Some portable devices facilitate temperature correction functions or active cooling elements, e.g., thermoelectric Peltier cooling. Radiation sources are more uniform throughout the NIR spectroscopy. Benchtop NIR spectrometers can operate with high stability and very good performance using conventional incandescent tungsten halogen bulbs. While thermal stability and electrical power of a few Watts needed for the operation of such source can be easily maintained in a benchtop instrument, in a handheld spectrometer these requirements may become challenging. Conveniently, current technology enables implementing tungsten halogen sources in miniaturized instruments.

8.1.2 Overview of the Technological Advancements in Miniaturized NIR Spectrometers

The instrumentation in spectroscopy and spectrometry can be divided into benchtop spectrometers, operational only in a laboratory setting, and autonomous spectrometers that can be deployed and used on-site. Commonly accepted classification of deployability of the instrumentation distinguishes the transportable, 'suitcase-type' and handheld spectrometers [6]. The criterion is based on the equipment weight, and the first class groups autonomous instruments weighting over 20 kg, which means they are deployable on field while mounted in a car. Next, an intermediate 'suitcase' format corresponds to equipment with weight of several kg. The handheld spectrometers are designed to be carried and operates by hand, and typically weigh under 1 kg [6]. Such classification should be understood broadly, as fieldable instrumentation in mass spectrometry (MS), nuclear magnetic resonance (NMR, relaxometry), or elemental (atomic) spectroscopy such as X-ray fluorescence (XRF) or laser-induced breakdown spectroscopy (LIBS) is available as well. Figure 8.4 demonstrates the progress of the miniaturization in different fields of spectroscopy and spectrometry [7]. The physical principles of NIR spectroscopy make it very suitable for miniaturization of the instrumentation, and NIR spectrometers achieved an outstanding progress in compact technology. While very compact sensors emerged recently for

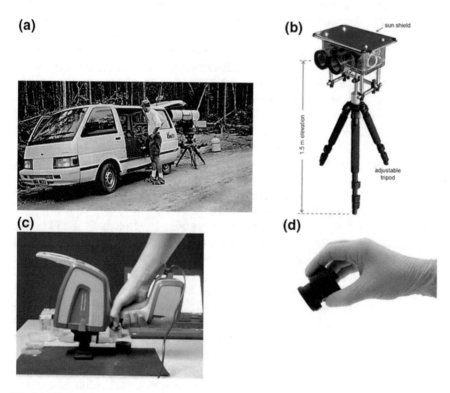

Fig. 8.4 Exemplary compact instrumentation in spectroscopy and spectrometry with different levels of portability. **a** Car-deployable long-path reflective FT-IR and GC–MS. **b** Portable sensor based on tunable diode-laser absorption spectroscopy (TDLAS). **c** Handheld attenuated total reflection (ATR) FT-IR spectrometer (Agilent 4300). **d** Ultra-miniaturized NIR spectrometer MicroNIR 1700. Panels in the figure reproduced; **a** from Ref. [24] with permission (Elsevier Open Access license); **b** from Ref. [25] (CC-BY 4.0 license); **c** from Ref. [26] (CC-BY 4.0 license)

some other techniques, e.g., fluorescence, NIR spectroscopy is superior in chemical specificity and applicability to a broad range of sample types. The most recent years have led to emerging new class of spectrometers, miniaturized NIR devices that can reach weigh of sub-100 g, and ultra-miniaturized sensors that are compact enough to be built-in directly into a smartphone device. Several ultra-miniaturized NIR spectrometers appeared in the past decade. This new-generation devices are USB powered or operate under own power source (usually Li-ion battery) and are intended for easy use. Such spectrometers often come with software designed for easy and rapid operation by non-expert personnel. Moreover, spectra measurement by these spectrometers is often much more rapid than in the case of benchtop instrument.

Sensor miniaturization has a critical importance for several practical applications [8]. The first breakthrough to practical availability of handheld NIR instruments was micro-optoelectronic-mechanical systems (MEMS) and miniature diode-array detectors (DAD). The first handheld, all-in-one portable (1.5 kg) NIR spectrometer was introduced commercially by Polychromix 2006, and the instrument is nowadays known as microPHAZIR by Thermo Fischer Scientific. This design was engineered with MEMS wavelength selector and tungsten light source. The next noteworthy step into further miniaturization was made in 2012 by JDS Uniphase (currently ViaviSolutions, Milpits, CA, USA) with the MicroNIR instrument. A very compact size was achieved through using a linear variable filter (LVF) element for the wavelength selection together with a 128-pixel InGaAs DAD. This multi-channel design also enabled very quick scanning, with less than 1 s required to measure the entire spectrum [9, 10]. In 2016, Texas Instrument presented an optical engine with the dimensions of 33 $\times$ 29 $\times$ 10 mm in their DLP NIRscan product. It incorporated a digital micromirror device (DMD) principle with diffraction grating system. The current advances in the miniature technology make it feasible to fully integrate a NIR spectrometer into smartphone in the near future (Fig. 8.5) [11].

For most practical applications that adopt portable spectrometers of all kinds, a successful design should possibly achieve the following characteristics; (i) handheld format/high level of miniaturization, (ii) ruggedness (e.g., no moving parts, resilience against external conditions, temperature, etc.), (iii) affordability, and (iv) straightforward applicability in routine analysis (e.g., rapid measurement, easy handling by non-expert personnel). It may be safely stated that, in most points, NIR spectrometers are particularly suitable for miniaturization. However, as it is known, NIR spectral analysis is not straightforward and typically requires supervision by expert personnel. Significant efforts are being directed nowadays to make NIR analysis by handheld devices more accessible for non-trained operators. Some of the proposed solutions, e.g., blackbox operation, 'factory' chemometric calibration procedures, cloud computing, still leave much to be desired and are discussed by the NIR spectroscopic community [12].

Nonetheless, advancements made over the past decade enabled spectroscopic measurements in previously unattainable scenarios and create new opportunities for innovative applications in wide field of science and industry. Therefore, portable instrumentation forms a significant breakthrough in NIR spectroscopy and may become the key advancement for widespread of this technique in forthcoming time.

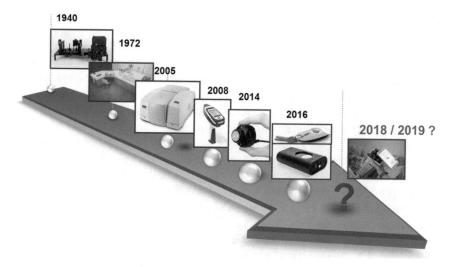

Fig. 8.5 Miniaturization of NIR spectrometers

8.2 The Principles of the Technology Underlying Miniaturized NIR Spectroscopy

The design of handheld NIR spectrometers mostly deciding about their working characteristics is the detector and monochromating (i.e., wavelength selection) technology [6]. There is much less variety in the technology used for light sources here, although certain notes need to be made below.

8.2.1 Light Sources

In general, there exist two types of light sources in commercially available miniaturized spectrometers. Most devices employ the conventional principle of tungsten light bulb (e.g., microPHAZIR, MicroNIR), although adopted for the needs of highly autonomous spectrometers. This primarily involves optimizations toward low power consumption. To maintain control over the thermal stability and output of these sources, manufacturers often recommend to perform frequent reference scans, which are not problematic as these devices often rapid scanning. Another used technology are light emitting diodes (LEDs). Semiconductor-based sources offer unparalleled power effectiveness, affordability, and compact dimensions. However, they suffer from narrow emission spectrum, which makes such sensors practically useful mostly in Vis/SW-NIR region (e.g., SCiO device).

8.2.2 Detectors

The silicon detectors are entry-level solutions that enable constructing low cost 1D and 2D array sensors. Multiplied Si detector elements can be fitted with own filter each, with each such element tuned toward measuring its own channel (i.e., wavelength region). Therefore, they are suitable for constructing very simple and inexpensive multichannel detectors. However, Si detectors yield inferior S/N parameter and because of a cut-off at ca. 1000 nm (10,000 cm^{-1}) are limited to operate in a narrow spectral region of visible/short-wave NIR (Vis/SW-NIR). Silicon-based detectors offer practical advantages, e.g., low power consumption. There exist two major types of such detectors, complementary metal–oxide–semiconductor (CMOS) and charge-coupled device (CCD), with CMOS requiring lower power consumption [6]. Charge-coupled device (CCD) is a silicon-based photon detector. When light strikes the chip, it directly induces as a small electrical charge in each cell of the photosensor. The cell is an analog circuitry, and the charge is amplified, converted into a digital value, and the output registered. Commercial success and wide-spread use of CMOS and CCD technology have brought down the price per unit of such detectors. However, silicon photodetectors offer inferior sensitivity toward sensing the NIR wavelengths. Therefore, for higher S/N, better performing detectors are preferable. Here, indium–gallium–arsenide (InGaAs) detector may be considered state of the art, with excellent sensitivity at wavelengths longer than ca. 1050 nm, superior S/N and scan time. The detector noise varies with temperature, which has been a problem in some earlier designs. Temperature stabilization by thermoelectric cooling was proved to be helpful in this regard [9]. Temperature correction functions have been also introduced in newer designs, e.g., MicroNIR 2200.

8.2.3 Wavelength Selectors

Wavelength selector can be considered the most critical element for the design of a miniaturized spectrometer. There is a large variety of the available solutions in this regard. Micro-electro-mechanical systems (MEMS; in combination with micro-optics: micro-opto-electro-mechanical systems, MOEMS or optical MEMS) are in-silicon microscaled mechanical devices manufactured similar to integrated circuitry. This technology advanced together with the progress made in semiconductor industry enabling the assembly of extremely miniaturized moving parts. MEMS technology can be used to implement few different wavelength selector principles in microscale, e.g., Hadamard mask, digital micromirror, Michelson and Fabry–Perot interferometers. Thus, grating-based monochromators for 'dispersive-like' and interferometers for Fourier transform (FT) spectrometers can be manufactured. It has become fairly popular solution for miniaturized NIR spectrometers with a number of MEMS-based devices proposed in the last 20 years.

MEMS element appeared in the first commercial handheld NIR spectrometer, microPHAZIR (known earlier as PHAZIR). In that case, a programmable Hadamard mask principle was implemented through the MEMS chip, which contained a large number of electronically actuated bars resembling a piano keyboard (Fig. 8.6a). A Hadamard-based spectrometer employs one or two masks, to record more than one wavelength at a time. Such optical configuration achieves both the Jacquinot and the multiplex advantages. In the single mask variant, the encoding mask selects half of the resolution elements and directs the light onto a single element detector. An inverse

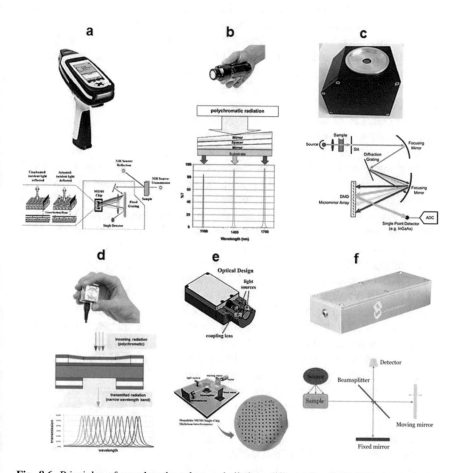

Fig. 8.6 Principles of wavelengths selectors built into different handheld NIR spectrometers: **a** MEMS Hadamard mask—microPHAZIR, Thermo Fisher Scientific, Waltham, USA; **b** LVF—MicroNIR Pro ES 1700, VIAVI, Santa Rosa, USA; **c** MEMS DMD—implementation of DLP NIRscan module, Texas Instruments, Dallas, USA; **d** MEMS Fabry–Perot interferometer—NIRONE Sensor S, Spectral Engines, Helsinki, Finland **e** MEMS Michelson interferometer—NeoSpectra, Si-Ware, Cairo, Egypt; **f** MEMS Michelson interferometer with a large mirror—nanoFTIR NIR, SouthNest Technology, Hefei, China. Reproduced from Ref [12]. (CC-BY 4.0 license)

Hadamard transform resolves the spectrum from the collected wavelengths [13]. Hadamard principle allows to employ a cost-effective single-pixel photodetector. For example, in microPHAZIR, NIR beam emitted from a low-power tungsten bulb is focused on a fixed grating that serves as the dispersive element. The keys of the MOEMS chip are actuated successively, reflecting the selected wavelengths onto a single-pixel InGaAs detector. This design enabled a rapid scanning capability (<10 s), good S/N, and a reasonable optical resolution of 11 nm. However, the device operates in a rather narrow wavelength region of 1596–2396 nm (6267–4173 cm^{-1}) [14, 15].

In early 2010s, the success of microPHAZIR leads to anticipation that MEMS spectrometers would rapidly dominate the market of miniaturized NIR sensors based on Hadamard principle. However, the subsequent progress was much less dynamic [6]. The limitations resulting from the size of the optics created issues with repeatability of operation and the ability of a MEMS comb actuator to drive the moving mirror; these factors outweigh the advantages of this technology, given its price. Nonetheless, commercial success of some handheld FT-IR (mid-IR) devices even paved the path for the appearance of FT-NIR spectrometers. NIR light sources are brighter, and detectors have a higher specific detectivity than those used in IR spectrometers; hence, mirror size is less of a constraint. For example, Thermo Fisher Scientific successfully scaled down the interferometer design, with a voice-coil and piston-bearing scheme, and a moving mirror of 1.2 cm diameter. A MEMS-based Michelson interferometer was commercialized by NeoSpectra, the division of Si-Ware Systems, with their FT-NIR miniaturized instrument (Fig. 8.6a).

The problem of maintaining a stable operation of MEMS element and the optical throughput of such devices is under constant development. Recent examples of refined designs include NIRONE sensors from spectral engines (Fig. 8.6d) and nanoFTIR NIR spectrometer from Hefei SouthNest Technology (Fig. 8.6f). The nanoFTIR NIR is a very recent sensor that uses a MEMS Michelson interferometer, in which in order to improve its light throughput efficiency, a large mirror in relation to the area of MEMS chip was implemented. Further, the spectrometer operates over the entire NIR wavelength region (800–2600 nm; 12,500–3846 cm^{-1}), which is a notable improvement over early MEMS-based sensors (Table 8.1). This is accompanied by a relatively high spectral resolution (6 nm), good S/N, and rapid scanning. Noteworthy, the design achieved significantly more compact dimensions (14.3 × 4.9 × 2.8 cm; weight 220 g) than any previous MEMS-based FT-NIR spectrometers.

Fabry–Perot interferometers are very suitable to serve as wavelength selectors in miniaturized spectrometers. The key element in such interferometer is Fabry–Perot filter consisting of two mirrors, either planar or curved, facing each other and separated by a distance d. Two variants exist, an etalon with fixed d, and the other with variable d. Interference condition in a Fabry–Perot interferometer is achieved through the standing wave effect between the two mirrors and division of a polychromatic light into several narrow wavelength bands. Important for miniaturized spectrometers, MEMS technology can be used to fabricate a fully programmable optical filter based on Fabry–Perot interferometer in microscale. This solution is implemented, e.g., NIRONE sensor series (Fig. 8.6d).

Table 8.1 Operating parameters of the selected handheld NIR spectrometers available on the market

Spectrometer	Spectral region		Spectral resolution (at wavelength)[a] [nm]
	[nm]	[cm^{-1}]	
microPHAZIR (Thermo Fisher Scientific)	1596–2396	6267–4173	11
MicroNIR Pro ES 1700 (VIAVI)	908–1676	11,013–5967	12.5 (at 1000) 25 (at 2000)
SCiO (Consumer Physics)	740–1070	13,514–9346	unknown[b]
NIRscan (Texas Instruments)	900–1700	11,111–5882	10
NIRONE sensors (Spectral Engines)	2000–2450	5000–4082	18–28[c]
NeoSpectra (Si-Ware Systems)	1350–2500	7407–4000	16 (at 1550)
nanoFTIR NIR (SouthNest Technology)	800–2600	12,500–3846	2.5 (at 1000) 6 (at 1600) 13 (at 2400)

[a] 'at wavelength' parameter listed if available in the datasheet provided by the vendor
[b] SCiO presents to the operator interpolated spectra with 1 nm data spacing, but the real resolution is considerably lower
[c] depending on the sensor implementation/factory configuration

Digital micromirror device (DMD) is a MEMS-based array of mirrors. DMD may primarily be used to lower the cost of miniature dispersive scanning spectrometers, and its key role is to enable replacing expensive micro-array detectors by a large single-pixel detector, which is a much more cost-effective solution. DMD element is used in DLP NIRscan module from Texas Instruments. The company's proprietary DLP technology is offered as two evaluation modules (EVMs): DLP NIRscan and DLP NIRscan Nano (Fig. 8.6c).

There exist microscaled technology solutions other than MEMS, which are suitable for the construction of wavelength selectors in miniaturized NIR spectrometers. For example, a linear variable filter (LVF) is an optical bandpass filter, in which through varying thickness of an optical coating, and thus, transparence against different wavelengths varies linearly across a wedged geometry of the filter. LVF technology is cost-effective, and compared with MEMS relies less on large-scale manufacturing, although requires to be used with less affordable array detectors. On the other hand, such configuration enables very rapid scanning capability. LVFs enable constructing very compact spectrometers with no moving parts, which improves the ruggedness of the spectrometer. A satisfying spectral resolution for real applications and low power consumption should also be noted. DMDs coupled with InGaAs array detectors are used in the line of NIR spectrometers, including specialized models aimed for process control, that were introduced to the market by VIAVI (Fig. 8.6b).

It needs to be highlighted, that the design of a miniaturized NIR spectrometer is a balance between the level of miniaturization, performance, and the economic cost. Through accepting certain limitations, very affordable devices are feasible.

For example, Consumer Physics designed a pocket-size spectrometer (SCiO), with dimensions of 67.7 × 40.2 × 18.8 mm and weight of 35 g, available at a very low price. This was accomplished through incorporation of a LED (IrED) light source and a silicon detector in the form of a 4 × 3 photodiode array with optical filters over the individual pixels. However, the device operates over a narrow Vis/SW-NIR spectral region (740–1070 nm; 13,514–9346 cm^{-1}) with a rather poor spectral resolution of ca. 28 nm because of just 12 resolution elements and sub-par S/N of the measured spectra.

8.3 Application and In-depth Evaluation of Performance Characteristics of Portable NIR Spectrometers

Variety of the technology solutions and miniaturization itself has a meaningful impact on the operating parameters and performance of handheld NIR spectrometers. The key characteristics such as the working spectral region, spectral resolution, sensitivity, and S/N, of such spectrometers differ from those available on benchtop instrumentation. These issues influence the applicability and analytical performance of miniaturized NIR spectrometers. It is now an active field of research from several scientific group to perform systematic evaluation studies of different handheld NIR spectrometers in variety of analytical applications.

Various approaches can be helpful in examining the analytical worthiness of miniaturized spectrometers. The most straightforward and definitive evaluation of the analytical accuracy is provided by the statistical errors of multivariate analysis, e.g., quantitative models constructed for prediction of the chemical contents or classification models for qualitative discrimination between samples. Correlation coefficients for regressions (e.g., by means of partial least squares), either for cross-validations or test-set validations, and root mean square errors (of cross-validation, calibration, estimation) deliver numerical values indicating the worthiness of a given spectrometer. However, these values are only valid for the given, particular application and are neither easily interpretable nor transferrable to other scenarios. As such, these are not sufficient, if one aims for comprehensive evaluation of the instrumentation or prediction how it should behave in other more or less similar scenarios. Therefore, some other approaches can be helpful in obtaining a more general overview of the concerning problem. Comparative measurements of the same sample sets on high-performing benchtop NIR spectrometers, with underlying reference analysis based on gold standard methods of analytical chemistry (e.g., chromatography coupled to mass spectrometry) are indispensable for establishing the performance level in best-case scenarios. Differences between the devices in the wavelength-dependent sensitivity levels can be easily visualized and assessed by performing 2D hetero-correlation analysis, in which spectra measured on different NIR spectrometers (e.g., miniaturized vs. benchtop) can be directly correlated. This approach should be repeated for different samples, as well as experimental conditions, to outline the performance profile of a given miniaturized spectrometer in a comprehensive way. Yet, to be able to fully predict even approximate performance in a given application, extensive and tedious analyses of numerous samples are necessary. To mitigate this imitation, quantum chemical simulations of NIR spectra are helpful. Firstly, detailed NIR band assignments enable linking the wavenumber regions that are established as meaningful in the chemometric models with the vibrational modes of the target molecule. This allows to interpret the regression vectors and to assess the sensitivity of a spectrometer toward specific modes. More importantly, through these steps, the ability to generalize the performance over a wider range of scenarios is given. For instance, the performance against certain classes of analytes can be predicted following their likeliness of appearing characteristic NIR bands within the sensitivity ranges of a given spectrometer. This can be obtained without the need to perform analyses of large number of analyses, or even measurement of standards. Furthermore, the influence of matrix effects on various vibrational modes can be obtained through quantum mechanical calculation of NIR spectra [10, 16, 17]. The scheme presenting the gains from combined approaches to comprehensive evaluation procedures of miniaturized spectrometers is presented in Fig. 8.7.

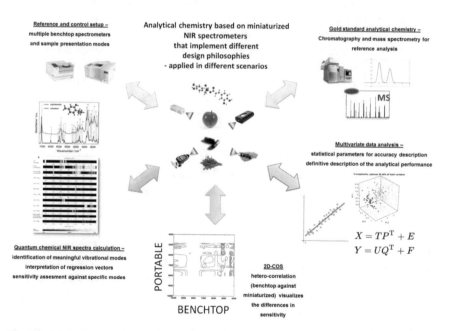

Fig. 8.7 Methods useful for comprehensive dissection and comparison of performance profiles of NIR spectrometers that are based on various design philosophies and differ in their operating characteristics

8.3.1 Example of an Application Where Differences Between Performances of Portable Instruments (Based on Different Designs) and a Benchtop Spectrometer Were Demonstrated

Validation of the performance of handheld NIR spectrometers in relation to a benchtop instrument in analyzing polyphenols in medicinal plant was conducted by Kirchler et al [10]. In that case, portable NIR sensors demonstrated high feasibility to predict concentrations of anti-oxidative active ingredients (rosmarinic acid and similar polyphenols) in *Rosmarinus officinalis*, folium. The operating characteristics of two handled spectrometers, microPHAZIR and MicroNIR 1700, were closely compared. These two devices differ significantly in the employed technology, i.e., wavelength selectors and detectors. The evaluation of the analytical performance of miniaturized devices was conducted through the comparison with a reference benchtop NIR spectrometer Büchi NIRFlex N-500. Sophisticated data analytical tools were employed to obtain exhaustive comparison, e.g., hetero-correlated 2D plots visualized the differences in the relative sensitivity toward different NIR absorption bands measured on the three spectrometers. In that study, the benchtop spectrometer yielded the best prediction of the rosmarinic acid content in the plant extracts. The moderate analyte concentration in the sample (less than 7%) and complex matrix, as well as several NIR wavenumber regions in which the molecule absorbs significantly, could have contributed toward the inferior capability of portable spectrometers, because of their limited spectral windows and resolutions.

8.3.2 Example of an Application Where an Ultra-miniaturized and Affordable NIR Spectrometer Performed Semi-comparably with a Benchtop Instrument

Wiedemair et al. have tested the performance of SCiO in comparison with Büchi NIRFlex N-500 for the analysis of protein content in millet samples [18] and the fat content in cheese samples [19]. As can be deduced from Tables 8.2 and 8.3, they found that the analytical performance of portable devices may considerably vary between different scenarios. Although clearly inferior in the former analytical problem (Table 8.2), in the determination of fat content in cheese (Table 8.3), the inexpensive SCiO sensor delivered the performance, evaluated by statistical values, comparable to the high-performing benchtop instrument. In this case, the analysis of major nutrient was more successful; presumably, because it was a more significant chemical constituent in the sample.

These examples demonstrate somewhat uneven performance profiles of miniaturized spectrometers, strongly depending on the particular analytical scenario.

Table 8.2 Millet analysis toward gallic acid equivalents (GAE) by benchtop and ultra-portable NIR spectrometers

Spectrometer	Sample form	R^2 (CV)	RMSECV/mgGAE/g	R^2 (TV)	RMSEP/mgGAE/g
NIRFlex N-500	Intact	0.953	0.365	0.940	0.467
	Milled	0.985	0.223	0.920	0.479
SCiO	Intact	0.876	0.601	0.814	0.806
	Milled	0.8240	0.743	0.782	0.840

RMSECV and RMSEP values resulting from PLS-R models for protein content (7–14% w/w) are used here as the parameters describing the prediction performance
CV cross-validated regressions; TV test set-validated regressions

Table 8.3 Cheese analysis toward gallic acid equivalents (GAE) by benchtop and ultra-portable NIR spectrometers

Spectrometer	Sample form	R^2 (CV)	RMSECV/mgGAE/g	R^2 (TV)	RMSEP/mgGAE/g
NIRFlex N-500	Intact	0.9726	1.5711	0.9431	1.8964
	Grated	0.9930	0.7845	0.9913	0.7676
SCiO	Intact	0.9801	1.2466	0.9838	1.1874
	Grated	0.9838	1.0527	0.9940	0.8194

RMSECV and RMSEP values resulting from PLS-R models for protein content (9–36% w/w) are used here as the parameters describing the prediction performance
CV cross-validated regressions; TV test set-validated regressions

Nonetheless, their potential is demonstrated even for challenging analysis of multi-constituent natural products where content variability and matrix effects are comparably stronger than most other types of samples. The discussed problem of performance in certain scenario is essential in various fields of research and analysis. Often, practical applications benefit largely from routine measurements being carried out on portable instrumentation, however, with calibration and validation controlled via benchtop spectrometer as the reference. Transfer of multivariate models between these devices is an essential feature and remains a focused lane of research. For instance, transferability of spectra and the resulting qualitative and quantitative calibrations, between benchtop and miniaturized spectrometers was explored by Hoffman et al. [20]. Recent literature demonstrates keen interest in the performance of handheld spectrometers in a variety of applications, e.g., material identification and product authentication [21]. Applicability and performance in quantitative analysis of miniaturized NIR spectrometers were also conducted in the context of pharmaceutical formulation [22]. Because of on-site capacity, miniaturized NIR spectrometers have very high potential for applications in the analysis of natural products. However, complex composition and demanding conditions in such analyses are challenging for accurate quantitative analysis. Wiedemair and Huck evaluated the performance of three different miniaturized NIR spectrometers in determining the antioxidant capacity of gluten-free grains [23].

8.4   The Conclusions and Prospects for Future

Over the past decade, a new generation of compact NIR instrumentation appeared and underwent a successively progressing level of miniaturization. These devices revolutionized several fields of application of NIR spectroscopy. However, in sharp contrast to the matured technology used in laboratory-scale benchtop spectrometers, the miniaturized NIR instrumentation available on the market is far less uniform as it implements various engineering principles. Therefore, the performance profiles of those devices differ considerably. The necessary design compromises, inevitably accepted to achieve a compact size, impose certain limits in the applicability of such devices. Furthermore, several miniaturized NIR sensors have been optimized toward low cost and ease of use by a non-expert operator. These circumstances rise considerable concerns on the miniaturization vs. performance factor in reference to scientific-grade benchtop spectrometers. It still remains a relatively shallowly explored problem. The aim of this chapter is to summarize the background of miniaturized technology in NIR spectroscopy and its impact on the existing and potential applications.

Rapidly increasing utilization of miniaturized NIR spectrometers in practical applications throughout a broad spectrum of fields is evident from the literature reports published over the last few years. It demonstrates the potential of portable NIR spectroscopy and its importance for modern analytical chemistry. Furthermore, this technology is one of the first methods of physicochemical analysis that aims to cross the barrier between industrial or professional and everyday life applications in general public. Notwithstanding, the impressive level of miniaturization often comes as a give-and-take against the accuracy and robustness in analytical sense. Furthermore, the diversity of the design principles implemented in portable NIR spectrometers affects their performance profiles in device-specific ways. Therefore, it is an active area of research to perform systematic evaluations of the analytical performance of various miniaturized NIR spectrometers available on the market, and to better direct future development of this technology.

References

1. A. Michelson, E.W. Morley, On the relative motion of the Earth and the luminiferous ether. Am. J. Sci. **34**, 333–345 (1887)
2. Q.S. Hanley, Fourier transforms simplified: computing an infrared spectrum from an interferogram. J. Chem. Educ. **89**, 391–396 (2012)
3. P.J. Jacquinot, The luminosity of spectrometers with prisms, gratings, or Fabry-Perot etalons. J. Opt. Soc. Am. **44**, 761–765 (1954)
4. P.B.J. Fellgett, On the ultimate sensitivity and practical performance of radiation detectors. J. Opt. Soc. Am. **39**, 970–976 (1949)
5. J. Connes, P. Connes, Near-infrared planetary spectra by Fourier spectroscopy. I. Instruments and results. J. Opt. Soc. Am. **56**, 896–910 (1966)
6. R.A. Crocombe, Portable spectroscopy. Appl. Spectro. **72**, 1701–1751 (2018)
7. C. Palmer, *Diffraction Grating Handbook (7th edition)* (Newport Corporation, 2014)

8. V. Wiedemair, M. De Biasio, R. Leitner, D. Balthasar, C.W. Huck, Application of design of experiment for detection of meat fraud with a portable near-infrared spectrometer. Curr. Anal. Chem. **14**, 58–67 (2018)

9. O.M.D. Lutz, G.K. Bonn, B.M. Rode, C.W. Huck, Towards reproducible quantification of ethanol in gasoline via a customized mobile NIR spectrometer. Anal. Chim. Acta **826**, 61–68 (2014)

10. G. Kirchler, C.K. Pezzei, K.B. Beć, S. Mayr, M. Ishigaki, Y. Ozaki, C.W. Huck, Critical evaluation of spectral information of benchtop versus portable near-infrared spectrometers: quantum chemistry and two-dimensional correlation spectroscopy for a better understanding of PLS regression models of the rosmarinic acid content in Rosmarini folium. Analyst **142**, 455–464 (2017)

11. P. Reinig, H. Grüger, J. Knobbe, T. Pügner, S. Meyer, Bringing NIR spectrometers into mobile phones. Proc. MOEMS Miniaturized Syst. XVII **10545**, 105450F (2018)

12. K.B. Beć, J. Grabska, H.W. Siesler, C.W. Huck, Handheld near-infrared spectrometers: Where are we heading? NIR News **31**(3-4), 28–35 (2020)

13. Y. Du, G.A. Zhou, A MEMS-driven Hadamard transform spectrometer. Proc. SPIE 10545, MOEMS Miniaturized Syst. XVII, **105450X**, (2018)

14. K. Pezzei, M. Watschinger, V.A. Huck-Pezzei, C. Lau, Z. Zuo, P.C. Leung, C.W. Huck, Infrared spectroscopic techniques for the non-invasive and rapid quality control of Chinese traditional medicine Si-Wu-Tang. Spectro. Eur. **28**, 16–21 (2016)

15. C.K. Pezzei, S.A. Schönbichler, C.G. Kirchler, J. Schmelzer, S. Hussain, V.A. Huck-Pezzei, M. Popp, J. Krolitzek, G.K. Bonn, C.W. Huck, Application if benchtop and portable near-infrared spectrometers for predicting the optimum harvest time of Verbena officinalis. Talanta **169**, 70–76 (2017)

16. K.B. Beć, J. Grabska, C.G. Kirchler, C.W. Huck, NIR spectra simulation of thymol for better understanding of the spectra forming factors, phase and concentration effects and PLS regression features. J. Mol. Liq. **268**, 895–902 (2018)

17. J. Grabska, K.B. Beć, C.G. Kirchler, Y. Ozaki, C.W. Huck, Distinct difference in sensitivity of NIR versus IR bands of melamine to inter-molecular interactions with impact on analytical spectroscopy explained by anharmonic quantum mechanical study. Molecules **24**, 1402 (2019)

18. V. Wiedemair, D. Mair, C. Held, C.W. Huck, Investigations into the use of handheld near-infrared spectrometer and novel semi-automated data analysis for the determination of protein content indifferent cultivars of Panicum miliaceum L. Talanta **205**, 120115 (2019)

19. V. Wiedemair, D. Langore, R. Garsleitner, K. Dillinger, C.W. Huck, Investigations into the performance of a novel pocket-sized near-infrared spectrometer for cheese analysis. Molecules **24**, 428 (2019)

20. U. Hoffmann, F. Pfeifer, C. Hsuing, H.W. Siesler, Spectra transfer between a fourier transform near-infrared laboratory and a miniaturized handheld near-infrared spectrometer. Appl. Spectro. **70**, 852–860 (2016)

21. H. Yan, H.W. Siesler, Identification of textiles by handheld near infrared spectroscopy: protecting customers against product counterfeiting. J. Near Infrared Spectro. **26**, 311–321 (2018)

22. H. Yan, H.W. Siesler, Quantitative analysis of a pharmaceutical formulation: performance comparison of different handheld near-infrared spectrometers. J. Pharm. Biomed. Anal. **160**, 179–186 (2018)

23. V. Wiedemair, C.W. Huck, Evaluation of the performance of three hand-held near-infrared spectrometer through investigation of total antioxidant capacity in gluten-free grains. Talanta **189**, 233 (2018)

24. B.A. Eckenrode, Environmental and forensic applications of field-portable GC-MS: an overview. J. Am. Soc. Mass Spectrom. **12**, 683–693 (2001)

25. J. Zhang, C.C. Teng, T.G. van Kessel, L. Klein, R. Muralidhar, G. Wysocki, W.M.J. Green, Field deployment of a portable optical spectrometer for methane fugitive emissions monitoring on oil and gas well pads. Sensors **19**, 2707 (2019)
26. C. Hutengs, B. Ludwig, A. Jung, A. Eisele, M. Vohland, Comparison of portable and benchtop spectrometers for mid-infrared diffuse reflectance measurements of soils. Sensors **18**, 993 (2018)

Chapter 9
NIR Optics and Measurement Methods

Akifumi Ikehata

Abstract What type of components does a NIR spectrometer consist of? How do these parts determine the performance of the instruments? The measurement targets of NIR spectroscopy span a wide variety from transparent liquids to opaque solid samples, and as described in Chap. 8, the NIR spectrometers are characterized by a wide variety of device specifications and shapes. Consequently, what are the criteria for choosing a spectrometer? In the first half of this chapter (9.1), the basics of the optics that comprise the NIR spectrometer, such as the light source, spectroscopic element, and detector, are explained. This will allow the reader to understand the specifications, that control the functions of the spectrometer. Next, in the latter half of this chapter (9.2), the measurement method is explained for each sample form, namely liquid, solid, and paste. The most characteristic feature of NIR spectroscopy is the use of diffuse reflected light, and the "interactance" method, which is a unique application. It can be inferred that diffuse reflectance method contributes to the expansion of the range of sample forms that are measurable by NIR spectroscopy.

Keywords Light sources · Spectrometers · Detectors · Sample cells · Interactance · Transflectance

9.1 Optics

9.1.1 Device Configuration

Near-infrared (NIR) spectrometers are composed of a light source, a sample optical system, a spectrometer, and a detector. The configuration of NIR spectrometers is largely confined to two types, as shown in Fig. 9.1a and b. In type (a), the sample is positioned after the spectrometer and thereby irradiated with monochromatic light. This configuration is often used for standard desktop UV–VIS spectrometers. In type

A. Ikehata (✉)

Food Research Institute, National Agriculture and Food Research Organization (NARO), 2-1-12 Kannondai, Tsukuba 305-8642, Japan

e-mail: ikehata@affrc.go.jp

© Springer Nature Singapore Pte Ltd. 2021

Y. Ozaki et al. (eds.), *Near-Infrared Spectroscopy*,

https://doi.org/10.1007/978-981-15-8648-4_9

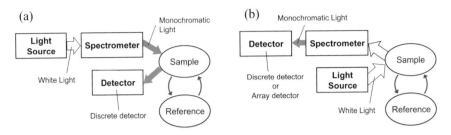

Fig. 9.1 Device configuration of NIR spectrometers. **a** Monochromatic light irradiation, and **b** white light irradiation

(b) spectrometers, the sample is positioned just after the light source and irradiated with white light. This configuration is often used for NIR applications, and achieves high-speed detection by use of an array detector. However, type (a) is better than (b) from the viewpoint of preventing damage to the sample by irradiation with light. A sample and a reference are measured alternately, and a spectrum is calculated from the ratio.

9.1.2 Near-Infrared Light Sources

9.1.2.1 Thermal Radiation

A light source that uses thermal radiation generated by heating a filament by passing an electrical current through it (resistive heating) is the most popular light source in the NIR region. Halogen lamps using tungsten filaments are inexpensive, constitute a stable thermal radiation source, and are widely adopted in NIR spectrometers on the market. Halogen lamps emit light with high brightness across the visible-to-infrared range by a reaction called the halogen cycle. To stabilize the halogen cycle, not only the filament but also the inner wall of the lamp must be heated to a sufficiently high temperature. Since the temperature is very high, care must be taken to prevent failure of the seal and deterioration of the socket. The emission spectrum of a halogen lamp is generally explained by Planck's law, and has an emission peak of approximately 1 μm. Higher filament temperatures result in higher brightness and larger emission intensities of short wavelengths. Heated nichrome wire (wavelength 2–5 μm) and Glover (silicon carbide; Wavelength 1–50 μm), which are often used as light sources for mid-and far-infrared wavebands, are also used in the NIR waveband. Since emission characteristics follow Planck's law, it is necessary to use a high temperature by passing a large amount of current to produce the NIR waveband.

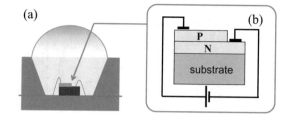

Fig. 9.2 a A schematic picture of an LED and **b** the *n*- and *p*- type semiconductor layers

9.1.2.2   LED

A light-emitting diode (LED) consists of a pn junction, in which a p-type semiconductor and an n-type semiconductor are joined (Fig. 9.2). When a forward voltage is applied, holes and electrons recombine, and excess energy is emitted as light. Although the emission bandwidth of LEDs is limited to about 100–200 nm, LEDs are inexpensive and can be driven with low power, so are useful for reducing the size, complexity cost of devices. The most common LED that emits NIR light is a gallium arsenide (GaAs) semiconductor, which has an emission peak at 940 nm. These are used in communication applications, such as remote control devices. For other wavelengths, GaAlAs LEDs emit shorter wavelengths (850, 880 nm) and InGaAs LEDs emit longer wavelengths (1300, 1550 nm).

A medical instrument called a pulse oximeter, used mainly in hospitals, can measure blood oxygen saturation non-invasively. The device mounts two single LEDs emitting either red or NIR light, corresponding to deoxidized and oxidized hemoglobin, respectively. The pulse oximeter is a successful example of cost reduction and miniaturization with LEDs.

9.1.2.3   Laser Diode

A laser diode (LD) is a semiconductor laser having the same structure as an LED but enables laser oscillation by forming a population inversion in a medium and causing a stimulated emission. A LD emits a pulse wave or continuous wave (CW) of monochromatic coherent photons and LDs are divided into several types, depending on the structure of the active layer (core layer) sandwiched between the *n*- and *p*-type semiconductor layers. Thus, LDs may be classified as Fabry–Perot type (FP type), distributed feedback type (DFB type), and vertical cavity surface emitting type (VCSEL). While the wavelength width of LEDs is about 100 nm, LDs have high monochromaticity, and the wavelength width is only a few nm (Fig. 9.3). Laser diodes with wavelengths that are often used in communication, such as 780, 850, 1310, and 1550 nm, are readily available. There are products with high outputs, exceeding 10 W in multimode output, which are used for processing and pumping light sources of solid-state lasers, as shown below.

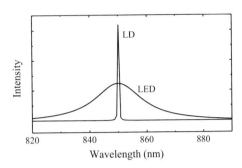

Fig. 9.3 Typical
light-emission characteristics
of an LD and an LED

9.1.2.4 Solid-State Laser

Solid-state lasers are lasers with a solid-state medium. The most representative NIR solid-state laser is Nd:YAG laser. An Nd:YAG laser oscillates very strong light of 1064 nm by using LDs as pump light. The titanium (Ti) sapphire laser is a tunable laser whose output wavelength can be freely changed across the range 700–1000 nm by introducing a wavelength selection element, such as an acousto-optic tunable filter (AOTF), into the resonator. Owing to the fluorescence characteristics of the Ti-sapphire crystal, the strongest light is emitted at around 800 nm. An argon ion laser or a Nd:YAG second-harmonic laser (532 nm) is used as the pump light. The Ti-sapphire laser excels at ultrashort pulse generation, but it can also emit CW light.

9.1.2.5 Supercontinuum Light

When ultrashort pulsed light with high peak power is introduced into a nonlinear optical material, continuous coherent light over a wide band can be generated by various nonlinear optical effects. This is called supercontinuum (SC) light and can be used as a white laser light source. The generation of SC light was first reported in 1970, [1] and the development of photonic crystal fibers in recent years allowed the development of a SC light pulse with a kW of peak power. The output wavelength range depends on the wavelength of the excitation light source. Supercontinuum light sources covering the wavelength range of approximately 500–2500 nm are on the market. Commercially available SC light sources are designed to extract light with a fiber optic, and the total output power is often about several mW. Since the SC light source is a coherent light source, like other lasers, interference due to scattered light from a rough surface (speckle) can occur.

9.1.3 Spectroscopic Elements

9.1.3.1 Prism

A prism is a classic feature of spectroscopes and is still sometimes used today. Because the refractive index of the prism differs depending on the wavelength (this is called "wavelength dispersion" or simply "dispersion"), incident light is emitted in different angles. Many optical materials highly disperse NIR light, so that good wavelength resolution is possible. However, the internal transmittance itself is inadequate. Consequently, most dispersion-type NIR spectrometers currently on the market use diffraction gratings to replace prisms, as described below.

9.1.3.2 Diffraction Grating

A diffraction grating is a dispersive element that diffracts white light into constituent wavelengths. The prototype was made by Fraunhofer in 1814, [2] and it has been widely commercially available since 1945 [3]. The spectral characteristics of a prism depend on its physical properties (refractive index), while the characteristics of a diffraction grating are determined only by the geometric structure. In general, a reflective diffraction grating, called a blaze-type grating (or an echelette grating), contains fine grooves engraved at equal intervals and in parallel, on a planar surface, is often used (Fig. 9.4). There are many commercially available gratings coated with aluminum or gold on a replica surface that is molded with resin. The fine blaze is produced by photoresist and ion beam irradiation. When light is incident at an angle α with respect to the normal line to the surface of the grating, light diffracted at each grating slope is strengthened or weakened under specific conditions. The condition that light of a specific wavelength, λ, is diffracted at an angle, β, is expressed by the following grating equation:

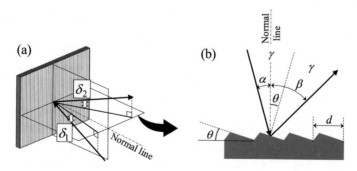

Fig. 9.4 a Schematic picture of a blaze-type diffraction grating. **b** The surface structure of a blaze-type grating showing the single reflection diffraction

$$d \cos \delta (\sin \beta + \sin \alpha) = m\lambda \qquad (9.1)$$

Here, m, is called the order number and represents the integer 0, 1, 2, ... δ is the angle that the incident or reflected beam makes with the plane perpendicular to the grooves, and these angles are matched with one another for blaze-type gratings.

$$\delta_1 = \delta_2 = \delta \qquad (9.2)$$

The enhanced diffracted light repeatedly appears in the angular direction depending on the order number. The number of grooves per mm, N, is the most important parameter determining the performance of the grating. Commercially available blaze gratings often have $N = 600$ or 1200 grooves/mm. Increasing the number of grooves reduces the groove spacing and increases the resolution of different wavelengths. Furthermore, the edge angle (blaze angle, θ) of the groove is an important value in the design of spectrometers. This is defined as the angle at which the m-th order diffracted light can be obtained with high reflection efficiency. The wavelength of the light at this angle is called the blaze wavelength. The blaze wavelength and blaze angle define the basic performance of diffraction gratings. When performing NIR spectroscopy, the blaze wavelength should be in the NIR region. The reflection efficiency rapidly decreases at wavelengths shorter than the blaze wavelength, but gradually decreases at longer wavelengths. The wavelength range of the diffraction grating is designed to be one to twice the blaze wavelength. The configuration of the spectrometer using a diffraction grating will be explained in detail in the next section.

9.1.3.3   Fourier Transform

Fourier transform (FT) spectroscopy is the mainstream technique in the mid-infrared region as FT-IR. An FT spectrometer specialized for the NIR region is also commercially available, called an FT-NIR.

The FT spectrometer measures the interference of light beams divided in two by use of a double-beam interferometer. Here, we will explain the principle of a double-beam interferometer using the Michelson interferometer (Fig. 9.5) as an example. The interferometer consists of a half mirror and two plane mirrors. One of the plane mirrors can move along the optical axis. Light emitted from the light source is collimated by the collimator, the beam reflected by the half mirror (HM) goes to the fixed mirror (M1), and the transmitted beam goes to the movable mirror (M2). The beams reflected by M1 and M2 are transmitted through, and reflected from, the opposite surface of the half mirror and combined again. The FT spectrometer detects the intensity of the combined wave with the detector (D) while changing the position of M2. The measured intensity increases or decreases with respect to the position x of the movable mirror, that is, produces an interference waveform, $F(x)$, called an interferogram. When the light source emits white light, the interferogram is multiplied by the wavenumber distribution, $B(k)$. In the FT spectrometer, the

Fig. 9.5 Michelson
interferometer forming the
basis of FT spectrometers

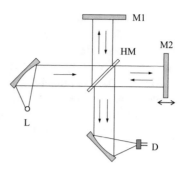

interferogram is Fourier-transformed by a built-in computer, and converted into a distribution, B (k), that is, the corresponding spectrum. Note that this spectrum is a function of wavenumber, k, not wavelength, λ. Owing to this measurement principle, the output of FT spectrometers is a spectrum of wavenumber (cm^{-1}).

In an actual interferometer, the movable range of M2 is finite. When the Fourier transform is performed with a limited range of integration, ringing occurs in the spectrum, so the transformation is performed by multiplying the window functions, whose ends of their integration range decrease smoothly. This window function is called an apodizing or tapering function, and this integration operation is called apodization. The maximum movable distance of the moving mirror is inversely proportional to the wavenumber resolution. In the FT calculation (discrete FT), the number of data points must be a factorial of 2. Therefore, the wavenumber resolution must be set as 1, 2, 4, 8, $16\,cm^{-1}$, ... In the NIR region, the spectral resolution is often set across the range 8–$32\,cm^{-1}$. In FT spectroscopy, the position of the movable mirror does not correspond to a certain wavelength, but the spectrum is obtained by Fourier transform of the interferogram. In other words, it is a spectrophotometer that measures all wavelengths simultaneously. This is synonymous with multiplex processing in the field of signal processing, and is extremely advantageous in improving the signal-to-noise ratio (S/N) (Fellgett advantage) [4]. However, it is difficult to determine which has lower noise, the grating type, or FT type. As explained in the next section, the noise level of a grating spectrometer is the same for all the wavelengths. In contrast, noise is superimposed on the interferogram signal in Fourier-type spectrometers. Consequently, the noise level at a specific frequency decreases the S/N at the corresponding wavelength, i.e., longer wavelengths are influenced by low-frequency noise. In addition, the effect of the slit width of FT spectrometers is smaller than that of dispersion-type spectrometers. Therefore, an FT spectrophotometer is a bright and high-throughput optical system (Jacquinot advantage) [5]. In order to accurately read the position of the movable mirror, a laser with a known wavelength is incident on the same optical axis as the observation light and is detected by another detector. Since the built-in He–Ne laser is stable, the 7th digit (632.9914 nm) of the oscillation wavelength is unchanged. Thus, the spectrum obtained with the FT spectrophotometer also has a high wavenumber accuracy of 7 digits (Connes advantage) [6].

Fourier transform spectrometers with excellent features were relatively large and expensive in the early years of development. Consequently, they were often used only in research laboratories. However, because the high wavenumber resolution is not always necessary in the NIR region, the movable range of the movable mirror can be reduced and miniaturization is now progressing. In recent years, devices with a single-chip interferometer based on micro-electro-mechanical systems (MEMS) technology have been released, and portable FT-NIR spectrometers using this device are also commercially available [7, 8].

9.1.3.4 Bandpass Filters

The filters discussed here are bandpass filters that transmit only light of a specific wavelength, such as interference filters and liquid crystal filters.

(1) Interference filter

Interference filters have a simple structure, in which a dielectric thin film is sandwiched between two semitransparent films. Interference filters are suitable for an inexpensive and robust system. The dielectric thin film is constructed using low refractive index fluorides, including MgF_2 and CaF_2, and high refractive index oxides, including TiO_2, and Al_2O_3, laminated together, as shown in Fig. 9.6. The principle of wavelength selection by the interference filter is the same as that of the thin-film Fabry–Perot interferometer.

If the thickness of the low refractive index (n) film, d is equivalent to a half wavelength for the desired wavelength, then transmitted wave and multiply reflected waves coincide in phase resulting in a wave having an amplitude of twice or more. This is constructive interference. At the same time, other wavelength components are attenuated by destructive interference. In addition, when the filter is tilted with respect to the incident light by the angle θ, the transmitted wavelength can be finely adjusted. The transmitted light of wavelength, λ, satisfies:

Fig. 9.6 a Schematic diagrams of an interference filter. **b** An enlarged view of the layer unit of dielectric films

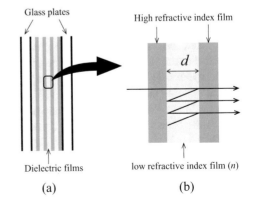

Glass plates

High refractive index film

d

Dielectric films

low refractive index film (n)

(a)   (b)

$$m\lambda = 2nd\cos\theta \tag{9.3}$$

Here, m is called order, and $m = 1$ or 2 is normally selected for NIR light. The total transmittance of an interference filter is given as:

$$T = \frac{(1-R)^2}{(1-R)^2 + 4R\sin^2(\delta/2)} \tag{9.4}$$

where R and δ represent the intensity of reflectance and the phase difference, respectively. δ is expressed by the following equation:

$$\delta = \frac{4\pi nd\cos\theta}{\lambda} \tag{9.5}$$

Since many interference waves are also generated by phase-shifted light, filters for blocking them are laminated on interference filters. The full width at half maximum (FWHM) of the spectrum of the transmitted light is given by the following equation:

$$\text{FWHM} = \frac{1-R}{R^{1/2}}\frac{c}{2\pi nd\cos\theta} \tag{9.6}$$

When reflectance, R, is increased, wavelength resolution is improved. However, simultaneously, loss due to the absorption of the dielectric film itself and reflection at the filter surface also increase, resulting in a decrease in transmittance. The effective transmittance of most products is about 70%. The band width of the filter is wider for longer wavelengths, the actual FWHM is about 10–30 nm. Those with transmission characteristics tailored to the commonly used laser oscillation wavelengths, including 780, 850, 1064, and 1550 nm, are inexpensive and easy to obtain.

Examples of NIR spectrometers equipped with interference filters include portable fruit sugar meters, moisture meters, and on-line meters that have been commercialized, mainly for specific analysis targets.

(2)   Variable filter

If the dielectric film of an interference filter has a taper, the transmitted wavelength can be changed depending on the irradiation position on the film. A spectrometer that utilizes this in the direction of rotation of a disk shape film is called a circular variable filter (CVF). The wavelength can be continuously swept by rotating the CVF around the central axis and using the light transmitted through a certain point. A filter whose thickness varies in the linear direction of a rectangular dielectric film is called a linear variable filter (LVF). Combining an LVF with an image sensor makes it possible to create an ultra-compact spectrometer without a mechanical component (Fig. 9.7). However, MEMS technology enabled a thickness control of a gap between reflectors of an interference filter, which is known as a MEMS-based Fabry–Perot interferometer (FPI) (Fig. 9.8). The MEMS-FPI realizes high-speed hyperspectral

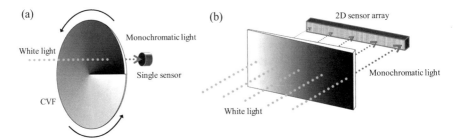

Fig. 9.7 a Circular variable filter (CVF) and **b** linear variable filter (CVF). The graduations show the thickness of the interference filters

Fig. 9.8 A schematic picture of a cross section of MEMS-FPI

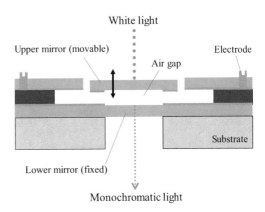

imaging because of its fast on-axis wavelength selection [9]. The detailed performance of commercially available variable filters will be described in Chap. 8 (New trend in instrumentation of NIR spectroscopy—miniaturization).

(3) Liquid crystal tunable filter

A liquid crystal tunable filter (LCTF) is an optical element that can electrically change the transmission wavelength. By laminating two birefringent plates having inclined optical axes in ±45° between two polarizers arranged in parallel, a phase difference (retardation) is caused, thereby transmitting a specific wavelength. This is called the Lyot filter. Such LCTFs use nematic liquid crystal instead of one birefringent plate. By applying an electric field to the liquid crystal, the same effect as rotating a uniaxial crystal around the optical axis can be obtained, and the birefringence can be changed. The transmittance wavelength is proportional to the birefringence. The high-speed response of liquid crystals results in high-speed wavelength switching, with a response time of about 50 ms. Wavelength resolution can be improved by connecting multiple sets of polarizers, birefringent plates, and liquid crystals, although the FWHM is 10 nm or more even for high-resolution products. Since LCTFs have a wide effective area, with a diameter of 20 mm, some spectral imaging systems apply them in combination with an area image sensor. Control of

Fig. 9.9 A geometry of
Bragg's diffraction by AOTF

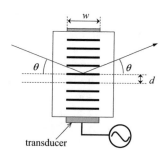

transducer

the orientation of the sample is important because an LCTF uses the polarization of
light, which changes according to sample orientation.

(4)   Acousto-optic tunable filter (AOTF)

An acousto-optic tunable filter (AOTF) is a spectroscopic device capable of wavelength selection by external electrical modulation. An AOTF can select wavelengths
at high speed without using mechanical parts and it does not generate high-order
diffracted light. Although it is classified as a filter, it works as a transmissio-type
diffraction grating (see also volume phase holographic grating). As shown in Fig. 9.9,
a transducer (piezoelectric element) is brought into close contact with a medium such
as a TeO_2 crystal, and an ultrasonic signal is generated in the medium by applying
an AC signal to the transducer. When the ultrasonic wave becomes a standing wave,
a periodic structure of density appears in the medium. The dense part corresponds
to a periodic change in refractive index and works as a diffraction grating for light.
The fringe pitch, that is, the ultrasonic wavelength, d, has the following relationship
with frequency, f, of the AC signal applied to the transducer:

$$\frac{v}{f} = d \qquad\qquad (9.7)$$

where v is the velocity of sound in the medium. The spacing of the diffraction grating
can be changed by modulating the frequency of the ultrasonic waves. An AOTF is
designed to realize Bragg's diffraction condition, whereby the angles of incidence
and diffraction are equal, θ. The Bragg's condition satisfies the following equation:

$$2d \sin \theta = m\lambda \qquad\qquad (9.8)$$

Hence, the wavelength of light, λ, can be switched by changing the grating space,
d, while fixing the diffraction angle θ. Thus, AOTF works as a tunable bandpass
filter. Many commercially available AOTFs have a range of sweep wavelengths of
about 1000 nm. The wavelength resolution of an AOTF is lower than that of a blazed
diffraction grating and the FWHM is about several nm to several tens of nm in
the visible-to-NIR region. Although wavelength resolution is low, AOTFs have the
advantage of their high-speed electric sweep, and it is often used for spectral imaging,

similar to LCTF, in the NIR region. The wavelength switching speed depends only on the dwell time of the ultrasound. The higher the speed of sound, the faster the wavelength switching speed, but the diffraction efficiency (the ratio of diffracted light intensity to incident light intensity) is inversely proportional to the square of speed, v. As a result, the use of a slow speed of sound is useful for AOTFs. The acoustic velocity of a longitudinal wave along the surface of (001) surface in a TeO_2 crystal is 4260 m s^{-1}, and the transverse wave along (110) surface is very slow at 616 m s^{-1}. Consequently, AOTFs with a TeO_2 crystal maintain their diffraction efficiency by use of transverse waves, and achieve a wavelength switching speed within 10 ms. When used with a thermal radiation source, the refractive index of the AOTF medium will change by warming, making it impossible to select the correct wavelength. For this reason, it is desirable to use a cold filter or apply an electrical correction method [10].

9.1.4 Detector

Since it is difficult to cover the entire NIR region with a single detector, it is necessary to select from the detectors shown below, depending on the wavelength band of interest [11].

9.1.4.1 Silicone Photo Diode (Si)

Photovoltaic silicon (Si) photodiodes can be used in the NIR region close to visible light (700–1100 nm: often called short NIR). Si photodiodes are suitable for quantitative analysis because of their wide dynamic range. The short NIR region is especially suitable for samples with high water content, so Si photodiodes are often used in sugar-content meters of fruits. Since Si photodiodes respond to visible light, it is necessary to use a wavelength cut-off filter to avoid indoor lighting and sunlight.

9.1.4.2 Indium Gallium Arsenide (InGaAs)

Indium gallium arsenide (InGaAs) is a photovoltaic device with a *pn* junction. The sensitivity band varies depending on the composition ratio of In and Ga. The larger the In:Ga ratio, the more sensitive it is to longer wavelengths. The standard sensitivity range is 900–1600 nm, but it is possible to extend this to 2600 nm. The *pn* junction type has a response speed of GHz, and high-speed measurement is therefore possible.

9.1.4.3   Lead Sulfide (PbS)

Lead sulfide (PbS) is a type of photoconductive element that decreases in resistance when light is incident on it. PbS is often used in NIR spectrometers because of its sensitivity across the range 900–2500 nm. However, since the decrease in resistance varies with the area exposed to light, it is necessary to design it so that the beam's diameter does not fluctuate depending on the sample. Lead selenide (PbSe) is also a NIR detector that can be used across the 1500–4500 nm range. In both cases, pink noise (1/f noise) occurs when temperature rises. For this reason, it is recommended to remove low-frequency components using an optical chopper. However, it is more effective to speed up the wavelength sweep when used for NIR spectroscopy. High-speed photovoltaic pn junction compounds of indium arsenide (InAs) and indium antimonide (InSb), having the same wavelength range as PbS and PbSe, are also available.

9.1.4.4   Photomultiplier Tube

In general, photomultiplier tubes (PMTs) are effective only for shorter wavelengths in the NIR region because the photoelectric effect cannot be generated by long wavelength light. The PMTs using a GaAs photocathode can be used across the range 500–800 nm and PMTs utilizing In/InGaAs semiconductors cover the NIR range up to 1700 nm. Since PMT was originally used for detection of faint light, PMTs are rarely used for quantitative analysis.

9.1.4.5   Image Sensor

An image sensor is a multi-channel detector with multiple sensors. An array of photo-diodes arranged in a row is called a linear image sensor, and a two-dimensional array spread on a plane is called an area image sensor (in contrast, an element with a single sensor is called a discrete semiconductor). There are charge coupled device (CCD) and complementary metal–oxide–semiconductor (CMOS) image sensors, and these devices have different coupling methods. As described above, a dispersive spectro-scope can distribute light of different wavelengths to different angles. Therefore, an image sensor can simultaneously capture a spectrum of a certain wavelength range without separating the wavelengths individually, using a slit. A spectrometer that takes out one wavelength at a time using a discrete detector is called a monochro-mator, and a spectrophotometer that measures multiple wavelengths simultaneously is called a polychromator. A device using an image sensor does not require mechan-ical parts, so it can be made robust and downsized. Since the two-dimensional area image sensor is used for a camera, it can be applied to spectral imaging in combination with LCTFs and AOTFs (Chap. 22).

9.1.5 Other Optical Materials

9.1.5.1 Transparent Materials for NIR Light

Materials transparent in the visible region are often transparent in the NIR region. Although ordinary glass, such as BK7, can be used in the short wavelength NIR range, because it contains O–H groups due to impurities, it exhibits absorption near 1400 and 2200 nm. For longer NIR wavelengths, fused silica (fused quartz), which is pure silicon dioxide glass, is most often used. However, fused quartz also contains O–H groups, depending on its purity. To avoid this problem, infrared grade high-purity quartz should be used. In terms of durability, sapphire is a very good material. However, attention should be paid to the possibility of loss and interference due to reflection, because sapphire has a refractive index of 1.7 or higher, even in the NIR region. CaF_2 and MgF_2 crystals used in the mid-infrared region are completely transparent in the NIR region, but they are rarely used because of problems of mechanical strength except, in cases where the low refractive index is essential.

9.1.5.2 High- and Low-Reflection Materials

The most popular mirror in the visible range has an aluminum-coated surface, but the reflectivity of Al decreases near 850 nm. Therefore, gold is used as a standard reflector in the NIR region. The reflectivity of gold consistently exceeds 96% in the NIR region (700–2500 nm). To prevent reflection, black materials tend to be used. However, even objects that appear black in the visible spectrum may reflect NIR light. For example, black alumite that is often used for optical mounts has a reflectance of nearly 50% in the NIR region, and this may cause stray light. Carbon black paint without lacquer or black flock paper can achieve lower reflectivity than black alumite. Black bakelite (phenolic resin) is also a good low-reflection material. An anti-reflective cloth with flocked fibers dyed black having a reflectance of only 2% is available at camera stores.

9.1.5.3 Polarizer

In normal use, a polarizer made by monoaxially stretched dye polymer film is easy to use in a large area. It can transmit light with electric field amplitude perpendicular to the stretching direction. However, transmittance is about 30–40% due to the dye. When using a laser, the polarizing film may be destroyed, so a polarizer based on the birefringence of the crystal is required. A typical Rochon prism polarizer made from quartz or MgF_2 is available in the NIR region.

9.2   Measuring Methods of NIR Spectroscopy

NIR spectroscopy can be used to analyze various kinds of samples such as liquids, solutions, suspensions, pastes, solids, fibers, powders, and gases. Moreover, it is used to assess a variety of bulk and raw materials along with biomedical samples. Therefore, different types of measuring methods are employed in NIR spectroscopy [12–14]. Both the selection of the measuring method as well as that of the pretreatment method and measuring conditions are crucial. The most characteristic feature of practical NIR spectroscopy is the measurement of a large number of samples in a series to create a calibration model for a wide range of target objective variables such as concentration. In this case, one must carefully choose the methods and conditions, and operate it in the same manner each time to obtain good prediction results. Sample selection is another important point for NIR measurements. Although this knowledge of experimental design is essential, it is not treated in this book. Reference [12] provides an outline for sample selection. In this section, the measurement methods, sample pretreatment methods, and measuring conditions in NIR spectroscopy are described.

9.2.1   Outline of NIR Measuring Methods

NIR measuring methods can be divided into two, namely, transmittance and reflectance. Figure 9.10 shows the schematics of the transmittance and reflectance method. The former can be employed for only clear liquids or thin solid samples such as solutions, suspensions, films, fibers, powders, and gases [12–14]. Conversely, the reflectance method can be used even for bulk materials such as whole small fruits, tablets, and human fingers. In addition, the Beet–Lambert law can be assumed to hold in the transmittance measurement but not in the reflectance measurement. The transmittance method can be further divided into two methods, namely regular transmittance and transflectance.

The reflectance method can be employed for solid samples such as suspensions, solids, cloths, grain, powders, and pastes, and it is based on reflection from the rough surfaces of the scatterers. A variety of bulk materials, such as tablets, agricultural

Fig. 9.10 Schematic of **a** transmittance method and **b** reflectance method. Top: reference measurement, bottom: sample measurement

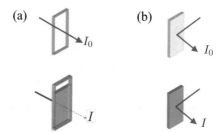

products, foods, polymers, and the human body, act as targets in reflectance measurements. As shown in Fig. 9.11, the incident ray is partially reflected by the surface and partially transmitted beneath the surface. The latter ray is refracted and reflected many times on the surfaces of particles (milled grain, microcrystals, emulsions, cells, etc.), and a part of it goes outside the scatterer. This is called the diffuse reflection (DR) light and is comparable to the transmittance light because the DR ray travels inside the absorptive material. As the rays jump out in different directions from the object surface, a gold-evaporated integrating sphere (Fig. 9.12a) is often used to efficiently collect them. In the case of rapid measurements, a detector is placed at the off-angle side of the regular reflection such that it is not in contact with the sample (Fig. 9.12b). Note that, in DR measurements, the sample functions as a small reflector itself. The theory of DR is described in detail in Chap. 3.

For both transmittance and reflectance methods, it is necessary to carry out a reference measurement (background measurement). For this purpose, the intensity, I_0, of the transmittance or reflectance light of the reference must be measured (Fig. 9.10). The intensity, I_0, can be measured before every sample or every convenient set of samples in a routine analysis. Next, the intensity of the light of the sample, I, is measured. Then, the absorbance, $A = -\log(I/I_0)$, of the sample is evaluated. In the case of reflectance measurement reflectivity, $R = I/I_0$, is obtained, and the absorbance

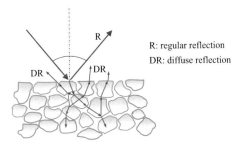

Fig. 9.11 Schematic of rays reflected from the surface of a material and within a material

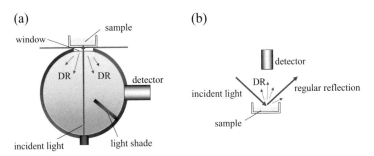

Fig. 9.12 **a** Gold-evaporated integrating sphere and **b** non-contact measurement arrangement

of the reflectance is expressed as $\log(1/R)$. Although NIR spectrometers for practical use often output only a single absorbance spectrum, I_0 and I are independently measured by single beam measurements in the apparatus. Note that the absorbance in NIR spectroscopy is the ratio of two single beam measurements.

NIR spectroscopy rarely uses attenuated total reflection (ATR) optics. Such ATR optics are used to suppress absorption that is too strong, for example, mid-IR. Thus, ATR is normally useless in the NIR. However, it may be applied if one needs to increase absorption. A method has been proposed in which a thin film of gold or metal oxide is formed in the ATR configuration and the absorption signal is enhanced by surface plasmon resonance [15]. This may be especially useful when only a small amount of sample can be used for measurement.

9.2.1.1 Clear Liquids and Solution Samples

For clear liquids and solutions, the transmittance method is used, and for such cases, the Beer–Lambert law can be applied except for dense solutions. It is the most commonly employed method for a variety of spectroscopic techniques. When the transmittance method is applied for a liquid sample, the selection of window materials and optimum path length of a cell becomes imperative. In the selection of a window material, its usable wavelength region, refractive index, and solubility in solvents must be considered. Glass or quartz (fused silica) is most often employed as a window material for the NIR region; however, they are not suitable for alkaline solutions. The thickness of a cell will be discussed later. Various kinds of liquid transmittance cells are commercially available, out of which three examples are displayed in Fig. 9.13. A cell with a fixed path length is commonly used, so that it is easy to perform precise quantitative analysis. Moreover, the effect of adsorption of a sample onto a cell wall, which is often a problem in IR spectroscopy, is negligible in the case of NIR spectroscopy because of the relatively long path length.

Another type of cell suitable for clear liquids and solutions is the transflection cell shown in Fig. 9.14. In this cell, the transmitted light is reflected back from a

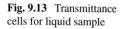

Fig. 9.13 Transmittance cells for liquid sample

(a)   (b)

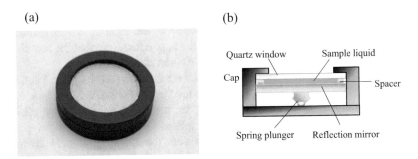

Fig. 9.14 **a** Transflection cell and **b** schematic

mirror and hence termed as "transflection". A transflection cell can be used as a DR-type instrument wherein the path length of the light is double the cell thickness. For practical purposes, transflectance is suitable for transmittance measurement because the cell is capable of being overhauled and cleaned, as the materials for the reflection plate employed are mostly ceramic, aluminum, stainless steel, and gold. This type of cell can also be used as a flow cell.

In the cases of liquid and solution samples, one must be careful about the effect of temperature. As already explained in Sect. 4.2 (Fig. 4.3), the NIR spectrum of water is sensitive to temperature; hence, the cell is often inserted into a thermostatted cell holder. For clear liquids and solutions, the sample pretreatment is, in general, not necessary; however, it is better to remove light scattering components from the solution as per the situation.

9.2.1.2   Suspensions

For suspensions, the measurement type should be selected based on the turbidity of a sample. A highly turbid sample such as milk scatters most of the incident light and transmits a small part. In such a case, the DR method is effective. In contrast, if the turbidity of a sample is low, transmittance or transflectance measurement is better for obtaining efficient absorption spectra. In the NIR measurement of suspensions, it is recommended to stir the sample well and measure its homogeneous part.

9.2.1.3   Solids

There are numerous types of solid samples. Typically, the DR method is employed for opaque solid samples. For thin transparent films such as thin polymer films, the transmittance methods are useful. In DR measurement of solid samples, it is necessary to avoid regular reflection from the surface as much as possible, because the intensity is very high, but the information obtained from it is relatively low. To solve this problem, interactance method is often used for NIR measurements

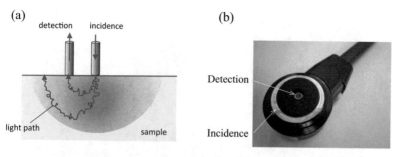

Fig. 9.15 a Interactance method and **b** an actual example

for various solid samples and bulk materials. Figure 9.15a depicts an interactance probe. It avoids regular reflection from the surface of a sample. In the interactance method, the irradiation and detection parts are separated physically, and thus, only DR light from the inside of the sample can reach the detector. Usually, a probe that integrates both the irradiation and detection parts is pushed directly onto the sample. Figure 9.15b shows an example of an interactance probe of a stationary NIR spectrometer.

When the distance between the irradiation and detection parts is small, the light that passes statistically through the shallow part of the sample can be detected. Alternatively, when the distance is large, the DR light that enters the deep interior of the sample can be collected. Therefore, an interactance probe provides a simple method to explore the depth profile of absorbance by changing the distance between the two parts. One of the representative applications of the interactance method is fNIRS, which enables noninvasive investigation of brain activity. The fNIRS monitors the changes in blood flow by delivering NIR light to the cerebral cortex by separating an irradiation probe and a light reception probe by several centimeters.

9.2.1.4 Powders and Grains

As emphasized before, DR method is used for powders, particles, and whole grains, and the corresponding cells are commercially available (Fig. 9.16). Before loading the powder and particle samples into the cell, homogenization is generally required to ensure that the samples have a uniform particle size because the effect of scattering on an NIR spectrum varies with the change in the particle size, particle shape, and its surface state. Proper pretreatment enables high precision in quantitative analysis and reproducibility in qualitative analysis of powders and particles. Grinders and mills equipped with a screen of an appropriate mesh size are commercially available for the pretreatment of NIR measurement.

For loading powder or particle samples, the packing density is significant because it affects the absorptivity and scattering conditions. To ensure constant and reproducible packing density, samples with equal weights should be loaded into the cell.

Fig. 9.16 Sample cells for powders and grains

Thus, sample setting is a key practice. In addition, to obtain a position average spectrum, some NIR instruments have a rotating or sliding cell option.

9.2.1.5   Pastes

Various kinds of pastes, such as ground meat, mayonnaise, fermented soybean paste (miso), and dough, are subjected to NIR measurement. Commercially available cells for the DR measurement of powders and grains are also useful for pastes. If it is possible to create a smooth surface, a paste sample packed in a polyethylene bag can be used as a measuring object. Smooth viscose paste samples, e.g., mayonnaise and butter, can be assessed using the transmittance method with a thin quartz cell comprising a sandwich of two quartz plates and a spacer of known thickness.

9.2.2   Sample Pretreatments and Measurement Conditions

As previously described, sample pretreatment is often necessary for NIR measurements. Some of them were already mentioned above, and examples of sample pretreatments are presented in Chap. 15 (agricultural products), Chap. 16 (woods and soils), Chap. 17 (sugarcane), Chap. 18 (pharmaceutical applications), Chap. 20 (medical applications), Chap. 21 (polymers), and Chap. 22 (on-line analysis). Various pretreatment methods have been used in NIR measurements, and a number of instruments and equipment are commercially available. Pretreatment methods may be divided into grinding, slicing, cutting, shredding, juicing, and homogenizing. For some samples, moisture control is also important. Some of these are described in different chapters. In addition, a detailed explanation about the pretreatments is provided in Refs. [12, 13].

Table 9.1 Molecular absorption coefficient of water at different wavelengths and optimum cell thicknesses

Wavelength/nm (Wavenumber/cm^{-1})	970 (10,300)	1450 (6897)	1930 (5180)
ε/dm^3 mol^{-1} cm^{-1}	0.0038	0.257	1.07
Pathlength for $A = 2$ (mm)	94.7	1.4	0.336
Cell thickness (mm)	10	1	0.3

9.2.2.1 Optimization of Light Pathlength

To obtain stable spectra, measurements should be performed within a dynamic range of the detector. As a general guideline for NIR absorption, it is highly desirable to maintain absorbance below 2 (or transmittance above 1/100). To satisfy this condition, one must optimize the light path length or devise a reference. In the transmittance measurement of transparent liquids, an optimum cell path length should be selected. In the IR spectral measurement, it is not easy to maintain a constant light path length, but in the NIR measurement, it is possible to use a cell with a mm size path length. Further, the absorbance of water can be a good reference for the selection of a cell. Table 9.1 summarizes the relationship between the peak wavelength of water in the NIR region and its absorption coefficient. The optimal path length for each peak is listed in the table. As the molar absorption coefficient, ε, of water is known, it is possible to determine the optimal light path length from the Beer–Lambert Law, $A = \varepsilon c l$. In the case of opaque liquid samples, the ray cannot pass straight in a sample, and thus, the practical light path length becomes longer than l. Moreover, the absorbance increases (the transmittance decreases) due to the diffusion of a significant part of the light. As the effects of the elongation of path length and scattering are difficult to remove, a practical method for maintaining absorbance below 2 is to perform NIR measurements by changing the sample thickness and reference materials.

9.2.2.2 Selection of a Reference

A reference is an external standard, which is used for comparing absorbance with transmittance or reflectance of an incident light. There are some NIR instruments housing references but if they are not available, the reference must be properly selected depending on the purpose.

9.2.2.3 References for Transparent Samples

In the case of a transparent sample that does not contain any scattering matter such as a solid transparent film, or a clear solution, one can measure a reference spectrum with nothing in the sample holder. For quantitative analysis of the concentration of solute in a solution using a UV–VIS spectrometer, a cell with the solvent can be used as a reference (blank sample). However, most solvents show strong absorption in the NIR region and often overlap with the absorption of the target solute. Therefore, for the NIR measurement of a clear solution, nothing or an empty cell should be placed in the sample chamber as a reference.

9.2.2.4 References for Opaque Samples

It is not easy to choose a reference for an opaque sample, i.e., when DR is applicable. For NIR measurements, it is highly desirable to cancel the contribution of the scattering component. To ensure this, the scattering coefficient of a reference should be similar to that of the sample, but the reference should not show absorbance in an objective wavelength region. For these reasons, polytetrafluorethylene (PTFE), ceramic plate, and gold are normally used as reference materials. Figure 9.17a shows the single beam intensity of PTFE, ceramic, and gold plates. It is noted that ceramic and gold have high reflectivity, but the reflectivity is low for PTFE. PTFE has negligibly small absorption in the NIR region but microcrystalline acts as its scattering component. Figure 9.17b compares the NIR DR spectra of brown rice (unpolished rice) measured using gold, ceramic plate, and PTFE references. Notably, PTFE yields a good spectrum without a rising baseline; however, the best reference for quantitative analysis and discriminant analysis is another matter. One has to determine the best reference from the precision of the results.

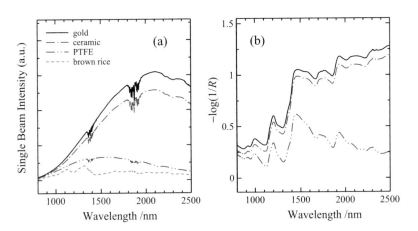

Fig. 9.17 **a** Single beam intensity profiles of reflection from gold, ceramic, and PTFE plates, **b** NIR DR spectra of brown rice measured using gold, ceramic, and PTFE plates as a reference

References

1. R.R. Alfano, S.L. Shapiro, Observation of self-phase modulation and small-scale filaments in crystals and glasses. Phys. Rev. Lett. **24**, 592–594 (1970). https://doi.org/10.1103/PhysRe vLett.24.592

2. J. Fraunhoffer, Kurtzer Bericht von the Resultaten neuerer Versuche über die Gesetze des Lichtes, und die Theorie derselbem. Ann D Phys **74**, 337–378 (1823)

3. C. Palmer, *Diffraction Grating Handbook*, 7th edn. (Newport Corporation, Irvine CA, 2014)

4. P.B. Fellgett, On the ultimate sensitivity and practical performance of radiation detectors. J. Opt. Soc. Am. **39**, 970–976 (1949). https://doi.org/10.1364/JOSA.39.000970

5. P.R. Griffiths, H.J. Sloane, R.W. Hannah, Interferometers versus monochromators: Separating the optical and digital advantages. Appl. Spectrosc. **31**(6), 485 (1977). https://doi.org/10.1366/000370277774464048

6. P. Connes, How light is analyzed, in *Laser and Light*, ed. by A. Schawlow (Freeman, San Francisco CA, 1969), p. 35

7. J. Antila et al., MEMS- and MOEMS-based near-infrared spectrometers, in *Encyclopedia of Analytical Chemistry* ed. by R.A. Meyers (Wiley Online, 2014)

8. Y.M. Sabry, D. Khalil, T. Bourouina, Monolithic silicon-micromachined free—space optical interferometers onchip. Laser Photonics Rev. **9**(1), 1–24 (2015). https://doi.org/10.1002/lpor.201400069

9. T. Kääriäinen, P. Jaanson, A. Vaigu, R. Mannila, A. Manninen, Active hyperspectral sensor based on MEMS Fabry-Pérot interferometer. Sensors **19**(9), 2192 (2019). https://doi.org/10.3390/s19092192

10. N. Saito, S. Wada, H. Tashiro, Dual-wavelength oscillation in an electronically tuned Ti:sapphire laser. J. Opt. Soc. Am. B **18**(9), 1288–1296 (2001). https://doi.org/10.1364/JOSAB.18.001288

11. A. Rogalski, History of infrared detectors. Opto-Electron Rev. **20**(3), 279–308 (2012). https://doi.org/10.2478/s11772-012-0037-7

12. P. William, Sampling, sample pretreatment, and sample selection, in *Handbook of Near-infrared Analysis*, ed. by D.A. Burns, E.W. Ciurczak (CRC Press, New York, 2007), pp. 267–296

13. S. Kawano, Sampling and sample presentation, in *Near-infrared spectroscopy-principles, instrumentations, applications*, ed. by H.W. Siesler, Y. Ozaki, S. Kawata, H.M. Heise (Wiley-VCH, Weinheim, 2002), pp. 115–124

14. Y. Ozaki, J. Berry, Sampling techniques in near-infrared transmission spectroscopy, in *Handbook of Vibrational Spectroscopy*, ed. by J.M. Chalmers, P.R. Griffiths (Wiley, West Sussex, 2002), pp. 953–959

15. A. Ikehata, T. Itoh, Y. Ozaki, Surface plasmon resonance near-infrared spectroscopy. Anal. Chem. **76**(21), 6461–6469 (2004). https://doi.org/10.1021/ac049003a

Chapter 10
Hardware of Near-Infrared Spectroscopy

Tsutomu Okura

Abstract The hardware of near-infrared (NIR) spectroscopy is almost the same as UV-VIS and infrared spectrometer except the wavelength area. However, the high SN ratio and stability of the instruments are required for a quantitative analysis by NIR spectroscopy, because of the smooth and dull absorption peaks of the NIR spectral shapes. These are realized by the hardware and computer technologies and are special features of the hardware of NIR spectroscopy. It is important to understand what they are when you use or design a near-infrared spectrometer. These aspects of the technologies are described. Instrumental difference also is an important problem in NIR spectroscopy where a calibration is used to predict contents of the matter. In this Chapter, not only the method to avoid the instrumental difference, but also the sources of the instrumental are described. To decrease the instrumental difference, it is crucial to understand why and how the instrumental difference is generated. The information described in this chapter will help you design a new NIR instrument, and a designing process with less effort is also described.

Keywords $1/f$ noise · Grating · Spectrometer · Hadamard · Instrumental difference

10.1 Noise Reduction Technology of the NIR Spectrometer

Karl Norris successfully evaluated agricultural products using near-infrared (NIR) spectroscopy around 1970 [1]. Near-infrared absorption was first discovered by Abney and Festing in 1881. However, it took 100 years to develop practical NIR applications.

It is said that the electronics and computer technologies played an important role in Karl Norris' success. However, it actually was a fight against the noise in the spectra.

T. Okura (✉)
Soma Optics, Ltd., 23-6 Hirai, Hinode-Machi, Nishitama-Gun, 190-0182 Tokyo, Japan
e-mail: tokura@somaopt.co.jp
URL: http://www.somaopt.co.jp

© Springer Nature Singapore Pte Ltd. 2021
Y. Ozaki et al. (eds.), *Near-Infrared Spectroscopy*,
https://doi.org/10.1007/978-981-15-8648-4_10

Karl Norris specialized in the field of electronics and was unfamiliar with spectroscopy, which is based on the wavelength and height of the absorption peak. His background enabled him to easily adopt a new scanning method and statistics to eliminate the noise from the spectra, thus enabling the detection of subtle spectral changes. This would have been difficult to achieve using an infrared spectrometer, which generated the spectra on a chart paper recorder using a pen. The experiments performed to eliminate noise were successful, and using statistics, information on the ingredients could be retrieved from the NIR spectra [2].

10.1.1 Noise and NIR Spectroscopy

Figure 10.1 shows the reflectance spectra of beef meat obtained using the Foss XDS analyzer. In Fig. 10.1, abs refers to the absorption, which is the logarithm of the reflectance R (%), as shown below in Eq. 10.1.

$$\text{abs} = -\log(R/100) \tag{10.1}$$

The intensity of the 928 nm fat absorption peak increases with increasing fat. Although it is difficult to confirm fat absorption peak in the lean spectra (Fig. 10.2a), the second derivative spectrum clearly indicates the presence of the fat peak (Fig. 10.2b).

The vertical scale of the spectrum shown in Fig. 10.2a is in (abs), while that of the spectrum in Fig. 10.2b corresponding to the second derivative is in (μabs).

Figure 10.3 shows the spectrum to which 20 μabs noise was added in the simulations. It is difficult to differentiate between Figs. 10.2a and 10.3a. However, the second derivative spectra shown in Figs. 10.2b and 10.3b are completely different. Even a low noise of 20 μabs, which is not easily discernible, deteriorated the second derivative spectrum illustrated Fig. 10.3b.

Fig. 10.1 Reflectance spectra of beef meat

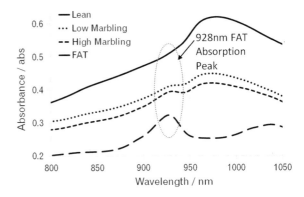

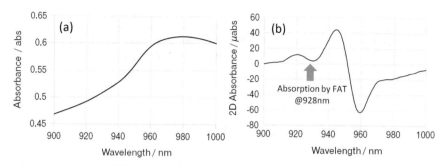

Fig. 10.2 Lean reflectance of beef (**a**) and its second derivative (**b**)

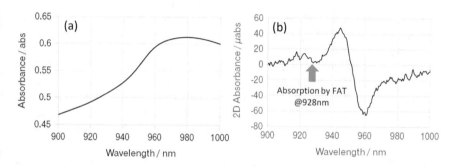

Fig. 10.3 Lean reflectance of beef with 20 μabs noise (**a**) and its second derivative (**b**)

The presence of noise affects the estimation of the ingredients, which depends on the sample characteristics. The effects of noise on the ratio of performance to deviation (RPD) [3] of three samples are shown on Fig. 10.4.

RPD of approximately 2.5 is obtained when the noise level is 20 mabs for the sugar in a tomato, 500 μabs for the protein in polished rice, and 50 μabs for the protein in unhulled rice. The required noise limits of the instrument are different

Fig. 10.4 RPD versus noise corresponding to different samples

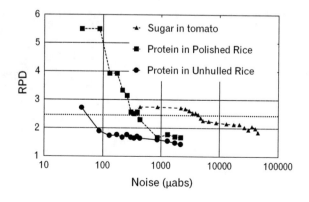

depending on the nature of the target of measurements. A typical high-grade NIR spectrometer has a low noise level of less than 20 μabs. Such a low noise level can be realized using high-grade A/D conversion instead of a pen recorder, and the new fast repeat scanning (FRS) method developed by Karl Norris.

10.1.2   Noise Reduction Using the FRS Method

The noise reduction technology used for an instrument, before the advent of digital technology until 1970, involved slow speed measurements and noise reduction using the time constant of the amplifier circuit. For example, the UV-VIS or Raman spectrometers employed in that era took a long time (>10 min or sometimes hours) to measure the sample, and the spectra were generated using a long time constant.

White noise with flat frequency characteristics can be reduced by employing slow scanning speeds and longer time constants. However, the actual noise is composed of white noise and *1/f* noise [4], as shown in Fig. 10.5, which has larger amplitudes at low wavelengths. The sources of the *1/f* noise include temperature and time, which influence many factors such as the sensor, optical parts, mechanism, and light source of the NIR spectrometer. The actual noise has a larger amplitude at lower frequencies than the *1/f* noise.

Karl Norris employed a method with FRS to avoid the effect of high noise levels at low frequencies. For example, when the wavelength scanning speed is 100× faster, the signal frequency is 100× higher, which reduces the noise due to the *1/f* noise. However, the frequency bandwidth of the signal is then 100× wider, which would require 100 × averages to obtain the same noise level as white noise.

The noise can be reduced by using a large number of averages. This method is called FRS, which cannot be realized using an analog instrument. The NIR spectrometers available in the market have a high-speed scanning mechanism of around 0.2 s per scan.

Fig. 10.5 Noise—frequency characteristics

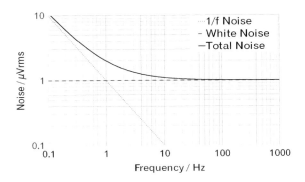

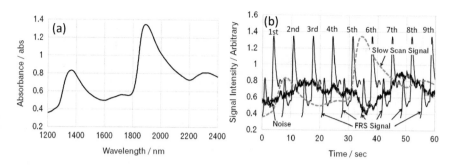

Fig. 10.6 The NIR spectral shape (**a**) and measured signals (**b**)

A typical waveform of the NIR spectrum is shown in Fig. 10.6a. The noise and signals measured using the slow scan and FRS methods over a 1 min duration are shown in Fig. 10.6b.

The signals and noise measured using the slow single scan and FRS methods are shown in Fig. 10.7.

In the spectrum obtained with the slow single scan, as shown in Fig. 10.7a, the noise is reduced by smoothing, similar to the time constant circuit. In Fig. 10.7b, nine repeated data with noise measured using the FRS method are averaged. These results are shown in Fig. 10.8, which indicate that the FRS scan method has a lower noise level than the slow single scan.

Karl Norris succeeded in retrieving the spectra with low noise using the FRS method and could estimate the ingredients of a material using statistics. This method can be realized only using digitalized technology. Almost all the wavelength scanning NIR spectrometers available in the market employ the above-discussed FRS method.

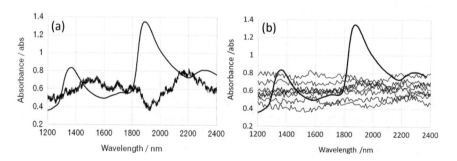

Fig. 10.7 The signals and noise measured using the slow single scan (**a**) and FRS (**b**) methods

Fig. 10.8 Spectra measured using the slow single scan and FRS methods

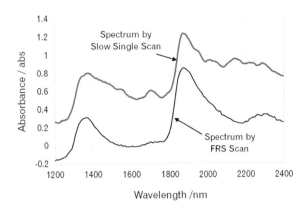

10.1.3 Noise Reduction in a Linear Array Spectrometer

Many NIR linear array spectrometers containing a linear array and grating spectrometer have been widely used for onsite ingredient measurements. A linear array spectrometer measures all the wavelengths simultaneously to ensure low noise levels. Most of the noise sources in a linear array spectrometer are intrinsic to the sensor.

The photoelectrons produced by light in a pixel of the linear array sensor are accumulated in a capacitor connected to each pixel. The number of accumulated photoelectrons in the capacitor is proportional to the incident light intensity. The noise in the linear array includes the circuit and quantum noises. Quantum noise is caused by the photoelectric effect. The number of photoelectrons produced by the incident light is proportional to the light intensity, and the noise is $\sqrt{N}$ when the averaged photoelectron number is N. The signal-to-noise ratio (SNR) (Eq. 10.2) can then be defined as follows [5].

$$\text{SNR} = \frac{N}{\sqrt{N}} = \sqrt{N} \tag{10.2}$$

Although the noise $\sqrt{N}$ increases at large N, the SNR improves. This indicates that high intensity light is required to achieve a good SNR in a NIR spectrometer.

The maximum number of photoelectrons is limited by the size of the capacitor. This is called saturation exposure, which is expressed in terms of the number of electrons (Me-). As shown in Fig. 10.9, when the saturation exposure is large, the capacitance of the linear array is large along with an improved SNR. A linear array with a large saturation exposure is called a "deep well."

The saturation exposure ranges from 0.03×10^6 to 1000×10^6 (Table 10.1). In a NIR spectrometer, a saturation exposure of $500–1000 \times 10^6$ is preferred to obtain a good SNR. As shown in Table 10.1, the saturation exposure of a CCD linear array is smaller, while that of the NMOS linear arrays is larger. Recently, CMOS linear arrays with large saturation exposures of around 1000×10^6 have been developed.

Fig. 10.9 Signal and saturation exposure of a linear array

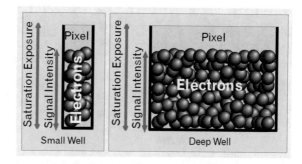

Table 10.1 Saturation exposure of a linear array

Linear array type		Saturation exposure ($\times 10^6$)
Silicon Linear array	CCD	0.03 ~ 0.6
	CMOS	0.08 ~ 900
	NMOS	31 ~ 312
InGaAs linear array		30

The SNR can be improved by averaging repeated measurements even when a linear array with a small saturation exposure is used. However, in this case, noise from the circuit is added with each measurement. Hence, a linear array with a large saturation exposure is preferred, and the exposure time should be selected such that a signal close to the saturation exposure is obtained.

When a "deep well" linear array is used, the light intensity should be high to obtain sufficient signal. Therefore, the optical design that supplies light of high intensity to the linear array detector should be considered.

10.1.4   Noise Caused by Wavelength Accuracy and Repeatability

In the IR or Raman spectroscopic techniques, where the peak position and peak height of spectral absorption are evaluated, the wavelength accuracy is approximately a third or fifth of the wavelength resolution. In NIR spectroscopy, though all the peaks are not sharp and the wavelength resolution is around ten nm, a high wavelength accuracy is still required to observe small subtle changes in the spectrum.

The spectrum shown in Fig. 10.10a (reflectance spectrum of a leaf) has a small absorption peak at 1728 nm, with its magnified plot shown in Fig. 10.10b. The peak height of the second derivative is approximately 20 μabs. The original spectrum has a slope of 1000 μabs/nm around 1728 nm. A 0.005 nm change in the wavelength around the 1728 nm position will cause a 5 (1000 × 0.005) μabs change. A higher wavelength accuracy is required where the spectrum has a large slope.

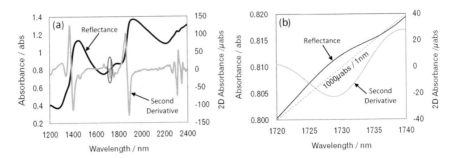

Fig. 10.10 The effect of the spectral slope on the wavelength accuracy, the reflectance of a leaf and its second derivative (**a**), and magnified graph around 1728 nm (**b**)

A laser is incorporated to calibrate wavelength with high accuracy in a Fourier-type spectrometer. A high-resolution rotary encoder is used to detect the grating rotation angle with high accuracy in a wavelength scanning-type grating spectrometer.

10.2 Grating Spectrometer

When considering the in-plane optics of a spectrometer, the off-plane angle δ is set to zero with the first order of diffraction, as shown in Fig. 10.11.

The relationship between the incident angle α, diffracted angle β, and wavelength λ can be expressed by Eq. 10.3 using the groove density N of the grating (see 5.2) [6].

$$\lambda = \frac{10^6}{N} \cdot (\sin\alpha + \sin\beta) \qquad (10.3)$$

λ Wavelength (nm)
N Groove density (mm^{-1})

Fig. 10.11 The principle behind diffraction grating

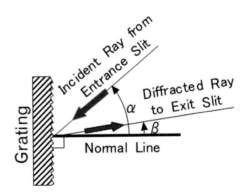

α Incident angle

β Diffraction angle.

The optics of the grating spectrometers can be examined based on Eq. 10.3.

10.2.1 Wavelength Scanning Grating Spectrometer

In a plane grating spectrometer, the parallel light beam on a grating is diffracted along the direction determined by Eq. 10.3. The diffracted parallel light is collimated into the exit slit. The wavelength can be scanned by rotating the grating.

(a) Optical Mount of the Plane Grating Spectrometer

The Czerny-Turner mount spectrometer shown in Fig. 10.12 is the commonly used optical mount [7]. The incident light from the entrance slit is reflected by the spherical mirror to form a parallel beam, which enters the plane grating. The light diffracted from the grating surface is collimated into the exit slit by the second spherical mirror. The wavelength of the spectrometer can be calculated using Eq. 10.3, producing Eq. 10.4 as shown below.

$$\lambda = \frac{2}{N} \cdot \cos\gamma \cdot \sin\theta \cdot 10^6 \tag{10.4}$$

Fig. 10.12 The Czerny-Turner mount spectrometer

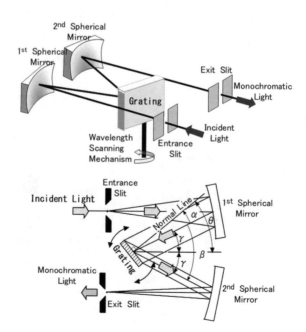

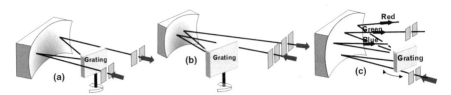

Fig. 10.13 Various optical mounts of a spectrometer **a** Ebert, **b** Littrow, and **c** Fastie Ebert

where

λ Wavelength (nm)

N Groove density of the grating (mm^{-1})

γ Deviation angle from the center axis (rad)

θ Rotation angle of the grating (rad).

Various other types of optical mounts have been developed, as shown in Fig. 10.13.

A grating spectrometer has stray light depending on the location of the optical axis on the principal plane [8]. The spectrometer should be designed such that any stray light is effectively eliminated.

(b) Concave Grating with Constant Interval Grooves—Seya-Namioka Mount

A spectrometer using a concave grating with constant interval grooves does not require any other additional optics (Fig. 10.14). This results in a simple instrument with minimized optical loss. In the Seya-Namioka mount [6, 9], the optical configuration can be calculated using Eqs. 10.5 and 10.6. The wavelength can be determined using Eq. 10.4.

$$2\gamma = 70.5° \tag{10.5}$$

$$R = R' = r \cdot \cos 35.25° \tag{10.6}$$

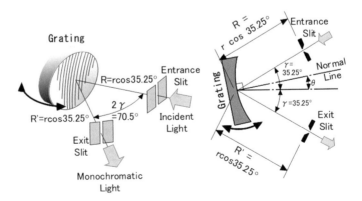

Fig. 10.14 Seya-Namioka mount

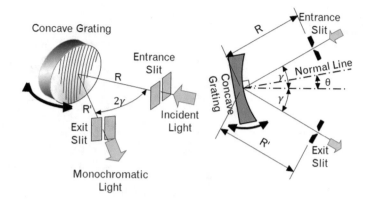

Fig. 10.15 Constant deviation mount using a concave grating

r Radius of curvature of the concave grating (mm)
R Entrance slit position (mm)
R' Exit slit position (mm).

The wavelength is calculated using Eq. 10.4.

(c) Concave Grating of Constant Deviation

Using a concave grating with grooves at unequal intervals, a spectrometer with a constant deviation angle of 2γ can be realized (Fig. 10.15). The specific angle γ, and focal points R and R' are determined based on the design of this grating [6]. Although the groove interval is not constant, the nominal groove density of the grating N is used in Eq. 10.4 to determine the wavelength.

10.2.2   Spectrometer with a Linear Array Detector

A spectrum can be measured by deploying a linear array detector at the exit slit position of a grating spectrometer [6]. This type of spectrometer has a compact size and enables rapid measurement. The NIR spectrometer can be realized using this type of spectrometer for specific samples.

(a) Linear Array Spectrometer with a Plane Grating

The focal point of each wavelength at the exit slit of a grating spectrometer should be on a flat plane to detect the spectrum without blurring, as the detector plane of a linear array is flat.

   In a grating spectrometer where all the optical axes are on the principle plane, such as the Czerny-Turner, Ebert, or Littrow mounts shown in Fig. 10.13, the positions of the grating collimating spherical mirror, and linear array determine the flatness of

Fig. 10.16 The super flat
condition for a linear array
grating spectrometer

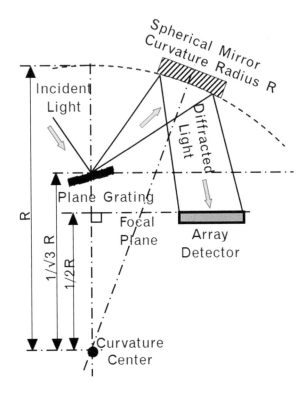

the focal plane. The grating position shown in Fig. 10.16, which is $1/\sqrt{3} \cdot R$ from
the center of the radius curvature of the collimating spherical mirror, is known as a
super flat condition [7] that achieves the flattest focal plane.

Astigmatism perpendicular to the slit image is generated in the Fastie-Ebert spectrometer (Fig. 10.13c) where the optical axis is out of the principle plane. However,
by deploying the linear array just above the grating, the image plane becomes approximately flat in a narrow wavelength band without any stray light. Thus, this mount
is appropriate for specific NIR applications.

The configuration parameters of the Czerny-Turner linear array spectrometer
(Fig. 10.17), i.e., the focal length, grating position, angle γ, and angle ϑ are
determined by Eq. 10.4 using the center wavelength λ_0, as shown in Eq. 10.7.

$$\lambda_0 = 2/N \cdot \cos\gamma \cdot \sin\theta \cdot 10^6 \tag{10.7}$$

The angles β_1 and β_2 corresponding to λ_{min} and λ_{max}, respectively, can be
determined using Eqs. 10.8 and 10.9.

$$\lambda_{min} = \frac{10^6}{N} \cdot (\sin(\theta - \gamma) + \sin\beta_1) \tag{10.8}$$

Fig. 10.17 Czerny-Turner linear array spectrometer

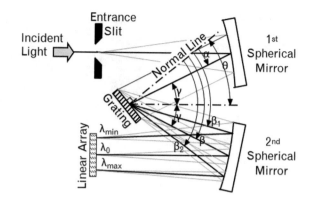

$$\lambda_{max} = \frac{10^6}{N} \cdot (\sin(\theta - \gamma) + \sin \beta_2) \tag{10.9}$$

All configurations of the spectrometer can be determined based on the above calculations.

(b) Concave Grating with Constant Interval Grooves

When a concave grating with constant interval grooves is used, the incident light through an entrance slit on the Rowland circle [6] is diffracted and focused back onto the same Rowland circle (Fig. 10.18). The Rowland circle has a diameter R, which is the radius of curvature of the concave grating and is in contact with its surface.

Fig. 10.18 A concave grating mount for a linear array spectrometer

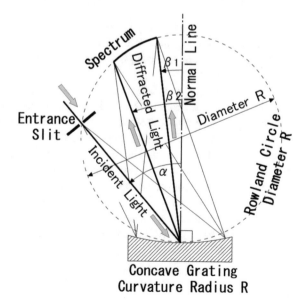

Fig. 10.19 Flat field concave grating for a linear array spectrometer

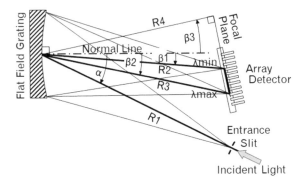

The relationship between α, $\beta 1$, and $\beta 2$ can be described using Eqs. 10.7–10.9.

The spectra can be measured by deploying a linear array on the Rowland circle. Some blurring is inevitable as the focal plane of the spectra is on the Rowland circle, while the linear array sensing area is on the flat plane.

(c) Flat Field Concave Grating

By adopting grooves with uneven/unequal intervals using holographic technology, a flat field concave grating with a flat spectral focal plane can be realized [6] (Fig. 10.19).

The configuration of the spectrometer ($R1$, α, $R4$, $\beta 3$) can be determined based on the grating design. The other parameters ($\beta 1$, $\beta 2$) can be calculated using Eqs. 10.7–10.9.

(d) Volume Phase Holographic (VPH) Grating

A grating with high efficiency is preferred for NIR spectrometer, which requires a high intensity light signal to decrease the noise. The VPH grating, which is based on Bragg diffraction, has a high efficiency of approximately 90% in the narrow wavelength band [10].

The VPH grating is a transmissive-type grating that is fabricated using holographic technology. The interference fringes generated by a laser are recorded three-dimensionally in a photosensitive material sandwiched between glass plates. The three-dimensional periodic modulation in the material causes Bragg diffraction, which occurs when $\alpha = \beta$. This enables an efficiency of >90% and produces the characteristics of the narrow wavelength range based on its thickness. Therefore, this grating is used around the Bragg condition. The anomaly that appears in a plane grating is not observed.

The Bragg condition is calculated using Eqs. 10.10 and 10.11. Angles β_1 and β_2 corresponding to $\lambda_{\min}$ and $\lambda_{\max}$, respectively, are calculated using Eqs. 10.12 and 10.13 (Fig. 10.20).

$$\alpha = \beta \qquad (10.10)$$

Fig. 10.20 Optics of a VPH grating

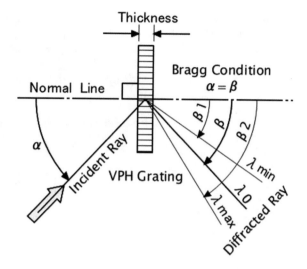

$$\lambda_0 = 2 \cdot \frac{10^6}{N} \cdot \sin \alpha \qquad (10.11)$$

$$\lambda_{min} = \frac{10^6}{N} \cdot (\sin \alpha + \sin \beta_1) \qquad (10.12)$$

$$\lambda_{max} = \frac{10^6}{N} \cdot (\sin \alpha + \sin \beta_2) \qquad (10.13)$$

Higher-order diffraction does not exist in Bragg diffraction. Therefore, a cut filter is not necessary in a VPH spectrometer to eliminate second and higher order diffraction. The VPH grating is easy to handle due to its sandwich structure. The surface of the VPH grating can be wiped or polished. The typical optical alignment of a VPH grating spectrometer is shown in Fig. 10.21.

10.2.3   Hadamard Spectrometer

In a Hadamard spectrometer, an image of the spectrum is focused onto the exit slit position of a grating spectrometer. The light passing through a multi-aperture plate deployed at the exit slit position is detected by a single detector (Fig. 10.22). This measurement is repeated more than n times for different multi-aperture plates. The signal intensity of the n signals corresponding to the different multi-aperture plates can be expressed in terms of Eq. 10.14.

Fig. 10.21 Optical mount of
a VPH grating spectrometer

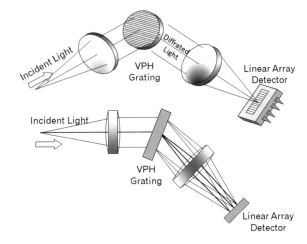

Fig. 10.22 Hadamard
spectrometer

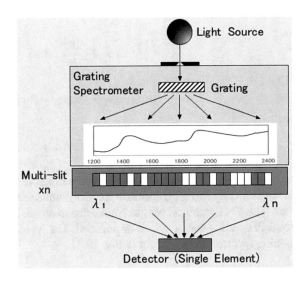

$$I_k = \sum_{i=1}^{n} P_i^k \cdot I_i \qquad (10.14)$$

where

I_k n-summed detector signal corresponding to the different multi-apertures
P_i^k Multi-aperture pattern
I_i Spectral component.

Equation 10.14 can be rewritten as Eq. 10.15. The spectral shape can be calculated from the *n* signals and *n* aperture patterns [11]. This is called the Hadamard spectrometer.

$$\begin{pmatrix} M_1 \\ \vdots \\ M_k \end{pmatrix} = \begin{pmatrix} P_1^1 & \cdots & P_1^k \\ \vdots & \ddots & \vdots \\ P_n^1 & \cdots & P_n^1 \end{pmatrix} \cdot \begin{pmatrix} I_1 \\ \vdots \\ I_k \end{pmatrix} \text{ or } M = P \times I \quad (10.15)$$

In Eq. 10.15, matrix I can be calculated using P^{-1}, which is the inverse matrix of s P (Eq. 10.16).

$$I = P^{-1} \times M \quad (10.16)$$

When noise e is present in the measurements, the signal can be expressed in terms of Eq. 10.17.

$$M = P \times I + e \quad (10.17)$$

The result of Eq. 10.15 would then be Eq. 10.18, as shown below.

$$I = P^{-1} \times (M - e) = P^{-1} \times M - P^{-1} \times e \quad (10.18)$$

Finally, the noise in the spectrum would be $P^{-1} \times e$, which can be minimized using a multi-aperture pattern. This is known as Hadamard transform spectroscopy. Although this method was proposed around 1970, it did not become popular because it was difficult to realize a dynamic n multi-aperture mechanism. However, the MEMS DLP element has been used as a multi-aperture since the advent of MEMS technology around 1990. This type of spectrometer has been available since 2015 [12]. Using the MEMS DLP element, users can design their own multi-aperture patterns for the desired applications.

10.2.4 Wavelength Resolution and Measurement Interval

(a) Wavelength Resolution of a Grating Spectrometer

A wavelength resolution of five to ten nm is sufficient for a NIR spectrometer due to the broad absorbance peak of the NIR spectra. The light intensity attained by the grating spectrometer is proportional to the square of the wavelength resolution. The lower the wavelength resolution of a NIR spectrometer, the higher is the signal intensity, which results in a good SNR. Therefore, the minimum possible low wavelength resolution should be selected. However, as explained later in Sect. 10.4, a lower wavelength resolution might cause greater instrumental differences induced by the spectral response of the spectrometer. Thus, both of SNR and instrumental difference should be considered.

As the grating spectrometer for NIR spectroscopy has a wavelength resolution of five to ten nm, the wavelength resolution of a spectrometer is decided mainly by the wavelength dispersion of the grating. Other factors such as the aberration of the optics, optical surface accuracy, and insufficient adjustment of the focal point increase the wavelength resolution of spectrometer. These factors will cause differences in the wavelength resolution even between the same spectrometer designs, and the instrumental differences will be larger depending on the different wavelength resolutions. The difference between the wavelength resolutions of the instruments should be within 10% to avoid any instrumental differences.

The wavelength resolution resulting from the dispersion caused by the entrance and exit slits can be described as follows [6, 13].

The dependence of the position deviation on the entrance slit and wavelength can be calculated by differentiating Eq. 10.3. The relationship between the wavelength and incident angle α or slit position can be expressed using Eq. 10.19. The slit wavelength width corresponding to the mechanical slit width can be expressed using Eq. 10.20.

$$\frac{d\lambda}{d\alpha} = \frac{10^6}{N} \cdot \cos\alpha \cdot \frac{d\lambda}{dw_{in}} = \frac{10^6}{N \cdot f} \cdot \cos\alpha \qquad (10.19)$$

$$\omega_{in} = \frac{10^6}{N \cdot f} \cdot \cos\alpha \cdot W_{in} \qquad (10.20)$$

ω_{in} Entrance slit wavelength width
W_{in} Mechanical entrance slit width
f Focal length of the spectrometer.

At the exit slit, the relationship between the wavelength and diffracted angle or slit position can be expressed in terms of Eq. 10.21, while the exit slit wavelength width caused by the mechanical slit width can be expressed by Eq. 10.22.

$$\frac{d\lambda}{d\beta} = \frac{10^6}{N} \cdot \cos\beta \cdot \frac{d\lambda}{dw_{exit}} = \frac{10^6 d \cdot \cos\alpha}{N \cdot f} \qquad (10.21)$$

$$\omega_{exit} = \frac{10^6 \cdot \cos\alpha}{N \cdot f} \cdot W_{exit} \qquad (10.22)$$

ω_{exit} Exit slit wavelength width
W_{exit} Exit slit mechanical width.

The shape of the total slit function of the spectrometer will be trapezoidal, with the top side and base line given by $\omega_{in} - \omega_{exit}$ and $\omega_{in} + \omega_{exit}$, respectively (Fig. 10.23b). When ω_{exit} is small, the shape in Fig. 10.23a is obtained, while a triangle shape is obtained when $\omega_{exit} = \omega_{in}$ (Fig. 10.23c).

In a wavelength scanning grating spectrometer, the $\omega_{exit} = \omega_{in}$ condition is selected to achieve the highest intensity at the same wavelength resolution.

In a linear array grating spectrometer, the pixel width is assumed to be the exit slit width and $\omega_{exit} \cdot \omega_{in}$. The slit function is illustrated in Fig. 10.23b.

Fig. 10.23 Slit function of
the spectrometer

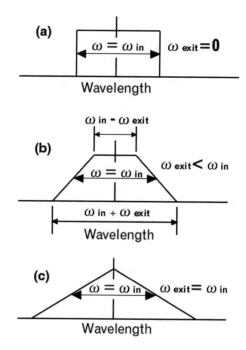

(b)   Wavelength Interval for Measurement

Selecting the right wavelength interval for a measurement is important to obtain the
right spectral shape. Thus, the spectral shape is influenced by the wavelength interval
[13].

When the wavelength interval is larger than the wavelength resolution, some spectral components between the measurement points will not be measured. When the
interval is too narrow, some spectral components will be overlapped in the measurement. The following important aspects should be considered to enable an even/equal
measurement of all the spectral components.

(b-1)   Wavelength Scanning Grating Spectrometer

The shape of the slit function of a wavelength scanning grating spectrometer is
usually triangular with a half bandwidth (HBW) of ω, as shown in Fig. 10.23.

When the wavelength interval $\Delta\lambda$ is ω (Fig. 10.23b) or $\frac{\omega}{n}$ (Eq. 10.23), all the
spectral components will be measured evenly.

$$\Delta\lambda = \frac{\omega}{n} \qquad\qquad (10.23)$$

$\Delta\lambda$ Wavelength interval
ω HBW of the slit function
n Integer 1, 2, …

When the wavelength interval $\Delta\lambda \neq \frac{\omega}{n}$ (Fig. 10.23a, c), the spectral components will not be measured evenly (Fig. 10.24). However, a large n in Eq. 10.23 will result in a smaller error.

(b-2) Linear Array Grating Spectrometer

The entrance slit wavelength width ω_{in} is always larger than the pixel width $\omega_{\text{pix}-w}$. The slit function of the signal from a pixel of the linear array in a spectrometer cannot be triangular. This is shown in Fig. 10.25, which is the same as Fig. 10.23b, where the exit slit wavelength width ω_{exit} is replaced by the pixel width wavelength $\omega_{\text{pix}-w}$.

This slit function is lined-up with the interval of the pixel interval wavelength width $\omega_{\text{pix}-i}$. When the entrance slit wavelength width ω_{in} is n times the pixel interval wavelength width $\omega_{\text{pix}-i}$ (Eq. 10.24), all the spectral components are evaluated evenly. The relationship between the mechanical size of the entrance slit W_{in} and pixel interval $W_{\text{pix}-i}$ is also described in Eq. 10.24.

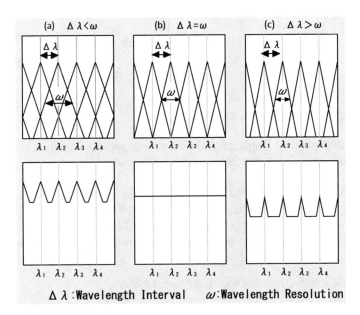

Fig. 10.24 Wavelength interval and wavelength resolution

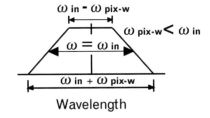

Fig. 10.25 Slit function of a pixel

$$\omega_{\text{in}} = n \cdot \omega_{\text{pix}-i} \cdot W_{\text{in}} = n \cdot \frac{\cos \beta}{\cos \alpha} \cdot W_{\text{pix}-i} \tag{10.24}$$

However, the wavelength widths of the slit and pixel vary according to the wavelength, and it is impossible to realize the condition defined in Eq. 10.24 at all wavelengths. Therefore, this condition can be realized around the center wavelength region.

(c)   Calculation of the Absorbance Data

At the wavelength λ_j, which is determined by the hardware of the spectrometer, the detector signal intensity is acquired for the white reference $W(\lambda_j)$ and sample $X(\lambda_j)$. The wavelength interval of the measurement is decided by the encoder in a wavelength scanning-type spectrometer and by a pixel in a linear array spectrometer and is not a constant round number, i.e., 1 nm or 0.5 nm.

Using the data $W(\lambda_j)$ and $X(\lambda_j)$ at the wavelength determined by the hardware, the reflectance, transmittance, or absorbance values at even wavelengths should be calculated. The process used for the calculations is described in the following subsections. The sequence of the calculations is important for avoiding the instrumental difference.

(c-1)   Calculation of the Absorbance

Before processing the data, the transmittance or reflectance $D(\lambda_j)$ should be calculated according to Eq. 10.25.

$$D(\lambda_j) = \frac{X(\lambda_j)}{W(\lambda_j)} \tag{10.25}$$

λ_j Wavelength used for the measurement
$W(\lambda_j)$ Signal intensity of the white reference
$X(\lambda_j)$ Signal intensity of the sample
$()$ Reflectance or transmittance.

(c-2)   Calculation of the Wavelength Spectra

Using the calculated reflectance or transmittance $D(\lambda_j)$, the data $S(\lambda_n)$ are calculated using Eq. 10.26 at each even wavelength, i.e., 1000, 1001, and 1002 nm.

$$S(\lambda_n) = \frac{\sum_{j=f}^{g} \left[(\Delta - |\lambda_n - \lambda_j|) \cdot D(\lambda_j) \right]}{\sum_{j=f}^{g} (\Delta \cdot D(\lambda_j))} \tag{10.26}$$

$S(\lambda_n)$ Reflectance or transmittance at any wavelength λ_n (nm)
$D(\lambda_j)$ Reflectance or transmittance at the wavelength λ_j (nm) determined by the hardware
Δ Wavelength resolution used in the calculations (nm)

j Measurement number (a smaller number corresponds to a shorter wavelength)

f Minimum measurement number that satisfies the condition $\left(\lambda_n - \lambda_f\right) < \Delta$

g Maximum measurement number that satisfies the condition $\left(\lambda_g - \lambda_n\right) < \Delta$.

Based on the above calculations, the reflectance or transmittance at any wavelength λ_n with the wavelength resolution Δ can be acquired. The actual wavelength resolution following these calculations will be a convolution of the slit function with the wavelength resolution ω, and the triangular shape with the wavelength resolution Δ. If ω is very small compared to Δ, the actual wavelength resolution will be close to Δ. If ω and Δ are almost identical, then the total wavelength resolution will be approximately $1.43 \times \Delta$.

(c-3) Preprocessing of the Spectral Data

The absorbance $A(\lambda_n)$ of the spectrum $S(\lambda_n)$ can be calculated using Eq. 10.27 if required.

$$A(\lambda_n) = -\log(S(\lambda_n)) \qquad (10.27)$$

Using the spectral data $A(\lambda_n)$ or $S(\lambda_n)$, the calibration can be established using statistics. Various data preprocessing steps, such as smoothing, can be applied prior to the statistical analysis to obtain good calibrations.

10.3 Designing a NIR Spectrometer for Special Materials

The main application of NIR spectroscopy is to measure the ingredients of a certain material, which is an easy and non-destructive method used onsite. An appropriate method of designing such a spectrometer is explained in the following sections. The performance of the desired instrument should be examined before commencing the design. It is particularly critical to assess the permissible noise levels with respect to the desired instrument.

When an instrument is designed without knowledge of the allowable noise levels for a particular application, the instrument should be repeatedly improved until satisfactory calibration is achieved. Generating calibrations repeatedly to evaluate the instrument is a waste of time and money.

The efficient design process of NIR spectroscopy with minimum waste is explained in Fig. 10.26. The process involves three steps, namely the first test measurement, the second step to determine the specification, and the third step involving the final manufacture of the instrument.

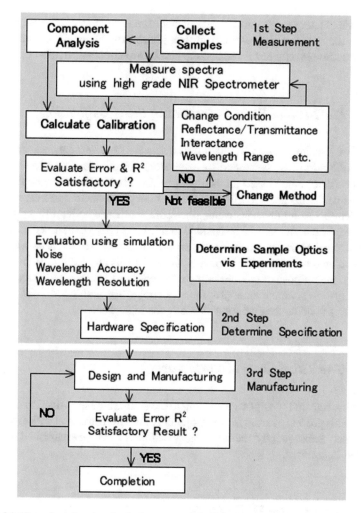

Fig. 10.26 Flow chart showing the design process of a NIR spectrometer

10.3.1   The First Step: Test Measurement

At least thirty samples, which have various ingredient values, are prepared and the spectra are measured using a high-grade laboratory-type NIR spectrometer. The true ingredient values of the samples are also analyzed using other official/established methods. The calibration is retrieved from the spectra and ingredient values using multivariate statistical analysis.

If the calibration performance is not satisfactory, the conditions (optics, wavelength range etc.) are changed and measurements are repeated until a satisfactory calibration is established.

Once satisfactory calibration is established, the second step is implemented to determine the specifications of the instrument hardware using this calibration data. The NIR method is replaced if satisfactory results are not obtained after many trials.

10.3.2   The Second Step: Determining the Specification

Noise of various amplitudes is added to the acquired spectral data via simulations and the calibrations are generated for each noise amplitude. The allowable noise level can be determined based on the relationship between the noise amplitude and calibration performance (R^2, SEP and RPD). Other performance parameters such as the wavelength resolution and wavelength accuracy can also be determined similarly.

The sampling optics should be decided in this step. Using experimental sampling optics to determine the specification, the samples are measured to check whether the optics generated the appropriate spectral shape for this application. It is important to note that the purpose of the measurement is not just to obtain the correct spectral shape but also to generate an appropriate spectral shape with good SNR, which in turn provides good calibration for estimating the ingredients.

10.3.3   The Third Step: Manufacturing

The NIR instrument is designed according to the determined specification and manufactured. This process ensures the development of a NIR instrument with satisfactory performance. Subsequently, the satisfactory calibration can be established using this developed instrument.

10.4   Instrumental Differences

10.4.1   Effect of Instrumental Differences

The main application of NIR spectroscopy is not to study the molecular structure but to measure the ingredients of a material. Therefore, many NIR instruments are being employed at various sites for manufacturing or agriculture. Even when the same type of instruments and same calibration are used, the measured ingredients differ due to a phenomenon known as the instrumental difference.

In UV-VIS or infrared spectroscopy, the main purpose of the measurement is to obtain the spectral waveform, and a small difference in the spectral shape does not present a major problem. However, in NIR spectroscopy, a small subtle difference in the spectral shape results in large differences in the predicted ingredient values, thus presenting a significant problem.

It is necessary to establish a calibration for each instrument when the instrumental difference is large. However, generation of the calibration requires substantial cost and time, which is a wasteful process. It is important to understand the factors causing the instrumental differences to find an effective solution. Instruments with small differences can be effectively designed by understanding the underlying causes.

10.4.2   Instrumental Differences Caused by the Sampling Optics

The spectral shape of the light reflected or transmitted by the sample depends on the physical configuration of the sample optics. During the manufacture of the sample optics, the angle of the optical axis and position of the optical parts should be precisely adjusted. The distance between the irradiated and observed areas should be considered to generate the identical spectra when manufacturing the interactance optics.

Furthermore, each lamp has light intensity angular characteristics due to the filament shape, and each spectrometer also has angular sensitivity dependence. Thus, when the lamp or entrance slit is collimated onto the sample, the angular characteristics of the lamp and spectrometer influence the spectral shape. The non-collimated optics and fiber optics are effective for eliminating the spectral shape difference, although the light intensity reduces.

10.4.3   Instrumental Differences Caused by the Spectral Sensitivity and Slit Function

Based on the spectral sensitivity of the spectrometer $H(\lambda)$ and slit function $\Delta(a)$, the measured signals for the sample $x(\lambda_0)$ and reference $w(\lambda_0)$ can be expressed by Eqs. 10.28 and 10.29 [14].

$$x(\lambda_0) = \frac{\int_{-D}^{+D} X(\lambda_0 - a) \cdot H(\lambda_0 - a) \cdot \Delta(a) \cdot da}{\int_{-D}^{+D} \cdot \Delta(a) \cdot da} \tag{10.28}$$

$$w(\lambda_0) = \frac{\int_{-D}^{+D} W(\lambda_0 - a) \cdot H(\lambda_0 - a) \cdot \Delta(a) \cdot da}{\int_{-D}^{+D} \Delta(a) \cdot da} \tag{10.29}$$

where

$x(\lambda_0)$ Sample signal intensity

$w(\lambda_0)$ Reference signal intensity
λ_0 Wavelength of the spectrometer
$X(\lambda_0)$ Sample spectrum
$W(\lambda_0)$ Reference spectrum
a Wavelength shift from λ_0
$H(\lambda)$ Spectral sensitivity of the spectrometer
$\Delta(a)$ Slit function of the spectrometer
D Integral width of the slit function.

Using Eqs. 10.28 and 10.29, the reflectance (or transmittance) spectrum $R(\lambda_0)$ can be calculated (Eq. 10.30) [15].

$$R(\lambda_0) = \frac{x(\lambda_0)}{w(\lambda_0)} \cdot W(\lambda_0) = \frac{\int_{-D}^{+D} X(\lambda_0 - a) \cdot H(\lambda_0 - a) \cdot \Delta(a) \cdot da}{\int_{-D}^{+D} W(\lambda_0 - a) \cdot H(\lambda_0 - a) \cdot \Delta(a) \cdot da} \cdot W(\lambda_0)$$

(10.30)

Equation 10.30 can be expressed as Eq. 10.31 when the reference spectrum $W(\lambda_0)$ is constant within the integral width D.

$$R(\lambda_0) = \frac{\int_{-D}^{+D} X(\lambda_0 - a) \cdot H(\lambda_0 - a) \cdot \Delta(a) \cdot da}{\int_{-D}^{+D} H(\lambda_0 - a) \cdot \Delta(a) \cdot da}$$

(10.31)

It is obvious from Eq. 10.31 that the reflectance $R(\lambda)$ is influenced by the spectral sensitivity $H(\lambda)$ and slit function $\Delta(a)$ [15].

Variations in $H(\lambda)$ within the integral width $\pm D$ cause instrumental differences. Variations in $H(\lambda)$ can mainly be attributed to the detector and grating. These optical elements should not exhibit steep/sharp changes within the integral width $\pm D$. The linear array has an etaloning effect that causes periodic sensitivity variations with respect to the wavelength. This effect differs depending on the manufacturing lot and is a significant factor producing of the instrumental difference.

When the variation of $H(\lambda)$ within the integral width $\pm D$ is small and considered constant, the slit function $\Delta(a)$ will be the only source affecting the reflectance spectral shape.

The slit function is a wavelength response of the spectrometer with respect to monochromatic light and is determined by the slit width and wavelength dispersion at the entrance and exit slits in the case of a grating spectrometer. The width of a slit function, which is equal to the wavelength resolution, increases due to aberrations in the optical system, mirror surface imprecision, and insufficient adjustment of the spectrometer optics. The mirror surface precision should be within $1/4 \cdot \lambda$. The adjustments should be checked by measuring the sharp spectral lines from a mercury or argon discharge lamp. The wavelength resolution of the grating-type NIR spectrometer should be adjusted to <0.1 nm of the master instrument to avoid instrumental difference.

10.4.4   *Considerations to Avoid Instrumental Differences*

The following factors should be considered to minimize the instrumental difference [15].

- The characteristics of the sample optics should be made approximately equal via adjustments.
- The wavelength resolution of each spectrometer should be adjusted within 0.1 nm.
- The spectral sensitivity of the spectrometer should be as flat as possible.
- A wide wavelength resolution generates low noise levels; however, the instrumental difference increases.
- The sequence of calculations is important (see Sect. 10.2.4c). The reflectance should be calculated first.

10.4.5   *Standardization Methods for the Calibration*

To compensate for the instrumental difference, the calibration or predicted results should be modified. The modification process is called standardization. Several standardization methods have been reported [16].

(a)   **The SBC Method**

The value predicted by a slave instrument using the calibration established by a master instrument can be corrected using the slope and bias correction (SBC) method [17]. As shown in the scatter plot in Fig. 10.27, the relationship between the predicted values from the master and slave instruments can be calculated.

The spectra of about 30 samples should be measured by the master and slave instruments. Subsequently, the same calibration is applied to both the instruments to obtain the predicted value. Using the predicted values from the master and slave

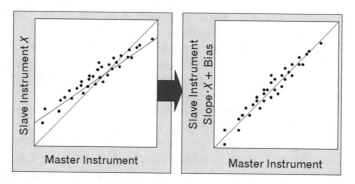

Fig. 10.27 The slope and bias correction (SBC) method

Fig. 10.28 The Shenk
method

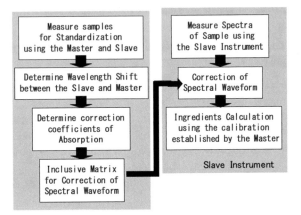

instruments, the relationship between these two instruments (slope and bias) is determined using regression analysis. These values are then applied to the value predicted by the slave instrument, as defined in Eq. 10.32.

$$A_K = \text{Slope} \cdot \sum K_{\lambda_i} \cdot I_{\lambda_i} + \text{Bias} \qquad (10.32)$$

A_k Corrected predicted value from the slave instrument
$K_{\lambda i}$ Calibration by the master instrument
$I_{\lambda i}$ Sample spectrum obtained using the slave instrument.

This method can be applied when the instrumental difference is small.

(b) Shenk Method

Shenk and Westerhaus proposed this method in a US patent [18] in 1991 (Fig. 10.28). Though this method is old, it is nevertheless important.

The principle idea is to identify a function that modifies the spectra from the slave instrument to match that of the master instrument. Subsequently, the calibration of the master instrument can be used for the slave instrument.

A minimum of 30 samples are measured using both the master and slave instruments. It is desirable that the samples cover all the features of the target.

Based on the correlation between the absorbance at λ_i of the master and λ_j of the slave, the relationship between λ_i of the master and λ_j of the slave can be acquired. Using these results, the wavelength of the slave instrument can be corrected.

Based on the absorbance values of the master and slave instruments at each corrected wavelength λ_j, the correction factors for the slope and bias corresponding to the absorbance value at each wavelength can be retrieved. Using the correction matrix that includes all the corrections for the wavelength and absorbance, the spectra of the slave instrument match those of the master. Subsequently, the ingredients can be calculated using the slave spectra and calibration established by the master.

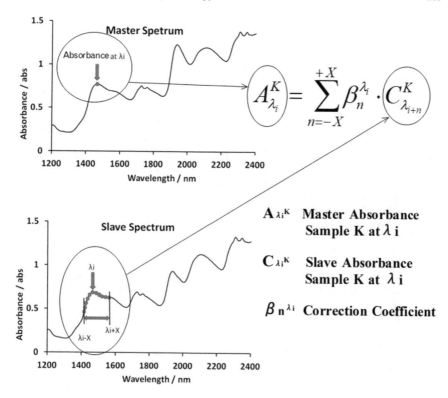

Fig. 10.29 The PDS method

(c) **PDS Method**

The piecewise direct standardization (PDS) method is similar to the Shenk method. In this method, the wavelength and absorbance are corrected using multivariate analysis [19] (Fig. 10.29).

Using the absorbance values of K samples from the master and slave instruments, the $A_{\lambda_i}^K$ of the master at wavelength λ_i can be calculated based on the slave absorbance values ($C_{\lambda_i}^K$) in the wavelength band λ_{i-X} to λ_{i+X} using the coefficient ($\beta_n^{\lambda_i}$). The coefficient ($\beta_n^{\lambda_i}$) can be calculated from the MLR, PCA, or PLS.

The direct standardization (DS) method involves using all the wavelength data of the slave instrument.

In the calculation of the above corrections, it is important to ensure the repeatability of the measurements for this standardization.

References

1. I. Ben-Gera, K.H. Norris, Direct spectrophotometric determination of fat and moisture in meat products. J. Food Sci. **33**(1), 64–67 (1968)
2. T. Davis, Happy 90th birthday to Karl Norris, Father of NIR technology. NIR News **22**(4), 3–16 (2011)
3. K.A. Sudduth, J.W. Hummel, Near-infrared spectrophotometry for soil property sensing, in *Proceedings of the SPIE Conference on Optics in Agriculture and Forestry*, vol. 1836 (Albuquerque, America, 1993), pp. 14–25
4. F.N. Hooge, 1/f noise sources. IEEE Trans. Electron Devices **41**(11), 1926–1935 (1994)
5. W. Schottky, Über spontane Stromschwankungen in verschiedenen Elektrizitätsleitern. Annalen der Physik (in German) **57**(23), 541–567 (1918)
6. C. Palmer, *Diffraction Grating Handbook*, 7th edn. (Newport Corporation, 2014)
7. K. Kudo, Optical properties of plane-grating monochromator. J.O.S.A. **55**(2), 150–161 (1965)
8. C.M. Penchina, Reduction of stray light in in-plane grating spectrometers. Appl. Opt. **6**(6), 1029–1031 (1967)
9. T. Namioka, Theory of concave grating III Seya-Namioka Monnochromator. J.O.S.A. **49**(10), 951–961 (1959)
10. S.C. Barden, J.A. Arns, W.S. Colburn, J.B. Williams, Volume-phase holographic gratings and the efficiency of three simple volume-phase holographic gratings. Astron. Soc. Pac. **112**, 809–820 (2000)
11. J.A. Decker, M. Harwit, Experimental operation of a Hadamard spectrometer. Appl. Opt. **8**(12), 2552–2554 (1969)
12. J. Xu, H. Liu, C. Lin, Q. Sun, SNR analysis and Hadamard mask modification of DMD Hadamard Transform Near-Infrared spectrometer. Opt. Commun. **383**(15), 250–254 (2017)
13. C.L. Sanders, F. Rotter, The spectroradiometric measurement of light sources. CIE 063 (1984). ISBN: 978-963-7251-23-8
14. K.S. Seshadri, R.N. Jones, The shapes and intensities of infrared absorption bands—a review. Spectrochim. Acta **10**, 1013–1085 (1963)
15. T. Okura, S. Piao, S. Kawano, Difference of predicted values by near-infrared spectrometers caused by wavelength resolution. J. Light Visual Environ. **38**, 29–36 (2014)
16. E. Bouveresse, B. Campbell, Transfer of multivariate calibration models based on near-infrared spectroscopy, in *Handbook of NIR Analysis*, 3rd edn, ed. by D. Burns, E. Ciurczak (CRC Press, FL33487, U.S.A, 2007), pp. 231–243
17. B.G. Osborne, T.J. Fearn, Collaborative evaluation of universal calibrations for the measurement of protein and moisture in flour by near infrared reflectance. Int. J. Food Sci. Technol. **18**, 453–460 (1983)
18. J.S. Shenk, M.O. Westerhaus, Optical instrument calibration system. U.S. Patent No. 4866644A, (1991) Sept. 12
19. E. Bouveresse, D.L. Massart, Improvement of the piecewise direct standardization procedure for the transfer of NIR spectra for multivariate calibration. Chemometr. Intell. Lab. Syst. **32**, 201–213 (1996)

Chapter 11
Time-of-Flight Spectroscopy

Tetsuya Inagaki and Satoru Tsuchikawa

Abstract This chapter summarizes the principle and application of time-of-flight (TOF) NIR spectroscopy, which can evaluate the contribution of scattering and absorption of light in samples simultaneously. In order to construct robust calibrations for organic materials by NIR spectroscopy, it is important to evaluate and understand the spectral contribution from light absorption (absorption resulting from harmonics or overtones of the fundamental absorptions of molecular vibrations) and light scattering (mainly due to the cellular structure). In this chapter, we introduce the principle of TOF-NIR spectroscopy and some applications to agricultural, medical area, and forest products.

Keywords Time-resolved spectroscopy · Time-of-flight spectroscopy · Spatially resolved spectroscopy · Absorption coefficient · Reduced scattering coefficient

11.1 Introduction

In the past three decades, many researchers have paid attention to the potential use of NIR spectroscopy as a practical use for the detection of organic compounds in materials. In fields such as agriculture, food, pharmaceuticals, medical, paper, and polymers, there is a strong interest in NIR spectroscopy because of its nondestructiveness, accuracy, quick measurement, and easy operation. The measurement of NIR spectra (i.e., detection of NIR light from the sample) is done by transmittance (including interactance) or diffuse reflectance mode. The transmission method is desirable for detecting internal information in large quantities of material, although the optical information from the diffuse reflectance spectrum is limited to the surface of the sample. It is particularly important to proceed with the development of NIR transmission devices for detecting the internal properties of high moisture fruits, vegetable products, or thick wood products.

T. Inagaki (✉) · S. Tsuchikawa
Graduate School of Bioagricultural Sciences, Nagoya University, Nagoya, Japan
e-mail: inatetsu@agr.nagoya-u.ac.jp

© Springer Nature Singapore Pte Ltd. 2021
Y. Ozaki et al. (eds.), *Near-Infrared Spectroscopy*,
https://doi.org/10.1007/978-981-15-8648-4_11

Behavior of transmitted or diffuse reflected light from an agricultural, forest product or human body (i.e., highly scattering media) is strongly affected by both physical and chemical properties of the tissues, making it complicate to examine the optical characteristics of the tissue in detail and to evaluate the sample constituents accurately. Especially in the 500–1100 nm wavelength range for most biological media, the scattering coefficient is much higher than the absorption coefficient. Although many studies have reported that the chemical, physical, and mechanical properties of biological material can be predicted by NIR diffuse reflectance/transmittance spectroscopy with the aid of statistical methods (i.e., chemometrics), such chemometric NIR approaches have some disadvantages. First, the contribution of the light absorption and scattering phenomena in acquired spectra cannot be explained independently. Second, the construction of a calibration model, which is usually not transferable among instruments, requires a considerable amount of spectral and objective data. Additionally, the light scattering contribution to NIR spectra is significant when the material has complex cellular structure result in the high scattering of light. In order to construct robust calibrations for organic materials by NIR spectroscopy, it is of importance to independently evaluate the spectral contribution from light absorption (absorption resulting from harmonics or overtones of the fundamental absorptions of molecular vibrations) and light scattering (mainly due to the cellular structure and refractive index mismatch at the boundary).

In order to understand such a complex phenomena of light propagation in organic materials, many researchers have given attention to time-of-flight (TOF) or time-resolved (TR) spectroscopy using short pulses of light emission and observe the reflected or transmitted light as a function of time in nano or pico order. A time-resolved measurement, or time domain system, could provide the TOF information of the detected light. In the TOF approach, tens of picosecond light pulses are usually injected into the tissue, usually using a suitable optical fiber.

The intensity of light pulse propagates through the tissue is detected at a certain distance from the injection point (Fig. 11.1). It is also possible to examine biological tissue using the transmission approach. In this approach, the source and detector fibers are placed on opposite sides of the tissue. Time domain intensity of photons propagated into the tissue, known as the photon distribution of time of flight (DTOF), results delayed, broadened, and attenuated because of the scattering and absorption of light inside the diffusive medium. Although it is possible to estimate the optical properties of materials by spatially resolved technique (SR) or spatial frequency domain technique, it is considered that TR technique is more accurate in the measuring of optical properties.

Patterson et al. [1] proposed the usefulness of the time-resolved reflectance and transmittance spectroscopy for the noninvasive measurement of tissue optical properties theoretically. They developed a model based on the diffusion approximation of radiative transfer, which yielded an analytical expression for pulse shape in terms of the interaction with a homogeneous slab, for the determination of optical properties [i.e., absorption coefficient (μ_a) and reduced scattering coefficient (μ_s')] in tissue. μ_s' is defined as

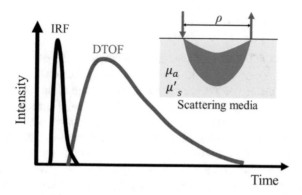

Fig. 11.1 Summary of TR NIR spectroscopy. The injected pulse (IRF: instruments response function) propagate into scattering media with absorption coefficient and reduced scattering coefficient. The detected light intensity at certain distance from injected point (ρ) with time domain broadens due to light scattering (DTOF: distribution of time-of-flight)

$$\mu'_s = (1 - g)\mu_s$$

where μ_s is the linear scattering coefficient, and g is the mean cosine of the scattering angle. A value of $g = 1$ represents forward scattering, while $g = 0$ represents isotropic scattering. Under the assumption that $\mu_a \ll \mu'_s$ (i.e., high scattering media), the diffusion of a photon can be considered to be in a random walk of step size $1/\mu'_s$, where each step involves isotropic scattering. Patterson et al. solve the diffusion equation using Green's function with two assumptions; 1. All the incident photon are initially scattered at the depth of $z_0 = 1/\mu'_s$ and 2. diffuse photon rate at the physical boundary between tissue and non-scattering medium would be 0. They successfully express the reflectance and transmittance ratio with the function of distance from light source and time. The usefulness of the function they reported has been proven by many researches. However, Leonardi and Burns [2] investigated quantitative measurements in scattering media on the basis of TOF spectroscopy with analytical descriptors. They found that experimental analysis from time-resolved profiles is efficient in estimating absorption and scattering coefficients. Numerical methods such as adding-doubling and Monte Carlo (MC) methods simulating light propagation in biological tissues are also often used.

Many researches revealed that the determination of μ_a and μ'_s in agricultural and food products can be used for the evaluation of chemical and physical properties in samples [3–5]. In the field of medical science, time domain method is expected to develop optical tomography techniques, which can be used for the noninvasive detection of cancer [6] or the changes of hemoglobin concentration associated with neural activation in human brain [7]. In this chapter, we introduce the principle of TOF-NIR spectroscopy and some applications to agricultural, medical area and forest products.

11.2 Measuring Apparatus

For the detection of DTOF data of samples, measuring system should contain 1. laser source, 2. photodetector, and 3. time-resolved system. Torricelli et al. summarized [4] the evolution history of TR spectroscopic components, i.e., in the first generation since the early 1990s, the laser from gas, dye, or solid state laser was used as light source and detected DTOF profiles using microchannel plate photomultiplier (PMT) with electronic chain for time-correlated single photon counting (TCSPC) with NIM module. The TR spectroscopic components at the second generation, 2000–2010, semiconductor laser heads with external RF driver was used as light source and detected by compact metal channel dynode PMT with TCSPC electronic board system. Now it is possible to use the supercontinuum fiber laser with powerful emission with broad wavelength range as light source and detected by hybrid PMT with time-to-digital converters module with USB controller. The distance from light injection points to receiver (ρ in Fig. 11.1) should be optimized according to the range of μ_a and μ_s' of the samples. Estimated value of attenuation is in the order of 10^6–10^8, and temporal dynamic span is over a 1–10 nm range when the sample has general optical properties in the NIR region, i.e., $\mu_a = 0 - 0.05$ cm^{-1}, $\mu_s' = 5 - 25$ cm^{-1} and $\rho = 1 - 3$ cm.

11.3 Data Analysis

After obtaining the time-resolved light intensity signal, μ_a and μ_s' are estimated by fitting the TR data obtained with the analytical solution of the diffusion equation by the nonlinear inverse algorithm. Convolution between the theoretical TR reflectance with the IRF is calculated at first in order to take the broad shape of the IRF and then used to fit the experimental TR reflectance curve. As mentioned in introduction, solution of diffusion equation can be applied only if the scattering is dominant compared to absorption in media, and radiance is detected at a sufficient larger distance from the injection point, i.e., distance should be much longer than one mean free path $1/\mu'$. MC simulation is also used to model the light propagation in biological tissue because of its flexibility and simplicity to simulate photon propagation processes in arbitrary shapes with complex boundary conditions or spatial localization. For that simulation, the photon propagation in the turbid medium is traced until it exits the sample surface or is absorbed. The movement of photons from one photon tissue interaction to the next photon tissue interaction is described by a probability function using the optical properties of the tissue. Repeat these processes for a large number of photons to estimate the photon distribution in the tissue. Detailed explanation for the data analysis with fitting or MC simulation can be found in book [8].

11.4   Application of TOF-NIRS to Agricultural Science

The important parameters for fruits or vegetables (i.e., maturity, quality parameters, and defect) can be detected by measuring the optical parameters as Lu et al. reviewed [5]. The couple of most important quality attributes to different fruits are firmness and soluble solids content (SSC). Many studies demonstrate that both μ_a and μ'_s have relation to hardness, SSC, and skin color. However, it was generally reported that μ_a was suitable for predicting these quality parameters. This may be due to the fact that pigments or other chemical compositions would change during maturation with changes of cell structure directly affecting hardness. In most of the cases, the combination of μ_a and μ'_s was found to improve the prediction of fruit maturity and quality parameters. The prediction of firmness and SSC by TR technology was widely reported for apple, kiwifruit, mango, nectarine, peach, and pear. Because physiological disorders often cause the changes in the chemical and structural properties of the fruits product result in the change of μ_a and μ'_s, it is possible to detect the defect in the fruits by observing the optical properties in fruits. Especially, TR technology can be used to detect internal browning and internal bleeding of apples, nectarines, plums, thanks to its ability to penetrate tissue deeper inside the fruits. Determined bulked absorption coefficients of fruits in the spectral regions of 500–1850 nm were largely dominated by the water in the NIR range and fruit-specific pigments in the visible range. The differences in μ'_s behavior (μ'_s decrease exponentially with the increase of wavelength) between fruits, cultivars and tissue type are related to microstructural differences, such as differences in cellular structure and porosity. μ'_s values are reported for various fruits ranging from 0 to 20 cm^{-1}.

11.5   Application of TOF-NIRS to Medical Science

One of the research areas where TR investigation is most actively conducted is the medical science. Diffuse optical tomography in NIR region at the range of wavelength from 700 to 1000 nm is proven to have the potential for noninvasive diagnoses of tissue oxygenation and thyroid cancers. As a first step toward properly designing devices, interpreting diagnostic measurements or planning therapeutic is to identify the accurate optical properties of a tissue. Following research might be the use of optical properties determined to describe the light transportation and absorption. Jacques [9] summarized 1. μ_a of various tissues in terms of the average hemoglobin concentration or some similar properties and 2. μ'_s with the parameters (a, b), or alternatively $(a', f_{\text{Rayleigh}}, b_{\text{Mie}})$ which explain the μ'_s variation with wavelength change. The $\mu'_s(\lambda)$ were expressed by the equation $\mu'_s = a\left(\frac{\lambda}{500(\text{nm})}\right)^{-b}$ or $\mu'_s =$
$a'\left(f_{\text{Rayleigh}}\left(\frac{\lambda}{500(\text{nm})}\right)^{-4} + \left(1 - f_{\text{Rayleigh}}\right)\left(\frac{\lambda}{500(\text{nm})}\right)^{-b_{\text{Mie}}}\right)$ where the λ is wavelength.
In the later equation, scattering is described in terms of the separate contribution

of Rayleigh and Mie scattering. Author explained that the equations are good for use in predicting behavior of light propagation or diffusion within the 400–1300 nm wavelength range. Author summarized also the mean values of coefficient a and b (skin: $a = 46.0$ cm^{-1}, $b = 1.421$, brain: $a = 24.2$ cm^{-1}, $b = 1.611$, breast: $a = 16.8$ cm^{-1}, $b = 1.055$, bone: $a = 22.9$ cm^{-1}, $b = 0.716$, other soft tissues: $a = 18.9$ cm^{-1}, $b = 1.286$, other fibrous tissues: $a = 27.1$ cm^{-1}, $b = 1.627$, fatty tissue: $a = 18.4$ cm^{-1}, $b = 0.672$). Fujii et al. [6] investigated the effects of three factors (trachea, refractive index mismatch at the boundary of trachea tissue, and neck organs other than the trachea [spine, spinal cord, and blood vessels]) on light propagation in the neck by 2D time-dependent radiative transfer equation. After they constructed an anatomical model of human neck from MR image, they performed segmentation of the MR image and recognized the pixel corresponding to organs of the human neck: the trachea, spine, spinal cord, and blood vessels. They simulated the light propagation in anatomical human neck models by numerical method and MC simulation. They showed that reflection and refraction at the trachea tissue interface significantly effect on the light intensities in the region between the trachea and the front of the neck surface. So, it is necessary to take into account the refractive index mismatch at the trachea tissue interface. Hoshi summarized the use of TR system for clinical monitoring of tissue oxygenation [7].

11.6 Application of TOF-NIRS to Forest Products

As many reviews and manuscripts about application of TR spectroscopy for medical and food product science are published, TR spectroscopic application for these research area is briefly explained in previous chapter. In present chapter, the use of TR spectroscopy for the determination of optical properties in wood is explained in detail.

Wood is a natural material widely used in construction because of its versatility and strength. As wood is a biomaterial, there are significant variations in wood properties (e.g., density, moisture content, grain angle) between species and even among the same species. From the point of view of quality assurance in industry, nondestructive measuring and control of the mechanical, physical, and chemical properties of wood are strongly desired. The light scattering in wood is especially complex because of the complex cellular structure in wood. Softwood mainly possesses a tracheid structure, arrayed along the longitudinal direction; whereas, hardwood structures have wide variation of cell structure (e.g., tracheids, vessels, libriform wood fibers, or ray cells). The optical properties of wood are significantly affected also by the water retained in cell walls or cell lumens.

Some groups reported the use of TR diffuse reflectance spectroscopy to determine the optical properties of wood. D'Andrea et al. [10] decided μ_a and μ'_s in the wavelength range of 700–1040 nm of two wood species treated in different conditions (dry wood, wet wood, and degraded wood) by TR spectroscopy with two orientations of the optical fiber (i.e., the emitted and detected fibers are set perpendicular or parallel to wood grain orientation) and obtained many interesting results. They reported that the μ_s' (10–200 cm^{-1}) was much larger than the μ_a (0.05–1.00 cm^{-1}) for all wood samples. μ_s' spectra were almost constant over the measured wavelength ranges. It was also found that μ_s' highly depends on the wood species (μ_s' value differs between silver fir and sweet chestnut wood greatly). μ_s' of wet wood was significantly small compared to dried wood because the refractive index mismatch between the wood cell wall substance and water in the pores is much smaller than that between wood cell and air. D'Andrea et al. also evaluated the moisture content of wood using the μ_a and found a high relationship between moisture content and the μ_a at a specific wavelength [11]. Kienle et al. investigated the origin of scattering in wood by comparing the light propagation in the microstructure of silver fir measured experimentally to simulation modeled by MC method [12]. They determined μ_s' (wet wood: 1.79 mm^{-1}, dried wood: 6.68 mm^{-1}) due to tracheids by solving Maxwell's equation. They also determined μ_{s-iso}', which is the scattering coefficient due to all other scattering media (rough border between the lumen and wood cell substance, pits, ray cells), calculated by fitting measured light propagation to simulated data. The light scattering in wood is significantly complex as wood is a hygroscopic, heterogeneous, cellular, and anisotropic material. Although the wood samples were regarded as a homogenous material when the μ_s' values were estimated using TR spectroscopic method, in fact, the scattering properties highly depended on the wood species and fiber direction because the cellular structure, which caused multiple light reflections at the boundary between cell wall and air (water), significantly differs between wood species (i.e., hardwood has various cell arrangements like ring-porous, diffuse-porous, radial-porous, and figured-porous). Kitamura et al. tried to determine true μ_a and μ_s' values of wood cell wall substance itself in order to construct the robust calibrations wood properties by NIR spectroscopy [13]. They expected that the μ_a and μ_s' values of the cell wall substance are identical or similar between species because the density of wood cell wall itself is about 1.4–1.5 g cm^{-3} regardless of species (wood density depends on the ratio of pore and cell wall volume in wood). As the density of cell wall is identical between species, it is thought that the factor affecting the optical properties might be the concentration ratio of the three main polymers in the cell wall (cellulose, hemicellulose, and lignin). As there was no specific absorption band at the wavelength used in their study (846 nm) as shown by Hans et al. [14], it implied that the concentration ratio of the cellulose, hemicellulose, and lignin does not strongly affect μ_a. In order to decide the true optical parameters of wood cell wall, Douglas fir wood samples were immersed in hexane, toluene, or quinolone and saturated with them to minimize the multiple light reflections at the boundary between pore cell wall substance in wood. TR transmittance result of organic liquid saturated wood samples was fitted to the diffusion approximation equation to decide μ_a and μ_s'. μ_s' showed the minimum value when the wood was saturated with toluene because the refractive index of toluene is close to wood cell wall substance. In the toluene saturated wood sample, Fresnel reflection (1) at small particle, (2) between lumen and wood cell wall substance for all cell types (tracheid, ray cells, vessel), and (3) at the rough border are minimized because refractive index mismatch between toluene and cell wall was very small. The optical parameters of wood cell wall substance calculated taking into account the volume fraction of wood cell wall substance were $\mu_a = 0.030$ mm^{-1} and $\mu_s' = 18.4$ mm^{-1}. Konagaya et al. [15] fully investigated the effect of boundary between cell wall substance and air or water (refractive index mismatch) on the μ_s'. They investigated optical properties of drying wood with the moisture contents ranging from 10 to 200% by TR spectroscopy. They divided the source of light scattering into two factors, 1. scattering from large scatters (i.e., scattering diameter is much larger than the wavelength λ of the light), which can be described by geometric optics, expressing light propagation in terms of rays, and 2. scattering from small scatters (i.e., when the scattering diameter is the same as or less than λ), where Mie theory ($\approx\lambda$) or Rayleigh theory ($<\lambda$) is applicable. They revealed the contribution of scattering source at each stage of wood drying (constant rate, initial part of the first decreasing rate, later part, and second decreasing rate period). Scattering from dry pores dominated during the constant drying rate period, and the drying process of smaller pores dominated during the period of decreasing drying rate. The surface layer and interior of the wood exhibit different moisture states, which affect the scattering properties of the wood. The light propagation in wood complex cell structure is simulated by MC method taking into account the light reflection and transmission at the boundary between wood cell wall substance and pore by Ban et al. [16]. They investigated the relation of wood texture parameters calculated from cross-sectional microscopic images of the 13 species of wood samples and μ_s' at 846 nm. They found that μ_s' has linear relation to the air-dry density ($R^2 = 0.56$), quadratical relation to the cell–wall area ratio ($R^2 = 0.76$), and exponentially relation to the median pore area ($R^2 = 0.54$). 85 percent of the variation in μ_s' between many wood species can be explained by these three parameters. They simulated the light propagation in wood using the measured cross-sectional microscopic image of wood. After they performed segmentation of the microscopic image and recognized the pixel corresponding to cell wall substance and air area, the simulations were performed in the MC code, MCVM. The refractive index mismatch at the boundaries is also considered to improve the precision of simulations in MCVM code. Figure 11.2 shows simulated photon propagation in (a) agathis, (b) yellow poplar, and (c) rubber wood. It is observed that photon spreads farther in wood cell wall woods through continuously connected cell walls. The high correlation of cell–wall area ratio and median pore area on μ_s' can be attributed to the thicker, more connected cell walls associated with large cell–wall area ratio and small median pore area. Accordingly, increasing the area ratio of the cell wall and decreasing the pore area increased the μ_s'.

11.7 Brief Explanation for SR Spectroscopy

Not only the TR spectroscopy, some techniques are used to determine the optical properties. SR technique was developed to understand light propagation in turbid media. Compared to TR spectroscopy, SR technique is well suitable for use in post-harvest applications thanks to its low instrumentation cost, easy implementation.

(a) Agathis	(b) Yellow poplar	(c) Rubber tree
μ': 10.4 mm^{-1}	μ': 14.2 mm^{-1}	μ': 30.9 mm^{-1}
ρ_{airdry} : 0.328 g/cm^3	ρ_{airdry} : 0.508 g/cm^3	ρ_{airdry} : 0.704 g/cm^3
r: 0.551	r: 0.721	r: 0.846
S: 145.0 µm^2	S: 13.1 µm^2	S: 8.3 µm^2

μ' : reduced scattering coefficient ρ_{airdry} : air dry density
r :Areal ratio of cell wall   S : Median value of pore area

Fig. 11.2 Light propagation in **a** agathis, **b** yellow poplar, and **c** rubber wood simulated by the Monte Carlo method

For the SR technique, the point light source is generally injected on sample and the spatial distribution surrounding the injected light is detected by optical fiber arrays or non-contact reflectance image, which can be implemented with fiber optic probe, monochromatic imaging, and hyperspectral imaging. Measured spatial distributions were fitted by the analytical equation derived by Farrell et al. [17] with an appropriate inverse algorithm.

11.8 New Measurement System Minimizing the Effect of Light Scattering.

It can be extremely difficult to obtain reliable calibrations with highly scattering medium by conventional NIR spectrometer if both of the scattering and absorption properties of the samples vary. For example, variations of light scattering with size or concentration of rubber particles in latex samples are complicate. The variation of scattering properties in sample makes it difficult to interpret the quantitative analyses and to construct linear calibrations (multiple linear regression, principal component regression, and partial least squares regression). Shimomura et al. established three-fiber-based diffuse reflectance spectroscopy (TFDRS) based on spatially resolved spectroscopy to estimate the sugar content in fruits or total hemoglobin concentration and tissue oxygen saturation in human tissues using wavelengths range of 800 nm to 1100 nm [18–20]. A geometry of the TFDR spectrometer is shown in Fig. 11.3. A continuous wave laser beam is emitted into the sample through an optical fiber.

Fig. 11.3 Geometry of
TFDRS

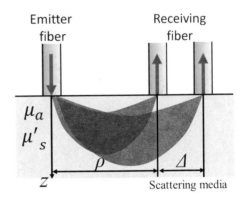

The spatially resolved light is collected by fibers that guide the light to the detector. The receiving fibers are aligned parallel to the distance of ρ and $\rho + \Delta$ apart from the emitting fiber. Shimomura et al. showed that a new physical parameter γ, which is calculated using the ratio of light intensity detected at distances ρ and $\rho + \Delta$, is independent of the optical path length and showed good linear relation to the analyte material concentration. One of the advantages of this method is that it is not necessary to measure reference signal. Shimomura et al. further developed a new system using three laser diodes at wavelengths of 911, 936, and 1055 nm, which were found as the best combination of wavelength to predict the sugar content of fruit. The sugar content in apples in the range of about 9–15 Brix was greatly predicted with high accuracy (Brix is an approximation of dissolved solid content in samples, representing the relation to a solution as a percentage of mass). They also constructed a small and cheap handheld commercial device employing NIR-LED and Si detector. Inagaki et al. showed this kind of technique using the wavelength range 850–1060 nm, which is a good method to decide the quality in highly scattering media, natural rubber latex samples [21]. They showed parameter γ has a strong linear relation to total solid content in latex (range 0.3–0.6 g g^{-1}) with a coefficient of determination value of 0.98 and root mean square error for total solid content of 0.014 g g^{-1}. Although the NIR spectra measured by conventional transmission or reflectance spectroscopy were highly affected by the scattering coefficient in the sample, simulation results in that study showed that the effects of scattering in the samples on γ can be reduced.

References

1. M.S. Patterson, B. Chance, B.C. Wilson, Time resolved reflectance and transmittance for the noninvasive measurement of tissue optical properties. Appl. Opt. **28**(12), 2331–2336 (1989)
2. L. Leonardi, D.H. Burns, Quantitative measurements in scattering media: photon time-of-flight analysis with analytical descriptors. Appl. Spectro. **53**(6), 628–636 (1999)

3. O.H.A. Nielsen, A.A. Subash, F.D. Nielsen, A.B. Dahl, J.L. Skytte, S. Andersson-Engels, D. Khoptyar, Spectral characterisation of dairy products using photon time-of-flight spectroscopy. J. Near Infrared Spec. **21**(5), 375–383 (2013)
4. A. Torricelli, D. Contini, A.D. Mora, E. Martinenghi, D. Tamborini, F. Villa, A. Tosi, L. Spinelli, Recent advances in time-resolved NIR spectroscopy for nondestructive assessment of fruit quality. Chem. Eng. Trans. **44**, 43–48 (2015)
5. R. Lu, R.V. Beers, W. Saeys, C. Li, H. Cen, Measurement of optical properties of fruits and vegetables: a review. Postharvest Biol. Tec. **159**, 111003 (2020)
6. H. Fujii, Y. Yamada, K. Kobayashi, M. Watanabe, Y. Hoshi, Modeling of light propagation in the human neck for diagnoses of thyroid cancers by diffuse optical tomography. Int. J. Numer. Meth. Bio. **33**(5), 1–12 (2017)
7. Y. Hoshi, Hemodynamic signals in fNIRS. Prog. Brain Res. **225**, 153–179 (2016)
8. F. Martelli, S.D. Bianco, A. Ismaelli, D. Zaccanti, Light propagation through biological tissue and other diffusive media.Theory, solutions, softw. (2009)
9. S.L. Jacques, Optical properties of biological tissues: a review. Phys. Med. Biol. **58**(11), R37–R61 (2013)
10. C. D'Andrea, A. Farina, D. Comelli, A. Pifferi, P. Taroni, G. Valentini, R. Cubeddu, L. Zoia, M. Orlandi, A. Kienle, Time-resolved optical spectroscopy of wood. Appl. Spectro. **62**(5), 569–574 (2008)
11. C. D'Andrea, A. Nevin, A. Farina, A. Bassi, R. Cubeddu, Assessment of variations in moisture content of wood using time-resolved diffuse optical spectroscopy. Appl. Opt. **48**(4), 87–93 (2009)
12. A. Kienle, C. D'Andrea, F. Foschum, P. Taroni, A. Pifferi, Light propagation in dry and wet softwood. Opt. Express **16**(13), 9895–9906 (2008)
13. R. Kitamura, T. Inagaki, S. Tsuchikawa, Determination of true optical absorption and scattering coefficient of wooden cell wall substance by time-of-flight near infrared spectroscopy. Opt. Express **24**(4), 3999–4009 (2016)
14. G. Hans, R. Kitamura, T. Inagaki, B. Leblon, S. Tsuchikawa, Assessment of variations in air-dry wood density using time-of-flight near-infrared spectroscopy. Wood Mater. Sci. Eng. **10**(1), 57–68 (2015)
15. K. Konagaya, T. Inagaki, R. Kitamura, S. Tsuchikawa, Optical properties of drying wood studied by time-resolved near-infrared spectroscopy. Opt. Express **24**(9), 9561–9573 (2016)
16. M. Ban, T. Inagaki, T. Ma, S. Tsuchikawa, Effect of cellular structure on the optical properties of wood. J. Near Infrared Spec. **26**(1), 53–60 (2018)
17. T.J. Farrell, M.S. Patterson, B. Wilson, A diffusion theory model of spatially resolved, steady-state diffuse reflectance for the noninvasive determination of tissue optical properties in vivo. Med. Phys. **19**(4), 879–888 (1992)
18. Y. Shimomura, S. Miki, T. Tajiri, H. Tanaka, Noninvasive measurement of absolute hemodynamic components in human tissue using three-fiber-based diffuse reflectance spectroscopy. in *2009 IEEE LEOS Annual Meeting Conference Proceedings* (2009), pp. 274–275
19. Y. Shimomura, T. Okada, Development of nondestructive measurement technique for fruits sugar content with near-infrared laser diodes operating at three different wavelengths. Rev. Laser Eng. **33**(9), 620–625 (2005)
20. Y. Shimomura, T. Takami, Y. Ichimaru, K. Matsuo, R. Hyodo, New measurement technique that uses three near infrared diode lasers for nondestructive evaluation of sugar content in fruits. Proc. SPIE 5739, Light-Emitting Diodes: Res. Manuf. Appl. IX **5739**, 145 (2005)
21. T. Inagaki, D. Nozawa, Y. Shimomura, S. Tsuchikawa, Three-fibre-based diffuse reflectance spectroscopy for estimation of total solid content in natural rubber latex. J. Near Infrared Spec. **24**(4), 327–335 (2016)

Chapter 12
Method Development

Benoît Igne⬤, Gary McGeorge⬤, and Zhenqi Shi⬤

Abstract A general framework for method development based on the analytical quality by design process is presented and applied to the development of near-infrared spectroscopic methods. The framework is particularly well suited to secure stakeholder alignment, setting appropriate expectations and ensuring that resources are spent appropriately. After setting method goals and expectations and confirming feasibility, a risk assessment is performed to identify all the factors that could affect the method. The method is then developed with the intention to mitigate the impact of those risks. The result is a robust method that can be tested and validated if required by the regulatory environment of use. Aspects of method lifecycle are also discussed as method development is only a part of the process of successfully using near-infrared spectroscopic methods in routine commercial applications. Aspects of interface to the process, sample set selection, model optimization, system suitability, and performance monitoring are discussed in the context of building robust methods. The analytical quality by design framework can significantly streamline method development and lifecycle management efforts to ensure a successful deployment and long-term value generation from a NIR spectroscopic method. Continuous improvement ensures method performance over the useful life of the method.

Keywords Near-infrared spectroscopy · Analytical quality by design · Method development · Method lifecycle

B. Igne (✉)
Vertex Pharmaceuticals Inc, 50 Northern Avenue, Boston, MA 02210, USA
e-mail: Benoit_Igne@vrtx.com

G. McGeorge
Bristol Myers Squibb, 1 Squibb Drive, New Brunswick, NJ 08901, USA
e-mail: gary.mcgeorge@bms.com

Z. Shi
Eli Lilly and Company, Indianapolis, IN 46285, USA
e-mail: shi_zhenqi@lilly.com

© Springer Nature Singapore Pte Ltd. 2021
Y. Ozaki et al. (eds.), *Near-Infrared Spectroscopy*,
https://doi.org/10.1007/978-981-15-8648-4_12

12.1 Introduction

An analytical method is a collection of documents and procedures describing how an analytical signal is collected and processed, how information is generated from the signal, and how it is reported. It is not only the chemometrics model that takes spectra as inputs to provide predictions or classifications, but also a description of how the samples are measured (frequency, instrument configuration, and operation), how the instrument performance is monitored (hardware calibration, frequency of recalibration and system suitability), how the model was built, tested, validated (if applicable), and will be maintained throughout its lifecycle, and what is done with the output (reporting, link to informatics systems). In this chapter, the general procedure for method development and lifecycle management will be discussed through the concept of analytical quality by design [1, 2] (AQbD) developed by the pharmaceutical industry, but applicable to all analytical fields of use of near-infrared spectroscopy (NIRS).

12.2 General Procedure for Method Development, Validation, and Lifecycle

Near-infrared spectra are usually too complex for directly establishing a relationship between an absorbance at a particular wavelength and a parameter of interest. While for clear liquids Beer's law may be directly applicable, most samples will exhibit diffuse reflectance or transmittance. The resulting spectra will differ in path lengths and require the development of empirical models. These models can be supervised or unsupervised depending on the intended use. The development of these models must follow a rigorous methodology to ensure that the resulting analytical method meets its intended purpose. The analytical quality by design framework presents an approach to building, testing, validation, and maintaining a NIR method.

Practitioners of near-infrared spectroscopy will be very familiar with the process described in Fig. 12.1 corresponding to a generic flow diagram of how a supervised model is built, tested, and validated if required by the regulatory environment of use. The model is critical, but only a part of an analytical method and the general framework of method development will be discussed in the next section. However, general principles need to be explored before focusing on the AQbD framework. The following paragraphs present generalities about model development.

As indicated in Fig. 12.1, a calibration set, corresponding to relevant variability for which the model will need to account during operational deployment (i.e., variability in chemical and physical parameters expected to be encountered), is regressed against the "true" (known or measured) quantity in the parameter of interest. The range of sample variability included in the model will dictate how robust a model is. If the model does not span an appropriate range (what the model is expected to encounter during deployment), the model usefulness will be limited to the variability included

Calibration Process

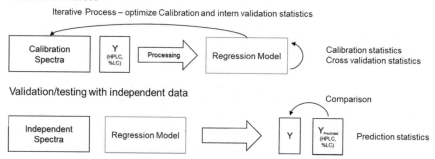

Fig. 12.1 Generic method development, testing, and validation workflow

in calibration. In addition, there is always error in the determination of the reference values, and both the accuracy and precision of the reference data should be determined to understand the impact on the NIR method as the error in the reference method will impact the performance of the chemometrics model.

Prior to regression, the spectra are often preprocessed to remove unwanted variance and highlight specific information through variable range selection and pretreatment methods. The choice of regression method will depend on the nature of the signal. The data collected by most spectrometers are highly collinear and will require the use of regression techniques that can handle the collinearity. The most utilized methods are based on latent variable extraction techniques [3, 4] (partial least-squares regression (PLSR), principal component regression (PCR)). These algorithms find the main directions of variance in the data and correlate them to the reference values (PCR) or the main directions of covariance between the spectra and the reference values (PLSR). However, for well-defined systems, Beer's law can be used through classical least-squares [4] (CLS) regression or more computationally intensive techniques such as artificial neural networks (ANN) or support vector machines (SVM) [5].

After regression, the model stability and performance are internally tested with cross-validation techniques. Numerous cross-validation approaches have been developed, but, in general, they remove a part of the calibration samples, redevelop the model without these samples, and subsequently predict the excluded samples. The operation is repeated until all samples have been used to test the model. Care should be taken to select the right approach for cross-validation. Leave-one-out cross-validation is a very simple approach, where only one sample is taken out at a time; but it will tend to be over-optimistic. Alternatively, if using block cross-validation, an entire source of variability could be inadvertently removed, and the resulting cross-validation error would be overly inflated resulting in a loss of confidence that the model will perform as required. Random block and venetian blinds are alternative sample selection techniques that can provide a more realistic estimate of the model performance. The selected approach should consider the specificities of the samples in the calibration set.

Once the calibration algorithm and testing approach are chosen, an iterative process of sample selection, variable selection, and spectral preprocessing is initiated, with the goal of obtaining the most suitable model for the intended purpose. Once that model has been developed, it should be tested, and if necessary validated, against independent data to confirm its performance.

The workflow displayed in Fig. 12.1 should however only be applied as a result of a careful consideration of the method purpose and expected performance and numerous activities should take place before a single model is built. The frequency at which spectra are collected should ensure that the information generated by the method meets the intended goals of the analytical method and should be driven by the measured process variability and the analytical instrument capability. The most accurate model may not prove useful if not associated with a method that measures the right information. For instance, an instrument known to be affected by environmental conditions that cannot be controlled effectively (e.g., temperature) but referenced only at the beginning and end of a week-long campaign may not allow the method to perform appropriately. Another consideration is related to the error of the reference method. If the desired error of the NIRS method is significantly lower than the error of the reference method, the model may not be able to meet expectations. Finally, if variability in raw materials are expected beyond what can be included in the calibration set (i.e., year-to-year variability of natural products), resulting in a significant change in the spectra, the model will lack robustness and need to be updated. If collecting new samples for reference analysis and model update is not feasible, the validity of the method will be limited.

For these reasons, following the analytical quality by design framework can be of significant help to set the parameters of the method and ensure that the expectations are set before undergoing a long and costly method development activity.

12.3 Analytical Quality by Design (AQbD)

12.3.1 Introduction to AQbD

The concept of analytical quality by design arises from the quality by design (QbD) concept documented by the International Council for Harmonisation of Technical Requirements for Pharmaceuticals for Human Use (ICH) [6]. Quality by design is defined as "a systematic approach to development that begins with predefined objectives and emphasizes […] understanding and […] control, based on sound science and quality risk management". A process developed using the QbD framework is well understood and delivers quality product within a design space identified during method development. For analytical methods, AQbD allows for a "well understood, fit for purpose, and robust method that consistently delivers the intended performance throughout its lifecycle" [2]. Figure 12.2 presents the components of AQbD. Each item will be discussed below.

Fig. 12.2 Components of
the analytical quality by
design framework

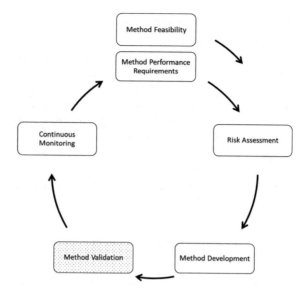

12.3.2   Analytical Target Profile

The analytical target profile (ATP) defines the intent of the method: What is the method aiming to measure? How should the method work? What will be done with the outputs of the method? It is a document where the method developers and the stakeholders agree on what the method needs to deliver. The criteria may be in terms of figures of merit (accuracy, precision, robustness, …), in terms of lifecycle (how is the method supposed to work and be updated on the long run?) and in terms of deployment (who will use the method and in which condition). For instance, a method built on a laboratory instrument and targeted for a manufacturing application may not meet the performance and deployment criteria; the sample properties may change between laboratory and plant, or the expert users in the laboratory will not be in charge of all the activities affecting the instrument once deployed.

Table 12.1 presents an example of figures of merits and corresponding high-level performance criteria a hypothetical method would need to satisfy to meet the needs of a hypothetical application. Refer to ICHQ2(R1) [7], regulatory guidance [8, 9], or standards [10] for definitions. Examples of statistical tests that can be used to evaluate these figures of merit are available in regulatory guidance and standards [11].

Thus, the ATP should be the document that sets the expectations and the direction of work for the technical team. However, it is a living document that is periodically evaluated. It is not because a method does not meet the criteria after development and testing that it may not be fit for purpose. As the ATP is initially drafted several months or years ahead of completion, it can be difficult to determine all the criteria and their acceptance limits.

Table 12.1 Example method performance category and high-level corresponding criteria

Category	Criteria
Accuracy	Fit for purpose based on the specifications and requirements of the method.
Precision	Repeat measurements error is acceptable with and without repositioning, by operators, at different concentration levels.
Specificity	The NIRS procedure should be able to assess unequivocally the analyte in the presence of other components, which may vary
Linearity	Model predictions are linear over the range of interest.
Range	Covers a relevant range of the parameter of interest to allow for the determination of non-conforming materials.
Robustness	Robust to chemical and physical variables, sampling and sample preparation, and variations in procedure parameters.

12.3.3 Feasibility Study

NIR models can often be expensive to develop depending upon the application, and it is beneficial to consider a staged approach to development whereby smaller feasibility studies explore the potential method. Often used as an input to, and completed in parallel to the ATP, a feasibility study can be used to de-risk an application before significant resources are deployed. However, an ill-designed or too limited feasibility study will provide over-optimistic results and further risks of feasibility studies are provided below.

The feasibility study should confirm elements that will become critical when moving forward with method development:

- Is the targeted analytical tool suitable for the measurement? When a feasibility study finds that NIRS is not the best tool for the job, other analytical tools shall be considered.
- What is the nature and stability of the sample? Developing a method for a sample that degrades or reacts upon sampling for reference analysis may be challenging.
- What is the appropriate means of sampling? While optimization of the scale of scrutiny and sample presentation can be completed during method development, the feasibility study should try to identify the factors that will affect the method. For example, if the feasibility study for an in-line measurement is performed on a benchtop system after manual sampling, then only a portion of the ATP has been answered; i.e., NIRS is a viable technology, or NIRS provides specificity, etc. Often, the sampling conditions (representativity, matrix effect of moving powder, reproducibility of sampling, etc.) can cause the method to not meet ATP criteria.
- What is the error of the laboratory reference values? A reference method with 5% error may not allow the development of a NIRS method with less than 5% error.
- What is the process variability? If significant variability in the process is expected (i.e., the process is developed at the same time as the analytical method), uncertainty can exist on what variability in raw materials, environmental, and processing

conditions will exist during routine use of the method and impact the performance of the model. Having a clear picture of the expected variability will provide insights into the scope of application and the variables/properties that may affect method robustness.

Needless to say, however, that comprehensively addressing these questions may not be feasible at an early stage. By its nature, a feasibility study cannot explore all the sources of variability that a model will need to handle during routine use. For instance, exposure to limited sources of variability may provide over-confidence in the performance of a method. A feasibility study designed to not address known risks will not provide the relevant outcomes, specifically for robustness.

Progressing from a feasibility study to method development is a business decision, and the scientific team should weigh the costs of progressing without all the answers versus performing a more extensive feasibility study (whose data could be used in the method development) and advance with more certainty. The feasibility study should also be used to potentially update the ATP and gain endorsement from stakeholders if factors have been identified that will change the desired performance criteria.

12.3.4   Risk Assessment

The risk assessment is the process of identifying and scoring the causes that could have an impact on the method. The result of a risk assessment is a formal document identifying the risks that exist to the method and how they will be controlled, accepted, mitigated, or avoided.

There are many ways to conduct a risk assessment. But, the general principles are as follows. First, all the stakeholders should be identified and convened together to conduct the exercise. These members should represent management (if appropriate), the process engineers, the method developers, and the method users. Getting the right membership to the risk assessment will ensure that all the relevant risks are identified. Second, the team should list all the possible failure modes that could affect the method performance. Figure 12.3 presents an example of a risk identification process also known as a fishbone diagram for a hypothetical method aimed at predicting the content of an active ingredient in-line. As presented, the risk identification should cover all relevant causes of impact to the model performance. Third, the risks should be categorized or scored by the team. If using a scoring system, each risk is assessed one at a time, for criteria such as the probability of occurrence, the severity if it occurred, and the probability of detection. Setting scores for the various risks will require team members to use a priori information from previous projects, feasibility studies, or experiences. Some risks may be scored high at first because at the time of the risk assessment, limited information may be available and then re-scored as new knowledge becomes available. On the contrary, a risk could have been determined to be low after the feasibility study but may prove to require controls or mitigation

upon method development. Finally, plans are put in place to address those with the
highest scores. They can be categorized as:

- Controlled variables.
  - Procedural control (i.e., lamp warm up may have a high risk because running
    the instrument when the lamp is not stable will affect the results. However, it
    can be proceduralized that the lamp needs to be on for a set time prior to use)
  - Physical control (i.e., sample will be presented in a holder to ensure repro-
    ducibility)
- Experimental variables. These variables should be explored during method
  development and will result in the risks being:
  - Accepted: variability during use will not significantly affect the method (i.e.,
    environmental pressures and humidity for well-controlled instruments)
  - Avoided: variability of the sample will be removed prior to analysis (i.e.,
    milling to a target particle size, drying to a constant moisture level)
  - Mitigated: variability will be built in the chemometrics model (i.e., if the sample
    varies in moisture, removing the -OH absorption bands from the spectra prior to
    modeling can limit the method from being sensitive to moisture variability, or
    if the sample particle size is expected to change, a design of experiment should
    be used to ensure that variability is presented to the model and preprocessing
    methods are employed to reduce their impact on the analytes(s) of interest and
    the remaining impact is maintained in the model as within-model variability
    (i.e., measured with the model diagnostic Hotelling's T^2) or out-of-model
    variability (i.e., measured with the model diagnostic Q-residual)

However, mitigating or avoiding all risks is impossible. Resource and time
constraints will push the scientific team to make trade-offs when building a model.
The result will be models that are robust to certain degrees to most but not all sources
of variability that the method could encounter in the future. It is also important
to recognize that identifying a risk does not mean that it can be fully managed or
addressed.

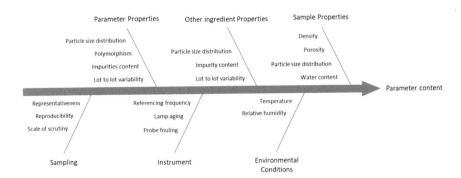

Fig. 12.3 Examples of risks identified for an in-line pharmaceutical method

As a result, some risk will be passed onto the method users because model robustness needs to be monitored throughout lifecycle management. They need to accept risk factors that may arise from previously unseen sources of variability as well as factors that were simply not identified *a priori*. These unforeseen sources of variability and their potential impacts on the method will be handled through the method lifecycle management process and a plan should be in place to maintain robustness of the model throughout its lifecycle.

12.3.5 Method Development

When reaching this stage of development, a significant amount of information is known about the desired performance criteria (analyte target profile), the suitability of the sampling approach and analytical technique (feasibility study), and the factors that may affect the model (risk assessment). These learnings are now taken into account to optimize the sampling methodology, the sample presentation to the analytical instrument, the data collection (to ensure measurement representativity), etc. It is essential that these factors are set prior to spectral data collection for the calibration, test and if applicable validation sets. Any change to instrument configuration and collection parameters should be assessed for impact on the method. This is particularly important in cases where the calibration set is collected with a different experimental setup, e.g., off-line calibration for in-line use. Such an example may occur when in-line calibration samples cannot be prepared and presented in a representative manner.

12.3.5.1 Sample Set Membership Considerations

The sample membership of the calibration set, test set(s), and potential validation set should be carefully designed. Specifically, the validation set should be completely independent from the calibration and test sets, should represent expected variability from the process, and should challenge the model across an appropriate range that the model is intended to cover. The test set(s) should be representative of variability expected during normal operating conditions. The calibration set should be built from the experimental variables identified in the risk assessment. If sample particle size, density, batch-to-batch or seasonal variability, etc., is expected to change, that information should be included in the model. If too much variability is included, the model may suffer from a lack of accuracy at the expense of robustness. If the process is expected to be highly variable, local chemometric methods segmenting large ranges of variability into smaller segments for model development may be better suited [12].

The model should be qualified for its use irrespective of the application, and the requirements for qualification (or validation) will vary across industries. In the pharmaceutical industry, it is necessary to satisfy ICH Q2(R1) requirements for method validation in addition to the relevant guidelines (e.g., EMA and FDA guidelines on

NIR) [7]. The ASTM E55.01 committee has also issued several documents on the topic [10, 11]. Proper testing of any model requires independence of the calibration and representative samples used for performance evaluation.

12.3.5.2 Sample Variability or Origin

Samples for building a NIR spectroscopic model may come from natural sources or manufacturing sources. In cases where it is difficult to design specific samples for model development, care should be taken to ensure that the calibration set is well balanced and not over representing a particular source of variability that may artificially influence the predictions.

The modeling team should ensure that all relevant variability is included in the calibration set. Algorithms such as the Kennard and Stone [13] sample selection approach can help reduce redundancy in spectra presenting the same variability for large datasets but the team must understand the data included and excluded, and rationalize the sample selection.

When feasible, particularly for the chemical and pharmaceutical industries, artificial samples can be produced at small or pilot scales. Designs of experiments are often used to derive a suitable spectral space by varying the chemical and physical properties in the samples to represent normal operating ranges of the process and that the model should be expected to suitably handle. Figure 12.4 provides an example of such a design (a 3-factor 2-level full factorial design with a center point) where the

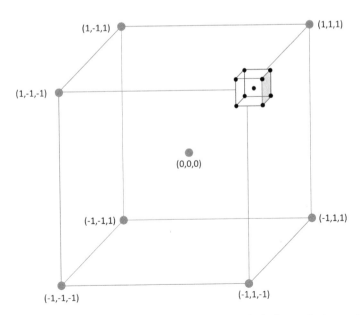

Fig. 12.4 Example of full factorial design commonly used for designing synthetic samples

chemical content in a sample is varied along with the concentration of other ingredients. There are many other types of designs that can be used (central composite designs, fractional factorial designs, mixture designs, spiral designs, D, and I optimal designs) and attempts at comparing their performance have been published [14, 15]. In situations where the samples are expected to vary both in chemical and physical properties, designs can be augmented through exposure to the conditions (i.e., calibration set is scanned multiple time after exposition to various moisture environments to build moisture robustness). Nested designs can also be employed where at each chemical design point (i.e., Fig. 12.4, point (1, 1, −1)) a design for other parameters is built. This approach is very common for pharmaceutical oral solid dosage forms such as tablets where the model is expected to be robust to tablet thickness and density variability.

The use of artificial samples is very attractive as it allows the rapid development of models for a fraction of the cost of running a manufacturing line. However, care must be taken to ensure the representativity of the samples designed at a scale different to what the model is expected to encounter during commercial use. In situations where the difference between small- and large-scale production samples is significant and cannot be accommodated for by spectral preprocessing, much of the work may not be relevant and a new approach to model building may need to be designed. Attempts to bridge the difference in sample matrix between small and large scale for powder mixing have been published [16].

12.3.5.3 Model Optimization

NIR spectroscopy relies on the use of multivariate regression methods such as PCR, PLS, CLS, ANN, or SVM to relate the absorbance values with the reference values. The description of each method and situation of use is beyond the scope of this chapter, and readers should refer to the chemometrics chapter, but all these methods will generate a set of coefficients that can be used to relate a new spectrum to its value in the parameter of interest as outlined in Fig. 12.1. Once a method is selected, much of the model optimization relies on determining the calibration set membership, the wavelength or wavenumber range, the pretreatment of the spectra, and setting the model complexity (i.e., number of factors for PCR and PLS, network structure and neuron numbers for ANN, kernel type for SVM).

The optimization is an iterative empirical process and should be performed to meet the method requirements outlined in the ATP (Table 12.1). Below are some elements to consider:

- Variable range: it should be optimized to ensure specificity and robustness. For instance, the development of a model for moisture should use spectral ranges at 1450 nm and or 1920 nm corresponding to the -OH absorption band. But the overlap of chemical absorption bands in the NIR range usually means that large variable ranges are included in the model.

- Preprocessing: used to reduce irrelevant variance, enhance relevant variance and/or linearize the relationship between the spectral data and the parameter of interest. Preprocessing options should be selected to address the particular needs of the model based on the sample matrix. For instance, if a difference in particle size exists across the samples, varying scattering intensities could be mitigated by standard normal variate [17] or multiplicative scatter correction [18].
- Model complexity: over- or under-determined models can significantly impact robustness, and the model complexity should be set so that the model is able to handle future variability while still delivering adequate accuracy.

12.3.5.4 Sources of Method Errors

The final method error is the combination of several sources of error that not only have the potential to impact the model accuracy but also the measurement suitability. Sources of error to consider include the sampling error, the error of the laboratory, the error due to instrument variability, etc.

Ensuring the measurement is representative of the process being analyzed is critical to the quality of the method outputs. There are several items to consider: the scale of scrutiny (the volume of sample analyzed by the instrument), the spectral collection frequency, the sampling method, and the process variability. If a spectrometer analyzes the entire sample volume but the sample is collected in a way that does not represent the variability of the process, the model error may be low (meeting the predefined criteria for figures of merit) but method error would be high. If the sampling is unbiased, the measurement volume is appropriate, but the frequency of measurement is low, the manufacturing process may not be appropriately monitored. Ensuring that the sample measured is representative of the process is critical to a method success.

As a secondary analytical method, NIRS relies on the determination of the parameter(s) of interest by a primary method (i.e., scale, chromatography and spectroscopy, etc.) for model building. The larger the error in the reference methodology (error of the laboratory), the larger the error of the NIRS model.

Another element related to the error of the laboratory is the quantity being measured as reference values (or the unit in which it is expressed) and its relationship with what NIRS measures. In a publication, the impact of selecting the unit of the parameter of interest (volume fraction vs. weight fraction) was shown to result in nonlinearities with the authors commenting on the fact that the sensitivity of NIRS to volume fraction over weight is distorting the established concept of artificial design of experiments based on weight [19]. The work was done on liquid samples but should apply to other forms of samples. Authors argue that the nonlinearity generated by using the "wrong" unit cannot be fully accommodated for by spectral preprocessing methods.

Instrumental error should also not be under-estimated. While modern instruments are highly reproducible with nearly noise-free detectors, variability in lamp intensity can be observed as a function of time and care should be taken to ensure a model is built over the range of intensity and several lamps (as lamp-to-lamp variability can be expected), and if necessary, limits on lamp intensity should be set. If the instrument is in a space where the ambient temperature will vary, variability in the spectra can be observed. Finally, if the instrument is deployed in-line or on-line, reproducibility of the positioning of the probe or probe fouling can have a significant effect on the resulting spectra. This is also the case for sample holders for at-line or off-line measurements. Any change in how a sample is presented to the sampling system will increase the error of the measurement.

12.3.6 Method Testing and Validation

As discussed in the introduction, a model is only a part of an analytical method. Method testing and potential validation does not solely mean applying independent spectra to the model but also ensuring the spectrometer is deployed as intended and that the model outputs are provided to the reporting system. Testing should be performed on relevant samples (independent from calibration and if possible, from different lots of materials) to evaluate the impact of the risks identified in the risk assessment on the method performance. If new factors affecting the method are identified during testing or if the experimental work proved to not adequately address some of the risks previously identified, the method and model may need to be updated. Evaluation of any new update to a model should be performed with independent samples. If the method performs as expected, the method development effort can be concluded.

12.3.7 Referencing, System Suitability, and Performance Monitoring

During routine use, the method must be used within the conditions it was built for, the suitability of the instrument must be confirmed, and the performance of the method must be monitored. Ensuring that the instrument, sampling, data collection, etc., are consistent between method development and method use is critical. While the samples will change, procedures should be in place to ensure that all the variables that could affect the method performance are controlled. If there is a significant difference between the conditions where the method was built and the routine manufacturing conditions, the method may need to be updated. That scenario will be discussed in a next section.

12.3.7.1   Referencing

One potential difference between method development and method use is the requirement for referencing the spectrometer. Off-line or at-line methods will usually call for referencing either before each sample (this is very common when the reference standards are built in the instrument) or at a set frequency that has been proven to sustain method performance. However, for on-line or in-line deployments, other considerations are at play. If the instrument is located in a manufacturing suite or technical space and these environments are expected to change in thermal conditions during use, the stability of the lamps and detectors may be affected. In such case, the referencing of the instrument (white reference (e.g., 99% reflectance standard) and potentially dark reference) may be required more frequently. The chosen rate/periodicity should ensure suitable spectrometer performance.

For in-line applications, it may not be possible to remove and clean the probe for referencing. Off-the-shelf commercial systems allow for the probe to be automatically removed from the process, cleaned and referenced. In addition, probes and spectrometer systems are commercially available with built-in referencing that allows the system to avoid drifts for extended periods of use (several days). But regardless of how the referencing is performed, it must be done at a relevant frequency to ensure that any change in the analytical system is accommodated for.

12.3.7.2   System Suitability

Turning on a spectrometer, letting it warm up, and collecting a reference should not be the only actions performed prior to use. Instruments usually follow a scheduled performance qualification program either completed by the vendor or the user, but these could be completed once a year or once every few months. Additional tests are necessary prior to use and during use to confirm that the instrument and sampling system are performing as intended when compared against acceptance criteria and these should be trended against historical values. These tests, often called system suitability tests or on-going performance verifications, will often use traceable standards to confirm that the equipment is meeting acceptance criteria and can be used to generate data as per the method procedure. System suitability is specifically discussed in guidelines and standards [7, 20].

12.3.7.3   Method Performance Monitoring

With chromatographic methods, it is typical to run a known working standard reference sample before and periodically throughout a sample set run to confirm that, over the duration of the run, the system performed as intended. In addition, these measurements are used to generate the model calibration curve at the time of use. However, as described earlier, typical NIR methods rely on *a priori* established calibration curves, and future samples will be projected onto that model. To ensure that the multivariate regression models are still valid, diagnostic metrics need to be routinely collected and trended. The calculation of diagnostic values such as the Hotelling's T^2 and the Q-residuals will confirm for every spectrum that it appropriately belongs to the model space. This will provide confidence that the spectrometer, sample, and environment are all producing a spectrum that is as required for the model to perform as developed, tested, and potentially validated.

Tracking and trending of the diagnostic values are essential to the successful deployment of a NIRS-based method. Statistical limits can then be set to raise alarms when a measurement appears to be different or tending toward process limits. The limits for the diagnostics can be set by incorporating the assessment into the method qualification or in parallel with initial batch experience during the first several production runs. These diagnostics will determine whether the sample belongs to the model space (Hotelling's T^2) and whether there is excessive unmodelled variance (Q-residual). Note that it is possible for a prediction to be within process limits but for the diagnostics to flag the sample as being "different" or an outlier. If a sample is determined as exceeding an outlier limit, it should not be assumed however that the sample is different or that the instrument is not functioning properly. If justified by consistent out of specification results beyond the expected statistical limits, an investigation of the system, the process, and the materials used should be undergone to determine the root cause. If it is determined that the sample was appropriate and that no instrument issue occurred, an update to the method may be considered.

12.4 Method Lifecycle

Maintaining method performance years after method development is the current challenge that NIRS practitioners are facing to ensure the successful use of NIRS, regardless of the field of application. A number of scenarios can be envisioned that may affect a method such as instrument changes and replacement (either caused by optical part replacement, instrument failure or transfer of the process to a new manufacturing site), process changes (process wear and tear and process improvements), and raw materials changes (long-term lot-to-lot variability and sourcing changes). It is also important to systematically monitor risk factors identified during the risk assessment until they are deprioritized or eliminated. Alternatively, when new risks arise, it may be necessary to either control the risks or update the model to handle that new variability. Finally, periodic assessment against the reference method should be performed to demonstrate the method is still performing adequately. This may be done even if predictions and diagnostics do not show any new concerns. In a regulated environment such as the pharmaceutical industry, it shall be noted that periodic assessment does not mean periodically performing a verification exercise against the original validation dataset, i.e., re-validation. Re-validation is only needed when a model update takes place.

If it is determined that an update to the model is required, the following approaches could be used:

- Optimizing the existing model (preprocessing, variable range, number of latent variables) to accommodate for the new sources of variability
- Update the calibration set to add a limited number of samples representing the missing sources of variability
- Re-build the model

In situations where the process is transferred to a different facility and/or analytical instrument, ensuring that the model performance before and after transfer is practically equivalent is critical. This can be done as part of method development by including samples from multiple instruments in the calibration, test and validation sets (robust models across units) or through the use of calibration transfer methods [21]. If calibration transfer cannot be achieved (resulting error on the secondary unit is too high), model re-build may be necessary.

12.5 Additional Considerations for Multipoint Systems

All the concepts discussed in this chapter are applicable to any NIR spectroscopic system. However, some additional considerations should be noted for multipoint (imaging) systems. Without discussing the details of how images are constructed, NIR imaging systems usually consist of arrays of sensing elements that can be considered as juxtaposed single point spectrometers. There are numerous NIR systems where the sample is measured at multiple points, but the resulting light is usually directed toward a single detector channel. On the contrary, imaging systems or specialized systems such as those employed in spatially resolved spectroscopy will detect light coming back to different detector elements, as being measured at different locations on the sample. This spatial information is key to imaging but also brings some additional points to consider regarding sample membership and how to construct a representative calibration dataset.

A single-point spectrometer assumes that the collected spectrum is representative of the area interrogated (scale of scrutiny). When collecting the equivalent sample for reference analysis, care is taken to ensure that the volume sampled for reference analysis corresponds to the variability measured by the NIR measurement. In multipoint systems, this assumption is not correct. While there will be a degree of correlation between neighboring sampling points, homogeneity of the sample cannot be assumed at the scale of measurement, particularly when using microscopic or wide-angle lenses. It is actually the desired intent: to investigate the distribution of physical and/or chemical properties within the area of interest via spectral information from pixel to pixel (or sampling location to sampling location).

What makes imaging an ideal tool for the investigation of sample spatial distribution also makes it a challenge for model development. In many cases, it is no longer correct to assume that the percentage of light reflected by a 99% reflectance standard is actually homogeneous across the detector array; it is no longer correct to assume that a wavelength standard will have the same composition at all pixel locations; and it is certainly incorrect to assume that the bulk content of a sample will be directly assignable to the individual spectrum from individual pixels from the image array of that same sample. In other words, if a sample containing a binary mixture of 92–8 for component A to B, the local aspect of using detector arrays does not result in every single pixel containing the signal of 92% of A and 8% of B. A distribution of spectra representing different levels of content for a particular analyte of interest is obtained for a particular spatial area. In addition, due to the photons' "random walk", the spectrum collected from a single pixel is a weighted averaged signal from adjacent pixels.

However, this has not stopped scientists from building successful applications and deploying them in commercial settings. There are two common approaches to building models for multipoint systems. The first simply consists of taking the mean of an image as a representative spectrum for the sample being analyzed by the reference method. The resulting spectrum (average of potentially thousands of spectra) will have a different signal to noise ratio to the individual pixel spectra but be relevant for model building. The second approach is to divide the sample in smaller areas for reference analysis and binning (averaging) spectra spatially so a bin, smaller than the entire image, can be associated to a local value of the property of interest. Readers should refer to the chapter on imaging for more information about how these techniques are used.

12.6   Summary

NIRS is a very powerful technology. It can determine in a matter of seconds the content of a particular attribute of interest in a complex sample matrix without preparation, in-line and in real-time. It is sensitive to both the physical and chemical make-up of the sample which can make any deployment challenging. The Analytical Quality by Design framework can significantly streamline method development and lifecycle management efforts to ensure a successful deployment and long-term value generation from a NIR spectroscopic method. The process of stating the method goals and requirements, performing feasibility study(ies), and a risk assessment before method development will help address knowledge gaps and set expectations before significant resources are spent. The result is a robust method that can be tested and validated if required by the regulatory environment of use. Continuous improvement through method lifecycle management ensures method performance over the useful life of the method.

References

1. P. Borman, P. Nethercote, M. Chatfield, D. Thompson, K. Truman, The application of quality by design to analytical methods (2007)
2. G.L. Reid, J. Morgado, K. Barnett, B. Harrington, J. Wang, J. Harwood, D. Fortin, Analytical Quality by Design (AQbD) in Pharmaceutical Development. American Pharmaceutical Review (2013)
3. S. Wold, M. Sjöström, L. Eriksson, PLS-regression: a basic tool of chemometrics. Chemometr. Intell. Lab. Syst. **58**(2), 109–130 (2001)
4. E.V. Thomas, D.M. Haaland, Comparison of multivariate calibration methods for quantitative spectral analysis. Anal. Chem. **62**(10), 1091–1099 (1990)
5. F. Chauchard, R. Cogdill, S. Roussel, J.M. Roger, V. Bellon-Maurel, Application of LS-SVM to non-linear phenomena in NIR spectroscopy: Development of a robust and portable sensor for acidity prediction in grapes. Chemometr. Intell. Lab. Syst. **71**(2), 141–150 (2004)
6. ICH, *Pharmaceutical Development* Q8(R2) (2009)
7. ICH, *Validation of Analytical Procedures* Q2(R1) (1996)
8. USFDA, *Development and Submission of Near Infrared Analytical Procedures* (2015)
9. European Medicine Agency, *Guideline on the Use of Near Infrared Spectroscopy by the Pharmaceutical Industry and the Data Requirements for New Submissions and Variations* (2014)
10. ASTM, Standard Guide for Multivariate Data Analysis in Pharmaceutical Development and Manufacturing Applications, in *E2891–13, ASTM International: West Conshohocken*, PA (2013)
11. ASTM, Standard Practice for Validation of the Performance of Multivariate Online, At-Line, and Laboratory Infrared Spectrophotometer Based Analyzer Systems, in *D6122-19b, ASTM International: West Conshohocken*, PA (2019)
12. W.S. Cleveland, S.J. Devlin, Locally weighted regression: An approach to regression analysis by local fitting. J. American Stat. Assoc. **83**(403), 596–610 (1988)
13. R.W. Kennard, L.A. Stone, Computer aided design of experiments. Technometrics **11**(1), 137–148 (1969)
14. R.W. Bondi Jr., B. Igne, J.K. Drennen Iii, C.A. Anderson, Effect of experimental design on the prediction performance of calibration models based on near-infrared spectroscopy for pharmaceutical applications. Appl. Spectrosc. **66**(12), 1442–1453 (2012)
15. O. Scheibelhofer, B. Grabner, R.W. Bondi, B. Igne, S. Sacher, J.G. Khinast, Designed blending for near infrared calibration. J. Pharm. Sci. **104**(7), 2312–2322 (2015)
16. S. Mohan, W. Momose, J.M. Katz, M.N. Hossain, N. Velez, J.K. Drennen, C.A. Anderson, A robust quantitative near infrared modeling approach for blend monitoring. J. Pharm. Biomed. Anal. **148**, 51–57 (2018)
17. R.J. Barnes, M.S. Dhanoa, S.J. Lister, Standard normal variate transformation and de-trending of near-infrared diffuse reflectance spectra. Appl. Spectrosc. **43**(5), 772–777 (1989)
18. P. Geladi, D. MacDougall, H. Martens, Linearization and scatter-correction for near-infrared reflectance spectra of meat. Appl. Spectrosc. **39**(3), 491–500 (1985)
19. H. Mark, R. Rubinovitz, D. Heaps, P. Gemperline, D. Dahm, K. Dahm, Comparison of the use of volume fractions with other measures of concentration for quantitative spectroscopic calibration using the classical least squares method. Appl. Spectrosc. **64**(9), 995–1006 (2010)
20. ASTM, Standard Guide for Risk-Based Validation of Analytical Methods for PAT Applications. In E2898–14 (2014)
21. J.J. Workman Jr., A review of calibration transfer practices and instrument differences in spectroscopy. Appl. Spectrosc. **72**(3), 340–365 (2018)

Part IV
Applications

Chapter 13
Overview of Application of NIR Spectroscopy to Physical Chemistry

Mirosław A. Czarnecki, Krzysztof B. Beć, Justyna Grabska, Thomas S. Hofer, and Yukihiro Ozaki

Abstract Near-infrared (NIR) spectroscopy is a powerful tool in studies of physicochemical properties of various kinds of samples. In particular, NIR spectroscopy contributed considerable to our understanding of intermolecular interactions (e.g. hydrogen bonding), molecular structure, solvent effect, clustering, phase transitions, kinetics. Because of mechanical and electrical anharmonicity of molecular vibrations, NIR spectra provide unique information not available from the other spectral regions. On the other hand, to elucidate useful information from NIR spectra, more sophisticated methods of data analysis than those applied in mid-infrared (mid–IR, MIR) region are necessary. This chapter presents selected examples demonstrating the variety of problems in the field of physical chemistry that have been studied by NIR spectroscopy.

Keywords Physical chemistry · NIR spectroscopy · Anharmonicity · Hydrogen bonding · Molecular structure · Intermolecular interactions · Solutions

M. A. Czarnecki
Faculty of Chemistry, University of Wrocław, F. Joliot-Curie 14, 50-383 Wrocław, Poland
e-mail: miroslaw.czarnecki@chem.uni.wroc.pl

K. B. Beć (✉) · J. Grabska
Institute of Analytical Chemistry and Radiochemistry, Leopold-Franzens University, CCB-Center for Chemistry and Biomedicine, Innrain 80/82, 6020 Innsbruck, Austria
e-mail: Krzysztof.Bec@uibk.ac.at

J. Grabska
e-mail: Justyna.Grabska@uibk.ac.at

T. S. Hofer
Institute of General, Inorganic and Theoretical Chemistry, Leopold-Franzens University, CCB-Center for Chemistry and Biomedicine, Innrain 80/82, 6020 Innsbruck, Austria
e-mail: t.hofer@uibk.ac.at

Y. Ozaki
Department of Chemistry, School of Science and Technology, Kwansei Gakuin University, 2-1 Gakuen, Sanda, Hyogo 669-1337, Japan
e-mail: ozaki@kwansei.ac.jp

© Springer Nature Singapore Pte Ltd. 2021
Y. Ozaki et al. (eds.), *Near-Infrared Spectroscopy*,
https://doi.org/10.1007/978-981-15-8648-4_13

13.1 Introduction

As discussed in the other chapters of this book, recent progress in instrumentation, in particular interest of instruments based on Fourier transform (FT) measurements significantly improved the accuracy of NIR spectra both in terms of wavenumbers and absorbance scale. The possibility of recording high accuracy spectra permitted to study fine effects and distinguished very small differences between individual spectra. Also, the rapid development of computational hardware and software stimulated progress in NIR spectroscopy and theoretical calculations of anharmonic spectra became an important tool for the understanding and interpretation of NIR spectra [1–4]. All these circumstances opened new possibilities and areas of applications in NIR spectroscopy. One of the most important fields of these applications is physical chemistry, which covers a variety of topics like molecular structure, intra- and intramolecular interactions (in particular hydrogen bonding), solvent effects, clustering, phase transitions, solution, kinetic studies, and so on [5].

The NIR region is very unique since it provides information not accessible from the other spectral regions [5]. This specificity results from the anharmonicity of molecular vibrations and the nonlinearity in the change of the dipole moments. Both phenomena influence the positions and intensities of bands originating from vibrational modes of different molecular fragments. Therefore, NIR spectra are a rich source of information on molecular structure and interactions. Since the overtones and combination modes are forbidden in the harmonic approximation, the corresponding bands are much weaker as compared with the fundamental ones. For these reasons, NIR spectroscopy is a very powerful tool for studies of highly absorbing samples like bulk materials, pure liquids, or even aqueous solutions. NIR spectra of bulk liquids can be recorded in commercially available cells of 1–10 mm width (pathlengths). Typically, these cells are made of quartz, which is water resistant. Hence, NIR spectroscopy can be employed for study of bulk water and aqueous solutions. However, due to very strong absorption from water, one has to use cells with shorter pathlengths (<1 mm). As can be seen in Fig. 13.1, both combination bands from bulk water are accurately recorded in a 0.1 mm quartz cell. In Fig. 13.1 are compared the spectra of neat 1–propyl alcohol and a 1:1 water/1–propyl alcohol mixture. It appears that the $\nu_2 + \nu_3$ combination band of water is particularly useful for spectroscopic studies since it is located in the region, which is free from the absorption of the other overtones and combination bands (5000–5300 cm^{-1}). In contrast, the $\nu_1 + \nu_3$ combination band of water (6800–7200 cm^{-1}) is overlapped by the first overtone of the alcoholic OH and the C–H combination bands [5].

In principle, NIR spectra do not reveal bands originating from carbon–carbon double and triple bonds. However, bands originating from C–H stretching groups attached to double or triple bonds are usually blue-shifted [5]. This shift is very well seen for the first and second overtones of the C–H stretching vibrations of cyclohexane and benzene (Fig. 13.2). This figure illustrates one more important property of NIR spectra. The first overtones of both compounds have a complex structure, since it results from the coexistence of the normal and local vibrations [6]. On the

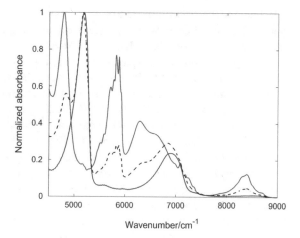

Fig. 13.1 Normalized NIR spectra of bulk water (blue solid line), bulk 1-propyl alcohol (red solid line) and an 1:1 water/1-propyl alcohol mixture (black dashed line) at 20 °C

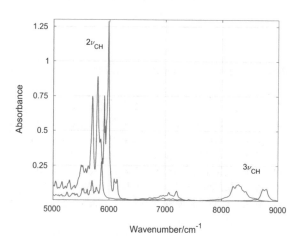

Fig. 13.2 NIR spectrum of liquid cyclohexane (blue) and benzene (red) at 20 °C

other hand, the structures of the second overtones (and higher) are relatively simple as these bands are due to more local vibrations [7, 8]. In some cases, absorption arising from the second overtone of the $C = O$ stretching vibration in the $5150-5050$ cm^{-1} region can be observed [9]. The position and intensity of this band appears to be very sensitive to solvation. Recently, the first identification of the bands due to the overtones of the $\nu(C \equiv N)$ in NIR spectra of simple nitriles has been reported [10]. Despite of their weak intensity, these bands can be a valuable source of information on the structure of liquid nitriles.

NIR spectra also contain rich physicochemical information on the sample [5]. These include not only the structural properties of the absorbing molecules but also a variety of other important features of matter and processes [5]. NIR spectroscopy plays a profound role in the exploration of hydrogen bonding, in which its sensitivity toward X–H stretching vibrations has been a key advantage [5, 11, 12]. In this context

NIR imaging spectroscopy should be mentioned as well since it demonstrates a strong potential in this area as well, e.g. when investigating processes involving solvent molecules or molecular interaction in polymers [13].

NIR spectroscopy has frequently been used when studying solution phase and various solvent effects [5]. Certain vibrations, e.g. $v(OH)$, undergo distinct spectral changes in response to change in the chemical environment (for instance change in the concentration level of solvent) [5]. Similar observations have been made for hydrogen-bonded complexes [14]. It has been noticed that the underlying mechanisms are non-trivial and involve an interplay of the anharmonicity in the vibrational potential and the nonlinearity of the transition dipole moments [2, 9]. NIR spectroscopy is indispensable in investigation of this phenomenon, as data on fundamental, first, second, and often third overtone bands are necessary. At the same time, conventional methods of spectral analysis fail to deliver decisive insights in this case and advanced tools of computational chemistry proofed necessary to reproduce and explain the changes occurring in mechanical and electrical anharmonicity in response to the changing environment [2, 9].

13.2 Hydrogen Bonding Studies

Vibrational spectra are sensitive markers of hydrogen bond interactions. The most evident proof of the presence of hydrogen bonding is a shift in the position of IR peaks originating from the groups involved in this interaction. The shift is often easily observable in vibrational spectra and enables a monitoring of hydrogen bond properties. The origin of this shift can be explained by using a relatively simple model of molecular vibrations, namely the classical harmonic oscillator. For a diatomic molecule with masses m and M, the reduced mass μ is given as:

$$\mu = \frac{Mm}{M + m} \tag{13.1}$$

This oscillator has a single vibration with the frequency v_{osc} expressed as:

$$v_{\mathrm{osc}} = \frac{1}{2\pi}\sqrt{\frac{k}{\mu}} \tag{13.2}$$

with k being the associated force constant. The frequency of vibration is proportional to the square root of the force constant specific to the bond. Hydrogen bond formation leads to changes in the effective force constant of the vibration, which is manifested as a shift of the corresponding band. Upon formation of a hydrogen bond, the oscillator may be subjected to two kinds of changes, depending on the type of vibration and the geometry of the bond (Fig. 13.3). Typically, the force constant of an X–H stretching vibration is red-shifted upon the formation of a hydrogen bond

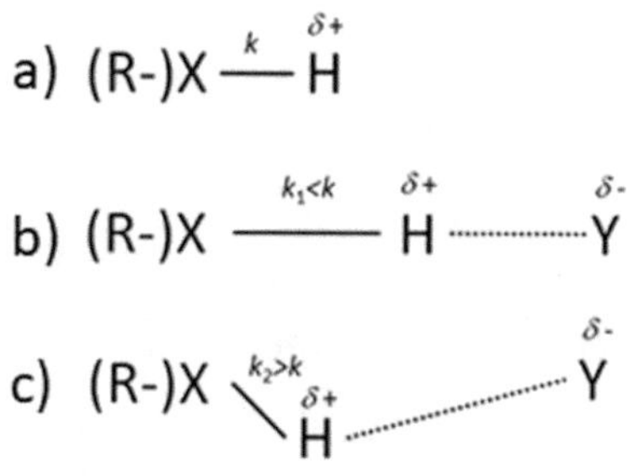

a) $(R-)X \xrightarrow{k} \overset{\delta+}{H}$

b) $(R-)X \xrightarrow{k_1 < k} \overset{\delta+}{H} \cdots\cdots\cdots \overset{\delta-}{Y}$

c) $(R-)X \underset{\overset{|}{H}}{\overset{k_2 > k}{\diagdown}} \overset{\delta+}{} \cdots\cdots\cdots \overset{\delta-}{Y}$

Fig. 13.3 Simplified scheme of the relation between the force constant of an X–H oscillator with respect to the formation of an X–H···Y hydrogen bond. **a** Non-bonded group. **b** X–H stretching vibration in the hydrogen-bonded group. **c** X–H bending vibration in the hydrogen-bonded group

(X–H···Y). This results from the attractive interaction between the positively charged H–atom and electron rich acceptor Y, which weakens the X–H bond and reduces the associated force required for its elongation [15]. On the contrary, the force constant of a bending vibration is effectively increased upon hydrogen bond formation since the X–H···Y bond is more rigid. As a result, a higher energy is required to induce angular deformations and the bending bands associated to the X–H bond are shifted to higher wavenumber (blue-shift). This is a simplified picture, however as it neglects a number of other factors influencing hydrogen bonding and the associated molecular vibrations. Comprehensive information on these phenomena may be found in the literature [15, 16]. It should be noted that the manifestation of hydrogen bonding in NIR spectra is different than in MIR spectra. The bands resulting from hydrogen-bonded species are strong in the fundamental but relatively weak in the overtone region. The stronger the hydrogen bonding, the more pronounced is this tendency. On the other hand, weakly bound and the free OH groups are more visible in the overtone region. Hence, an analysis of both spectral regions provides more comprehensive information on the hydrogen bonding properties. An overview of the current state of knowledge on this subject will be provided in Sect. 13.4 of this chapter.

The apparent manifestation of hydrogen bonding in NIR spectra attracted considerable attention since the 1950s [11]. As discussed earlier, MIR spectra of self-associating samples are dominated by broad absorption from the hydrogen-bonded species, whereas the absorption resulting from free and weakly bound groups is very weak [5, 17, 14]. An opposite situation is observed in NIR region. Figure 13.4 displays MIR and NIR spectra of neat *tert*-butyl alcohol. It can be seen that the absorption of the free OH group is not visible in the fundamental region, whereas the corresponding first overtone displays a prominent band. Most of models of association of alcohols are based on the knowledge of the population of the monomers [5, 18–20]. Therefore, examination of the dissociation of higher associates into monomers and smaller associates is more convenient by using NIR spectroscopy. In addition, the overtones of different hydrogen–bonded species are better separated as compared with the analogous fundamentals. As a result, NIR spectroscopy has been intensively applied for studies of a variety of hydrogen–bonded systems ranging from simple

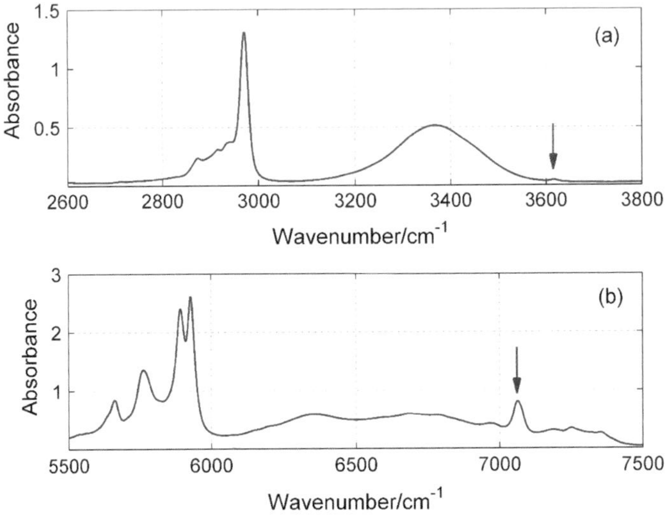

Fig. 13.4 MIR (**a**) and NIR (**b**) spectra of neat *tert*-butyl alcohol at 20 °C. The red arrows indicates the position of the band due to the free OH group

organic liquids to complex biological samples. Particular interest has been given to the examination of bulk water and aqueous solutions [20–22]. Also, self-association of alcohols and phenols was intensively explored by NIR spectroscopy [5, 19, 23].

The introduction of FT technique to NIR spectroscopy opened new possibilities in studies of hydrogen bonding. One of the first works showing a potential of FT–NIR spectroscopy was devoted to temperature-induced dissociation of fatty acids in the liquid phase [24]. The obtained high-quality spectra enabled to remove the contribution from the C–H combination bands and determine the intensity of the first overtone of the free OH as a function of the temperature. This way, it was possible to determine the population of the free OH groups and associated thermodynamical parameters such as ΔH and ΔS for the process of dissociation of the dimers into the monomers. As expected, the mean association number decreases upon elevation of the temperature. Later, the usefulness of the second overtone of the OH stretching mode for studies of the hydrogen bonding was demonstrated [5]. However, most works employ the first overtone since its intensity is significantly higher. NIR spectroscopic studies also explored the dissociation and thermodynamic properties of N–methylacetamide in the pure liquid state and CCl_4 solutions [5]. Another study of decan–1–ol in the pure liquid phase and CCl_4 solutions revealed that the bands of the first and second overtones associated to the free OH have a fine structure. As shown, this structure has a complex origin and is due to the rotational isomerism of the OH group (for the first time observed in NIR spectra) and the presence of the free terminal OH groups in linear associates [25]. Afterward, this assignment was confirmed by 2DCOS analysis of temperature-dependent NIR spectra of oleyl alcohol and numerous other studies reviewed elsewhere [5].

Self-association of a series of aliphatic alcohols in CCl_4 (0.01–1.00 M) has for the first time been examined by means of multivariate curve resolution (MCR) in the NIR region from 1900–2200 nm (5263–4556 cm^{-1}) [19]. The MCR method enabled to resolve NIR spectra into three components originating from monomers, linear aggregates, and cyclic aggregates. As shown, the size of the aggregates increases with increasing concentration. Due to steric hindrance, the branched alcohols tend to form smaller aggregates as compared to straight chain alcohols. It should be noted that the formation of the cyclic aggregates is more favorable than theô linear ones. As a result, at low to moderate concentrations of alcohols, the solutions are expected to consist mainly of the cyclic aggregates.

The examination of concentration and temperature-induced changes in NIR spectra of different alcohols lead to the conclusion that the degree of association in the saturated straight-chain 1–alcohols decreases with an increase in the chain length [26, 27]. For long-chain alcohols the hydrophobic interactions between the chains have a stronger effect on the aggregation than the OH···OH interactions. Similarly, an increase in the order of the alcohol also leads to a decrease in the extent of the self-association as a result of restricted accessibility of the OH group [18, 27, 28]. In spite of using FT–NIR spectroscopy coupled with 2DCOS it was not possible to identify any differences between the spectra of the optically active and racemic samples of octan–2–ol [28]. A possible explanation is that the relatively high degree of freedom of the OH group in octan–2–ol is reducing the chiral discrimination effect.

The combination of NIR spectroscopy with measurements of nonlinear dielectric effects provided further information on the hydrogen bonding and association of octyl alcohols in the pure liquid phase and CCl_4 solutions [29]. At low to moderate concentrations of the alcohol, nonpolar cyclic species dominate. An increase in the concentration of the alcohol leads to the formation of linear species showing a significant dipole moment, and this tendency is more pronounced in case of straight-chain alcohols. The branching in the vicinity of the OH group enhances the probability of the formation of cyclic associates.

Recently, Orzechowski and Czarnecki studied the association of 1–hexanol in n–hexane by using experimental (nonlinear dielectric effect, NIR spectroscopy) and theoretical (DFT) methods [30]. Numerical fitting of the dielectric data provided populations of all species present in the mixture (Fig. 13.5). At lower alcohol content, the cyclic species dominate, while at higher concentrations of 1–hexanol, the equilibrium shifts toward linear associates. The obtained results do not confirm common assumption that the cyclic tetramers are the most populated species in liquid alcohols. It appears that the most abundant cyclic species are trimers and population of the cyclic associates rapidly decreases with increasing size of the associates. An average size of the associates depends on the range of concentrations. This observation nicely explains differences between models of association obtained from various studies. Differences in the population of the free OH group obtained from dielectric and NIR measurements may suggest that the bands due to the free OH do not include contribution from free-terminal OH groups in the open associates. However, this controversial supposition needs further experimental verification.

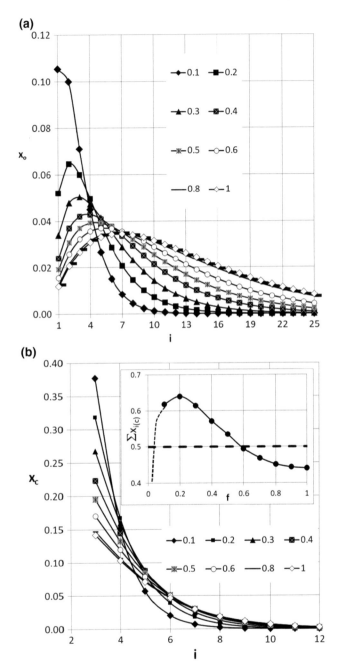

Fig. 13.5 Fraction of 1–hexanol included in the open (**a**) and cyclic (**b**) associates consisting of "i" molecules for different mole fractions of 1–hexanol in hexane. An insert shows the sum of mole fractions of molecules in cyclic associates as a function of the mole fraction of 1–hexanol. Reprinted with permission from Elsevier (Ref. [30])

Morisawa and Suga studied the effect of intermolecular interactions on intensity of the overtones of OH stretching vibrations of methanol, methanol–d_3, and tert-butanol-d_9 in n–hexane [31]. The authors determined the relative intensities of the free and hydrogen–bonded OH for the fundamental and overtone bands (v_{01}, v_{02}, v_{03}, and v_{04}). The obtained results suggest that variation in the dipole moment function of the OH group generated by hydrogen bond formation induces changes in the transient dipole moment. Dong et al. examined interactions in ethanol/water mixture by using NIR spectroscopy [32]. The application of a curve-fitting method enabled the identification of six different kinds of water species in the alcohol–poor region (<10%). The connection of the excess spectra and 2DCOS for mixtures with higher alcohol content (10–100%) revealed that the maximum of the alcohol–water interactions occur at an alcohol concentration of 40%. Further increase of the alcohol content resulted in self-association of ethanol molecules at the expense of the water–alcohol associates.

The structure of water has been also studied by NIR spectroscopy together with PCA, 2DCOS, and MCR. The studies are described in the other chapters of this book.

13.3   Anharmonic Effects in Vibrational Spectroscopy

NIR spectroscopy is a powerful tool to investigate the anharmonicity in molecular vibrations and its impact on vibrational spectra. Considerable attention has been paid to the interplay of hydrogen bonding and anharmonic effects [12]. It is known that the formation of a hydrogen bond induces a change in anharmonicity in the vibrations of the respective bonded groups. However, an intensive discussion of the extent of this change in anharmonicity took place over the years. For some hydrogen–bonded complexes, anharmonicity leads to a decrease in the fundamental excitation of an X–H bond, whereas the wavenumbers of overtones displays an increase. However, in many cases, it has been difficult to obtain quantitative insight into the anharmonicity and the associated coupling constants since experimental methods have a limited potential in this regard. Therefore, progress achieved in this area is strongly connected with quantum chemical calculation studies. Over the last decades, numerous investigations have been aimed at different types of hydrogen–bonded complexes. The anharmonicity of a moderately strong hydrogen–bonded HCN$\cdots$HF complex is relatively well studied. It was determined that a moderate increase by 27 cm^{-1} (from 90 to 117 cm^{-1}) in the anharmonic constant of the v(HF) mode occurs in the HCN$\cdots$HF complex as compared to non-bonded HF [12]. Investigations of alcohols enabled to draw comparative data on the change of anharmonicity induced to stretching and bending OH vibrations upon formation of hydrogen-bonded OH$\cdots$O complexes. The conclusion was drawn that the anharmonicity of the high–frequency stretching vibration is amplified and the magnitude of this effect is correlated to the strength of the hydrogen bond [12].

The theory of hydrogen bonding and the role of anharmonicity remained a matter of intense research. With the availability of advanced quantum mechanical calculations further progress in the understanding of the relation between hydrogen bonding and anharmonicity could be achieved. By applying the methodologies described in the chapter "*Introduction to Quantum Vibrational Spectroscopy*," detailed information on the change in the anharmonicity of the vibrational potential and the transition dipole moments upon the formation of hydrogen bonding can be obtained. A number of examples have been discussed in the recent literature, for instance, a systematic research by Futami and co-workers [3]. These investigations yielded deep insights into the discussed phenomenon by studying NIR and IR spectra of pyrrole, pyridine, and pyrrole–pyridine complex in solution phase (CCl_4). The foundation for the study was formed by the observation that the first overtone of the NH stretching vibration of a non-bonded pyrrole molecule appears as a well-resolved band at 6856 cm^{-1}; however, it is not observed for a pyrrole–pyridine complex. The theoretical calculations by Futami et al. used Johnson's reformulation of the Numerov approach and yielded detailed information on the vibrational levels and the dipole moment functions of the ν(NH) mode in a non-bonded pyrrole molecule as well as the pyrrole–pyridine complex. Those results explained why the 2ν(NH) transition of pyrrole is weakened upon the formation of a NH–N hydrogen bond in the pyrrole–pyridine complex. Firstly, reproduction of the shift observed for the experimental peak as well as the variation in its intensity was achieved. Further insight was obtained from the analysis of the one-dimensional vibrational wave functions of the two molecular systems. The conclusion was drawn that upon the formation of hydrogen bonding, the transition dipole moment diminishes while at the same time the overlap integral of the wave function is enhanced. This leads to a dramatic decrease in the intensity of the 2ν(NH) band of pyrrole to a level at which it is hardly detectable by the experiment. Therefore, the study revealed a significant decrease in the 2ν(NH) transition dipole moment upon formation of the pyrrole–pyridine complex. This in turn results in a remarkably weak intensity of the first overtone band observed experimentally for the hydrogen–bonded NH group [3].

The investigation of the changes caused by hydrogen bonding to the anharmonicity of vibrational potential and transition dipole moment has been continued with a number of different systems [3]. Those included complexes featuring NH–π hydrogen bonding such as pyrrole–ethylene and pyrrole–acetylene. This investigation revealed that the stabilization energy of NH–π hydrogen bond is almost two-thirds lower than the stabilization energy of a typical NH–N bond. Furthermore, the formation of NH–π hydrogen bonding induces a comparably small red-shift in the fundamental and first overtone of ν(NH) bands. It was concluded that the energy shift depends on the intermolecular force between the hydrogen–bonded molecules. On the other hand, Futami et al. also observed an increasing trend in the intensity of the ν(NH) fundamental absorption but a decreasing trend in the intensity of the ν(NH) first overtone band. Consequently, this observation made for NH–π hydrogen bonding remained in full agreement with earlier results for the pyrrole–pyridine case, in which NH–N hydrogen bonding is present.

Highly accurate calculations of vibrational levels and transition dipole moments is indispensable in reproducing fine effects observed in NIR spectra that are induced by the interactions with solvent molecules. The spectral manifestation of the interaction with solvent was investigated in detail by Futami et al. using similar methods [33]. Again, a pyrrole molecule was examined while the selection of solvents (n–hexane, CCl_4, $CHCl_3$, and CH_2Cl_2) was dictated by their gradually changing properties, being dielectric constant, polarity, and acidity. These solvents are suitable for spectral measurements in the NIR and IR regions where $\nu(NH)$ fundamental and the first overtone bands appear. A variation in the shift and intensity of the $\nu(NH)$ and $2\nu(NH)$ bands of pyrrole was observed depending on the solvent used. It was observed that the shift in the wavenumbers decreases in the following order of solvents: $CCl_4 > CHCl_3 > CH_2Cl_2$. At the same time, the absorption intensity of these two bands increases in the same order and is more pronounced for the fundamental than the overtone band. These trends correspond directly to the increasing order of the static solvent permittivity of the solvents being $CCl_4 > CHCl_3 > CH_2Cl_2$. The study suggested that the dependency of the solvent shift on the solvent permittivity results from the anharmonicity of the vibrational potential. However, the intensity variations result from changes in the slope of the dipole moment function (Fig. 13.6) [33]. Therefore, the study suggested that mechanical and electrical anharmonicity each have a distinct and non-trivial impact on the observed NIR spectra [33]. Interestingly, the spectral variability of NH stretching bands of pyrrole caused by solvent effects is quite different from the trends resulting from hydrogen bonding.

Further insights into the dependency X–H stretching vibrations of a solvated molecule with respect to the solvent permittivity were reported by Futami et al. in their examination of the HF molecule [34]. HF is an archetypal polar molecule and a simple electronic system suitable for application of more advanced theoretical methods. In addition to density functional theory (DFT) calculations on B3LYP/6–311 + +G(3df,3pd) level, the significantly more reliable coupled-cluster singles and doubles method (CCSD) in conjunction with the aug-cc-pVQZ basis set was applied as well. Both approaches utilize a solvent cavity model by means of SCRF/IPCM [34]. The study revealed that the vibrational potential and dipole moment function of a solvated HF molecule vary in accordance with the permittivity of the solvent. Another finding indicated that the absorption intensities of the fundamental increase proportionally to the permittivity. However, the intensities of the first, second, and third overtones do not increase continuously. In addition, the study demonstrated the accuracy of the applied quantum chemical approaches, with the DFT–B3LYP and CCSD methods leading to substantially different calculation results in the dependence of absorption intensities on static solvent permittivity [34].

The impact of solvent effects on the anharmonicity of the potential was further explored by Gonjo et al. [2]. They examined solvent effects in MIR and NIR spectra of phenol and its 2,6–dihalogenated derivatives with F, Br, and Cl atoms (Fig. 13.7). The experimental study revealed a characteristic pattern in the intensity of the consecutive OH stretching band of the fundamental, first, second, and third overtone. Moreover, it was deemed sensitive to the interaction with solvent molecules. Again, the same solvents as in previous studies were used, namely CH_2Cl_2, $CHCl_3$, CCl_4, and

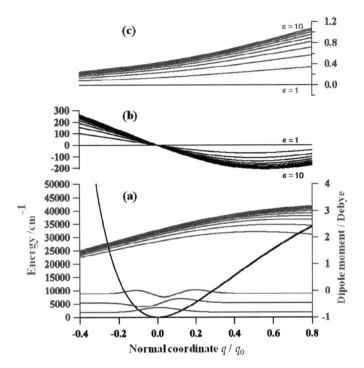

Fig. 13.6 **a** Dependences of the potential energy curve, dipole moment function ($\varepsilon = 1$ to 10) and wave function ($\varepsilon = 1$) of NH stretching mode on the relative permittivity. **b** Difference of the potential energy curve between the calculation result for permittivity of 1 and a variety of permittivity. **c** difference of the dipole moment function between the calculation result for permittivity of 1 and a variety of permittivity. Reprinted with permission from Ref. [33]. Copyright (2011) American Chemical Society

n–hexane. These solvents were shown to modulate the magnitude of the observed intensity variations [2]. A bandshift of ν(OH) bands (ν_{01}, ν_{02}, ν_{03}, and ν_{04}) from a gas state to a solution state (solvent shift) was observed and a linear relationship with the solvent was concluded (Fig. 13.7). Moreover, a solvent slope effect is observed for the ν(OH) mode of phenols. Comparison of the relative ν_{01}, ν_{02}, ν_{03}, and ν_{04} band intensities of ν(OH) measured in CCl_4, $CHCl_3$, and CH_2Cl_2 against the corresponding intensity in n–hexane demonstrated an increase in the fundamental and the second overtone but a decrease in case of the first and third overtones (Fig. 13.7). The slope of the solvent shift decreases in the order of phenol, 2,6–difluorophenol and 2,6–dichlorophenol while becoming larger with the increase in the solvent permittivity. To analyze the experimental spectra in a wide range of wavenumbers, covering visible, near-infrared, and infrared regions (15,600–2500 cm^{-1}), quantum chemical calculations capable of accurate reproduction of these transitions were applied. These required to determine the potential energy curve along the OH stretching coordinate q within the boundaries of -0.7 to $1.0\ q_0$ around the equilibrium q_0 with a fine step $0.02\ q_0$. In that case, q_0 denoted the atomic displacement vectors in Å along the

Fig. 13.7 Slopes of solvent shifts of phenol, 2,6–difluorophenol, 2,6–dichlorophenol and 2,6–dibromophenol in *n*–hexane, CCl$_4$, CHCl$_3$, and CH$_2$Cl$_2$. **a** Observed and **b** calculated B3LYP/6–311 ++G(3df,3pd) level. **c** Calculated at B3LYP/cc–pVTZ level. Reprinted with permission from Ref. [2]. Copyright (2011) American Chemical Society

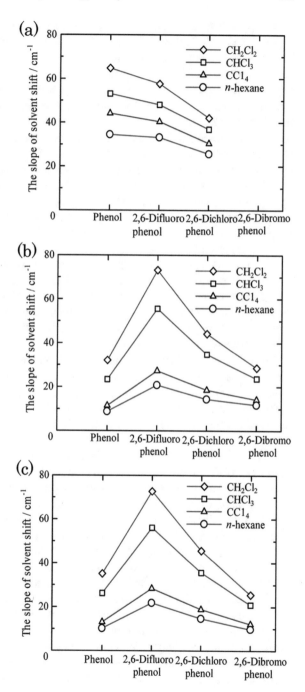

normal coordinate corresponding of ν(OH) mode. The potential energy in each point was obtained at the B3LYP/6-311++G(3df,3pd) level employing the IPCM solvation model. Subsequently, to obtain vibrational states, the one-dimensional Schrödinger equation was solved using Johnson's reformulation of Numerov's approach. This approach yielded the vibrational levels at the accuracy level exceeding 0.001 cm^{-1} with respect to the determined vibrational potential, and allowed to avoid approximations such as the fitting of a Morse function [2]. The corresponding transition intensities were obtained from the integrated absorption coefficient (km mol^{-1}, base e). The accurate calculation of wavenumbers and intensities of the ν(OH), 2ν(OH) and 3ν(OH) bands enabled a direct interpretation of the observed intensity variations as well as respective solvent dependency. The calculations reproduced the observed "parity" in the intensity of the ν(OH) bands ν_{01}, ν_{02}, ν_{03}, and ν_{04} in phenols as well as the respective solvent dependency. The parity effect was notably more prominent for phenol than for 2,6–dihalogenated phenols. It was concluded that this difference results from phenol having a stronger intermolecular hydrogen bonding in contrast to its derivatives that feature a weaker intramolecular hydrogen bond. By this reasoning, it was suggested that the intermolecular hydrogen bond between the OH group and the Cl atom is responsible for the observed tendencies [2]. The electrical anharmonicity in the system, manifested as a nonlinear dependence of the transition dipole moments with respect to the nuclear coordinates [35], may contribute to the parity effects observed by Gonjo et al. [2].

Most of the combined experimental and computational studies of anharmonic effects focused on X–H vibrations (e.g. N–H, O–H, and F–H) since the respective fundamental and overtone bands are relatively strong and well-resolved. Comparatively little knowledge is available about the anharmonicity of other kinds of vibrations and how they are influenced by the molecule's chemical surrounding. However, ν(C $=$ O) modes were investigated in the context of anharmonicity and solvent effects by Chen et al. [9]. The study focused on the IR and NIR spectral regions of acetone and 2–hexanone. As solvents, n–hexane, CCl$_4$, and CHCl$_3$, were used and the study considered vapor phase data for comparison as well. It was confirmed that the wavenumbers, absorption intensities, and oscillator strengths of the ν(C $=$ O) modes demonstrate a distinct solvent dependence. In case of the fundamental and the first overtone bands, the ν(C $=$ O) intensities were found to be significantly stronger than those of the ν(C–H) vibration. At the same time however, the ν(C $=$ O) and ν(C–H) bands were found to be comparable in terms of their intensity. Quantum chemical calculations reproduced the observed trends in integrated intensity upon going from the fundamental to the first overtone of the ν(C $=$ O), ν(O–H), ν(C–H), and ν(S–H) vibrations (Fig. 13.8). The combined theoretical and experimental results suggest that the weak intensity observed for the 2ν(C $=$ O) stretching overtone has a twofold cause. Low anharmonicity of the vibrational potential and a substantial reduction in the oscillator strength were suggested to contribute to this spectral effect [9].

As one may conclude from this chapter so far, studies focusing on solvent effects are of particular importance in physicochemical NIR spectroscopy. However, advanced computational approaches (e.g. Refs. [2, 9]), explicitly taking into account solvent effects may often be unsuitable in practice, e.g. for investigations of larger

Fig. 13.8 Experimental oscillator strength of the C = O stretching mode of acetone and 2–hexanone in n–hexane, CCl$_4$, and CHCl$_3$, along with the oscillator strengths of the O–H, C–H, N–H, and S–H stretching modes. Reproduced with permission from Ref. [9]. Copyright (2014) American Chemical Society

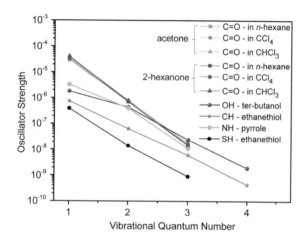

molecules in solution. Therefore, it is of particular importance to evaluate the accuracy vs. efficiency of different approaches. It is essential to know, when it is permissible to accept approximations in order to be able to perform reliable quantum chemical calculations of NIR bands of more complex systems. The impact of a relatively inert solvent, such as CCl$_4$, can often be reasonably approximated by an implicit solvation included in the calculations. For instance, the polarizable continuum model (PCM) is a commonly used method and was shown to improve the calculated spectrum in large number of cases [36]. However, inclusion of solvent effects has to be carefully considered. In spite of its low relatively permittivity of 2.228, CCl$_4$ may act as both a weak hydrogen bond acceptor as well as a halogen bond donor. The impact of these interactions significantly influences the vibrational behavior of a solvated molecule. Accordingly, explicit structural motifs formed between the solute and solvent molecules have to be taken into account. This problem has recently been investigated in detail based on anharmonic analysis using the Numerov and VPT2 methods (refer to the chapter *Introduction to Quantum Vibrational Spectroscopy*). The former approach yields highly accurate prediction of vibrational frequencies and is very useful to explore fine spectral effects. Although in practice limited to applications focused on smaller molecules, it may be used to benchmark the accuracy of the method intended to use for larger systems. The evaluation carried out by Schuler et al. [37] on the basis of methanol, phenol, and thymol molecules highlighted a consistent decrease in the ν(OH) and 2ν(OH) wavenumbers in the order vacuum > implicit solvation > explicit CCl$_4$ model (using one or two solvent molecules) in vacuum. The explicit approach provided better results as compared with implicit treatment of solvation effects. Harmonic and Numerov approaches showed no further improvement by placing the explicit solute–solvent model in an implicit solvation. However, in this example, Numerov approach yielded the best predictions with deviations from experiment of 5, 20, and 18 cm^{-1} in case of the fundamental bands and 10, 39, and 40 cm^{-1} for the first overtone of methanol, phenol, and thymol, respectively. This corresponds to errors being smaller than 0.5% in each case that can be attributed to the approximations inherent to the underlying electronic structure theory. Unfortunately, modeling an explicit solvation in CCl_4 dramatically increases the computational demand with each solvent molecules adding 74 electrons to the calculation. This is in contrast to the comparably inexpensive PCM implicit model. As benchmarked, it adds only miniscule cost to VPT2 calculations of NIR spectra [38]. Conveniently, the strongly local character of the $\nu(OH)$ mode opens a way for a feasible simplification applied within the Numerov approach. Accordingly, the r_{OH} distance can be varied only by changing the coordinate of the associated oxygen and hydrogen atom, thus serving as a simple approximation to the normal coordinate. Hence, the harmonic analysis step may be skipped in this approach. Interestingly, the VPT2 calculations by Schuler et al. [37] led to the best results with implicit solvation. This result confirmed previous findings by Beć et al. [39] which have not been benchmarked vs. higher level anharmonic computations, but instead were based on comparisons with the experimental spectra. Nevertheless, it is advised to examine carefully the frequencies predicted by VPT2 calculations in conjunction with an implicit solvation model. The studies discussed here reveal the need to take spectral shifts in the calculated frequencies into account, which depend on the chosen theoretical method and the examined solute as well.

Finally, one should highlight the key importance in spectroscopy of anharmonic approaches to molecules in aqueous environment [40]. As it was mentioned earlier, water serves as the essential medium for biochemical processes. Therefore, the detailed understanding of the vibrational spectra of hydrated molecules is crucial for progressing the potential of vibrational spectroscopy in the monitoring of biological samples. However, water creates a polar solvation environment with high permittivity, a high mobility of solvent molecules as well as directional interactions, e.g. change dipole interactions and hydrogen bonding. It is challenging to properly account for the related effects, which calls for reliable high-quality computational solvation treatments. The considerations toward feasible approaches to this problem have been recently reviewed [3]. Typical examples are sophisticated stationary point calculations incorporating implicit solvation models or explicitly considered solvent molecules, as well as applications of ab initio molecular dynamics (MD) [3]. Because of the high computational cost of the latter, one needs to consider the accuracy level necessary to describe inner and outer solvation layers and the impact of the necessary approximations on the vibrational analysis. Such considerations have been made by Lutz et al. in their methodological study of the vibrational spectrum of aqueous glycine [40]. The authors compared the spectra of hydrated glycine simulated using diverse approaches to solvation modeling with subsequent anharmonic treatments [40]. Simplified MD simulations indicated that an accurate quantum–mechanical treatment that is applied solely to the hydrated molecule, leads to an inadequate description of the hydration. An adequate account for the influence of hydration requires stepping beyond a simplified QM/MM scheme, in which the coupling is realized via empirical Coulombic and non-Coulombic interaction potentials. Further, MD simulations with electrostatic embedding offered only a slightly improved description but still did not reproduce the preferred charge configuration of the solvated glycine. In contrast, Hartree–Fock-based QMCF-MD simulation that included a QM treatment of the inner hydration shell showed an improved accuracy of the simulated vibrational spectrum. The authors reasoned that application of a higher ab initio level method such as perturbation theory to the latter scheme would likely yield an excellent agreement with the experimental spectrum. The study indicated the promising development directions toward the interpretation of the spectra of hydrated molecules [40].

13.4 Structural Information Derived from NIR Spectra

NIR spectra are a rich source of information on molecular structures, which are sensitive to the chemical environment and solvent effects. Sophisticated approaches are often required to effectively elucidate this information. Toyama et al. examined temperature-dependent spectral changes in the first overtone of the OH groups of alkane–α,ω–diols in the liquid and solid phase [41]. It appears that the spectra of alkane–α,ω–diols with an odd number of carbon atoms are similar in the pure liquid and solid states. On the other hand, the spectra of diols with an even number of carbon atoms reveal significant changes upon moving from the solid to the liquid phase. Hence, an analysis of NIR spectra confirmed the presence of the even–odd alternation in solid alkane–α,ω–diols. Liu et al. applied NIR spectroscopy coupled with chemometric methods for the examination of polymorphic transformations of oleic acid [42]. Temperature-dependent NIR spectra were resolved into independent spectral components by using alternating least squares (ALS) optimization. The obtained results demonstrate that the $\gamma \rightarrow \alpha$ transition is determined by the behavior of the COOH group, while the $\alpha \rightarrow \beta$ transition is due to the conformational changes of the acyl chain.

A significant advancement in elucidating structural information from NIR spectra is linked to the progress in the practical applications of anharmonic methods in computational chemistry [1]. The theory of NIR spectroscopy and selected applications is presented in another chapter (*Introduction to Quantum Vibrational Spectroscopy*). Here, a number of examples of using computational chemistry to increase the chemical structural specificity of NIR spectroscopy are outlined. General anharmonic approaches, such as VSCF or VPT2 (*Introduction to Quantum Vibrational Spectroscopy*), have been implemented in an almost a routine way in popular computational chemistry software packages. The computationally efficient VPT2 method has mostly been used in practical applications to NIR spectroscopy. Relatively accurate calculations of entire NIR spectra of medium-sized organic molecules have become feasible over the recent years [3] improving our understanding of this spectral range. In comparison with MIR or Raman spectroscopies, computational studies are of particular importance for NIR spectroscopy, due to the complex character of NIR spectra. Even relatively simple molecules have a large number of contributing bands as demonstrated by the calculated spectra [43]. As an example, we show the reconstructed spectra of vinylacetic acid. Figure 13.9 displays the individually modeled bands and the predicted theoretical lineshapes, which result via summation

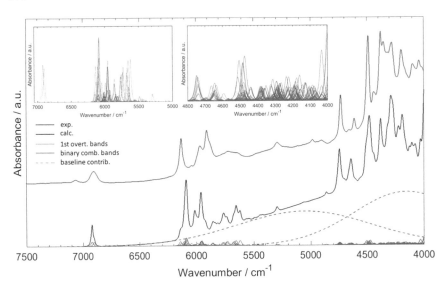

Fig. 13.9 Experimental and simulated NIR spectra of vinylacetic acid. All bands are depicted on a common scale; note the significant band overlapping. Reprinted with permission from Ref. [43]. Copyright (2017) American Chemical Society

of these bands, in comparison to the experimental spectrum, presented in a common intensity scale. To better identify specific details, the bands predicted in these two regions are additionally enlarged (Fig. 13.9). One should mention that the number of contributing bands rapidly increases with an increase in the size of molecule. Hence, NIR spectra of more complex molecules can be expected to show a huge number of underlying contributions steaming from overtones and combination bands.

Theoretical investigations also reveal that the majority of the meaningful NIR bands result from two–quanta vibrational transitions—the first overtones $\nu_{0\to2}$ and binary sum combinations $\nu_{00\to11}$. The probability of a transition decreases substantially for higher order excitations. Hence, three–quanta bands, i.e. second overtones $\nu_{0\to3}$ and ternary sum combinations $\nu_{000\to111}$, generally have a markedly lower intensity and are meaningful only in region of higher wavenumbers where the two–quanta bands are missing. For instance, the region above ca. $7200\ \mathrm{cm}^{-1}$ includes exclusively bands resulting from higher order transition [44, 45]. Besides, the contributions from higher order overtones and combination bands are important for molecules with heavy atoms only. It should be mentioned that apart from sum combinations, the vibrational spectra may include difference combinations as well; however, the probability of such transitions is extremely low. Unlike sum combinations, a difference combination involves a transition that takes place from an excited state, e.g. $\nu_{012\to001}$. At room temperatures, the majority of molecules populate the vibrational ground state and the associated difference combination bands are very weak.

Grabska et al. [44]. and Beć et al. [45] have systematically studied methanol and ethanol. They quantitatively estimated the relative contributions from different types of transitions to NIR spectra. This estimation is based on the ratio between the integrated intensities of the calculated bands, showing approximately only 19% (methanol) and 27% (ethanol) of the relative contribution to entire NIR region $(10,000\text{--}4000\ cm^{-1})$ originate from three–quanta transitions. Moreover, the three–quanta bands are numerous but very weak and tend to show a large degree of overlap. Thus, these bands are far less specific than the two–quanta counterparts. Therefore, it is sufficient to limit the calculations to the first overtones and binary combinations to predict NIR spectra with good accuracy. This conclusion has important practical implications, as for larger systems the requirement in terms of computational resources rises rapidly for the prediction of three–quanta transitions. Note that the results obtained for methanol and ethanol confirm the previous conclusions that bands resulting from two–quanta transitions are sufficient to explain the majority of features observed in NIR spectra [3, 4].

The theoretical reconstruction of the spectra enables robust band assignments, and the potential benefit is already evident in cases of relatively simple molecules such as methanol (Fig. 13.10) [3]. Similar improvements in the reconstruction of NIR spectra have been reported for a number of medium–sized molecules in diluted solutions. For instance, the simulations carried out by Grabska et al. [46] for the isomers of butyl alcohol (*n*–butanol, *sec*–butanol, *iso*–butanol and *tert*–butanol) reconstructed the differences in the respective NIR spectra between 6000 and 4000 cm^{-1} and reproduced the fine structure of NIR spectra in the regions from 5200 to 4600 and from 4500 to 4000 cm^{-1} (Fig. 13.11).

The rotational freedom of the OH group may lead to energetically distinguishable structures, called "rotational isomers." The OH group in various rotational isomers absorbs at different wavenumbers, but the corresponding bands are close to each other. Hence, the presence of the rotational isomers can be observed only in high-resolution spectra of diluted alcohols in inert solvents. MIR, NIR, and DFT studies of butyl alcohols in dilute CCl_4 solutions (0.01 M) revealed the presence of various rotational isomers in *n*–, *iso*- and *sec*–butanol (Fig. 13.12) [48]. The *trans* conformer is more favorable than its *gauche* counterpart, and as a result absorbs at lower wavenumbers. These studies revealed a minor effect of C–C dihedral rotation on the position of the first and second overtone of the OH group. The position of the OH bands due to rotational isomers primarily results from the order of the alcohol, while the relative population of a particular rotational isomer depends on the steric effects of the groups in α and β positions in relation to the OH group.

Highly accurate simulations are capable of reproducing the effect of conformational isomerism manifested also in the other NIR bands [46]. Isomers of butyl alcohol have very different conformational flexibility depending on the structure of the main chain. The number of stable rotational conformers is 14, 9, 5, and 1 for *n*–butanol, *sec*–butanol, *iso*–butanol, and *tert*–butanol, respectively. As shown in Fig. 13.11, the spectra of conformers differ noticeably throughout the entire NIR region, not only in vicinity of the $2v(OH)$ band. To reproduce NIR spectrum in detail, it is necessary to calculate the spectra of each form co-existing in the sample and mix them in accordance to the relative Boltzmann population. Grabska et al. [46] have performed a detailed conformational search for all butyl alcohols and estimated populations of

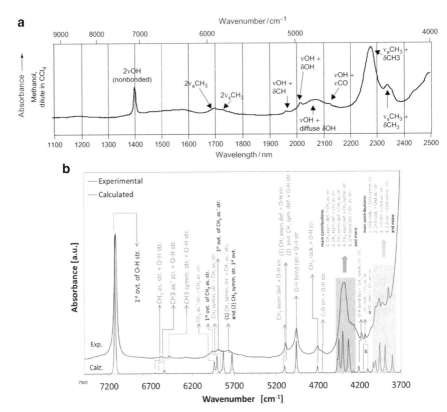

Fig. 13.10 Exemplary progress in the understanding of NIR spectra of methanol diluted in CCl$_4$ achieved via theoretical calculations of NIR spectra **a** band assignments of methanol obtained with classical spectroscopic methods. Reproduced with permission from Weyer and Lo [47], **b** the improvement achieved through anharmonic calculations (GVPT2//DFT–B2PLYP/SNST + CPCM). Reproduced from Ref. [39] with permission from the PCCP Owner Societies

all conformational isomers. Figure 13.11 displays the simulated spectra of each of the conformer weighted using the calculated Boltzmann terms, demonstrating the importance of taking the full conformational space into account. For instance, the band near 4050 cm^{-1} in the spectrum of *iso*–butanol (Fig. 13.11c) results from the second most abundant form Gg' (thin violet line in this figure), while the major conformer Gg (thin green line in this figure) does not contribute to this band in a significant way. The simulated spectra accurately reproduces the experimental shape of the 2ν(OH) band for butyl alcohols (Fig. 13.13) [46] and confirms the previous conclusions that it is an effect of conformational isomerism with respect to the rotation over C–O(H) bond [48]. The manifestation of conformational isomerism in NIR spectra has been reproduced for several other alcohols including ethanol, *n*–propanol, *n*–hexanol, and cyclohexanol as well as fatty acids [3, 43]. In case of particularly flexible molecules, it is essential to pre-screen their wide conformational space in order to select the most meaningful conformers for the more detailed spectra calculations.

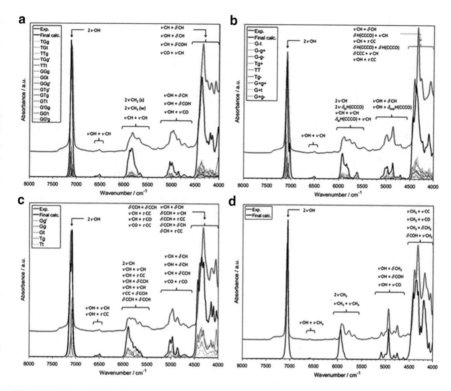

Fig. 13.11 Experimental and simulated (harmonic: B2PLYP/def2–TZVP; VPT2: B3LYP/SNST; CPCM) NIR spectra of 1–butanol (**a**), 2–butanol (**b**), iso–butanol (**c**), and tert–butyl alcohol (**d**). The contributions of the spectral lineshapes corresponding to conformational isomers presented as well (colored lines). Reprinted with permission from Ref. [46] Copyright (2017) American Chemical Society

Fig. 13.12 FT–NIR spectrum of $2\nu(OH)$ of n–butanol in 0.01 M CCl_4 solution at 20 °C together with resolved *trans* (right) and *gauche* (left) conformers

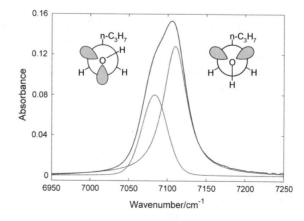

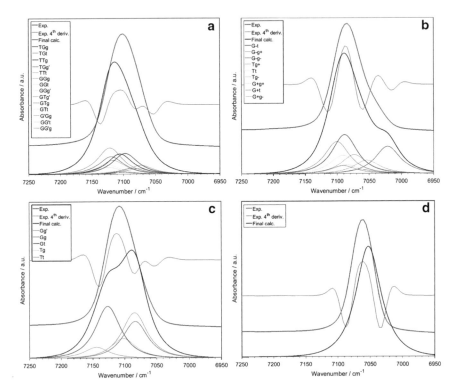

Fig. 13.13 Details of the first overtone band of OH stretching mode of *n*–butanol (**a**), *sec*–butanol (**b**), *iso*–butanol (**c**), and *tert*–butanol (**d**). Experimental spectrum, fourth derivative of the experimental spectrum, the calculated lineshape and contributions to NIR spectra originating from conformational isomers. Reprinted with permission from Ref. [46] Copyright (2017) American Chemical Society

For *n*–hexanol, where a selection of the 32 most stable forms out of 243 theoretically possible conformers resulted in the efficient modeling of the NIR spectrum [3]. High-level quantum mechanical calculations enhance the potential of NIR spectroscopy in conformational studies, as they allow to elucidate structural information even from weak and strongly overlapping bands. Moreover, such investigations are no longer limited to well-resolved bands (e.g. non-bonded 2ν(OH)), and may include broad spectral regions.

A reasonably accurate reproduction of NIR spectra was achieved for medium-size molecules like rosmarinic acid (Fig. 13.14) by Kirchler et al. [49]. The obtained band assignments (Table 13.1) was used to interpret the relationships between 2DCOS hetero–correlation contour plots and PLSR regression coefficients constructed from NIR spectra measured on different benchtop and miniaturized portable spectrometers. This example evidences the ability of obtaining detailed and reliable band assignments even for molecules with more than 40 atoms. However, the complex nature of NIR spectra resulting from the large number of overlapping bands decrease

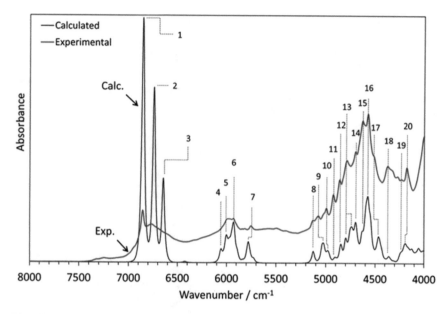

Fig. 13.14 Experimental (powder) and theoretical NIR spectrum of rosmarinic acid obtained in anharmonic GVPT2//DFT–B3LYP/N07D simulation. Band numbers correspond to those presented in Table 13.1. Reproduced from Ref. [49] with permission from The Royal Society of Chemistry

the utility of presenting band assignments in a conventional way (Table 13.1). In NIR spectra well-resolved bands seldom appear separated. Because of band overlap, e.g. as illustrated in Fig. 13.9, it is rarely possible to distinguish dominant vibrations that could be written in a simple tabular form. A far more suitable way to present the complex nature of NIR spectra are colormaps developed by Beć et al. [36]. Figure 13.15 shows an exemplary colormap for thymol. The NIR spectrum is represented in form of density maps of vibrational intensities in false-color code, reflecting the relative contribution as a function of wavenumber. This way, the band assignments can be visualized in an alternative way. In the case of thymol, this methodology permitted for observation of a distinct spectral pattern in response to the change of the sample state. The study based on thymol in polycrystalline, melted and solution (100 and 10 mg mL^{-1} in CCl$_4$) states revealed two spectral regions, in which the bands manifest relatively low sensitivity to the sample state. Spectra simulation evidenced that these two regions are populated by C–H and CH$_3$ stretching overtones and combinations (6000–5600 cm^{-1}) as well as combinations of CH$_3$ stretching and deformation modes, in addition of ring deformation (4490–4000 cm^{-1}). Interestingly, the bands with contributions from OH modes undergo significant spectral shifts and width changes. This result is reasonable since the OH group is much more sensitive to the environment as compared to the C–H and CH$_3$ groups.

Isotopic substitution is another powerful tool for the interpretation of MIR spectra. In particular, a selective deuteration of X–H groups (X = C, O, N) leads to noticeable

Table 13.1 Band assignments in NIR spectrum of rosmarinic acid, based on GVPT2//DFT–B3LYP/N07D calculation

	Wavenumber/cm^{-1}		Major contributions
	Exp.	Calc.	
1	6854.9	6853	2νOH (ar)
2	6767.2	6741	2νOH (ar)
3	~6680	6645	2νOH (carboxyl)
4	~6044	6056	2νCH (ar, aliph, in–phase)
5	5986.5	6001	2νCH (ar, aliph, opp.–phase); 2νCH (ar)
6	5929.7	5930	2νCH (ar); 2νCH (ar)
7	5752.5	5780	$[\nu_{as}CH_2, \nu CH$ (ar)$] + [\nu_{as} CH_2, \nu CH$ (ar)$]$;
8	5128.0	5126	$[\nu C = O, \delta_{ip}OH$ (carboxyl)$] + [\nu OH$ (carboxyl)$]$
9	5075.8	5027	$[\delta_{ring}, \delta_{ip}OH$ (ar)$] + [\nu OH$ (ar, para–)$]$; $[\delta_{ring}, \delta_{ip}OH$ (ar)$] + [\nu OH$ (ar, para–)$]$; $[\delta_{ring}] + [\nu OH$ (ar, para–)$]$
10	4994.9	4980	$[\delta_{ring}, \delta_{ip}OH$ (ar)$] + [\nu OH$ (ar, meta–)$]$; $[\delta_{ring}, \delta_{ip}OH$ (ar)$] + [\nu OH$ (ar, meta–)$]$
11	4923.8	4906	$[\delta_{ring}, \delta_{ip}OH$ (ar)$] + [\nu OH$ (ar, para–)$]$; $[\delta_{ring}, \delta_{ip}OH$ (ar)$] + [\nu OH$ (ar, para–)$]$
12	4860.0	4847	$[\delta_{ring}, \delta_{ip}OH$ (ar)$] + [\nu OH$ (ar, meta–)$]$; $[\delta_{ring}, \delta_{ip}OH$ (ar)$] + [\nu OH$ (ar, meta–)$]$
13	4788.3	4798	$[\nu CC] + [\nu OH$ (ar, para–)$]$; $[\nu CC] + [\nu OH$ (ar, para–)$]$; $[\nu CC] + [\nu OH$ (ar, meta–)$]$; $[\delta CCH$ (carboxyl)$] + [\nu OH$ (carboxyl)$]$; $[\delta CH$ (ar), $\delta_{ip}OH$ (ar)$] + [\nu OH$ (ar, para–)$]$; $[\delta CH$ (ar), $\delta_{ip}OH$ (ar)$] + [\nu OH$ (ar, para–)$]$
14	4701.0	4701	$[\delta CH$ (aliph)$] + [\nu OH$ (ar, meta–)$]$; $[\delta CH$ (ar), $\delta_{ip}OH$ (ar)$] + [\nu OH$ (ar, meta–)$]$
15	4629.4	4632	$[\delta CH$ (ar), $\delta_{ring}, \delta_{ip}OH$ (ar)$] + [\nu OH$ (ar, meta–)$]$; $[\delta CH$ (ar), $\delta_{ring}, \delta_{ip}OH$ (ar)$] + [\nu OH$ (ar, para–)$]$; $[\delta_{ip}OH$ (ar), δCH (ar), $\delta_{ring}] + [\nu OH$ (ar, para–)$]$

(continued)

Table 13.1 (continued)

	Wavenumber/cm^{-1}		Major contributions
	Exp.	Calc.	
16	4575.7	4757	[δCH (ar), δ_{ring}, δ_{ip}OH (ar)] + [νOH (ar, meta–)]; [δ_{ip}OH (ar), δCH (ar), δ_{ring}] + [νOH (ar, meta–)]
17	~4508	4465	[δ_{ring}, δ_{ip}OH (ar)] + [νCH (ar)]; [δ_{ring}] + [νCH (ar)]; [δ_{ring}, δ_{ip}OH (ar)] + [νCH (ar)]; [νC–O (carboxyl), δ_{ip}OH (carboxyl)] + [νOH (carboxyl)]; [δ_{ring}] + [νCH (ar)]; [δ_{ring}, δ_{ip}OH (ar)] + [νCH (ar)]
18	4372.3	4360	[δ_{ring}] + [νCH (ar)]; [δ_{ring}] + [νCH (ar)]
19	4233.3	4237	[δ_{sciss}CH$_2$] + [ν_{as}CH$_2$, νCH (ar)]; [δ_{sciss}CH$_2$] + [ν_{as}CH$_2$, νCH (ar)]
20	4179.4	4194	[δCH (aliph)] + [νCH (ar, aliph, opp.–phase)]; [δ_{sciss} CH$_2$] + [ν_sCH$_2$]; [δCH (aliph)] + [νCH (ar, aliph, in–phase)]

Arbitrary band numbers correspond to those presented in Fig. 13.14. Reproduced from Ref. [49] with permission from The Royal Society of Chemistry

shifts in the frequencies of the respective oscillators [50]. In contrast, due to band overlapping, the corresponding changes in the NIR spectra are less obvious. Moreover, combination modes may involve substituted and unsubstituted groups leading to more complex spectral pattern. For these reasons, spectra simulations provide a more comprehensive understanding of fine details in NIR spectra of deuterated samples. Grabska et al. [44]. applied anharmonic calculations to study methanol and all its possible deuterated species (CXXXOX; X = H, D). This study resulted in a number of novel conclusions including detailed band assignments in methanol (CH$_3$OH) and its major derivatives (CH$_3$OD, CD$_3$OH, CD$_3$OD). This way, it was possible to propose a complete picture of the NIR spectral patterns resulting from the isotopic substitution of the methyl and hydroxyl groups. In the same study [44], NIR spectra of several samples of CH$_3$OD obtained from different chemical suppliers have been measured. Isotopic impurities from co-existent OH/OD groups were easily identified based on the prominent 2ν(OH)/2ν(OD) peak. At the same time, the NIR spectra of CH$_3$OD showed the additional peaks that could not be assigned to CH$_3$OH, CH$_3$OD, CD$_3$OH, or CD$_3$OD. The origin of these peaks was explained by comparison with the spectra simulated for all kinds of unevenly substituted forms. This way the presence of CDHHOH in a commercial sample of CH$_3$OD could be demonstrated

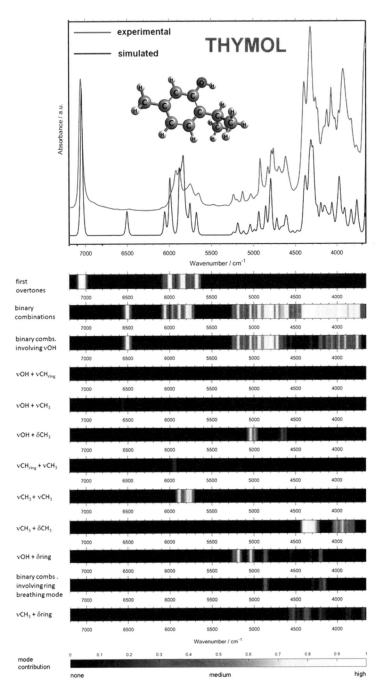

Fig. 13.15 Quantum mechanically calculated contributions to NIR spectrum of thymol (solution; 100 mg mL^{-1} CCl$_4$) presented in the form of false-color density maps for the selected vibrations. Reprinted with permission from Elsevier (Ref. [36])

[44]. This conclusion was possible to attain from the spectra simulations only, as the corresponding experimental spectra were not accessible.

Beć et al. [45] explored the effect of isotopic substitution on NIR spectra of ethanol and its derivatives by applying similar methodology further in detail. NIR spectra of these molecules are more complex than that of methanol, as a consequence of presence of the methylene group and co-existence of *gauche* and *trans* rotamers. Detailed band assignments for six ethanol isotopomers including CH_3CH_2OH, CH_3CH_2OD, CH_3CD_2OH, CD_3CH_2OH, CD_3CD_2OH, and CD_3CD_2OD were presented. In addition, the NIR spectra of CH_3CD_2OD and CD_3CH_2OD were theoretically predicted, since these samples are not commercially available. This way, the spectra simulations provide information not accessible from experimental studies. The examination of ethanol and its isotopomers provided an in-depth understanding of the effect of isotopic substitution on NIR spectra. Table 13.2 summarizes the relative contributions from two ($2\nu_x$ and $\nu_x + \nu_y$) and three–quanta ($3\nu_x$, $\nu_x + \nu_y + \nu_z$, and $2\nu_x + \nu_y$) transitions. The obtained results lead to the conclusion that the contributions from the CH_3 group appear to be more important than those from the CH_2 group. The isotopic substitution in the CH_3 group results in the most prominent intensity changes in the NIR spectra as compared to the changes due to the substitution of the other atoms. The bands resulting from three–quanta transitions appear to be more important for isotopomers of ethanol [45] than for derivatives of methanol [44].

Quantum mechanical calculations of NIR spectra proved to be very useful in explaining several other observations. For instance, the effect of baseline elevation appearing in the spectra of carboxylic acids is due to a significant red-shift and broadening of specific combination bands originating from hydrogen–bonded cyclic dimers. These findings were consistent in the studies of eight different systems [3]. However, further investigations are required to provide a decisive explanation of the mechanism underlying such selective shifts and broadening. Further, the manifestation of the C = C bond in aliphatic chains of long-chain fatty acids has been reproduced in simulation as well [51]. NIR spectra are highly sensitive to differences in the molecular structure as shown for *n*–hexanol, cyclohexanol, and phenol [52]. These structural differences induce prominent changes in the associated spectra that can be accurately reconstructed by theoretical approaches.

Spectra simulations are helpful in elucidating similarities and dissimilarities between the overtone and the fundamental regions. The fundamental bands of acetonitrile as well as the first, second, and third overtones, together with binary and ternary combination transitions, have been calculated by Lutz et al. with an attempt to benchmark a novel, highly correlated treatment of anharmonic spectra based on the CR–CC(2,3) method [52]. The examination of vibrational spectra of nitriles by quantum chemical calculations was essential for the successful elucidation of structural information [10]. The combined MIR, Raman, and NIR study of acetonitrile, acetonitrile–d_3, and trichloroacetonitrile suggested a distinct influence of the chemical environment on MIR and NIR spectra [10]. Further evidences of this effect were provided in a study of MIR and NIR spectra of polycrystalline spectra of melamine by Grabska et al. [53]. The explanation of different effects of the chemical environment on MIR and NIR spectra was obtained from the spectra simulation. IR

Table 13.2 Contributions (in %) from the first and second overtones as well as binary and ternary combinations into NIR spectra of ethanol isotopomers based on GVPT2//B2PLYP–GD3BJ/def2–TZVP//CPCM calculations)[a]

	10,00–4000 cm^{-1}				
	$2\nu_x$	$3\nu_x$	$\nu_x + \nu_y$	$\nu_x + \nu_y + \nu_z$	$2\nu_x + \nu_y$
CH_3CH_2OH	26.1	1.7	47.0	14.3	10.9
CH_3CH_2OD	18.0	2.2	51.2	17.4	11.1
CH_3CD_2OH	35.8	1.7	41.5	11.8	9.2
CD_3CH_2OH	40.9	1.2	32.6	15.1	10.1
CD_3CD_2OH	46.0	0.3	23.7	15.8	14.2
CD_3CD_2OD	43.1	2.0	19.2	17.8	17.9
CH_3CD_2OD	27.9	3.2	44.7	15.0	9.2
CD_3CH_2OD	36.0	2.5	35.5	10.8	15.2
	10,000–7500 cm^{-1}				
	$2\nu_x$	$3\nu_x$	$\nu_x + \nu_y$	$\nu_x + \nu_y + \nu_z$	$2\nu_x + \nu_y$
CH_3CH_2OH	0.0	39.7	0.0	22.3	38.0
CH_3CH_2OD	0.0	55.5	0.0	15.4	29.1
CH_3CD_2OH	0.0	43.9	0.0	30.5	25.6
CD_3CH_2OH	0.0	66.9	0.0	1.4	31.7
CD_3CD_2OH	0.0	0.0	0.0	43.0	57.0
CD_3CD_2OD	0.0	100.0	0.0	0.0	0.0
CH_3CD_2OD	0.0	69.9	0.0	16.7	13.4
CD_3CH_2OD	0.0	76.3	0.0	0.5	23.2
	7500–4000 cm^{-1}				
	$2\nu_x$	$3\nu_x$	$\nu_x + \nu_y$	$\nu_x + \nu_y + \nu_z$	$2\nu_x + \nu_y$
CH_3CH_2OH	26.5	1.2	47.6	14.2	10.5
CH_3CH_2OD	18.4	1.0	52.4	17.5	10.7
CH_3CD_2OH	36.0	1.4	41.8	11.7	9.1
CD_3CH_2OH	41.4	0.4	33.0	15.3	9.9
CD_3CD_2OH	46.0	0.3	23.7	15.8	14.2
CD_3CD_2OD	43.7	0.5	19.5	18.0	18.2
CH_3CD_2OD	28.4	2.0	45.5	15.0	9.1
CD_3CH_2OD	37.0	0.5	36.4	11.1	15.0

[a]The comparison is based on integrated intensity (cm^{-1}) summed over simulated bands, convoluted with the use of Cauchy–Gauss product function (details in the text) in relation to the total integrated intensity

Reproduced with permission from Ref. [45]

bands of melamine are more sensitive to the molecule's chemical neighborhood as compared to NIR bands. It was reasoned that neglecting the chemical environment as done in simple models (e.g. single molecule) largely reduces the accuracy in the description of those vibrations. In contrast, NIR bands seem to be less dependent on the chemical environment, in particular to the adjacent crystalline planes in the structure of melamine [53]. Comprehension of the relationships between MIR and NIR spectra is important from the point of view of spectroscopic studies and applications. Further investigations are necessary to provide a more complete picture; however, it is clear that spectral simulations play an important role in the exploration of this important aspect.

It is to be noted that new developments in anharmonic theories and applications are predominantly validated on the basis of MIR spectra, which are readily available and simpler for analysis [40]. However, results of these studies have important implications for NIR spectroscopy as well [3]. For instance, an in–depth understanding of the origin and nature of overtone and combination bands in the MIR range may be achieved [54]. Interestingly, also in MIR region, these bands are far more numerous than fundamental counterparts [54]. Typically, overtones and combinations bands are very weak, unless they are in resonance with the fundamental bands. In practical applications, calculations based on VSCF have a less favorable efficiency–to–accuracy ratio and a number of improvements have been proposed in recent years. These improvements increase the efficiency of the VSCF approach, e.g. by reducing the grid density for potential evaluations or by employing more efficient ways for determination of the electronic structure underlying of anharmonic vibrational analysis (e.g. resolution of the identity (RI) approximation in connection with Moller–Plesset second-order perturbation, i.e. RI–MP2). Interestingly, anharmonic (VSCF and VPT2) calculations have been used in connection with IR power spectra predicted by velocity autocorrelation of ab initio QMCF–MD and QM/MM–MD trajectories [40]. Such approaches can substantially increase the potential of interpretation of MIR spectra measured for highly labile systems, e.g. hydrogen–bonded molecules in aqueous solution [40].

13.5 Solution Chemistry

In the NIR region, molar absorption coefficients of many molecules are small, and thus, NIR spectroscopy is highly suitable for the investigation of hydrogen bonding in condensed phase. One can measure bands such as those due to water and solutions easily with high reproducibility. One can also reduce the effect of interface which often yields a serious issue in IR and ATR–IR spectroscopy. Using these advantages, Ikehata et al. [55] explored miscibility of solutions. They investigated the thermal phase behavior of triethylamine (TEA)–water mixtures which show phase separation with the lower critical solution temperature (LCST) type by NIR spectroscopy. They paid attention on a band shift of the first overtone of the C–H stretching modes of TEA and made a phase diagram of the mixtures. They originally thought that

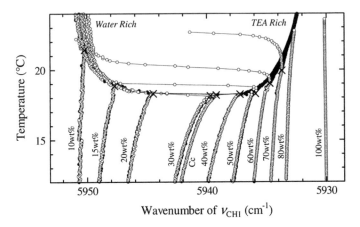

Fig. 13.16 Temperature-dependent changes of the wavenumber of the first overtone of CH_3 asymmetric stretching band (ca. 5930 cm^{-1}) in NIR spectra of triethylamine aqueous solutions with the concentration of 10, 15, 20, 30, 40, 50, 60, 70, and 80 wt%). Reprinted with permission from Elsevier (Ref. [55])

probably the shifts of the C–H bands occur due to the increase in the compression rate upon the phase separation. It has turned out that the observed shifts are bigger than the expected ones. Figure 13.16 shows the temperature-dependent changes of the wavenumber of the first overtone of CH_3 asymmetric stretching band (ca. 5930 cm^{-1}) in the NIR spectra of triethylamine aqueous solutions with the concentration of 10, 15, 20, 30, 40, 50, 60, 70, and 80 wt%. The largest shift was observed at the critical composition(Cc; 32.12 wt% The compression rate is the largest). In this way, Ikehata et al. [55] obtained the result which has a relation with density and partial molal volume. NIR spectroscopy is concerned with overtones, so that one can plot more detailed shifts than in IR spectroscopy. This result demonstrates that NIR spectroscopy is very useful for the observation of micro phase separation.

The analysis of hydrogen bonding based on concentration difference spectra and the specific attention on the vibrational modes of hydrophobic parts by Ikehata et al. [55] also suggested that even ethanol–water mixtures, which is a miscible solution, is microscopically in a state close to phase separation.

13.6 Summary and Future Perspective

In many aspects of physicochemical investigation, NIR spectroscopy is a powerful tool capable for providing unique insights that are not easily accessible from IR or Raman spectroscopy. NIR spectra consist of weak absorption bands that result from mechanical and electrical anharmonicity, which are not obscured by strong fundamental bands. The positions and intensities of these two kinds of bands are often affected in different ways, which creates a rich source of information on molecular structure and interactions. For these exclusive values, NIR spectroscopy has contributed remarkably to advancements accomplished in these fields over the last three decades. On the other hand, because of extensive band overlapping, NIR spectra are far more difficult for direct interpretation in comparison to fundamental transitions. Therefore, NIR spectroscopy in physicochemical research strongly depends on advanced methods of spectra analysis, e.g. chemometrics or 2DCOS tools. In the recent years, progress achieved in computational chemistry enabled the feasible simulation of NIR spectra by anharmonic quantum chemical calculations. This development dramatically increases the level of detail in band assignments, enables to follow up on fine spectral effects and to ascribe them to structural changes occurring at the molecular scale. Still, large molecules or some effects that are delocalized over a volume of the sample presently remain inaccessible for computational approaches due to the prohibitive cost associated to the anharmonic treatment. However, once these practical limitations are overcome, new insights, e.g. into solvent effects, molecular dynamics in large interacting systems, polymer properties, etc., can be achieved. With the continued progress in technology and quantum theory in the near future, it may be anticipated that the research focus in NIR spectroscopy will be strongly influenced in the coming years as well.

References

1. V. Barone, M. Biczysko, J. Bloino, M. Borkowska-Panek, I. Carnimeo, P. Panek, Toward anharmonic computations of vibrational spectra for large molecular systems. Int. J. Quantum Chem. **112**, 2185–2200 (2012)
2. T. Gonjo, Y. Futami, Y. Morisawa, M.J. Wójcik, Y. Ozaki, Hydrogen bonding effects on the wavenumbers and absorption intensities of the OH fundamental and the first, second, and third overtones of phenol and 2,6-dihalogenated phenols studied by visible/near-infrared/infrared spectroscopy. J. Phys. Chem. A **115**, 9845–9853 (2011)
3. K.B. Beć, J. Grabska, C.W. Huck, Y. Ozaki, Quantum mechanical simulation of near-infrared spectra. Applications in physical and snalytical chemistry. in *Molecular Spectroscopy: A Quantum Chemistry Approach.* ed by Y. Ozaki, M.J. Wójcik, J. Popp, (Weinheim, Grermany, Wiley-VCH, 2019) pp. 353–388
4. K.B. Beć, C.W. Huck, Breakthrough potential in near-infrared spectroscopy: spectra simulation. A review of recent developments. Front. Chem. **7**, 48 (2019)
5. M.A. Czarnecki, Y. Morisawa, Y. Futami, Y. Ozaki, Advances in molecular structure and interaction studies using near-infrared spectroscopy. Chem. Rev. **115**, 9707–9744 (2015)
6. L. Bokobza, Origin of near-infrared absorption bands. in *Near-Infrared Spectroscopy: Principles, Instruments, Applications*, ed. by H.W. Siesler, Y. Ozaki, S. Kawata, H.M. Heise, (Wiley-VCH Verlag GmbH, 2002), pp. 11–42
7. B.R. Henry, Use of local modes in the description of highly vibrationally excited molecules. Acc. Chem. Res. **20**, 429–435 (1987)
8. H.G. Kjaergaard, H. Yu, B.J. Schattka, B.R. Henry, A.W. Tarr, Intensities in local mode overtone spectra: Propane. J. Chem. Phys. **93**, 6239–6248 (1990)
9. Y. Chen, Y. Morisawa, Y. Futami, M.A. Czarnecki, H.S. Wang, Y. Ozaki, Combined IR/NIR and density functional theory calculations analysis of the solvent effects on frequencies and intensities of the fundamental and overtones of the C=O stretching vibrations of acetone and 2-hexanone. J. Phys. Chem. A **118**, 2576–2583 (2014)

10. K.B. Beć, D. Karczmit, M. Kwaśniewicz, Y. Ozaki, M.A. Czarnecki, Overtones of $\nu C\equiv N$ vibration as a probe of structure of liquid CH_3CN, CD_3CN, and CCl_3CN: combined infrared, near-infrared, and Raman spectroscopic studies with anharmonic density functional theory calculations. J. Phys. Chem. A **123**, 4431–4442 (2019)

11. C. Bourderon, C. Sandorfy, Association and the assignment of the OH overtones in hydrogen bonded alcohols. J. Chem. Phys. **59**, 2527–2536 (1973)

12. C. Sandorfy, Hydrogen bonding: How much anharmonicity? J. Mol. Struct. **790**, 50–54 (2006)

13. Y. Hu, J. Zhang, H. Sato, Y. Futami, I. Noda, Y. Ozaki, C-HOC hydrogen bonding and isothermal crystallization kinetics of poly(3-hydroxybutyrate) investigated by near-infrared spectroscopy. Macromolecules **39**, 3841–3847 (2006)

14. Sandorfy, C. Chapter 13 In *The hydrogen bond. Recent developments in theory and experiments.* (Schuster, P., Zundel, G., Sandorfy, C., Eds.), Amsterdam, North-Holland Publ. Co, 1976, pp 615–654

15. J. Joseph, E.D. Jemmis, Red-, blue-, or no-shift in hydrogen bonds: A unified explanation. J. Am. Chem. Soc. **129**, 4620–4632 (2007)

16. M.D. Struble, C. Kelly, M.A. Siegler, T. Lectka, Search for a strong, virtually "No-Shift" hydrogen bond: A cagemolecule with an exceptional OH…F interaction. Angew. Chem. Int. Ed. **53**, 8924–8928 (2014)

17. W.A.P. Luck, W. Ditter, Die assoziation der alkohole bis in überkritische bereiche. Ber. Bunsen-Ges. Phys. Chem. **72**, 365–374 (1968)

18. M. Iwahashi, M. Suzuki, N. Katayama, H. Matsuzawa, M.A. Czarnecki, Y. Ozaki, A. Wakisaka, Molecular self-assembling of butan-1-ol, butan-2-ol, and 2-methylpropan-2-ol in carbon tetrachloride solutions as observed by near-infrared spectroscopic measurements. Appl. Spectrosc. **54**, 268–276 (2000)

19. L. Stordrange, A.A. Christy, O.M. Kvalheim, H. Shen, Y. Liang, Study of the self-association of alcohols by near-infrared spectroscopy and multivariate 2D techniques. J. Phys. Chem. A **106**, 8543–8553 (2002)

20. Segtnan, V.H., Šašić, S., Isaksson, T., Ozaki, Y. Studies on the structure of water using two-dimensional near-infrared correlation spectroscopy and principal component analysis. *Anal. Chem.* 2001, 73, 3153 − 3161

21. Šašić, S., Segtnan, V. H., Ozaki, Y. Self-modeling curve resolution study of temperature-dependent near-infrared spectra of water and the investigation of water structure. *J. Phys. Chem. A* 2002, 106, 760 − 766

22. B. Czarnik-Matusewicz, S. Pilorz, J.P. Hawranek, Temperature-dependent water structural transitions examined by near-IR and mid-IR spectra analyzed by multivariate curve resolution and two-dimensional correlation spectroscopy. Anal. Chim. Acta **544**, 15–25 (2005)

23. R. Iwamoto, H. Kusanagi, Determination of the hydrate structure of an isolated alcoholic OH in hydrophobic medium by infrared and near-infrared spectroscopy. J. Phys. Chem. A **113**, 5310–5316 (2009)

24. M.A. Czarnecki, Y. Liu, Y. Ozaki, M. Suzuki, M. Iwahashi, Potential of Fourier transform near-infrared spectroscopy in studies of the dissociation of fatty acids in the liquid phase. Appl. Spectrosc. **47**, 2162–2168 (1993)

25. M.A. Czarnecki, M. Czarnecka, Y. Liu, Y. Ozaki, M. Suzuki, M. Iwahashi, FT-NIR study of dissociation of decen-1-ol in the liquid phase – I. Spectrochim. Acta A **51**, 1005–1015 (1995)

26. Czarnecki, M.A., Maeda, H., Ozaki, Y., Suzuki, M., Iwahashi, M. Resolution enhancement and band assignments for the first overtone of OH stretching modes of butanols by two dimensional near-infrared correlation spectroscopy. 2. Thermal dynamics of hydrogen bonding in n- and tert-butyl alcohol in the pure liquid states. *J. Phys. Chem. A* 1998, 102, 9117 − 9123

27. M.A. Czarnecki, Y. Ozaki, The temperature-induced changes in hydrogen bonding of decan-1-ol in the pure liquid phase studied by two-dimensional Fourier transform near-infrared correlation spectroscopy. Phys. Chem. Chem. Phys. **1**, 797–800 (1999)

28. M.A. Czarnecki, Near-infrared spectroscopic study of hydrogen bonding in chiral and racemic octan-2-ol. J. Phys. Chem. A **107**, 1941–1944 (2003)

29. M.A. Czarnecki, K. Orzechowski, Effect of temperature and concentration on self-association of octan-3-ol studied by vibrational spectroscopy and dielectric measurements. J. Phys. Chem. A **107**, 1119–1126 (2003)
30. K. Orzechowski, M.A. Czarnecki, Association of 1-hexanol in mixtures with n-hexane: dielectric, near-infrared and DFT studies. J. Mol. Liq. **279**, 540–547 (2019)
31. Y. Morisawa, A. Suga, Effects of intermolecular interactions on absorption intensities of the fundamental, and the first, second and the third overtones of OH stretching vibrations of methanol and t-butanol d_9 in n-hexane studied by visible/near-infrared/infrared spectroscopy. Spectrochim. Acta A **197**, 121–125 (2018)
32. Q. Dong, C. Yu, L. Li, L. Nie, D. Li, H. Zang, Near-infrared study of molecular interaction in ethanol-water mixtures. Spectrochim. Acta A **222**, 1–8 (2019)
33. Y. Futami, Y. Ozaki, Y. Hamada, M.J. Wójcik, Y. Ozaki, Solvent dependence of absorption intensities and wavenumbers of the fundamental and first overtone of NH stretching vibration of pyrrole studied by near-infrared/infrared spectroscopy and DFT calculations. J. Phys. Chem. A **115**, 1194–1198 (2011)
34. Y. Futami, Y. Morisawa, Y. Ozaki, Y. Hamada, M.J. Wójcik, Y. Ozaki, The dielectric constant dependence of absorption intensities and wavenumbers of the fundamental and overtone transitions of stretching vibration of the hydrogen fluoride studied by quantum chemistry calculations. J. Mol. Struct. **1018**, 102–106 (2012)
35. O. Golonzka, A. Tokmakoff, Polarization-selective third-order spectroscopy of coupled vibronic states. J. Chem. Phys. **115**, 297–309 (2001)
36. Beć, K.B., Grabska, J., Kirchler, C.G., Huck, C.W. NIR spectra simulation of thymol for better understanding of the spectra forming factors, phase and concentration effects and PLS regression features. *J. Mol. Liq.* 2018, 268, 895–902
37. M.J. Schuler, T.S. Hofer, C.W. Huck, Assessing the predictability of anharmonic vibrational modes at the example of hydroxyl groups – ad hoc construction of localised modes and the influence of structural solute–solvent motifs. Phys. Chem. Chem. Phys. **19**, 11990–12001 (2017)
38. K.B. Beć, M.J. Wójcik, T. Nakajima, Quantum chemical calculations of basic molecules: alcohols and carboxylic acids. NIR News **27**(8), 15–21 (2016)
39. K.B. Beć, Y. Futami, M.J. Wójcik, Y. Ozaki, A spectroscopic and theoretical study in the near-infrared region of low concentration aliphatic alcohols. Phys. Chem. Chem. Phys. **18**, 13666–13682 (2016)
40. O.M.D. Lutz, C.B. Messner, T.S. Hofer, L.R. Canaval, G.K. Bonn, C.W. Huck, Computational vibrational spectroscopy of glycine in aqueous solution—Fundamental considerations towards feasible methodologies. Chem. Phys. **435**, 21–28 (2014)
41. Y. Toyama, K. Murakami, N. Yoshimura, M. Takayanagi, Even-odd alternation of near-infrared spectra of alkane-$\alpha\omega$-diols in their solid states. Spectrochim. Acta A **197**, 148–152 (2018)
42. L. Liu, Y. Cheng, X. Sun, F. Pi, Numerical modeling of polymorphic transformation of oleic acid via near-infrared spectroscopy and factor analysis. Spectrochim. Acta A **197**, 153–158 (2018)
43. Grabska, J., Ishigaki, M., Beć, K.B., Wójcik, M.J., Ozaki, Y. Structure and near-infrared spectra of saturated and unsaturated carboxylic acids. An insight from anharmonic DFT calculations. *J. Phys. Chem. A* 2017, 121, 3437–3451
44. Grabska, J., Czarnecki, M.A., Beć, K.B., Ozaki, Y. Spectroscopic and quantum mechanical calculation study of the efect of isotopic substitution on NIR spectra of methanol. *J. Phys. Chem. A* 2017, 121, 7925–7936
45. Beć, K.B., Grabska, J., Huck, C.W., Czarnecki, M.A. Spectra–structure correlations in isotopomers of ethanol (CX_3CX_2OX; X = H, D): combined near-infrared and anharmonic computational study. *Molecules* 2019, 24, 2189
46. Grabska, J., Beć, K.B., Ozaki, Y., Huck, C.W. Temperature drift of conformational equilibria of butyl alcohols studied by near-infrared spectroscopy and fully anharmonic DFT. *J. Phys. Chem. A* 2017, 121, 1950–1961

47. L.G. Weyer, S.C. Lo, *Spectra-structure correlations in the near-infrared*, ed by J.M. Chalmers, P.R. Griffiths, Handbook of vibrational spectroscopy, vol. 3 (Chichester, Wiley, 2002)
48. M.A. Czarnecki, D. Wojtków, K. Haufa, Rotational isomerism of butanols: infrared, near-infrared and DFT study. Chem. Phys. Lett. **431**, 294–299 (2006)
49. Kirchler, C.G., Pezzei, C.K., Beć, K.B., Mayr, S., Ishigaki, M., Ozaki, Y., Huck, C.W. Critical evaluation of spectral information of benchtop vs. portable near-infrared spectrometers: quantum chemistry and two-dimensional correlation spectroscopy for a better understanding of PLS regression models of the rosmarinic acid content in Rosmarini folium. *Analyst* 2017, 142, 455–464
50. Chalmers, J.M., Griffiths, P.R., (Eds.) *Handbook of vibrational spectroscopy. Vol. 1*, John Wiley & Sons Ltd, 2002
51. Grabska, J., Beć, K.B., Ishigaki, M., Huck, C.W., Ozaki, Y. NIR spectra simulations by anharmonic DFT-saturated and unsaturated long-chain fatty acids. *J. Phys. Chem. B* 2018, 122, 6931–6944
52. O.M.D. Lutz, B.M. Rode, G.K. Bonn, C.W. Huck, The impact of highly correlated potential energy surfaces on the anharmonically corrected IR spectrum of acetonitrile. Spectrochimica Acta A **131**, 545–555 (2014)
53. Grabska, J., Beć, K.B., Kirchler, C.G., Ozaki, Y., Huck, C.W. Distinct difference in sensitivity of NIR vs. IR bands of melamine to inter-molecular interactions with impact on analytical spectroscopy explained by anharmonic quantum mechanical study. *Molecules* 2019, 24, 1402
54. Beć, K.B., Grabska, J., Ozaki, Y., Hawranek, J.P., Huck, C.W. Influence of non-fundamental modes on mid-infrared spectra. Anharmonic DFT study of aliphatic ethers. *J. Phys. Chem. A* 2017, 121, 1412–1424
55. A. Ikehata, C. Hashimoto, Y. Mikami, Y. Ozaki, Thermal phase behavior of triethylamine–water mixtures studied by near-infrared spectroscopy: band shift of the first overtone of the C-H stretching modes and the phase diagram. Chem. Phys. Lett. **393**, 403–408 (2004)

Chapter 14
Application of NIR in Agriculture

Baeten Vincent and Pierre Dardenne

Abstract NIR has been used for decades as an innovative technique in agriculture. There are many benefits, and today, researchers active in agronomy science are not the only ones using NIR extensively in their daily research but also breeders, farmers and agri-processors, using it as an efficient tool for the assessment of a large number of parameters and criteria including detection of contaminants. Undoubtedly, NIRS has demonstrated clear advantages in the analysis of soil, crops, forages, silages and faeces, but also for the analysis of agro-food products such as feed and dairy products. These analyses are no more conducted only at the laboratory level but go more and more to the sample. The new generation of instruments (portable and handheld devices) allow to perform the analyses at the field, farm, orchard or greenhouse level in order to get information to take the right decision at the right moment. This chapter aims to summarise some of these applications and attempts to give the trends of a selection of recently completed or current projects. Readers aiming to delve further into the potential of NIR in agriculture can refer to dedicated books (Williams and Norris in Amer Assn of Cereal Chemists, 312 p, 2001 [1]) or recent reviews (Baeten et al. in Handbook of food analysis, pp 591–614, 2015 [2], Dale et al. in Appl. Spectrosc. Rev. 48(2):142–159, 2013 [3], García-Sánchez et al. in Agricultural systems, pp 97–127, 2017 [4]).

Keywords Quality · Agriculture · Agro-food · Forage · Feed · Farm · Crop · Faeces

B. Vincent (✉)
Quality and Authentication of Products Unit, Department Knowledge and Valorization of Agricultural Products, Walloon Agricultural Research Centre (CRA-W), Gembloux, Belgium
e-mail: v.baeten@cra.wallonie.be

P. Dardenne
Henseval Building, Chaussée de Namur 24, 5030 Gembloux, Belgium
e-mail: dardennepaj@gmail.com

© Springer Nature Singapore Pte Ltd. 2021
Y. Ozaki et al. (eds.), *Near-Infrared Spectroscopy*,
https://doi.org/10.1007/978-981-15-8648-4_14

14.1   Introduction

Different areas will be considered in this short section regarding applications of NIR in agriculture: soil analysis, field analysis, forage and silage analysis, feed analysis, milk analysis, faeces and effluent analysis and orchard analysis. In order to help the readers to assess the potential of NIR agriculture, some practical results in terms of NIR equations are provided. The figures presented are those of the Belgian REQUASUD network. This network has been in place in the Walloon Region of Belgium (south part of the country) since 1989 and has the ambition of providing classical analyses and NIR analyses of premium quality for the agri-food sector. The network currently includes seven quality control laboratories performing routine analyses for the public and private sectors (www.requasud.be). The Walloon Agriculture Research Centre (CRA-W) manages the NIR instrument network consisting of 10 NIR spectrometers. A brochure in French explaining the organisation of the NIR network, the current achievements and future developments can be downloaded from the following address: http://www.requasud.be/wp-content/uploads/2017/07/brochure_requasud_spectrometrie_proche_infrarouge.pdf. In this chapter, it has been decided to provide R^2, SEC and RPD_{sec} of the NIR equations in use in the REQUASUD network. These figures could be considered to be a good estimation of the performance that could be achieved by NIR technology. It is important to underline that the performance figures have been obtained with multi-annual spectral databases covering a wide diversity in terms of origin and assembled in a NIR network equipped with one single type of benchtop instrument (Foss XDS). Moreover, reference values have been provided by different laboratories having high quality standards. Indeed, REQUASUD frequently organises interlaboratory studies to assess the quality of implementation of the Reference and NIR techniques. Frequent tailored trainings are provided to members of the NIR network.

14.2   Applications in the Field and Crop Analysis

Applications of NIR to field and soil analysis are well illustrated in the literature. These topics are also widely presented and discussed in NIR-focused conferences (e.g. ICNIRS = www.icnirs.org; IDRC = www.idrc-chambersburg.org; HELIOSPIR = www.heliospir.net). In the following section, the current status of application of the NIR technique to soil and crop analysis is given. Some ongoing research with their challenging objectives is briefly presented to give an idea of some of the forthcoming developments in these fields.

14.2.1 Soil Analysis by NIR—A Technique in Development

There is great interest for farmers having a good knowledge of the quality of their soil. Specifically, it is strategic and crucial to be able to adequately determine the physical and chemical characteristics and properties of the soils proposed and used for agricultural purposes. Based on different quality parameters, farmers will decide which crop to plant and which agricultural practices to follow in order to maximise the intrinsic value of their plots. Adequate management of inputs is of prime importance in order to maximise the benefits and to maintain the soil fertility. Soil analysis by classical techniques is tedious, time consuming and requires standardised methods using significant amounts of reagents, making such analyses particularly unfriendly for the environment. Several reviews have been published dealing with the perspectives offered by NIRS for soil analysis [5–7]. Chabrillat et al. reviewed the achievements and perspectives in soil mapping and monitoring based on imaging spectroscopy from air and space-borne sensors [8]. This review underlines that the next generation of hyperspectral satellite sensors could greatly help to meet the demand for global soil mapping and monitoring. Recently, Hutengs and co-authors have compared the performance of portable NIR and mid-infrared (MIR) devices for assessment of organic carbon in soils. Even though a comparison has not been made on the same sample batches, they conclude with the fact that handheld MIR gives significantly more accurate results for on-field analyses. This work clearly demonstrates that the prediction of soil properties (whatever the technique) is improved when the samples are dried and ground (RMSE = 0.155 for NIR) instead of being analysed in situ (RMSE = 0.243 for NIR). Ongoing projects also aim to develop the application of NIR to tackle the current challenges faced by soil analysis. For instance, the INDIGGES project (http://www.cra.wallonie.be/fr/indigges) aims to develop direct and indirect indicators to evaluate greenhouse gas emissions and carbon storage at the farm. In this project, the NIRS technique is tested for characterising both fresh and dried soils. NIR benchtop, portable, handheld and hyperspectral imaging devices are tested (Fig. 14.1). Another challenging issue where NIR approaches are interesting is the detection of foreign material in soil. An example is the detection, identification and quantification of macro- and micro-plastics in cultivated lands. Micro-plastics are emerging as persistent contaminants of increasing concern. They come from mulching film, sludge, wastewater irrigation and atmospheric deposition. It seems that the micro-plastics influence soil physio-chemistry and biota [9].

A key issue regarding soil analysis is the heterogeneity of samples proposed for classical or NIR analyses. It is crucial to take care and devote the required resources to be sure that the sample specimen analysed is representative of the soil for which we want to determine the physical and chemical properties.

Table 14.1 presents the performance of equations used in the REQUASUD network for soil analysis (2018 status). Only the parameters for which a RPD_{sec} is higher than 2 are presented, i.e. organic carbon content, nitrogen, CEC and clay content. The performance for soil from grasslands and lands under cultivation is given [6, 10]. The NIR analysis protocol for soil and the spectral database has been

Fig. 14.1 Illustrations of NIR soil measurements taken in the framework of the INDIGGES project. Portable VIS-NIR measurements taken directly in the field (*Source* CRA-W)

Table 14.1 Performance of equations implemented in the REQUASUD network for analysis of soil from grasslands and lands under cultivation

Grasslands								
Properties	N	Min	Max	Mean	SD	R^2	SEC	RPD
Total organic carbon % MS	8849	0.01	14.91	3.64	1.49	0.91	0.49	3.4
CEC (meq/100 g)	855	0.02	71.2	9.6	7.03	0.85	3.15	2.6
Nitrogen (g/kg)	1077	0.2	6.92	3.18	1.25	0.82	0.59	2.4
Clay % MS	210	2.56	57.7	18.52	8.23	0.82	4.12	2.3
Lands under cultivation								
Total organic carbon % MS	10,139	0.05	7.66	1.54	0.69	0.93	0.21	3.8
CEC (meq/100 g)	1228	0.48	44.3	12.01	4.3	0.81	2.47	2.3
Nitrogen (g/kg)	3240	0.17	9.31	1.61	0.75	0.92	0.25	3.6
Clay % MS	575	1.9	72.65	19.92	8.41	0.84	4.08	2.5

N—Number of samples in the spectral database; Min—Minimum; Max—Maximum; SD—Standard Deviation; SEC—Standard Error of Calibration; R^2—Coefficient of determination; RPD—Ratio of Performance to Deviation = SD_{ref}/SEC; DM—Dry Matter Basis; CEC—Cation Exchange Capacity
Source CRA-W, Adapted from [10]

built since 2011 by the University of Liège in collaboration with CRA-W. Different regression algorithms have been tested. The LOCAL approach with the use of PLS is the most appropriate [6]. An important public resource is the NIR spectral database developed in the framework of the European LUCAS initiative (https://esdac.jrc.ec. europa.eu/projects/lucas). In this initiative, about 20,000 topsoil samples have been collected in 25 European Union (EU) Member States with the goal of producing a European physical and chemical topsoil database with the aim to harmonise soil monitoring [11].

14.2.2 Crop Analysis—Direct Analysis in the Field or Laboratory Analysis to Support Farmers and Breeders

For the monitoring the crop before and during harvesting, NIR spectroscopy can be an interesting solution. Indeed, it can be used for optimising the harvest date and for crop management. NIR technology is used, among other things, for measuring moisture, production yield, nitrogen status of the crop and to monitor the occurrence of plant pests and diseases. Determining these chemical compositions and properties can be done directly in the field during farm operations or at the harvesting stage for optimal monitoring of crops throughout their life cycle. The objective is to support breeders or farmers in their management. It can also be done in the field to support breeder observations or farmer choices, as well as at the receipt stages of storage facilities or of industry. Today, there is a common effort by farmers, researchers, instrument manufacturers and farm advisory services to develop operational solutions for assessing optimal crop management, to optimise the use of inputs, to assure the best product quality and to maximise the financial benefits [2]. Classical NIR benchtop instruments are also used by researchers and breeders for routine analysis of the dried and ground aerial parts of the crop in order to determine key parameters such as nitrogen and carbon content [12]. Moreover, near-infrared microscopy (NIRM) has also been proposed as a rapid technique to predict the chemical composition (e.g. nitrogen content) of dried and ground materials when the material quantity is insufficient to perform analysis by classical instrumentation. It has been demonstrated with very small samples ($\ll 1$ g) of tomato (*Solanum lycopersicum* L.) leaf powder coming from experiments. The calibration model obtained for nitrogen content proved to be excellent, with a calibration coefficient of determination (R_c^2) higher than 0.9 and a ratio of performance to deviation (RPD$_c$) higher than 3. It appears that NIRM is a promising and suitable tool for a rapid, non-destructive and reliable determination of nitrogen content of tiny samples of leaf powder [13]. The use of the NIR hyperspectral imaging instrument for crop analysis seems to be an increasingly interesting approach as it provides spatial information in addition to chemical information from the spectral data. In that sense, this approach is being investigated to build phenotyping strategies useful in breeding programmes that focus on wheat varieties (e.g. PhenWheat project; http://www.cra.wallonie.be/fr/phenwheat), sugar beet varieties (e.g. BeetPhen project; http://www.cra.wallonie.be/en/beetphen) and potato varieties (e.g. First project; http://www.cra.wallonie.be/fr/first). Figure 14.2 shows the NIR hyperspectral imaging device used at CRA-W. Challenges are presentation of the device to the crop and the development of a robust protocol to calibrate and validate the system using spectral data not collected in the controlled environment of a laboratory.

The potential of NIR hyperspectral imaging spectroscopy and chemometrics for the discrimination of roots and crop residues extracted from soil samples has also been demonstrated. The study of these materials in different field conditions is important to identify suitable soil management practices for sustainable crop production. In

Fig. 14.2 Use of Hyperspectral NIR imaging for crop status monitoring (*Source* CRA-W)

order to eliminate the cumbersome hand-sorting step, avoid confusion between these elements and reduce the time needed to quantify roots, a protocol based on near-infrared hyperspectral imaging spectroscopy has been established. The best results have been achieved using a support vector machine to first discriminate the materials and then to quantify them in the soil samples [14]. The methodology has been used, for instance, to better understand the effect of tillage or fertilisation on root system development. Another interesting application of NIR to crops is the use of NIR hyperspectral imaging in the study of legume root systems. This technology has been used in the framework of several studies conducted on pea root systems. First, the suitability of this approach to quantify the mass of pea roots in root samples collected under pea–wheat intercropping has been demonstrated. Secondly, this analytical method was used to quantify leghaemoglobin in individual pea nodules. Fixation activity of the nodules is related to the concentration of this molecule in the pea nodules (Fig. 14.3; [15]).

14.3   Applications on Farm Products or Effluents

Different studies have proposed NIR technology to assess at the farm the control of feed, forages (fresh, dried and silages), milk and effluents [16]. The interest is in optimising costs and reducing the impact on the environment.

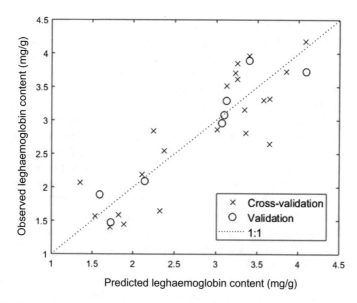

Fig. 14.3 Cross-validation (open circles) and validation (crosses) results of the PLS regression model. Leghaemoglobin content was measured with the cyanmethaemoglobin method and predicted on the basis of nodule NIR imaging spectra. Results are expressed in mg leghaemoglobin g^{-1} of fresh nodules. Leghaemoglobin was predicted with a RMSECV of 0.45 and a determination coefficient (R^2) of 0.74 (*Source* CRA-W)

14.3.1   An Efficient Tool to Assess Forage and Silage Quality for Precision Feeding

In the current economic (e.g. price volatility of main inputs and agricultural productions) and environmental (e.g. reduction of inputs, optimal valuation of farm production by maximising the use of productions and reducing the impact of effluents) context, the appropriate control of forage quality is of prime importance. In some region (e.g. Walloon Region of Belgium), feed produced at the farm contributes significantly (around 50%) to the feeding autonomy of the farm. Different types of forage are generally identified: green forages (i.e. grazed grass, whole plant maize, immature cereals and protein mixed crop); silage forages of grass, whole plant maize or beet pulp obtained by the application of a process to preserve wet forages through anaerobic lactic fermentation; dry fodder; artificially dehydrated and pelleted fodder and cereal/pea straws (Minet et al., to be published). One of the most important issues of forages is their high heterogeneity in terms of physical appearance and nutritional value. This heterogeneity is observed between different types, but also inside each class of forages making determination of forage quality essential in farm management. Sampling is a critical step for forage quality assessment whether analysed by classical techniques or NIR techniques. Samples must be as representative of the whole forage batch as possible regardless of its conditioning. When sampling has to be performed, several portions of the batch must be taken. It is advisable to collect at least 10 sub-samples (and even better 20) of 50 g each. Then, the sub-samples, also called primary samples, are mixed and transferred to a hermetic plastic bag, stored and transported at low temperature to the laboratory in order to avoid deterioration.

For the farmer, knowing the composition and nutrition value of their forage precisely and throughout the year enables feeding of the animals to be optimised for meeting the animals' requirements [10]. Economic losses can be avoided by ensuring that animals receive the diet that allows them to reach optimal milk production (for instance) without a risk of underfeeding or overfeeding. On the basis of forage quality, a farmer may either overestimate the nutritive value of feed and not cover his animals' needs, or underestimate it, with the risk of producing manure that is too rich and could potentially pollute the environment (Minet et al., to be published). Determining the chemical composition and nutritive value of feed ingredients produced at the farm is crucial. Even though several initiatives are being taken to perform it at the farm on fresh samples, this determination is usually done on previously dried and ground samples in an external laboratory using classical chemical methods or NIR methods. Several studies and reviews on the potential of NIR for assessing feeding values of forages exist. Generally, the LOCAL approach gives better results for the analysis of forages [17].

Table 14.2 presents the performance of equations used in the REQUASUD network (2018 status). A selection of parameters for which a RPD$_{sec}$ higher than 2 is presented, i.e. dry matter, proteins, cellulose, ash, digestibility of dry matter, digestibility of the organic matter and total soluble sugar. It is commonly accepted that most of the parameters that allow the farmer to estimate nutritional value of

Table 14.2 Performance of equations used in the REQUASUD network for analysis of grass forages

Grass forages								
Properties	N	Min	Max	Mean	SD	R^2	SEC	RPD
Dry matter	1877	88.84	97.49	93.16	1.44	0.78	0.68	2.1
Protein % MS	1877	4.45	31.26	15.49	5.26	0.98	0.76	6.9
Cellulose % MS	1465	11.27	41.10	26.18	4.97	0.95	1.11	4.5
ASH % MS	1989	3.44	16.66	10.05	2.20	0.85	0.86	2.6
Digestibility of dry matter (De Boever) % MS	1156	50.19	108.28	79.23	9.68	0.96	1.89	5.1
Digestibility of the organic dry matter (De Boever) % MS	1291	46.02	108.06	77.04	10.34	0.96	1.97	5.3
Total soluble sugar % MS	629	0.12	36.12	11.47	8.22	0.97	1.35	6.1

N—Number of samples in the spectral database; Min—Minimum; Max—Maximum; SD—Standard Deviation; SEC—Standard Error of Calibration; R_2—Coefficient of determination; RPD—Ratio of Performance to Deviation = SD$_{ref}$/SEC; DM—Dry Matter Basis
Source CRA-W, Adapted from [10]

forages can be determined by NIRS with relevant precision. The cost of this NIR determination is about one-tenth of the cost of determination by classical methods, and it is obtained in less time, which is more compatible with farm management requirements. Table 14.2 presents the performance for grass forages only [10, 17].

Today, with the evolution of technology, forage analysis can be performed at the farm with handheld NIR instruments and applied directly on wet samples. Several private companies have dedicated instruments for testing of forages and silages at farm (e.g. AURORA = http://www.grainit.it/en/portfolio-items/aurora-nir-analisi-dei-foraggi-in-stalla/) and some offer a full service to the farmer (e.g. NIR4FARM = https://www.abvista.com/Products/GB/NIR-4-Farm.aspx). Another new perspective is the use of NIR hyperspectral imaging to detect and discriminate grassland species in forage [3].

14.3.2   Determination of Key Parameters and Detection of Contaminants/Impurities in Feed

Today, for compound feed specialists, NIR spectroscopy is considered an essential analytical tool that can contribute greatly to quality and safety control and enhancement of their products. The technology has been implemented with success at different stages of feed production chains. This provides not only gains in speed of analysis but also larger analytical throughputs. For instance, NIR spectroscopy is used to characterise raw materials entering the factory and allows the production process to be optimised to assess the nutritive features of the different processed feeds. Networks of tens (even hundreds in some cases) of spectrometers are implemented in major feed companies that daily and routinely perform numerous determinations to assess the quality of feed ingredients, feed additives and compound feeds. Several reviews have addressed the application of NIR to feed analysis [18, 19].

Different parameters can be adequately predicted by NIR [20, 21]. Table 14.3 presents the performance of the equation used in the REQUASUD network to assess the quality of feed (2018 status). Only the parameters for which a RPD$_{sec}$ higher than 2 are presented, i.e. moisture, nitrogen, fat, cellulose, ash and starch [10].

In the feed area, NIR technology can be also relevant to detect contamination by plant, animal, mineral, chemical contaminants or any undesirable substances [22]. It has to be admitted that the use of NIR for detecting contaminants and undesirable substances in feed products is not widely practised. However, several studies have demonstrated the unique advantages of using this fingerprinting technique in the continuing effort to give stakeholders the means to check the safety of the feed chains [23]. Examples include the potential of NIR (NIR microscope and NIR hyperspectral imaging devices) for detection of animal protein in feed ingredients and compound feeds [24–26], detection of plant contaminants [21, 27], the detection of chemical contaminants such as melamine [28, 21], paper and plastic residues coming from packaging, assessment of the origin of feed ingredients [28–30] and the presence of

Table 14.3 Performance of equations used in the REQUASUD network for analysis of compound feeds

Compound feeds								
Properties	N	Min	Max	Mean	SD	R^2	SEC	RPD
Moisture	24,962	2.60	16.65	11.27	1.99	0.88	0.68	2.9
Proteins	23,734	7.10	62.10	20.91	8.66	0.97	1.39	6.2
Fat	8391	0.70	31.40	5.61	4.49	0.97	0.73	6.2
Fibre	5792	0.20	17.90	5.45	2.99	0.91	0.91	3.3
Ash	21,678	1.30	33.00	7.54	3.49	0.79	1.59	2.2
Starch	961	3.30	59.20	30.77	10.86	0.96	2.10	5.2

N—Number of samples in the spectral database; Min—Minimum; Max—Maximum; SD—Standard Deviation; SEC—Standard Error of Calibration; R^2—Coefficient of determination; RPD—Ratio of Performance to Deviation = SD_{ref}/SEC; DM—Dry Matter Basis
Source CRA-W, Adapted from [10]

insects [5]. A study conducted in a feed factory has also demonstrated the interest to use NIR technique coupled to a fibre optic probe to detect at the early stage non-conformity of feed ingredients [21]. In this study, issued from a EC project (Q-saffe output project = https://cordis.europa.eu/project/rcn/97821/factsheet/en), online spectrometer allows automatically and sequentially acquiring NIR spectra of sub-samples from incoming batch and detect if it differs to the spectra of the rest of the batch and to the spectra obtained from similar feed ingredient.

14.3.3 A Tool to Assess the Quality of Dairy Products and to Track Milk Quality in the Milking Parlour

Whereas NIR analysis of derived dairy products is common in the industry (for instance, determination of composition parameters and properties in cheese and butter), NIR analysis of milk is more anecdotal [31]. The main reason seems to be the fact that milk should be ideally measured in the transmission mode, and also that control of the temperature and homogenisation of the milk have to be properly addressed [22]. As far we know, only a few dedicated and appropriate instruments for milk analysis have been developed in the framework of research project and industrial initiatives [32], and only one including a temperature control system and homogenisation system has been commercialised (www.bruker.com). Milk is a complex matrix and contains many components such as lipids, proteins, carbohydrates and minerals in variable concentrations. Several authors have reviewed the potential of NIR in the analysis of milk and dairy products to assess the quality, discriminate the origin and detect adulteration [33, 34]. Quality analysis of dairy products relies mainly on manual sampling followed by chemical or physical measurements. This procedure uses laboratory methods characterised by a significant time lag between sample collection and generation of a result. One of the characteristics of this procedure is the fact that it does not permit interaction with the industrial process in order to instantly correct for deviations from target parameters of the process. In the framework of the Walloon Milkinir research project [32], a near-infrared (NIR) spectrometer-based system is used for online monitoring of milk quality during the process, allowing the milk quality of an individual cow to be monitored. Daily measurement of milk components individual cows could be a decisive tool for farm management and development of animal breeding or feeding programmes to produce milk with a specific milk composition.

14.3.4 Analysis of Faeces and Farm Effluent, A Way to Optimise Their Valuation

At the farm, NIR technology can be also used for the analysis of effluents and faeces. Farm activities produce organic residues, i.e. farm effluents and manure. These residues are of great interest to improve the fields as they are rich in phosphorus, potassium and nitrogen. Rational use of farm effluent and manure based on their intrinsic quality is interesting from the economic point of view. The challenge is the appropriate strategy in the preparation of the sample submit for analysis in order to take into account the high heterogenic nature of this product. Misuse can lead to reduction of soil fertility, environmental pollution and reduction of the farm's profitability.

NIR technique has been also proposed to analyse faeces in order to correlate spectral information to chemical composition or biological status of the diet. Development of models to determine quantitative and qualitative characteristics of grass and feed on the basis NIR spectra has been proposed [35]. It has been demonstrated, among other things, that this approach is relevant for estimating in vivo digestibility and voluntary intake of animals. Moreover, ruminants' diet composition in terms of plant species can be ranked using NIR data. The current work relates to the development of decision support tools for improving grazing management schemes based on NIR determination.

14.4 Applications in the Orchard and in the Fruit Sector

In the fruit sector and since beginning of the twentieth century, VIS and NIR techniques are becoming more and more widely adopted as a non-destructive technique to rapidly and cost effectively assess the quality of fruit. In production, harvesting, storage, processing and consumption of fruit, it is crucial to determine several quality parameters and criteria. A key issue in the analysis of fruit by NIR is appropriate sample presentation. Different studies have concluded that measurement in the reflection and interaction modes is more appropriate for fruit analysis. It is essential to report that the penetration depth for apples has been measured in the reflection mode and is about 4 mm for the 700–900 nm range and 2–3 mm for the 900–1900 nm range. In the transmission mode and in the 1400–1600 nm range, less than 1% of the initial intensity of the radiation goes through a 1 mm slice. The skin definitely poses a major barrier for the light entering the flesh of the apple, requiring a strict protocol for presentation of the sample to the instrument. This protocol will be adapted to the fruit analysed, the architecture of the instrument (mainly the relative position of the source/sources and detector/detectors). Several reviews summarise the potential of NIR for determining different parameters and criteria of fruit [36, 37]. A specific review has been dedicated to challenges and solutions for quality inspection for robotic fruit instrumentation [38].

Several parameters can be determined with enough precision to be used routinely. A number of authors have reported on the use of NIR spectroscopy to determine apple quality parameters, such as soluble solids, acidity, pulp firmness, maturity indexes, polyphenols and vitamin C [36]. Pissard et al. has shown that NIR technique can be used to determine sugar, vitamin C and total polyphenols contents [39]. This study, based on large spectral databases (between 1274 and 2646 depending on the parameter studied) built in the framework of breeding programmes and European projects, has demonstrated the high precision of models that can be achieved. Low standard error of prediction values, in addition to relatively high ratio to prediction (RPD) values, has been obtained especially for total polyphenol and sugar content (RPD values of 5.1 and 4.3 for polyphenol and sugar, respectively). These same authors have also studied the intra-fruit variability in apples using classical and NIR techniques [40]. This paper proposes and validates a protocol to analyse fruit based on reference analyses of a representative sample of the apple and NIR measurements collected at four points 45° from each other in the equatorial region of the fruit (i.e. apple). It has been demonstrated that there was little difference between the mean value at the four points and the mean value of the entire apple. The potential of NIR spectroscopy on fresh apples to determine the phenolic compounds and dry matter content in peel and flesh has been also studied [41]. In general, one of the challenges is the online analysis of intact fruit.

More and more handheld NIR devices are commercially available and proposed to analyse fruit. NIR uses under field conditions (i.e. orchard) have been limited for many years due to restrictions imposed by the size and low robustness of the instruments available. Recently, the development of new technologies used in the construction of NIR spectrometers and data acquisition strategies has enabled a significant reduction in size and cost of these instruments but often a decreased of the robustness of the methods developed [42, 43]. The challenge is to set up the right procedure to use the historical databases and calibration models, previously developed using benchtop spectrometers.

References

1. P. Williams, K. Norris, Near-infrared technology: in the agricultural and food industries. *Amer Assn of Cereal Chemists*, 2 edn (2001), 312 p.
2. V. Baeten, H. Rogez, J.A. Fernández Pierna, P. Vermeulen, P. Dardenne, Vibrational spectroscopy methods for the rapid control of agro-food products, in *Handbook of Food Analysis*, 3rd edn, vol. II, ed. by F. Toldra, L.M.L. Nollet (Chapter 32, 2015), pp. 591–614
3. L.M. Dale, A. Thewis, C. Boudry, I. Rotar, P. Dardenne, V. Baeten, J.A. Fernández Pierna, Appl. Spectrosc. Rev. **48:2**, 142–159 (2013)
4. F. García-Sánchez, L. Galvez-Sola, J.J. Martínez-Nicolás, R. Muelas-Domingo, M. Nieves, *Agricultural Systems*. Intechopen, open book, Chap. 5. https://doi.org/10.5772/67236 (2017), pp. 97–127
5. V. Bellon-Maurel, A. McBratney, Near-infrared (NIR) and mid-infrared (MIR) spectroscopic techniques for assessing the amount of carbon stock in soils—critical review and research perspectives. Soil Biol. Biochem. **43**(7), 1398–1410 (2011)
6. V. Genot, L. Bock, P. Dardenne, G. Colinet, Use of near-infrared reflectance spectroscopy in soil analysis. A review [L'intérêt de la spectroscopie proche infrarouge en analyse de terre (synthèse bibliographique)]. Biotechnol. Agron. Soc. Environ. **18**(2), 247–261 (2014)
7. M. Nocita, A. Stevens, B. van Wesemael, M. Aitkenhead, M. Bachmann, B. Barthès, E.B. Dor, D.J. Brown, M. Clairotte, A. Csorba, P. Dardenne, J.A.M. Demattê, V. Genot, C. Guerrero, M. Knadel, L. Montanarella, C. Noon, L. Ramirez-Lopez, J. Robertson, H. Sakai, J.M. Soriano-Disla, K.D. Shepherd, B. Stenberg, E.K. Towett, R. Vargas, J. Wetterlind, Soil spectroscopy: an alternative to wet chemistry for soil monitoring. Adv. Agron. **132**, 139–159 (2015)
8. S. Chabrillat, E. Ben-Dor, J. Cierniewski, C. Gomez, T. Schmid, B. van Wesemael, Imaging spectroscopy for soil mapping and monitoring. Surv. Geophys. **40**(3), 361–399 (2019)
9. S. Piehl, A. Leibner, M.G.J. Löder, R. Dris, C. Bogner, C. Laforsch, Identification and quantification of macro- and microplastics on an agricultural farmland. Sci. Rep. **8**(1), art. no. 17950 (2018)
10. O. Minet, F. Ferber, L. Jacob, B. Lecler, R. Agneessens, T. Cugnon, V. Decruyenaere, V. Genot, S. Gofflot, E. Pitchugina, V. Planchon, M. Renneson, G. Sinnaeve, B. Wavreille, P. Dardenne, V. Baeten, *La spectrométrie proche infrarouge: Une technologie rapide, précise et écologique pour déterminer la composition et la qualité des produits agricoles et alimentaires.* REQUASUD brochures (www.requasud.be), 31 p. (2018)
11. A. Stevens, M. Nocita, G. Tóth, L. Montanarella, B. van Wesemael, Prediction of soil organic carbon at the european scale by visible and near infrared reflectance spectroscopy. PLoS ONE **8**(6), e66409. https://doi.org/10.1371/journal.pone.0066409 (2013)
12. F.B. Abdallah, M. Olivier, J.P. Goffart, O. Minet, Establishing the nitrogen dilution curve for potato cultivar Bintje in Belgium. Potato Res. **59**(3), 241–258 (2016)
13. G. Lequeue, X. Draye, V. Baeten, Determination by near infrared microscopy of the nitrogen and carbon content of tomato (*Solanum lycopersicum* L.) leaf powder. Sci. Rep. **6**, art. no. 33183 (2016)
14. D. Eylenbosch, B. Bodson, V. Baeten, J.A. Fernández Pierna, NIR hyperspectral imaging spectroscopy and chemometrics for the discrimination of roots and crop residues extracted from soil samples. J. Chemometr. **32**(1), art. no. e2982 (2018a)
15. D. Eylenbosch, B. Dumont, V. Baeten, B. Bodson, J.A. Fernández Pierna, Quantification of leghaemoglobin content in pea nodules based on near infrared hyperspectral imaging spectroscopy and chemometrics. J. Spectral Imaging **7**(a9), 1–10 (2018b)
16. L.M. Dale, J.A. Fernández Pierna, P. Vermeulen, B. Lecler, A.C. Bogdan, F.S. Păcurar, I. Rotar, A. Thewis, V. Baeten, Hyperspectral imaging applications in agriculture and agro-food product quality and safety control: a review. J. Food, Agric. Environ. **10**(1), 391–396 (2012)
17. G. Sinnaeve, P. Dardenne, R. Agneessens, A choice for NIR calibrations in analyses of forage quality. JNIRS **2**(3), 163–175 (1994)
18. L. Chen, Y. Zengling, L. Han, A review on the use of near-infrared spectroscopy for analyzing feed protein materials. Appl. Spectrosc. Rev. **48**(7), 509–522 (2013)

19. V. Baeten, P. Dardenne, Applications of near-infrared imaging for monitoring agricultural food and feed products, in *Spectrochemical Analysis Using Infrared Multichannel Detectors*, ed. by R. Bhargava, I.W. Levin (Chapter 13, Blackwell Publishing, 2005), pp. 283–302

20. V. Baeten, P. Vermeulen, J.A. Fernández Pierna, P. Dardenne, From targeted to untargeted detection of contaminants and foreign bodies in food and feed using NIR spectroscopy. New Food **17**(3), 16–23 (2014)

21. J.A. Fernández Pierna, O. Abbas, B. Lecler, P. Hogrel, P. Dardenne, V. Baeten, NIR fingerprint screening for early control of non-conformity at feed mills. Food Chem. **189**, art. no. 16461, 2–12 (2015)

22. V. Baeten, J. Fernández Pierna, F. Dehareng, G. Sinnaeve, P. Dardenne, Regulatory considerations in applying vibrational spectroscopic methods for quality control, in *Applications of Vibrational Spectroscopy in Food Science*, vol. 2 (Wiley, 2010), pp. 595–607

23. J.A. Fernández Pierna, P. Vermeulen, O. Amand, A. Tossens, P. Dardenne, V. Baeten, NIR hyperspectral imaging spectroscopy and chemometrics for the detection of undesirable substances in food and feed. Chemometr. Intell. Lab. Syst. **117**, 233–239 (2012)

24. V. Baeten, C. von Holst, A. Garrido, J. VanCutsem, Renier A. Michotte, P. Dardenne, Detection of banned meat and bone meal in feeding stuffs by near-infrared microscopy. Anal. Bioanal. Chem. **382**, 149–157 (2005)

25. J.A. Fernández Pierna, V. Baeten, A. Michotte Renier, R.P. Cogdill, P. Dardenne, Combination of SVM and NIR imaging spectroscopy for the detection of MBM in compound feeds. J. Chemometr. **18**(7–8), 341–349 (2005)

26. Z. Yang, L. Han, J.A. Fernández Pierna, P. Dardenne, V. Baeten, Review of the potential of near infrared microscopy to detect, identify and quantify processed animal by-products. J. NIRS **19**(4), 211–231 (2011)

27. P. Vermeulen, J.A. Fernandez Pierna, H.P. van Egmond, P. Dardenne, V. Baeten, Online detection and quantification of ergot bodies in cereals using near infrared hyperspectral imaging. Food Addit. Contam. **29**(2), 232–240 (2012)

28. O. Abbas, B. Lecler, P. Dardenne, V. Baeten, Detection of melamine in feed ingredients by near infrared spectroscopy and chemometrics. J. Near Infrared Spectrosc. **21**(3), 183–194 (2013)

29. L.M. Dale, A. Thewis, C. Boudry, I. Rotar, F.S. Păcurar, O. Abbas, P. Dardenne, V. Baeten, J. Pfister, J.A. Fernández Pierna, Discrimination of grassland species and their classification in botanical families by laboratory scale NIR hyperspectral imaging: preliminary results. Talanta **116**, 149–154 (2013)

30. P. Vermeulen, M.-B. Ebene, B. Orlando, J.A. Fernández Pierna, V. Baeten, Online detection and quantification of particles of ergot bodies in cereal flour using near infrared hyperspectral imaging. Food Addit. Contam. Part A, **34** (8: 5th International Feed Conference), 1312–1319 (2017)

31. T. Troch, E. Lefébure, V. Baeten, F. Colinet, N. Gengler, M. Sindic, Cow milk coagulation: process description, variation factors and evaluation methodologies. A review. Biotechnol. Agron. Soc. Environ. **21**(4), 276–287 (2017)

32. H.N. Nguyen, F. Dehareng, M. Hammida, V. Baeten, E. Froidmont, H. Soyeurt, A. Niemöeller, P. Dardenne, Potential of near infrared spectroscopy for on-line analysis at the milking parlour using a fibre-optic probe presentation. NIR News **22**(7), 11–13 (2011)

33. T.M.P. Cattaneo, S.E. Holroyd, The use of near infrared spectroscopy for determination of adulteration and contamination in milk and milk powder: updating knowledge. J. Near Infrared Spectrosc. **21**(5), 341–349 (2013)

34. M. Kamal, R. Karoui, Using near-infrared spectroscopy in agricultural systems. Trends Food Sci. Technol. **46**(1), 27–48 (2015)

35. V. Decruyenaere, V. Planchon, P. Dardenne, D. Stilmant, Prediction error and repeatability of near infrared reflectance spectroscopy applied to faeces samples in order to predict voluntary intake and digestibility of forages by ruminants. Anim. Feed Sci. Technol. **205**, 49–59 (2015)

36. B.M. Nicolaï, K. Beullens, E. Bobelyn, A. Peirs, W. Saeys, K.I. Theron, J. Lammertyn, Postharvest Biol. Technol. **46**(2), 99–118 (2007)

37. W. Saeys, N.N. Do Trong, R. Van Beers, B.M. Nicolaï, Multivariate calibration of spectroscopic sensors for postharvest quality evaluation: a review. Postharvest Biol. Technol. **158**, 2019–110981 (2019)
38. B. Zhang, B. Gu, G. Tian, J. Zhou, J. Huang, Y. Xiong, Challenges and solutions of optical-based nondestructive quality inspection for robotic fruit and vegetable grading systems: a technical review. Trends Food Sci. Technol. **81**, 213–231 (2018)
39. A. Pissard, H. Bastiaanse, V. Baeten, G. Sinnaeve, J.-M. Romnee, P. Dupont, A. Mouteau, M. Lateur, Use of NIR spectroscopy in an apple breeding program for quality and nutritional parameters. Acta Hort. **976**, 409–414 (2013)
40. A. Pissard, V. Baeten, J.M. Romnee, P. Dupont, A. Mouteau, M. Lateur, Classical and NIR measurements of the quality and nutritional parameters of apples: a methodological study of intra-fruit variability. Biotechnol. Agron. Soc. Environ. **16**(3), 294–306 (2012)
41. A. Pissard, V. Baeten, P. Dardenne, P. Dupont, M. Lateur, Use of NIR spectroscopy on fresh apples to determine the phenolic compounds and dry matter content in peel and flesh. Biotechnol. Agron. Soc. Environ. **22**(1), 3–12 (2018)
42. J.A. Fernández Pierna, P. Vermeulen, B. Lecler, V. Baeten, P. Dardenne, Calibration transfer from dispersive instruments to handheld spectrometers (MEMS). Appl. Spectrosc. **64**(6), 644–648 (2010)
43. N.A. O'Brien, C.A. Hulse, D.M. Friedrich, F.J.V. Milligen, M.K. Von Gunten, F. Pfeifer, H.W. Siesler, Nondestructive measurement of fruit and vegetable quality by means of NIR spectroscopy: a review. Proc. SPIE **8374**(837404), 1–8 (2012)
44. J.A. Fernández Pierna, V. Baeten, P. Dardenne, Screening of compound feeds using NIR hyperspectral data. Chemometr. Intell. Lab. Syst. **84**, 114–118 (2006)
45. L.S. Magwaza, U.L. Opara, H. Nieuwoudt, P.J.R. Cronje, W. Saeys, B. Nicolaï, NIR spectroscopy applications for internal and external quality analysis of citrus fruit–a review. Food Bioprocess Technol. **5**(2), 425–444 (2012)
46. P. Vermeulen, J.A. Fernández Pierna, H.P. van Egmond, J. Zegers, P. Dardenne, V. Baeten, Validation and transferability study of a method based on near-infrared hyperspectral imaging for the detection and quantification of ergot bodies in cereals. Anal. Bioanal. Chem. **405**(24), 7765–7772 (2013)

Chapter 15
Applications: Food Science

Marena Manley and Paul James Williams

Abstract The combination of speed, accuracy and simplicity provided by NIR spectroscopy ensured its use as a preferred quality control tool in the food and beverage industries. These applications are increasingly simplified by the availability of readily available factory calibrations. A challenge receiving increasing attention is that of the detection of food adulteration, and a large effort is being made to evaluate NIR spectroscopy as a suitable method. The recent trend towards miniaturisation of NIR instruments contributes to the technology becoming portable and more affordable. The trust put into NIR spectroscopy as an effective analytical tool in the food industry will remain. In addition, investigations into new and innovative applications to the benefit of the food industry are seen on a daily basis.

Keywords Beer · Dairy · Cereals · Fish · Food authenticity · Fruit · Meat · Wine · Quality control · Vegetables

15.1 Introduction

Following agricultural applications, NIR spectroscopy is used as a quality control tool predominantly in the food industry [1]. The first use of infrared spectroscopy in foods was reported in the late 1930s when Ellis and Bath [2] determined the amount of water in gelatine. In 1975, Phil Williams replaced the Kjeldahl testing method for protein determination in the Canada Western Red Spring (CWRS) wheat programme with NIR spectroscopy [3]. This was the first-ever application of NIR technology in industry. Since then, NIR spectroscopy applications rapidly extended to now cover a wide range of food and beverage analyses [1, 4]. These include, apart from the initial wheat flour analysis, also other cereals and cereal products [4], meat and meat

M. Manley (✉) · P. J. Williams
Department of Food Science, Stellenbosch University, Private Bag X1, Matieland,
7602 Stellenbosch, South Africa
e-mail: mman@sun.ac.za

P. J. Williams
e-mail: pauljw@sun.ac.za

© Springer Nature Singapore Pte Ltd. 2021
Y. Ozaki et al. (eds.), *Near-Infrared Spectroscopy*,
https://doi.org/10.1007/978-981-15-8648-4_15

products [5], beer [6], wine [7], fruit and vegetables [8] as well as sensory properties of foods [9]. Commodities not considered in earlier years include cocoa beans [10], pistachio nuts [11], hazelnut kernels [12] and honey [13]. Investigations on olive oil [14] are still prominent.

Although NIR spectroscopy should theoretically only be applied to organic materials, of which the C–H, O–H and N–H bonds absorb in the NIR region, calibration models can also be developed to predict physical properties of samples. Wheat and maize kernel hardness are related to the particle size of the flour and meal, which can be measured due to the different size particles scattering the NIR light differently. It is also possible to measure the content of disolved salt (NaCl), which does not absorb in the NIR region, in food products [15]. This is because the presence of salt causes a shift in the water bands along the wavelength axis proportional to the salt concentration. The magnitude of the shift depends on the salt concentration which can be measured by NIR spectroscopy.

Food authenticity issues are a major concern in the industry and products or ingredients that are high in value are usually targeted [16]. Products with Protected Designation of Origin (PDO) or Protected Geographical Indication (PGI) which are usually produced at high costs are also prone to be adulterated with cheaper substitutes. Due to the complexity of the food matrix and its heterogeneous composition, it is difficult to identify adulterated products. The variability of the adulterations adds to this complexity. Adulteration has also become more refined in recent years. NIR spectroscopy has been evaluated for its potential to detect adulteration and/or confirm the authenticity of food products for many years. Manley and Baeten [17] recently provided an extensive review on the use of NIR spectroscopy in authenticity studies. NIR spectroscopy adulteration studies are expensive to perform due to the costs involved to collect suitable and large sample sets that include adequate variation. Because of this, many authenticity studies are performed on a limited number of samples, thus usually only demonstrating feasibility. There is a good likelihood of successful applications in industry, but such work is usually performed in-house and the results are not made available in the public domain.

In this chapter, quantitative and qualitative NIR spectroscopy food and beverage applications will be considered and briefly discussed.

15.2 Cereals and Cereal Products

The first application of NIR spectroscopy in the wheat industry dates back to 1975 when NIR spectroscopy was used to predict protein content in wheat [3]. Protein and moisture content measurements on wheat flour and whole grain are still one of the most widely used applications. Routine wheat applications have since extended to also include wheat hardness as well as ash content and starch damage. NIR spectroscopy is now commonly used as a rapid cereal quality control technique, and factory calibrations for a number of the measurements are readily available which can be purchased from NIR instrument manufacturers.

Distinguishing between wheat varieties based on their breadmaking quality was the first qualitative analysis performed in the 1980s [18]. Downey et al. [19] correlated wheat hardness with breadmaking quality, using a wheat hardness index, to differentiate between the wheat samples. Differences in particle sizes and the presence of inorganic additives enabled 97% correct classification of a range of commercial white flours, i.e. biscuit, self-raising, household, cake, bakers' and soda bread mix [20].

Wheat used for food applications comprised bread wheat (*Triticum aestivum*) and durum wheat (*Triticum durum*). The latter is used for pasta production and has different chemical and physical properties compared to bread wheat. In some European countries such as Italy, pasta is required to be produced using only durum wheat semolina and water. The addition of bread wheat results in a lower-quality product which would have inadequate resistance to cooking. Potential adulteration of durum wheat with bread wheat is thus of great concern. The potential to detect the addition of bread wheat flour to durum wheat flour was illustrated with uncertainties associated with the models to be about half of that of the official Italian method [4].

Although NIR spectroscopy is extensively used to quantify chemical composition in cereals (e.g. protein, moisture, oil), limited studies are available on cultivar discrimination and traceability of cereals [21].

15.3   Meat and Meat Products

NIR spectroscopy is extensively used to determine the content of meat components. The first NIR spectroscopy models developed included those which could predict intramuscular fat and moisture content. These could be predicted at excellent accuracies with low SEP results (0.18% for intramuscular fat; 0.37% for moisture) and high RPD values (9.17 for intramuscular fat; 7.21 for moisture) demonstrated [22]. Quantification of protein also with excellent prediction accuracies (SEP = 0.35%; RPD = 5.13) followed soon. NIR prediction of technological properties is more challenging as can be seen in SEP results and RPD values obtained for pH (0.05; 1.28), colour (0.42; 2.16) and water-holding capacity (WHC; 2.355; 1.27). Better results were obtained for pH when spectra were collected from intact meat samples. When minced meat was used, chemical composition predictions were more successful than for intact meat. Some success was achieved with more complex predictions such as ash content (SEP = 0.15%; RPD = 4.53). Adding the visible range enabled improved NIR spectroscopy predictions for colour. Water-holding capacity and drip loss measurements could, however, only be done with limited success thus far.

NIR spectroscopy measurement of sensory properties has not been successful due to intact meat samples not being homogenous, which is the main reason for poor predictions to date. The subjectivity of taste panels also contributes, in addition to inconsistent sample preparation. Consistent presentation of the sample to the instrument is also important and should receive the required attention when acquiring NIR spectra. A reasonable accuracy was obtained when beef tenderness was predicted with an SEP result of 0.35 and an RPD value of 3.82. NIR spectroscopy meat and meat product applications have been comprehensively reviewed by Prieto et al. [5].

Discrimination studies in meat receive continuous attention with the main issues being replacement of meat cuts of high value with cuts less costly. The potential of NIR spectroscopy to distinguish between different meat cuts has been considered extensively over the years. Discrimination between raw pork, chicken and turkey [4] as well as kangaroo and beef [23] have been demonstrated. McElhinney et al. [24] determined the lamb content mixed in raw minced beef. Classification accuracies of more than 85% were obtained when raw beef, lamb, pork and chicken were classified [4]. Meat properties such as intramuscular fat, fatty acids as well as muscle structure and type of muscle fibres would have contributed to the discrimination in these studies.

One of the earlier meat adulteration studies addressed the concern of selling frozen-then-thawed meat as fresh meat cuts [4], which is still evaluated today [25]. A more recent study also included assessment of mince beef adulteration [26] while the most recent study classified turkey meat products [27].

15.4 Fish and Fish Products

Quantification of moisture, fat and free fatty acids is the most common analysis performed with NIR spectroscopy on fish [28]. Because fish is highly perishable, microbial spoilage and freshness are important quality characteristics. Selling frozen-then-thawed fish as fresh is an important concern for the fish industry. Zhou et al. [29] demonstrated the use of NIR spectroscopy to determine freshness in fish flesh. As was the case for meat, heterogeneity of the fish samples resulted in NIR spectroscopy having limited potential for prediction of sensory properties. Subjectivity of the taste panel, which is the reference method in this case, could also have contributed to inaccurate predictions. The first application of the use of a small (miniature) handheld NIR device was demonstrated in fish authenticity studies [30] and is now increasingly evaluated [31] with the added advantage of onsite analysis.

15.5 Milk and Milk Products

Milk, a turbid and opaque liquid, is a challenging commodity to analyse with NIR spectroscopy. This is due to milk being a suspension containing fat globules and casein micelles which cause it to be a highly scattering medium [32]. NIR spectroscopy has initially only been used on low moisture products such as milk powders. Nowadays, it is widely used covering the entire range of dairy products. Holroyd [33] extensively reviewed the application of NIR spectroscopy in milk and milk products while Cattaneo and Holroyd [34] reviewed determination of adulteration and contamination in milk and milk powder. Chemical composition predictions in cheese, i.e. dry matter, fat and sodium chloride, could be done with accuracies suitable for routine analysis. RPD values of 6.0, 3.2 and 2.9 were obtained, respectively.

If is often difficult to develop calibration models in industry, due to the lack of variation between the samples. The compound to be measured would cover only a limited range. Filho and Volery [35] demonstrated how this can be overcome when they quantified the solids content using a 'broad-based' calibration including five different fresh cheeses with low, medium and high solids contents.

More recently, qualitative calibration model development has progressed considerably. González-Martín et al. [9] illustrated the power of NIR spectroscopy to predict sensory attributes of cheese. Texture measurements such as hardness, chewiness and creamy could be predicted with RPD values of 3.3, 2.7 and 1.6, respectively, with the hardness measurements suitable for routine analysis. Taste predictions resulted in RPD values of 1.6, 2.1, 2.3 and 2.6 for salty, buttery, rancid flavour and pungency, respectively. Volatile compounds could also be measured with reasonable accuracy, i.e. 2-nonanone (RPD = 3.4), acetaldehyde (RPD = 2.3), ethanol (RPD = 2.8), 2-heptanone (RPD = 2.8), 2-butanol (RPD = 2.1) and 2-pentanone (RPD = 2.0).

One of the most common methods of milk adulteration is the addition of water. Adulteration with melamine which is harmful when consumed is, however, of much greater concern. Melamine gives a false indication of increased protein content. The difficulty in using NIR spectroscopy as a method of analysis [36] is the low levels of melamine required to be detected.

15.6   Vegetable and Olive Oils

Sato et al. [37] performed the first NIR spectroscopy study on fats and oils. They suggested that a spectral library could be compiled which could then be used to check if the spectrum of an unknown sample matches any of the spectra in the database. They continued with a study in which they successfully distinguished between nine different types of vegetable oils, using principal component analysis (PCA) [37]. At the same time, Bewig et al. [38] illustrated the use of discriminant analysis and only four wavelengths to classify four different oils (cottonseed, peanut, soybean, canola). Similarly, Hourant et al. [39] demonstrated the use of selected wavelength ranges (1700–1800 and 2100–2400 nm) to classify seven vegetable oils.

The high value of extra virgin olive oil resulted in its potential adulteration with less costly oils [40]. Adulteration with inferior olive oils tends to be of concern, especially in olive oil producing countries. In contrast, addition of vegetable or seed oils seems to be of concern more likely in countries which produce these oils and import olive oils. The most important indicator of adulteration is the fatty acid composition of the oil [41]. Detection and quantification of the type of adulterant in virgin olive oil at an accuracy of 75% were demonstrated in an early study [42]. Using discriminant analysis, the authors subsequently correctly identified the type of adulterant in extra virgin oil with a

90% accuracy [42]. The level of adulteration could also be accurately predicted (±0.9% w/w).

Downey et al. [4] used SIMCA to classify authentic extra virgin olive oils from the same oils adulterated with sunflower oil. It was possible to detect adulteration at levels as low as 1% (w/w) as well as to predict the level of sunflower oil added using PLS regression (SECV = 0.8% w/w). They developed a model that could determine the level of sunflower oil adulterant with an accuracy suitable for industry use. Subsequent studies detected the adulteration of olive oils with a range of adulterants with very low error limits [43]. A recent study confirmed the use of NIR spectroscopy as a method to screen for adulterated olive oils [14]. When an unadulterated sample was also analysed, the level of detection was as low as 2.7% (w/w). Using SIMCA and without an unadulterated sample, the level of detection was less accurate (20%).

The use of handheld instruments has also been considered for oil analysis [44]. In spite of the handheld device only using the wavelength range of 950–1650 nm, lard adulteration in palm oil could be detected with a model accuracy of more than 0.95 using SIMCA. Using PLS regression gave even better results ($R^2 = 0.99$). As is the case with many adulteration studies, the sample set was limited, thus only demonstrating the feasibility of the application.

15.7   Fruit and Vegetables

One of the earliest fruit-related studies, detection of adulteration of orange juices, was reviewed by Shilton et al. [45]. As was the case with the early studies on oil, Shilton et al. [45] suggested the use of NIR spectroscopy as a 'fingerprint' technique rather than trying to predict specific constituent levels. In a subsequent study, however, LDA and PLS were used to classify apple juices up to 100% correctly, based on fruit variety [4].

The ability of NIR instruments, in association with chemometrics, to predict fruit and vegetable quality properties has been comprehensively reviewed [46]. Studies considered include dry matter content of onions, soluble solids content (SSC) of apples and water in mushrooms. Prediction of acidity was less accurate than predicting SSC due to NIR spectroscopy not being able to measure it directly, but based on its correlation with sugars. Similarly, fruit maturity could be predicted based on its correlation with sugar content and the microstructure of the fruit tissue. The microstructure of the fruit affects how the NIR light penetrates into and scatters within the fruit tissue which enables measurement of stiffness, internal damage as well as sensory attributes.

The successful measurement of changes in soluble solids and dry matter in individual mango fruit over time during ripening was demonstrated with a handheld device (950–1650 nm) [47]. The penetration depth of about 7.4 mm into the fruit tissue ensured representative sampling and contributed to the success of the developed models.

15.8 Honey

Initial NIR spectroscopy studies on honey mainly comprised the determination of its chemical composition [48]. Honey is a completely natural, high-value product comprising simple carbohydrates with a distinct sucrose–glucose–fructose profile and water. Cane invert sugar preparations are often prepared to mimic this profile and added to pure honey. Addition of such preparations is usually difficult to detect. Differences in floral species, maturity, environment, processing and storage techniques contribute to the natural variability of honey and thus the complexity of detecting honey adulteration.

The use of NIR spectroscopy and PLS-DA to detect adulterated honey has been demonstrated [49]. The addition of fructose and glucose could be detected with a high degree of success (99%). Similarly, the pure honey could be accurately identified (96%). The importance of temperature control during NIR analysis was, however, stressed. When SIMCA was evaluated as a discrimination technique [50], the adulterated honey could be 100% correctly identified compared to only 90% of the authentic honey. PLS-DA was shown as an effective classification method to discriminate South African from intentionally adulterated as well as imported honey [51]. Overall classification accuracies of between 93.3 and 99.9% were obtained. The handheld device evaluated in this study performed as accurately as the desktop instrument.

15.9 Tea

Growth in the functional food and bioactive ingredients markets encouraged the application of NIR spectroscopy within this field. McGoverin et al. [52] provided an extensive review on the quantification of bioactive compounds within food commodities such as tea.

Tea is of great interest due to its beneficial health properties. It is made from the processed leaves of *Camellia sinensis* and one of the most popular beverages consumed worldwide. Osborne and Fearn [53] reported one of the first discriminant studies on tea, distinguishing between black teas of differing sensory profiles. Grant et al. [54] showed a reliable classification of six teas differing in origins and taste. The effect of the grinding method when preparing and analysing powdered samples by NIR spectroscopy should always receive suitable attention [53]. Using SIMCA as a classification modelling technique, it was possible to identify four different tea varieties [55]. Manley et al. [56] quantified the major phenolic compounds, soluble solid content and total antioxidant activity of green rooibos (*Aspalathus linearis*), an indigenous South African herbal tea.

15.10   Coffee

The initial work on coffee, aimed at process control, illustrated the ability to distinguish between regular and decaffeinated coffee [57]. While this is no longer the focus of current work, at the time it showed the great potential of NIR spectroscopy. Today, the majority of commercially produced coffee is either Arabica or Robusta, with the former highly regarded for its improved sensory attributes. Since replacing the one with the other, or mixing/blending Arabica with Robusta is considered adulteration, detecting and quantifying this are important. Downey et al. [58] illustrated the capability of NIR spectroscopy to discriminate between pure and blends of Arabica and Robusta coffees. The coffee samples were either green or roasted and whole or ground beans, and a classification accuracy of 96.2% was achieved for the pure whole bean coffees. This was attributed to the caffeine content; it is well known that Robusta has a higher concentration than Arabica. When 50:50 blends were included in the model, lower accuracies of between 82.9 and 87.6% were, respectively, obtained for 20 green and 20 roast samples. A handheld device was successfully used for Arabica coffee grading, detecting the presence of peel/sticks, maize and Robusta coffee [59].

15.11   Wine and Distilled Alcoholic Beverages

There are numerous and diverse applications of NIR spectroscopy for wine analysis. Cozzolino et al. [7] predicted a number of phenolic compounds, simultaneously, in fermenting must and red wine. Most of the applications on wine focused on characteristics such as alcohol content, sensory and aromatic attributes and fermentation [4]. Cozzolino et al. [7] reviewed additional properties, such as the measurement of grape composition. Good to excellent RPD values were reported for total soluble solids (4.0), total anthocyanins (4.2), acidity and pH (2.8). In addition, measures such as alcoholic degree (5.7), total acidity (2.27), pH (2.4), glycerol (4.0), reducing sugars (10.3) and total sulphur dioxide (1.8) were reported for wine composition. Dambergs et al. [7] and Cozzolino et al. [7] were able to predict wine sensory quality, demonstrating the versatility of NIR spectroscopy. A problem often encountered with wine analysis and specifically when observing the fermentation process is the fact that the sample changes with time. In another study, Manley et al. [7] used NIR spectroscopy to measure sugar in grape must and to distinguish between samples based on their free amino nitrogen (FAN) content. Furthermore, the authors distinguished between Chardonnay wines and tables wines, based on their malolactic fermentation status and ethyl carbamate content, respectively.

Wine authenticity received considerable attention in the past [60]. Manley et al. [61] categorised four classes of rebate brandy; whereas, Pontes et al. [62] proposed a strategy to detect adulteration in whiskeys, brandies, rums and vodkas.

Since its first application for grape compositional analysis, there has always been a need to take the instrument to the sample, enabling analysis of grapes on the vine.

This is now possible due to technological advancement which led to the development of not only portable, but low-cost small handheld instruments [63].

15.12   Beer

Process control, with processing information continuously provided, is important during beer production, in addition to using good-quality raw materials. The main application in this field has been the development of models to select the best barley varieties to produce high-quality malt for beer production [6]. Investigations included genotype classification, mycotoxin detection and quantitative analysis of intact and ground grain for moisture, protein and β-glucan. In addition, intermediate products such as wort, extract and free amino nitrogen (FAN) were also investigated, while on the completed product real extract and ethanol were determined. Process analytical technology (PAT) is rapidly developing and becoming synonymous with product process optimisation strategies [7].

15.13   Aquaphotomics

Water plays a complex role in food systems and despite being studied extensively over many years, it is still not well understood. The term, aquaphotomics, has been introduced by Tshenkova [64] to describe the concept of approaching water as a multi-elemental system. Visible–NIR spectroscopy, being a powerful tool and source of spectral information, facilitated the establishment of this term. Aquaphotomics uses the information from water absorbance bands and patterns to provide knowledge of water structures and interactions between water and other components in, for example, a food system. The aim of aquaphotomics is thus to build up knowledge of and understand water absorbance bands over the entire electromagnetic spectrum in relation to functions of different biological systems. Bazar et al. [65] used aquaphotomics-based analysis to study honey adulteration. A difference in the water molecular structure of the honey and the added high fructose corn syrup (HFCS) was shown, with the honey samples containing a larger amount of highly organised water. Water matrix coordinates were assigned to the characteristic water bands of the honey with different levels of HFCS mixed into them. The variation of these coordinates describes the water spectral patterns of the different samples and can be visualised in aquagrams.

15.14   Conclusion

NIR spectroscopy has developed into a prominent analytical quality control tool in the food and beverage industries, due to its distinctive combination of speed, accuracy and simplicity. The requirement for the development of calibration models for each application and commodity is nowadays addressed by the availability of a number of factory calibrations readily available. The capabilities of NIR spectroscopy instrumentation are continually improving to maximise its performance, and the availability of small handheld instruments makes NIR spectroscopy portable and more affordable.

References

1. M. Manley, Near-infrared spectroscopy and hyperspectral imaging: non-destructive analysis of biological materials. Chem. Soc. Rev. **43**, 8200–8214 (2014)
2. J.W. Ellis, Alterations in the infrared absorption spectrum of water in gelatin. J. Bath J. Chem. Phys. 6, 723–729 (1938)
3. P. Williams, J. Antoniszyn, M. Manley, *Near-Infrared Technology: Getting the Best Out of Light* (AFRICAN SUN MeDIA, Stellenbosch, 2019)
4. T. Woodcock, G. Downey, C.P. O'Donnell, Better quality food and beverages: the role of near infrared spectroscopy. J. Near Infrared Spectrosc. **16**, 1–29 (2008) and papers therein
5. N. Prieto, O. Pawluczyk, M.E.R. Dugan, J.L. Aalhus, A review of the principles and applications of near-infrared spectroscopy to characterize meat, fat, and meat products. Appl. Spectrosc. **71**, 1403–1426 (2017)
6. V. Sileoni, O. Marconi, G. Perretti, Near-infrared spectroscopy in the brewing industry. Crit. Rev. Food Sci. Nutr. **55**(12), 1771–1791 (2015)
7. D. Cozzolino, R.G. Damsbergs, L. Janik, W.U. Cynkar, M. Gishen, Analysis of grapes and wine by near infrared spectroscopy. J. Near Infrared Spectrosc. **14**, 279–289 (2006) and papers therein
8. H. Wang, J. Peng, C. Xie, Y. Bao, Y. He, Fruit quality evaluation using spectroscopy technology: a review. Sensors **15**, 11889–11927 (2015)
9. M.I. González-Martín, P. Severiano-Pérez, I. Revilla, A.M. Vivar-Quintana, J.M. Hernández-Hierro, C. González-Pérez, I.A. Lobos-Ortega, Prediction of sensory attributes of cheese by near-infrared spectroscopy. Food Chem. **127**, 256–263 (2011)
10. S. Sunoj, C. Igathinathane, R. Visvanathan, Nondestructive determination of cocoa bean quality using FT-NIR spectroscopy. Comput. Electron. Agr. **124**, 234–242 (2016)
11. R. Vitale, M. Bevilacqua, R. Bucci, A.D. Magrì, A.L. Magrì, F. Marini, A rapid and non-invasive method for authenticating the origin of pistachio samples by NIR spectroscopy and chemometrics. Chemom. Intell. Lab. Syst. **121**, 90–100 (2013)
12. A. Pannico, R.E. Schouten, B. Basile, R. Romano, E.J. Woltering, C. Cirillo, Non-destructive detection of flawed hazelnut kernels and lipid oxidation assessment using NIR spectroscopy. J. Food Eng. **160**, 42–48 (2015)
13. M. Ferreiro-González, E. Espada-Bellido, L. Guillén-Cueto, M. Palma, C.G. Barroso, G.F. Barbero, Rapid quantification of honey adulteration by visible-near infrared spectroscopy. Talanta **188**, 288–292 (2018)
14. N. Vanstone, A. Moore, P. Martos, S. Neethirajan, Detection of the adulteration of extra virgin olive oil by near-infrared spectroscopy and chemometric techniques. Food Qual. Saf. **2**, 189–198 (2018)
15. T. Hirschfeld, Salinity determination using NIRA. Appl. Spectrosc. **39**, 740–741 (1985)

16. S. Lohumi, S. Lee, H. Lee, B.K. Cho, A review of vibrational spectroscopic techniques for the detection of food authenticity and adulteration. Trends Food Sci. Tech. **46**, 85–98 (2015)
17. M. Manley, V. Baeten, Spectroscopic technique: near infrared (NIR) spectroscopy, in *Modern Techniques for Food Authentication*, ed. by Sun, D.-W., 2nd edn (Elsevier, Oxford, 2018), pp. 51–102
18. M.F. Devaux, D. Bertrand, Discrimination of bread-baking quality of wheats according to their variety by near infrared reflectance spectroscopy. Cereal Chem. **63**(2), 151–154 (1986)
19. G. Downey, S. Byrne, E. Dwyer, Wheat trading in the Republic of Ireland: the utility of a hardness index derived by near infrared reflectance spectroscopy. J. Sci. Food Agric. **37**, 762–766 (1986)
20. A. Sirieix, G. Downey, Commercial wheat flour authentication by discriminant analysis of near infrared reflectance spectra. J. Near Infrared Spectrosc. **1**, 187–197 (1993)
21. D. Cozzolino, An overview of the use of infrared spectroscopy and chemometrics in authenticity and traceability of cereals. Food Res. Int. **60**, 262–265 (2014)
22. N. Prieto, R. Roehe, P. Lavín, G. Batten, S. Andrés, Application of near infrared reflectance spectroscopy to predict meat and meat products quality: a review. Meat Sci. **83**, 175–186 (2009)
23. H.B. Ding, R.-J. Xu, Differentiation of beef and kangaroo meat by visible/near-infrared reflectance spectroscopy. J. Food Sci. **64**(5), 814–817 (1999)
24. J. McElhinney, G. Downey, C. O'Donnell, Quantification of lamp content in mixtures with raw minced beef using visible near and mid-infrared spectroscopy. J. Food Sc. **64**(4), 587–591 (1999)
25. F. Huang, Y. Li, J. Wu, J. Dong, Y. Wang, Identification of repeatedly frozen meat based on near-infrared spectroscopy combined with self-organizing competitive neural networks. Int. J. Food Prop. **19**, 1007–1015 (2016)
26. C. Alamprese, M. Casale, N. Sinelli, S. Lanteri, E. Casiraghia, Detection of minced beef adulteration with turkey meat by UV–vis, NIR and MIR spectroscopy. LWT-Food Sci. Technol. **53**, 225–232 (2013)
27. D.F. Barbin, A.T. Badaró, D.C.B. Honorato, E.Y. Ida, M. Shimokomakia, Identification of turkey meat and processed products using near infrared spectroscopy. Food Control **107**, 106816 (2020)
28. D. Liu, X.A. Zeng, D.-W. Sun, NIR spectroscopy and imaging techniques for evaluation of fish quality—a review. J. Appl. Spectrosc. Rev. **48**, 609–628 (2013)
29. J. Zhou, X. Wu, Z. Chen, J. You, S. Xiong, Evaluation of freshness in freshwater fish based on near infrared reflectance spectroscopy and chemometrics. LWT-Food Sci. Technol. **106**, 145–150 (2019)
30. N. O'Brien, C.A. Hulse, F. Pfeifer, H.W. Siesler, Near infrared spectroscopic authentication of seafood. J. Near Infrared Spectrosc. **21**, 299–305 (2013)
31. S. Grassi, E. Casiraghi, C. Alamprese, Handheld NIR device: a non-targeted approach to assess authenticity of fish fillets and patties. Food Chem. **243**, 382–388 (2018)
32. D.J. Dahm, Explaining some light scattering properties of milk using representative layer theory. J. Near Infrared Spectrosc. **21**(5), 323–339 (2013)
33. S.E. Holroyd, The use of near infrared spectroscopy on milk and milk products. J. Near Infrared Spectrosc. **21**, 311–322 (2013)
34. T.M.P. Cattaneo, S.E. Holroyd, The use of near infrared spectroscopy for determination of adulteration and contamination in milk and milk powder: updating knowledge. J. Near Infrared Spectrosc. **21**, 341–349 (2013) and papers therein
35. P.A.D.A. Filho, P. Volery, Broad-based versus specific NIRS calibration: determination of total solids in fresh cheeses. Anal. Chim. Acta **544**, 82–88 (2005)
36. E. Domingo, A.A. Tirelli, C.A. Nunes, M.C. Guerreiro, S.M. Pinto, Melamine detection in milk using vibrational spectroscopy and chemometrics analysis: a review. Food Res. Int. **60**, 131–139 (2014)
37. T. Sato, Application of principal component analysis on near infrared spectroscopic data of vegetable oils for their classification. J. Am. Oil Chem. Soc. 71(3), 293–298 (1994) and papers therein

38. K.M. Bewig, A.D. Clarke, C. Roberts, N. Unklesbay, Discriminant analysis of vegetable oils by near-infrared reflectance spectroscopy. J. Am. Oil Chem. Soc. **71**(2), 195–200 (1994)
39. P. Hourant, V. Baeten, M.T. Morales, M. Meurens, R. Aparicio, Oils and fats classification by selected bands of near-infrared spectroscopy. Appl. Spectrosc. **54**(8), 1168–1174 (2000)
40. S. Armenta, J. Moros, S. Garrigues, M. De La Guardia, The use of near-infrared spectrometry in the olive oil industry. Crit. Rev. Food Sci. Nutr. **50**, 567–582 (2010)
41. E. Christopoulou, M. Lazaraki, M. Komaitis, K. Kaselimis, Effectivenes of determinations of fatty acids and triglycerides for the detection of adulteration of olive oils with vegetable oils. Food Chem. **84**, 463–474 (2004)
42. I.J. Wesley, F. Pacheco, A.E.J. McGill, Identification of adulterants in olive oils. J. Am. Oil Chem. Soc. **73**(4), 515–518 (1996)
43. A.A. Christy, S. Kasemsumran, Y. Du, Y. Ozaki, The detection and quantification of adulteration in olive oil by near infrared spectroscopy and chemometrics. Anal. Sci. **20**, 935–940 (2004)
44. K.N. Basri, M.N. Hussain, J. Bakar, Z. Sharif, M.F.A. Khir, A.S. Zoolfakar, Classification and quantification of palm oil adulteration via portable NIR spectroscopy. Spectrochim. Acta, Part A **173**, 335–342 (2017)
45. N. Shilton, G. Downey, P.B. McNulty, Detection of orange juice adulteration by near-infrared spectroscopy. Sem. Food Anal. **3**, 155–161 (1998)
46. E. Arendse, O.A. Fawole, L.S. Magwaza, U.L. Opara, Non-destructive prediction of internal and external quality attributes of fruit with thick rind: a review. J. Food Eng. **217**, 11–23 (2018)
47. E.J. Nascimento Marques, S.T. De Freitas, M. Fernanda Pimentel, C. Pasquini, Rapid and non-destructive determination of quality parameters in the 'Tommy Atkins' mango using a novel handheld near infrared spectrometer. Food Chem. **197**, 1207–1214 (2016)
48. M. García-Alvarez, J.F. Huidobro, M. Hermida, J.L. Rodríguez-Otero, Major components of honey analysis by near-infrared transflectance spectroscopy. J. Agric. Food Chem. **48**, 5154–5158 (2000)
49. G. Downey, V. Fouratier, J.D. Kelly, Detection of honey adulteration by addition of fructose and glucose using near infrared transflectance spectroscopy. J. Near Infrared Spectrosc. **11**, 447–456 (2003)
50. J.D. Kelly, C. Petisco, G. Downey, Potential of near infrared transflectance spectroscopy to detect adulteration of Irish honey by beet invert syrup and high fructose corn syrup. J. Near Infrared Spectrosc. **14**, 139–146 (2006)
51. A. Guelpa, F. Marini, A. Du Plessis, R. Slabbert, M. Manley, Verification of authenticity of South African honey and fraud detection using NIR spectroscopy. Food Control **74**, 1388–1396 (2017)
52. C.M. McGoverin, J. Weeranantanaphan, G. Downey, M. Manley, Review: the application of near infrared spectroscopy to the measurement of bioactive compounds in food commodities. J. Near Infrared Spectrosc. **18**(2), 87–111 (2010)
53. B.G. Osborne, T. Fearn, Discriminant analysis of black tea by near infrared reflectance spectroscopy. Food Chem. **29**, 233–238 (1988)
54. A. Grant, J.G. Franklin, A.M.C. Davies, Near infra-red analysis: the use of multivariate statistics for investigation of variables in sample preparation and presentation of tea leaf. J. Sci. Food Agric. **42**, 129–139 (1988)
55. Q. Chen, J. Zhao, G. Zhang, X. Wang, Feasibility study on qualitative and quantitative analysis in tea by near infrared spectroscopy with multivariate calibration. Anal. Chim. Acta **572**, 77–84 (2006)
56. M. Manley, E. Joubert, M. Botha, Quantification of the major phenolic compounds, soluble solid content and total antioxidant activity of green rooibos (*Aspalathus linearis*) by means of near infrared spectroscopy. J. Near Infrared Spectrosc. **14**, 213–222 (2006)
57. A.M.C. Davies, F. McClure, Near infrared analysis of foods. Near infrared analysis in the Fourier domain with special reference to process control. Anal. Proc. **22**, 321–322 (1985)
58. G. Downey, J. Boussion, D. Beauchêne, Authentication of whole and ground coffee beans by near infrared reflectance spectroscopy. J. Near Infrared Spectrosc. **2**, 85–92 (1994)

59. R.M. Correia, F. Tosato, E. Domingos, R.R.T. Rodrigues, L.F M. Aquino, P.R. Filgueiras, V. Lacerda Jr, W. Romão, Portable near infrared spectroscopy applied to quality control of Brazilian coffee. Talanta **176**, 59–68 (2018)

60. I.S. Arvanitoyannis, M.N. Katsota, E.P. Psarra, E.H. Soufleros, S. Kallithraka, Application of quality control methods for assessing wine authenticity: use of multivariate analysis (chemometrics). Trends Food Sci. Technol. **10**, 321–336 (1999)

61. M. Manley, N. De Bruyn, G. Downey, Classification of three-year old, unblended South African brandy with near infrared spectroscopy. NIR News **14**(5), 8–9, 11 (2003)

62. M.J.C. Pontes, S.R.B. Santos, M.C.U. Araújo, L.F. Almeida, R.A.C. Lima, E.N. Gaião, U.T.C.P. Souto, Classification of distilled alcoholic beverages and verification of adulteration by near infrared spectrometry. Food Res. Int. **39**, 182–189 (2006)

63. M. Alcalà, M. Blanco, D. Moyano, N.W. Broad, N. O'Brien, D. Friedrich, F. Pfeifer, H.W. Siesler, Qualitative and quantitative pharmaceutical analysis with a novel hand-held miniature near infrared spectrometer. J. Near Infrared Spectrosc. **21**, 445–457 (2013)

64. R. Tsenkova, Aquaphotomics: dynamic spectroscopy of aqueous and biological systems describes peculiarities of water. J. Near Infrared Spectrosc. **17**, 303–313 (2009)

65. G. Bazar, R. Romvari, A. Szabo, T. Somogyi, V. Éles, R. Tsenkova, NIR detection of honey adulteration reveals differences in water spectral pattern. Food Chem. **194**, 873–880 (2016)

Chapter 16
Wooden Material and Environmental Sciences

Te Ma, Satoru Tsuchikawa, and Tetsuya Inagaki

Abstract Near-infrared spectroscopy (NIRS) is suitable for both the qualification and quantification of organic properties associated with C–H, O–H, or N–H groups. There have been considerable efforts made toward proposing and developing various technologies and devices for the rapid and nondestructive measurement of various samples related to natural materials and environmental sciences. In this chapter, the utilizations of NIRS in the fields of wood material, soil, sediment, waste liquid, atmospheric gas detection, and archeological science will be explained through some representative studies.

Keywords Wood material · Soil · Sediment · Waste liquid · Archeological science

16.1 Introduction

Near-infrared spectroscopy (NIRS) is useful for the nondestructive measurement of various samples related to natural materials and environmental sciences. A typical example is wood material. The advantages of wood as a building material still outweigh other products (e.g., steel and brick) when looking at the environmental impact. It is naturally renewable and helps to mitigate climate change through carbon storage. However, there still is a limited extent in industry practice. The main reason is that wood has some drawbacks as a typical natural material such as deformation and cracking, caused by its heterogeneous structures and hygroscopic nature. Quality assessment of such characteristics just can get customers truly approbate and support and promote the development of the whole wood industry finally. Recently, there have been considerable efforts made toward proposing and developing various technologies, and NIRS makes great contributions to this field. NIRS also has been developed significantly at soil composition analysis since the 1990s, taking advantage of previous advances in agricultural instrumentation and chemometrics. Currently,

T. Ma · S. Tsuchikawa · T. Inagaki (✉)
Graduate School of Bioagricultural Sciences, Nagoya University, Furo-Cho, Chikusa, Nagoya 464-8601, Japan
e-mail: inatetsu@agr.nagoya-u.ac.jp

© Springer Nature Singapore Pte Ltd. 2021
Y. Ozaki et al. (eds.), *Near-Infrared Spectroscopy*,
https://doi.org/10.1007/978-981-15-8648-4_16

NIRS has been envisioned as a replacement for laboratory analysis in certain applications (e.g., soil carbon credit assessment at the farm level). Moreover, there have been many studies on the evaluation of lake sediment core composition changes by NIRS. Since the sediment core length increases when the sampling interval decreases, the need for a quick and inexpensive means of determining sediment composition became apparent. Finally, the utilizations of NIRS for waste liquid evaluation, i.e., wastewater, sewage, black liquor (a waste product from the crafting process in digesting pulpwood into paper pulp to remove lignin, hemicellulose, and other extracts from the wood and to release cellulose fibers), atmospheric gas detection, and archeological science will be explained in this chapter.

16.2 Wood

Figure 16.1 shows the IR and NIR spectra of *Chamaecyparis obtusa* (softwood) and *Zelkova serrata* (hardwood). Most absorption bands in the NIR region are corresponding to overtones or combinations of fundamental vibrations in the IR region. However, the absorption of NIR light in organic materials is very weak compared to

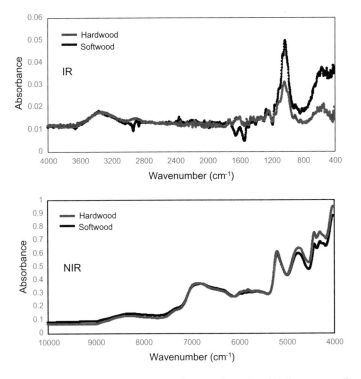

Fig. 16.1 IR and NIR spectra of *Chamaecyparis obtusa* (softwood) and *Zelkova serrata* (hardwood)

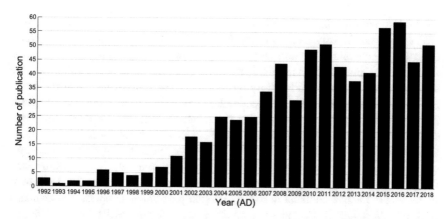

Fig. 16.2 Number of publications due to "NIR" and "Wood," searching by Web of Science

that of IR light; hence samples can be measured nondestructively. Another important advantage of NIRS is that many properties could be evaluated simultaneously which use the same NIR data. It is superior to sort wood, which generally is based on several criteria together.

However, as overtones or combinations of fundamental vibrations are multiply overlapped each other at the NIR range, measured spectra are "opaque spectroscopic information." They are generally analyzed with the aid of multivariate analysis, such as principal component analysis (PCA), partial least squares regression (PLS-R), to observe "useful material information." Some other NIR spectra collection and data analysis methods have also been developed since wood samples have much higher light scattering than light absorption. The spectral contributions of them were evaluated independently, for constructing robust calibration models of various wood properties without relying on complex multivariate statistical analysis.

Investigating NIRS for the quality assessment of pulp and paper has been going on for a number of years, Brikett and Gambino estimated pulp kappa number with NIRS approximately 30 years ago [1]. After that, the publication related to wood science and technology with NIRS dramatically increased. Figure 16.2 shows the number of research publications, including the keywords "NIR" and "Wood," searching by Web of Science. Nowadays, NIRS has already been exploited successfully by many paper-making companies to evaluate moisture content (MC) online. Below we introduce some representative studies separated into different wood properties.

16.2.1   Wood Chemical Composition

Wood is typically composed of three main chemical components: cellulose, hemi-cellulose, and lignin. Their percentages are roughly 45%, 25%, and 25%, respectively [2]. Wood also contains extractives that can be removed by solvents, which has been shown to play a large role in the protection of living tree [3]. The first NIR works were mostly focused on evaluating wood chemistry directly, especially the cellulose content [4–6] and then, shift to estimate wood lignin and extractives. Da Silva et al. presented an assessment of total phenolic compounds and extractive contents of mahogany wood rapidly by NIRS [7]. He and Hu indicated the benefits of FT-NIR to predict the lignin and extractive content of different wood species [8]. The validation results confirm that the selection of relevant wavenumbers and suitable data preprocessing methods produced values within tolerance levels. Lepoittevin et al. indicated that it is important to remove extractives before NIR spectra collection for the prediction of other wood chemistry traits [9]. Uner et al. utilized the genetic inverse least squares method for constructing the calibration models of lignin and extractive contents in milled Turkish pine wood samples. The standard error of calibration (SEC) and standard error of prediction (SEP) were 0.35% and 2.40%, respectively [10].

16.2.2 Wood Moisture Content

The molar absorption of water at the NIR range is 1/1000–1/10,000 compared to that of IR region [11]. Nevertheless, as the NIR range has rich light absorbance information of oxygen and hydrogen (O–H) structures, many researchers have invested NIRS to predict water within wood. In general, the state of water within wood can be categorized as either free or bound water. Here, free water is defined as liquid water located in the lumens and intercellular spaces of wood but without a chemical bond with the wood cell wall; whereas, the water attached by intermolecular forces between the major chemical components of wood cell walls is considered as bound water, which has profound effects on wood physical properties. For MC by mass (i.e., both free and bound water), Watanabe et al. compared the accuracy of NIRS with a commercial capacitance-type moisture meter for greenwood sorting purposes. Their experimental results showed the performance of NIR approach was better than the capacitance-type at predicting high moisture wood samples. Besides, compared to the capacitance-type moisture meter, the NIR method also has the advantage of measuring MC without the need for density correction [12]. However, Defo et al. suggested that NIR spectrometer may be less useful for the lumbers with significant gradients between core and surface layers [13]. For this limitation, Tham et al. recently highlighted the potential of NIRS combined with an industrial MC capacitance meter to predict MC from greenwood to oven-dried conditions [14]. Experimental results showed a good prediction accuracy (coefficient of determination (R^2) = 0.80, root mean square error of cross-validation (RMSECV) = 25.70%, and the ratio of percentage deviation (RPD) = 2.22) could be achieved at various sample thicknesses and wood species without density correction. It suggests that NIRS can be assisted by other techniques with higher transmission abilities, when measuring thick wood samples such as timber and lumber wood.

NIRS also could contribute to study bound water within wood, and it is a powerful tool for evaluating molecular water dynamics based on wavelength shift characteristics [15]. Inagaki et al. compared the variation of water adsorption between modern and archeological wood samples. The wavelength range of 1818–2128 nm due to the O–H first overtone was selected. Curve-fitting method was used to separate the baseline-corrected NIR difference spectra into three components that have different vibrational energy. Ma. et al. used PCA to characterize the variance of NIR spectral range of 1340–1610 nm due to the O–H second overtone after baseline correction. The data analysis results showed PC1 loading mainly correlates with wood water content by mass; however, the PC2 loading values contain the information about water–wood hydrogen structure interactions [16].

16.2.3   Wood Density

Density is a crucial parameter for wood strength and stiffness, which are critical considerations for a wooden structure. Most NIR calibration models were constructed based on light absorption differences caused by the three main chemical wood components (i.e., cellulose, hemicellulose, and lignin). Alves et al. calibrated the maritime pine and hybrid larch wood density measured by an X-ray densitometer with NIR spectra using PLS-R analysis [17]. The best PLS-R model could fulfill the requirements for the NIR model development and maintenance guidelines provided by the American Association of Cereal Chemists (AACC Method 39-00). Santos et al. also estimated the wood density of Portuguese Blackwood using NIRS combined with PLS-R analysis [18]. The RPD limit was 2.5, even though the number of spectra collected from each wood disk was only three. Fujimoto et al. examined the effect of MC on the accuracy of predicting wood density. They discussed the chemometric background for the potential to predict the wood density ($R^2 = 0.86 - 0.87$, SEP = 22 kg m^{-3}) at various moisture conditions [19].

Some studies also focused on light scattering caused by physical wood structure to predict density. Hans et al. measured seven softwood and hardwood species using time-of-flight NIRS (TOF-NIRS) which provides additional light scattering information. Then, curve-fitting procedure was used to separate absorption and reduced scattering coefficients. The square root of the adsorption/scattering ratio could correct the scattering effects in absorbance NIR spectra [20]. Ma et al. used spatially resolved spectroscopy (SRS) method, and a NIR imaging camera was utilized to catch the light scattering patterns on Douglas fir wood surface which illuminated by a concentrated halogen light source (Ø 1 mm). A steady-state diffusion theory model was applied to estimate light absorption and reduced scattering coefficients. The experimental results indicate that a few key wavelengths could achieve the prediction of subsurface density and grain direction without relying on multivariate statistical analysis [20]. Such an approach is worth pursuing further since it will contribute to the design of a low-cost measurement system and with robust prediction accuracy. Recently, Kitamura and Tsuchikawa [21] have shown the possibility of designing a cost-effective densitometer by a continuous NIR single wavelength laser source and an avalanche photodiode module. It could obtain a good calibration result with a conventional X-ray densitometer (RMSECV $= 0.046$ g cm^{-3}).

16.2.4 Wooden Anatomical Features

Wood has a porous three-dimensional structure. It is composed mostly of elongated cells that are parallel along the tree. A basic understanding of the wooden anatomical features is essential. Hein has developed NIR models to predict the microfibril angle (MFA) of Eucalyptus wood [22]. The RMSE between NIR predicted and X-ray diffraction derived values was 1.3°. Inagaki et al. demonstrated high-quality results when utilizing NIR to predict the fiber length of Eucalyptus solid [23]. Isik et al. examined NIRS to predict wood cell wall thickness, coarseness, air-dry density, MFA, and modulus of elasticity (MOE) of loblolly pine [24]. Furthermore, it suggested that NIRS can be utilized for screening loblolly pine progeny tests for surrogate wood traits.

16.2.5 Wood Mechanical Properties

The mechanical properties of wood are its fitness to resist outside forces. Knowledge of these properties is very important in the wood industry. However, conventional measurement methods are mostly destructive and time-consuming. Many studies have shown that wood mechanical properties could be evaluated by NIRS assisted by chemometrics. Wood density and the cellulosic feature are important in constructing prediction models from the viewpoint of the chemical absorption band. Horvath et al. utilized NIRS to predict the green modulus of elasticity (MOE) and green ultimate compression strength (UCS) of 1- and 2-year-old transgenic and wild-type aspen. Calibration results showed a well-predicted UCS ($R^2 = 0.91$, RMSEP $=$ 1.04 MPa) and green MOE ($R^2 = 0.78$, RMSEP $= 538$ MPa) [25]. Scimleck et al. examined to predict MOE, density, and modulus of rupture (MOR) simultaneously by diffuse NIR reflectance collected from the transverse surface of Pernambuco blocks [26]. Calibration results showed that the density prediction had the highest accuracy, followed by MOE, which results in MOR were pore. They suggested the presence of extractives may weaken the NIR-based calibration models. Watanabe et al. developed PLS-R-based calibration models for rapidly evaluating longitudinal growth strain (LGS) [27]. The LGS is one of the most important wood quality indices, and high levels easily cause end splitting. NIR spectra and LGS were measured from the peripheral locations of three Sugi green logs. The spectra with higher LGS tended to be lower absorbance may be caused by the chemical and physical properties related to the LGS. The calibration model achieved good accuracy (R^2 was 0.61 with a RMSEP of 0.015%). Kobori et al. and Sofianto et al. tested to measure NIR spectra from the

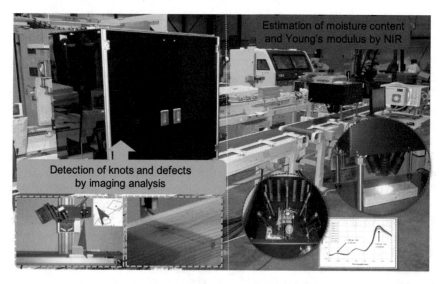

Fig. 16.3 Development of the spectrometer with linear sensor for estimation of moisture content and Young's modulus. Conveyor speed: 120 m/min [28, 29]

tangential surface of Hinoki [28] and Sugi [29] lumbers at a developed spectrometer with a speed of 120 m min^{-1} (Fig. 16.3). A diffraction grating linear sensor and high-intensity lighting were utilized in the measurement system. It showed fast speed and a sufficient prediction accuracy for the quality screening of wood lumber with the aid of PLS-R analysis, although the Sugi lumber samples cover more knots.

16.2.6 Wood Engineering Wood

Engineering wood, also called composite wood, human-made wood, is commonly composed of some kinds of wooden materials such as veneers, fibers, or particles together with adhesives. The advantage of engineering wood is that it can be designed to meet application-specific performance requirements. However, because the raw materials are continually changing during the compositing process, quality monitoring technique is also required to ensure its reliable performance. Maioli Campos et al. reported that NIRS combined with multivariate statistical analysis could be used for agro-based particleboards classification [30]. Rials et al. reported that good predictive models were generated for MOE, MOR, and internal bond of medium-density fiberboard (MDF) samples by NIRS with PLS-R analysis. Kohan et al. demonstrated the potential of NIRS to predict the ultimate tensile strength, tensile MOE, bending strength, and bending stiffness of strand feedstock [31]. Hein et al. showed the key role of adhesives, cellulose, and lignin for the mechanical and physical properties NIRS calibrations of agro-based particleboards [32].

16.2.7 Wood Modification and Degradation

Wood modification and degradation are other great topics. Wood is a hygroscopic natural material, and cell wall moisture adsorption or desorption occurs until an equilibrium moisture content (EMC) is reached in response to variation in surrounding relative humidity and temperature. Such sorption behaviors influence the stability of the wood dimension and mechanical properties. Hence, wood modification typically required before using. For example, when wood is treated with acetic anhydride, the hydroxyl groups of lignin, hemicelluloses, and cellulose are replaced with acetyl groups, and it has the advantages of dimensional stability, decay resistance to fungi. Schwanninger et al. reported that the chemical changes in wood due to acetylation could be monitored by NIRS [33]. Green et al. utilized NIRS for accessing wood decay in pine sapwood wafers [34]. Experimental results showed a strong correlation between NIR spectral data and sample mass loss, compression strength, and early stages of wood decay. Sandak et al. developed an FT-NIR-based methodology for estimating the biodegradation rate of recycled paper [35]. There were significant light absorption differences that correspond to C–H and O–H functional groups of cellulose. They found a good agreement between spectroscopic and reference methods (microscopy, mechanical testing, mycological tests). Inagaki et al. measured NIR reflectance spectra from the wood samples heated at 90, 120, 150, and 180 °C from 5 min to approximately 1.4 years. Kinetic analysis of principal component scores was useful to understand the chemical change in the thermally treated wood samples [36].

16.2.8 Wood Pulp and Paper

NIRS has been traditionally used in the quality analysis of pulp and paper. Downes et al. reported to predict Kraft pulp yield and cellulose content in Eucalyptus wood using NIRS [37]. Gigac and Fiserova indicated that NIR spectra could be used to predict the filler content, Kappa number, and strength properties of raw materials and paper [38]. Meder et al. compared the performance of laboratory and handheld NIR instruments in predicting Kraft pulp yield in standing trees (5 mm or 12 mm increment cores). The results showed the handheld NIR devices were also capable of predicting cellulose content and Kraft pulp yield [39]. Tyson et al. indicated that the tightly regulated pulping processes reduce the variability of the pulps, which affect on constructing physical and mechanical prediction models based on NIR spectra [40]. Yonenobu et al. [53] pointed out that NIRS was powerful in investigating the chemical condition of washi (literally "Japanese paper," which has played a vital role in Japanese culture since the early eighth century). This approach had obtained satisfactory results with conventional sugar analysis.

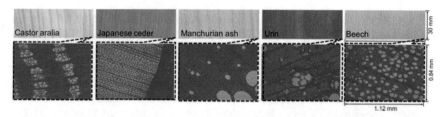

Fig. 16.4 Microscopic images of five representative wood sample species. Castor aralia and Manchurian ash are ring-porous hardwood, Japanese cedar is softwood, and Ulin and Beech are diffuse-porous hardwood, respectively. Each sample structure is unique from the other

16.2.9 Wood Species Classification

With the high diversity of species, it is of high importance to obtain accurate iden-tification. Figure 16.4 shows five representative wood species, including the three main types of wood: softwood, diffuse-porous hardwood, and ring-porous hardwood. The conventional identification approach is based on wood macroscopic charac-teristics. However, such methods are time-consuming and need full knowledge of wood anatomy. Hence, automatic identification systems are required in the fields of wood recycling and monitoring illegal logging protected tree species. Batista et al. explored NIRS as a potential option for the classification of several wood species. Experimental results showed that NIR spectra obtained from solid wood surfaces assisted with PLS discriminant analysis (PLS-DA) could achieve low identifica-tion errors [41]. Cooper et al. also pointed out that several factors may influence the NIRS performance, such as surface roughness, MC, and localized density differences [42]. Yang et al. classified softwood and hardwood by NIRS coupled with PLS-DA. They indicated that the differences of hemicelluloses and lignin components between softwood and hardwood species contributed much to the classification model [43]. Abe et al. were successful in the separation of two softwood species and indicated that light scattering might be useful for wood species classification purposes with advanced measurement systems [44]. Recently, Ma et al. evaluated the light scat-tering differences of five softwood and ten hardwood species based on NIR-SRS. They also encourage the observations that light scattering patterns in wood samples could be used for wood classification [45].

16.2.10 Imaging Analysis at the Field of Wood

Since the above wood properties are significantly varied between different regions of wood samples, NIR hyperspectral imaging (HSI) technique is a powerful approach that can provide not only spectral information but also including spatial informa-tion. It can provide a more detailed property analysis in every single annual ring. For example, Fernandes et al. measured wood density with a high spatial resolution of 79 μm by means of visible-NIR-HSI imaging [46]. The R^2 value between the present method and X-ray microdensitometer was 0.810 with an RMSE of 6.54 × 10^{-2} g cm^{-3}. The difference between latewood and earlywood (i.e., inside every single annual ring) was shown clearly. Lestander et al. applied NIR-HSI images to separate wood chips with elevated levels of extractives [47]. Meder et al. used HSI images for detecting the compression part of softwood. The compression wood is required to be detected at an early age since its higher proportion of lignin and lower cellulose easily, which causes further trucks. [48]. Kobori et al. [49] and Ma. et al. [16] successfully visualized MC distribution in wood by HSI techniques. Figure 16.5 shows the MC mapping results of three representative wood samples (Japanese cedar: softwood, Beech: diffuse-porous hardwood, and Manchurian ash: ring-porous hardwood), and water was preferentially retained in the latewood as the wood dries. Ma et al. evaluated the calibration between the SilviScan analysis system (FPInnovations, Vancouver, Canada) data and NIR-HSI imaging. The SilviScan system with a high spatial resolution provided the reference data for wood density with 25 μm resolution and for microfibril angle (MFA) with 1 mm resolution. Both the two important indexes were successfully mapped at a 156 μm spatial resolution [50]. Using the same HSI camera, Sofianto et al. successfully constructed a prediction model and

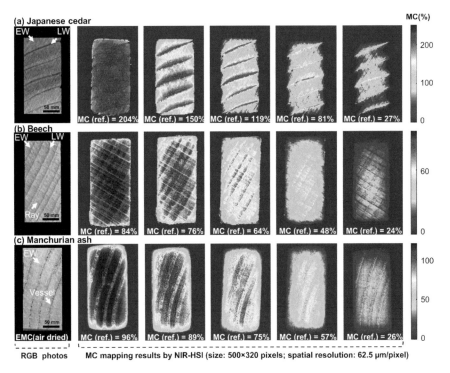

Fig. 16.5 Wood MC mapping result of three wood samples (Japanese cedar: softwood, Beech: diffuse-porous hardwood, and Manchurian ash: ring-porous hardwood), and water was preferentially retained in the latewood as the wood dries [16]

map the predicted modulus of elasticity (MOE) values. The results of MOE mapping could distinguish between a sound knot and a dead knot [51].

16.3 Soil

Soil is important to humans and all living things on the earth because it produces food, fiber, and energy. The ability of the soil to support these functions depends on its biological, chemical, and physical properties. NIRS for soil analysis started later than agricultural research. However, this research field has experienced a boom over the past couple of decades. NIRS can be used to estimate important properties in soil like salinity, moisture, total carbon, organic carbon, total N, or clay minerals. The assigned bands due to clay minerals (kaolin doublet, smectite, illite, carbonate) and organics (aromatics, amine, alkyl asymmetric–symmetric doublet, carboxylic acids, amides, aliphatics, methyls, phenolics, polysaccharides, carbohydrates) were summarized by Rossel et al. [52]. For the quantifications of soil parameters, visible (VIS) region spectra have been generally measured with NIR region. The light absorption by soil organic matter in the visible region is wide but more clear [53]. There are various reports suggesting that VIS–NIR is superior to built accurate calibration for organic matter than NIR alone [54]. Soil organic carbon (SOC) is an important property of soil quality that affects the type of organic compounds in the soil and the physical properties of the soil, therefore frequently estimated by VIS–NIR calibrations. The prediction results reported until 2010, summarized by Stenberg et al. [55], showed that the RMSEP for the prediction of SOC or total C was in the range of 2.5–9.0 mg g^{-1}. The variation of R^2 and RMSEP is due to the soil types, soil classes, and range of concentrations. Properties of which the calibration is reported are moisture, clay minerals, texture, pH, plant nutrients, and contaminant (i.e., heavy metals). Some metals, which are not absorbed by NIR light, can be detected because of covariation with spectrally active components.

16.4 Sediment

The nondestructiveness and rapidity of NIRS are particularly advantageous for the measurement of lake sediments and marine sediments. Lake sediments are derived from autochthonous production, substances derived from the lake itself, or allochthonous material (basin or aerial) and provide information about the region and its environmental history. Sediment is formed continuously in all lakes, and its composition depends on the biogeochemistry of the lake and the chemistry of the basin. Thus, lake sediments reflect the lake's average biogeochemistry, allowing integrated measurements of lake chemistry variables estimated based on sediment characteristics. Sediments have not only large spatial and temporal variations in physical and chemical properties but also show elemental flow between this compartment, hydrosphere, and atmosphere. Sediment records provide much information about the history of a body of water. This information can be used to reconstruct historical changes in water quality and to identify the natural state of waters prior to impacting the human body, thereby setting goals for lake restoration and restoration activities. The important thing for climatologists is the coherence between climate change and carbon cycling. Several studies have shown that most lakes in the world are net sources of CO_2 to the atmosphere and that the CO_2 emission from lakes is proportional to the input and lake mineralization of terrestrial organic carbon. Distribution and retention of heavy metals in lake sediments are also critical to understand biogeochemical processes in aquatic systems and lake management. Moreover, the investigation of marine sediments is also important for paleoceanographic and paleoclimatic views. A quick, inexpensive, high spatial resolution, and reproducible method for determining various chemistry of sediments like NIRS would be very useful for many environmental research and monitoring programs related to the aquatic system.

Malley et al. [56] reported the prediction of Cd, Cu, Zn, Pb, Ni, Mn, and Fe with an R^2 value of 0.86, 0.63, 0.91, 0.93, 0.81, 0.88, and 0.93, respectively, from the NIR reflection spectra of sandy and highly organic littoral sediments from a Precambrian Shield lake (37-ha surface area) in northwestern. In this research, cores were taken by scuba diver using a 5 cm plexiglass tube inserted by hand. After cutting the core (6–14 cm) into 1-cm-thick sections, samples were freeze-dried and sieved thorough no. 10 sieves before NIR measurement. Although there is no absorbance band due to these meatal in the NIR region, they attribute the reason for the high accuracy of heavy metal by NIR spectra to the correlation between heavy metals and organic matter containing protein, cellulose, and oil. Inagaki et al. [57] predicted values for water content, total nitrogen, total organic carbon, total sulfur, Al_2O_3, S/Al_2O_3, Fe_2O_3/Al_2O_3, Sc/Al_2O_3, Cu/Al_2O_3, and Zn/Al_2O_3 with coefficients of determination for cross-validation of 0.73, 0.89, 0.88, 0.73, 0.92, 0.81, 0.82, 0.75, 0.82, and 0.82, respectively, from the sediments samples from almost 20-m-depth cores, covering approximately the last 10,000 years in Lake Ogawara, Japan. They concluded NIR absorbances of organic matter contributed to the calibration and interpreted this absorbance primarily describes the ligands in the highly organic samples. Kleinebecker et al. [58] reported the acceptable to the excellent prediction for total and NaCl-extractable concentrations of Al, Ca, Fe, K, Mg, N, Na, P, S, Si, and Zn as well as oxalate-extractable concentrations of Al, Fe, Mn, and P in sediment samples collected from core samples (0–10 cm) using piston sampler. They collected the samples from 191 locations distributed over the Netherlands. Air-dried samples were screened through a sieve with 2 mm mesh wire before NIR measurement in their study. The prediction of total C, CO_3^{2-}, N, P, and diatoms 47-cm-long freeze core from the deepest point in Lake Arendsee, Mecklenburg Plain in northern Germany by NIR spectroscopy was reported by Malley et al. [59]. Total organic carbon (TOC) prediction in 400 core samples from Lake Suigetsu, Japan, was reported by Pearson et al. [60]. Korsman et al. [61] investigated the spatial variance in the NIR spectral data from 165 surface sediments samples from a northern Swedish humic, mesotrophic lake, and revealed that water depth and organic matter account for 20 and 16%, respectively, of the variance in the NIR absorbance data.

They regarded this result, indicating that NIR analysis might become a valuable complementary tool to traditional sediment characterization.

16.5 Wastewater

Because the advantage of NIRS including the optical path length can be longer and no reagent is required for such measurement, some researches attempt to build accurate calibration models for the estimation of some pollution degrees in wastewater (i.e., sewage, wastewater from sugar refinery processing, and black liquor). Ding et al. [62] reported the accurate prediction model for the mixture samples tributyl phosphate and methyl iso-butyl ketone in aqueous solutions over the concentration range of 1 ± 160 ppm with SEP for MIBK of 3.82 ppm. They used the C–H combination bands in the range of 5000–4000 cm^{-1} measured in transmittance mode (2 mm path length) with a spectrometer equipped with the liquid nitrogen cooled InSb detector. Pan et al. [63] showed NIRS could be used for the prediction of chemical oxygen demand (COD) in sugar refinery wastewater with validation SEP of 25.0 mg L^{-1} using transmittance absorbance spectra (2 mm path length) in the range of 780–1100 nm. The COD is an indicative measure of the amount of oxygen that can be consumed by reactions in a measured solution. They optimized the wavelength for the prediction by MWPLS. Quintelas et al. [64] investigated the feasibility of NIR transmittance spectroscopy for the quantification of pollutants, like pharmaceuticals, in wastewater. Two hundred seventy-six samples obtained from an activated sludge wastewater treatment process were analyzed in the range of 200–14,000 cm^{-1} (transmittance mode, 0.7 mm pathlength). They obtained an adequate calibration curve for the prediction of ibuprofen, sulfamethoxazole, 17β-estradiol, and carbamazepine with coefficients of determination around 0.95. Lindgrerz et al. [65] reported it is possible to monitor the delignification process during a laboratory Kraft cook on softwood by NIR transmission measurement of the black liquor. They showed a good calibration result for Klason lignin content in the pulps and pulp yield. Some researchers conducted to monitor the pollution degree which was reported [66–68]. The COD values in wastewater samples collected in wastewater treatment plants were well predicted [66] with an RMSEP value of 19 mgO$_2$ L^{-1}. Many organic matter index [68] and volatile fatty acids, bicarbonate alkalinity, and total and volatile solids content [67] were also predicted from NIR spectra of sewage sludge.

16.6 Atmospheric Gas Detection

The development of effective gas detectors is essential because many gases can be harmful to humans or animals. NIR-tunable diode laser spectroscopy has the potential for the development of effective and inexpensive detectors for monitoring of gas species as Martin summarized [69]. For example, Scott et al. developed the airborne laser infrared absorption spectrometer II (ALIAS-II), including a lightweight mid-infrared absorption spectrometer based on cooled lead salt-tunable diode laser sources. The chemical species such as long-lived tracers N_2O and CH_4 and chemically active species HCl and NO_2 could be measured precisely [70]. Durry et al. used commercial distributed- feedback InGaAs laser diodes for the monitoring of CH_4 and H_2O [71].

16.7 Archeological Science

Archeological science is to develop techniques for the analysis of archeological materials. Because most archeological materials are rare, nondestructive ways are required. Yonenobu et al. [72] compared the NIR spectra of modern hinoki cypress (*Chamaecyparis obtusa*) and antique ones from the upright pillars of an old building with a construction date estimated to be around A.D. 750. They reported that the hemicellulose and holocellulose decrease, whereas lignin increased relatively with aging for -1300 years under atmospheric conditions without fungal hyphae or some kind of beetles attack by checking the difference second-derivative NIR spectra. Sandak et al. [73] also evaluate the archeological wood samples by NIRS. They examined five pedunculate oak (*Quercus robur* L.) pieces of the archeological wood collected from the waterlogged sites in Poland. The range of waterlogged period of these samples was 700–2700 years. They measured moisture content, density, cellulose, holocellulose, lignin, extractive contents, crystallinity, and degree of polymerization by the traditional method and constructed a good calibration curve for lignin and cellulose. They also showed oak samples representing several degradation levels are grouped and clearly separated from each other by PCA score from NIR spectra. Linderholm et al. [74] measured the NIR spectra of rock paintings and local lithology background in Scandinavian Stone Age rock paintings site Flatruet, Härjedalen, Sweden using field-based NIR spectrometer. They showed that, although there was a large spread in the spectra of both background and red paint objects, PLS-DA for NIR spectra can separate the background and paintings. Their group used a hyper spectral image to identify the animal bone materials in complex sieved soil sediments matrics from archeological evacuation in northern Scandinavia [75]. They took NIR hyperspectral image of elk bone and a sieved sediment fraction and identified the presence of bones, even including variate states of preservation. They further proposed a new methodology-based NIRS for studying stratigraphy and depth profiles in archeological excavations [76]. The soil samples were collected from a 0.8-m-deep stratigraphy of a Neolithic site that was analyzed by NIRS and hyperspectral measurement. It was shown the NIRS combined with multivariate analysis could be useful for finding soil horizon traits.

16.8 Conclusion

As shown above, NIR applied research has attracted considerable attention recently at the fields of wooden materials and environmental sciences due to its rapid measurement and nondestructive sampling and low-cost characteristics. Another significant advantage is that many properties could be evaluated simultaneously.

Meanwhile, basic research also has been proceeded actively to make sure prediction model robustness. It is very important to clarify the spectroscopic background and know the limitation of NIRS. Sometimes, a "bridge" research between theory and practice is also required.

References

1. M.D. Birkett, M.J.T. Gambino, Estimation of pulp kappa number with near-infrared spectroscopy. Tappi J. **72**(9), 193–197 (1989)
2. E. Sjostrom, *Wood Chemistry: Fundamentals and Applications* (Gulf Professional Publishing, 1993)
3. A.M.M. Alves, R.F.S. Simões, C.A. Santos, B.M. Potts, J. Rodrigues, M. Schwanninger, Determination of Eucalyptus globulus wood extractives content by near infrared-based partial least squares regression models: comparison between extraction procedures. J. Near Infrared Spectrosc. **20**(2), 275–285 (2012)
4. D.B. Easty, S.A. Berben, F.A. DeThomas, P.J. Brimmer, Near-infrared spectroscopy for the analysis of wood pulp: quantifying hardwood-softwood mixtures and estimating lignin content. Tappi J. **73**(10), 257–261 (1990)
5. J.A. Wright, M.D. Birkett, M.J.T. Gambino, Prediction of pulp yield and cellulose content from wood samples using near infrared reflectance spectroscopy. Tappi J. **73**(8), 164–166 (1990)
6. L. Wallbäcks, U. Edlund, B. Norden, I. Berglund, Multivariate characterization of pulp using solid-state 13C NMR, FTIR, and NIR. Tappi J. **74**(10), 201–206 (1991)
7. A.R. Da Silva, T.C.M. Pastore, J.W.B. Braga, F. Davrieux, E.Y.A. Okino, V.T.R. Coradin, J.A.A. Camargos, A.G.S. Do Prado, Assessment of total phenols and extractives of mahogany wood by near infrared spectroscopy (NIRS). Holzforschung **67**(1), 1–8 (2013)
8. W. He, H. Hu, Rapid prediction of different wood species extractives and lignin content using near infrared spectroscopy. J. Wood Chem. Technol. **33**(1), 52–64 (2013)
9. C. Lepoittevin, J.P. Rousseau, A. Guillemin, C. Gauvrit, F. Besson, F. Hubert, D. Da Silva Perez, L. Harvengt, C. Plomion, Genetic parameters of growth, straightness and wood chemistry traits in Pinus pinaster. Ann. For. Sci. **68**(4), 873–884 (2011)
10. B. Üner, İ. Karaman, H. Tanrıverdi, D. Özdemir, Determination of lignin and extractive content of Turkish Pine (Pinus brutia Ten.) trees using near infrared spectroscopy and multivariate calibration. Wood Sci. Technol. **45**(1), 121–134 (2011)
11. S. Tsuchikawa, H. Kobori, A review of recent application of near infrared spectroscopy to wood science and technology. J. Wood Sci. **61**(3), 213–220 (2015)
12. K. Watanabe, S.D. Mansfield, S. Avramidis, Application of near-infrared spectroscopy for moisture-based sorting of green hem-fir timber. J. Wood Sci. **57**(4), 288–294 (2011)
13. M. Defo, A.M. Taylor, B. Bond, Determination of moisture content and density of fresh-sawn red oak lumber by near infrared spectroscopy. For. Prod. J. **57**(5), 68–72 (2007)
14. V.T.H. Tham, T. Inagaki, S. Tsuchikawa, A novel combined application of capacitive method and near-infrared spectroscopy for predicting the density and moisture content of solid wood. Wood Sci. Technol. **52**(1), 115–129 (2018)

15. V.H. Segtnan, Š. Šašić, T. Isaksson, Y. Ozaki, Studies on the structure of water using two-dimensional near-infrared correlation spectroscopy and principal component analysis. Anal. Chem. **73**(13), 3153–3161 (2001)

16. T. Ma, T. Inagaki, S. Tsuchikawa, Rapidly visualizing the dynamic state of free, weakly, and strongly hydrogen-bonded water with lignocellulosic material during drying by near-infrared hyperspectral imaging. Cellulose. https://doi.org/10.1007/s10570-020-03117-6

17. A. Alves, A. Santos, P. Rozenberg, L.E. Pâques, J.P. Charpentier, M. Schwanninger, J. Rodrigues, A common near infrared-based partial least squares regression model for the prediction of wood density of Pinus pinaster and Larix × eurolepis. Wood Sci. Technol. **46**(1–3), 157–175 (2012)

18. A.J.A. Santos, A.M.M. Alves, R.M.S. Simões, H. Pereira, J. Rodrigues, M. Schwanninger, Estimation of wood basic density of Acacia melanoxylon (R. Br.) by near infrared spectroscopy. J. Near Infrared Spectrosc. **20**(2), 267–274 (2012)

19. T. Fujimoto, H. Kobori, S. Tsuchikawa, Prediction of wood density independently of moisture conditions using near infrared spectroscopy. J. Near Infrared Spectrosc. **20**(3), 353–359 (2012)

20. G. Hans, R. Kitamura, T. Inagaki, B. Leblon, S. Tsuchikawa, Assessment of variations in air-dry wood density using time-of-flight near-infrared spectroscopy. Wood Mater. Sci. Eng. **10**(1), 57–68 (2015)

21. R. Kitamura, S. Tsuchikawa, Construction of a novel densitometer that utilizes a near-infrared laser system with Douglas fir (Pseudotsuga menziesii). Wood Mater. Sci. Eng. **10**(1), 69–74 (2015)

22. P.R.G. Hein, Estimating shrinkage, microfibril angle and density of Eucalyptus wood using near infrared spectroscopy (2012)

23. T. Inagaki, M. Schwanninger, R. Kato, Y. Kurata, W. Thanapase, P. Puthson, S. Tsuchikawa, Eucalyptus camaldulensis density and fiber length estimated by near-infrared spectroscopy. Wood Sci. Technol. **46**(1–3), 143–155 (2012)

24. F. Isik, C.R. Mora, L.R. Schimleck, Genetic variation in Pinus taeda wood properties predicted using non-destructive techniques. Ann. For. Sci. **68**(2), 283–293 (2011)

25. L. Horvath, I. Peszlen, P. Peralta, S. Kelley, Use of transmittance near-infrared spectroscopy to predict the mechanical properties of 1- and 2-year-old transgenic aspen. Wood Sci. Technol. **45**(2), 303–314 (2011)

26. L.R. Schimleck, J.L.M. De Matos, J.T. Da Silva Oliveira, G.I.B. Muniz, Non-destructive estimation of pernambuco (Caesalpinia echinata) clear wood properties using near infrared spectroscopy. J. Near Infrared Spectrosc. **19**(5), 411–419 (2011)

27. K. Watanabe, K. Yamashita, S. Noshiro, Non-destructive evaluation of surface longitudinal growth strain on Sugi (Cryptomeria japonica) green logs using near-infrared spectroscopy. J. Wood Sci. **58**(3), 267–272 (2012)

28. H. Kobori, T. Inagaki, T. Fujimoto, T. Okura, S. Tsuchikawa, Fast online NIR technique to predict MOE and moisture content of sawn lumber. Holzforschung **69**(3), 329–335 (2015)

29. I.A. Sofianto, T. Inagaki, K. Kato, M. Itoh, S. Tsuchikawa, Modulus of elasticity prediction model on sugi (Cryptomeria japonica) lumber using online near-infrared (NIR) spectroscopic system. Int. Wood Prod. J. **8**(4), 193–200 (2017)

30. A.C. Maioli Campos, P.R.G. Hein, R.F. Mendes, L.M. Mendes, G. Chaix, Near infrared spectroscopy to evaluate composition of agro-based particleboards. BioResources. **4**(3), 1058–1069 (2009)

31. N.J. Kohan, B.K. Via, S.E. Taylor, Prediction of strand feedstock mechanical properties with near infrared spectroscopy. BioResources **7**(3), 2996–3007 (2012)

32. P.R.G. Hein, A.C.M. Campos, R.F. Mendes, L.M. Mendes, G. Chaix, Estimation of physical and mechanical properties of agro-based particleboards by near infrared spectroscopy. Eur. J. Wood Wood Prod. **69**(3), 431–442 (2011)

33. M. Schwanninger, B. Stefke, B. Hinterstoisser, Qualitative assessment of acetylated wood with infrared spectroscopic methods. J. Near Infrared Spectrosc. **19**(5), 349–357 (2011)

34. B. Green, P.D. Jones, D.D. Nicholas, L.R. Schimleck, R. Shmulsky, Non-destructive assessment of Pinus spp. Wafers subjected to Gloeophyllum trabeum in soil block decay tests by diffuse reflectance near infrared spectroscopy. Wood Sci. Technol. **45**(3), 583–595 (2011)

35. A. Sandak, I. Modzelewska, J. Sandak, Fourier transform near infrared analysis of waste paper with the addition of cereal bran biodegraded by Ascomycetes fungi. J. Near Infrared Spectrosc. **19**(5), 369–379 (2011)
36. T. Inagaki, Y. Asanuma, S. Tsuchikawa, Selective assessment of duplex heat-treated wood by near-infrared spectroscopy with principal component and kinetic analyses. J. Wood Sci. **64**(1), 6–15 (2018)
37. G.M. Downes, C.E. Harwood, J. Wiedemann, N. Ebdon, H. Bond, R. Meder, Radial variation in Kraft pulp yield and cellulose content in Eucalyptus globulus wood across three contrasting sites predicted by near infrared spectroscopy. Can. J. For. Res. **42**(8), 1577–1586 (2012)
38. J. GlGAC, M. Flserova, Identification of semichemical fluting properties by application of near infrared spectroscopy. Wood Res. **56**(2), 189–202 (2011)
39. R. Meder, J.T. Brawner, G.M. Downes, N. Ebdon, Towards the in-forest assessment of Kraft pulp yield: comparing the performance of laboratory and hand-held instruments and their value in screening breeding trials. J. Near Infrared Spectrosc. **19**(5), 421–429 (2011)
40. J.A. Tyson, L.R. Schimleck, A.M. Aguiar, J.I. Muro Abad, G.D.S.P. Rezende, O.M. Filho, Development of near infrared calibrations for physical and mechanical properties of eucalypt pulps of mill-line origin. J. Near Infrared Spectrosc. **20**(2), 287–294 (2012)
41. J.W.B. Braga, T.C.M. Pastore, V.T.R. Coradin, J.A.A. Camargos, A.R. da Silva, The use of near infrared spectroscopy to identify solid wood specimens of swietenia macrophylla0 (Cites Appendix II). Iawa J. **32**(2), 285–296 (2011)
42. P.A. Cooper, D. Jeremic, S. Radivojevic, Y.T. Ung, B. Leblon, Potential of near-infrared spectroscopy to characterize wood products 1. Can. J. For. Res. **41**(11), 2150–2157 (2011)
43. Z. Yang, B. Lü, A. Huang, Rapid identification of softwood and hardwood by near infrared spectroscopy of cross-sectional surfaces. Spectrosc. Spectr. Anal. **32**(7), 1785–1789 (2012)
44. H. Abe, K. Watanabe, A. Ishikawa, S. Noshiro, T. Fujii, M. Iwasa, H. Kaneko, H. Wada, Simple separation of Torreya nucifera and Chamaecyparis obtusa wood using portable visible and near-infrared spectrophotometry: differences in light-conducting properties. J. Wood Sci. **62**(2), 210–212 (2016)
45. T. Ma, T. Inagaki, M. Ban, S. Tsuchikawa, Rapid identification of wood species by near-infrared spatially resolved spectroscopy (NIR-SRS) based on hyperspectral imaging (HSI). Holzforschung **73**(4), 323–330 (2018)
46. A. Fernandes, J. Lousada, J. Morais, J. Xavier, J. Pereira, P. Melo-Pinto, Measurement of intra-ring wood density by means of imaging VIS/NIR spectroscopy (hyperspectral imaging). Holzforschung **67**(1), 59–65 (2013)
47. T.A. Lestander, P. Geladi, S.H. Larsson, M. Thyrel, Near infrared image analysis for online identification and separation of wood chips with elevated levels of extractives. J. Near Infrared Spectrosc. **20**(5), 591–599 (2012)
48. R. Meder, R.R. Meglenb, Near infrared spectroscopic and hyperspectral imaging of compression wood in Pinus radiata D. Don. J. Near Infrared Spectrosc. **20**(5), 583–589 (2012)
49. H. Kobori, N. Gorretta, G. Rabatel, V. Bellon-Maurel, G. Chaix, J.M. Roger, S. Tsuchikawa, Applicability of Vis-NIR hyperspectral imaging for monitoring wood moisture content (MC). Holzforschung **67**(3), 307–314 (2013)
50. T. Ma, T. Inagaki, S. Tsuchikawa, Calibration of SilviScan data of Cryptomeria japonica wood concerning density and microfibril angles with NIR hyperspectral imaging with high spatial resolution. Holzforschung **71**(4), 341–347 (2017)
51. I.A. Sofianto, T. Inagaki, T. Ma, S. Tsuchikawa, Effect of knots and holes on the modulus of elasticity prediction and mapping of sugi (Cryptomeria japonica) veneer using near-infrared hyperspectral imaging (NIR-HSI). Holzforschung **73**(3), 259–268 (2018)
52. R.A.V. Rossel, T. Behrens, Using data mining to model and interpret soil diffuse reflectance spectra. Geoderma **158**(1–2), 46–54 (2010)
53. M.F. Baumgardner, L.R.F. Silva, L.L. Biehl, E.R. Stoner, Reflectance properties of soils. Adv. Agron. **38**(C), 1–44 (1986)
54. T. Udelhoven, C. Emmerling, T. Jarmer, Quantitative analysis of soil chemical properties with diffuse reflectance spectrometry and partial least-square regression: a feasibility study. Plant Soil **251**(2), 319–329 (2003)

55. B. Stenberg, R.A. Viscarra Rossel, A.M. Mouazen, J. Wetterlind, *Visible and Near Infrared Spectroscopy in Soil Science*, 1st edn (Elsevier Inc., 2010)
56. D.F. Malley, P.C. Williams, Use of near-infrared reflectance spectroscopy in prediction of heavy metals in freshwater sediment by their association with organic matter. Environ. Sci. Technol. **31**(12), 3461–3467 (1997)
57. T. Inagaki, Y. Shinozuka, K. Yamada, H. Yonenobu, A. Hayashida, S. Tsuchikawa, A. Yoshida, Y. Hoshino, K. Gotanda, Y. Yasuda, Rapid prediction of past climate condition from lake sediments by near-infrared (NIR) spectroscopy. Appl. Spectrosc. **66**(6), 673–679 (2012)
58. T. Kleinebecker, M.D.M. Poelen, A.J.P. Smolders, L.P.M. Lamers, N. Hölzel, Fast and inexpensive detection of total and extractable element concentrations in aquatic sediments using near-infrared reflectance spectroscopy (NIRS). PLoS ONE **8**(7), e70517 (2013)
59. D.F. Malley, H. Rönicke, D.L. Findlay, B. Zippel, Feasibility of using near-infrared reflectance spectroscopy for the analysis of C, N, P, and diatoms in lake sediments. J. Paleolimnol. **21**(3), 295–306 (1999)
60. E.J. Pearson, S. Juggins, J. Tyler, Ultrahigh resolution total organic carbon analysis using Fourier Transform Near Infrared Reflectance Spectroscopy (FT-NIRS). Geochemistry. Geophys. Geosyst. **15**(1), 292–301 (2014)
61. T. Korsman, M.B. Nilsson, K. Landgren, I. Renberg, Spatial variability in surface sediment composition characterised by near-infrared (NIR) reflectance spectroscopy. J. Paleolimnol. **21**(1), 61–71 (1999)
62. Q. Ding, B.L. Boyd, G.W. Small, Determination of organic contaminants in aqueous samples by near-infrared spectroscopy. Appl. Spectrosc. **54**(7), 1047–1054 (2000)
63. T. Pan, Z. Chen, J. Chen, Z. Liu, Near-infrared spectroscopy with waveband selection stability for the determination of COD in sugar refinery wastewater. Anal. Methods **4**(4), 1046–1052 (2012)
64. C. Quintelas, D.P. Mesquita, E.C. Ferreira, A.L. Amaral, Quantification of pharmaceutical compounds in wastewater samples by near infrared spectroscopy (NIR). Talanta **194**(1), 507–513 (2019)
65. T. Lindgren, U. Edlund, Prediction of lignin content and pulp yield. Nord. Pulp Pap. Res. J. **13**(1), 76–80 (2007)
66. A.C. Sousa, M.M.L.M. Lucio, O.F.B. Neto, G.P.S. Marcone, A.F.C. Pereira, E.O. Dantas, W.D. Fragoso, M.C.U. Araujo, R.K.H. Galvão, A method for determination of COD in a domestic wastewater treatment plant by using near-infrared reflectance spectrometry of seston. Anal. Chim. Acta **588**(2), 231–236 (2007)
67. J.P. Reed, D. Devlin, S.R.R. Esteves, R. Dinsdale, A.J. Guwy, Performance parameter prediction for sewage sludge digesters using reflectance FT-NIR spectroscopy. Water Res. **45**(8), 2463–2472 (2011)
68. L. Galvez-Sola, J. Morales, A.M. Mayoral, C. Paredes, M.A. Bustamante, F.C. Marhuenda-Egea, J.X. Barber, R. Moral, Estimation of parameters in sewage sludge by near-infrared reflectance spectroscopy (NIRS) using several regression tools. Talanta **110**(15), 81–88 (2013)
69. P.A. Martin, Near-infrared diode laser spectroscopy in chemical process and environmental air monitoring. Chem. Soc. Rev. **31**(4), 201–210 (2002)
70. D.C. Scott, R.L. Herman, C.R. Webster, R.D. May, G.J. Flesch, E.J. Moyer, Airborne Laser Infrared Absorption Spectrometer (ALIAS-II) for in situ atmospheric measurements of N_2O, CH_4, CO, HCL, and NO_2 from balloon or remotely piloted aircraft platforms. Appl. Opt. **38**(21), 4609–4622 (1999)
71. G. Durry, G. Megie, Atmospheric CH_4 and H_2O monitoring with near-infrared InGaAs laser diodes by the SDLA, a balloonborne spectrometer for tropospheric and stratospheric in situ measurements. Appl. Opt. **38**(36), 7342–7354 (1999)
72. H. Yonenobu, S. Tsuchikawa, Near-Infrared Spectroscopic. **57**(11), 1451–1453 (2003)
73. A. Sandak, J. Sandak, M. Zborowska, W. Pra, Near infrared spectroscopy as a tool for archaeological wood characterization. J. Archaeol. Sci. **37**(9), 2093–2101 (2010)
74. J. Linderholm, P. Geladi, C. Sciuto, Field-based near infrared spectroscopy for analysis of Scandinavian Stone Age rock paintings. J. Near Infrared Spectrosc. **23**(4), 227–236 (2015)

75. J. Linderholm, J.A. Fernández Pierna, D. Vincke, P. Dardenne, V. Baeten, Identification of frag-mented bones and their state of preservation using near infrared hyperspectral image analysis. J. Near Infrared Spectrosc. **21**(6), 459–466 (2013)
76. J. Linderholm, P. Geladi, N. Gorretta, R. Bendoula, A. Gobrecht, Near infrared and hyper-spectral studies of archaeological stratigraphy and statistical considerations. Geoarchaeology **34**(3), 311–321 (2019)

Chapter 17
Information and Communication Technology in Agriculture

Eizo Taira

Abstract Near-infrared spectroscopy (NIRS) enables rapid and nondestructive analyses of the components of an object. NIRS has many applications in agriculture. In particular, it has been combined with information and communication technology (ICT) to facilitate the management of equipment and data, thereby significantly expanding its application range. This chapter discusses the application of ICT, especially in sugarcane production.

Keywords Information and communication technology (ICT) · Network · Geographic information system (GIS) · Unmanned aerial vehicles (UAV)

17.1 NIR Network System

Near-infrared spectroscopy (NIRS) can provide desired data through calibration or discrimination models. NIR instruments can be used to rapidly measure data related to product management, grading, and pricing. An Internet-connected NIR network system enables the efficient and effective collection, analysis, and use of these data [1]. Further, an NIR network system comprising several NIR devices can not only collect data but also perform device diagnostics, calibration transfer, and management of several calibration models simultaneously. Such systems are, therefore, very important tools for system managers because they allow end users to perform all measurements without setting up a calibration model for the given NIR device and measuring conditions. NIR network software systems are provided either by NIR manufacturers or by other developers.

NIR network systems afford significant benefits to users. For example, Taira et al. used an NIR network system to set up sugarcane quality payment [2]. In Japan, sugarcane is cultivated on ten islands in the southern region. Further, 16 sugarcane factories are located across 14 islands. Both users and distributing companies face some transportation and maintenance issues on these islands when using NIR instruments.

E. Taira (✉)
University of the Ryukyus, Nishihara, Japan
e-mail: e-taira@agr.u-ryukyu.ac.jp

© Springer Nature Singapore Pte Ltd. 2021
Y. Ozaki et al. (eds.), *Near-Infrared Spectroscopy*,
https://doi.org/10.1007/978-981-15-8648-4_17

Further, NIR instrument operations such as calibration transfer, database management, and measurement settings require considerable time and effort. To solve these problems, 16 NIR instruments connected through a networked system have been introduced in each sugarcane factory for operations such as updating calibration models and performing database maintenance.

As these NIR instruments were distributed across distances of 1000 km over these islands, an effective monitoring and control system was necessary. A computer network was the best solution for this problem. The network's control center was located at the sugar association on Okinawa Main Island, where monitoring and control operations were performed. Each local system was connected to the control center through the Internet. The network system worked very well and was easy to operate. The network system operator could conduct daily checks and address any minor problems experienced by the slave instruments. Further, sugar content measurement results collected from all the slave systems could be monitored daily.

Such network systems are advantageous when a calibration manager needs to update the calibration model on all instruments. Once a sample has been measured using all instruments, the user can correct biases arising due to slight instrument differences by using a "repeatability file". Taira et al. showed the calibration result for Pol in cane (sugar index for payment) with and without a repeatability file [2].

These calibration models showed lower pooled standard error (P-SE) and pooled bias than the no repeatability file models, with the first-derivative standard normal variate (1DSNV) pre-treatment showing the lowest root mean square errors of prediction (RMSEPs). The network system can be used to apply calibration models to all instruments in all regions. Furthermore, this system can be used to estimate nutrient compositions for supporting fertilization operations (Fig. 17.1).

17.2   Assisting Smart Agriculture in Sugarcane Production

In Japan, growers' sugarcane prices are based on the sugarcane quality. Therefore, growers try to maintain the unit yield and preserve and/or increase the sugar content. Quality data, such as sugar content and yield, for all sugarcane fields are automatically collected by the quality payment system. Thus, this system functions as a big data collection system.

Watanabe et al. investigated the nutrients present in sugarcane juice to identify the key factors affecting sugarcane quality [3]. Juice analysis over a 3-year period showed that potassium (K^+) and chloride (Cl^-) were the most abundant cation and anion in the juice, respectively, and that both negatively correlated with the sucrose concentration. Further, K^+ and Cl^- concentrations varied significantly depending on the production area. Traditionally, most growers could not obtain information about the soil and plant chemistry on their farm, and for decades, they simply applied fertilizers in an unscientific way. Even for sugarcane production, fertilizers, such as the potassium fertilizer KCl, were applied without confirming the soil condition and chemical compositions. When typhoons strike these islands, many farms suffer soil

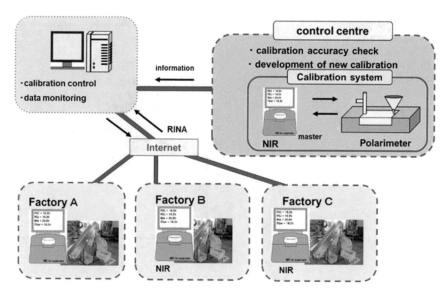

Fig. 17.1 NIR network system for sugarcane payment in Japan

depletion because of physical and salt stress. Moreover, these islands do not have mountains, and therefore, water resources for irrigation are limited; water resources, including life water, are only available from underground sources. Fertilizers and irrigation are crucial to the sustainability of agriculture; therefore, plant diagnosis is a key technique for improving the sugarcane quality and ensuring sustainability.

Several studies have applied NIRS to plant nutritional diagnosis in the field of agriculture [4–6]. Taira et al. reported calibration performances for basic nutrients, including nitrogen, phosphorus, and potassium, using a NIR instrument for payment [7].

Simple diagnoses as mentioned above as well as those of the sugar content revealed a negative correlation between the sugar and potassium contents, as shown in the scatter plot in Fig. 17.2. These results were obtained using NIRS and were similar to those obtained through laboratory methods. Such results can be obtained in the payment system simultaneously without requiring additional cost and effort. NIRS provides similar results to those of other conventional chemical analysis methods. These results could assist, for example, fertilizer application on farmlands. NIRS also has a big potential to find use in cane farming and farmers' decision makings.

17.3 Combining NIR System with a GIS

A geographic information system (GIS) is an effective tool for analyzing the spatial characteristics of quality data. Through a GIS, low-quality sugarcane and low-yield

Fig. 17.2 Relationship
between sugar content and
potassium content in cane
juice measured by NIRS in
Minami Daito Island, Japan

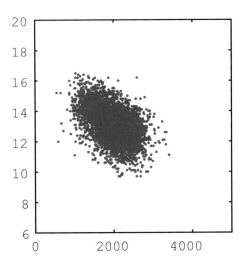

fields can be visualized easily. Such visualization is important for accurately fertil-
izing sugarcane fields. Figure 17.3 shows a map of sugarcane fields in Minami Daito
Island, Japan. This island is located 400 km east of Okinawa, Japan, and is heavily
dependent on its sugar industry. All polygons represent agricultural farmlands, and
sugarcane is cultivated on 87% of these farmlands. This map shows farmlands where

Fig. 17.3 Farming
evaluation a using
combination of NIR and GIS
in Minami Daito Island,
Japan

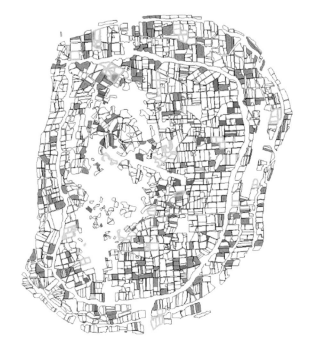

the sugar content decreased from 2012 to 2013, with red and blue indicating sugar content of less than and more than 14%, respectively. However, the potassium content in cane juice increased in all farmlands. This indicates that potassium and the sugar content are negatively correlated, that is, reducing potassium in the soil or adding an organic fertilizer can increase the sugar content [8]. This evaluation system has been used not only for cane payment but also as a low-cost diagnostic system to improve field management. The NIR database contains information on all cane farmers in Japan and could be extended to other farm management applications.

17.4   On-Site Analysis for Agriculture

In sugarcane production, estimating the cane quality and quantity prior to harvest is key for optimizing harvest scheduling and supply chain management. Doing so would contribute to increasing profitability for growers and sugar production factories. A previous study reported a calibration model for analyzing sugarcane stalks and evaluating components such as sugar, moisture, and fiber contents [9–11]. It used a portable NIR instrument for the direct scanning of cane stalks (Fig. 17.4). Further, it investigated the optimum integration times and preprocessing techniques. The best model for Pol and fiber had coefficients of determination of the prediction set (r^2) of 0.84 and 0.81 and RMSEPs of 1.2% and 0.63%, respectively. These models could be used for screening, and, therefore, they help breeders perform rapid measurements and potentially monitor sugarcane quality in the field to aid breeding programs. Moreover, the measured results could enable the estimation of harvest times and the early detection of diseases.

Fig. 17.4   Direct scanning of cane stalk using portable NIR instrument (P-TF1, HNK Engineering Co., Ltd., Iwade, Japan)

Fig. 17.5 UAV equipped with multispectral camera (left) and flight planner (right) (HG Robotic Co., Ltd., Bangkok, Thailand)

NIRS is also being used in remote sensing systems on unmanned aerial vehicles (UAV). Studies have shown that the properties of agricultural fields can be easily estimated from infrared images [12, 13]. UAV could be used to acquire the canopy reflectance to explore correlations between desired crop parameters. For example, a UAV equipped with multispectral cameras and operated with a customizable flight planner (Fig. 17.5) and an automatic controller as well as analysis software for mapping, interpreting, visualizing, and reporting the quantity and quality of sugarcane in fields was used as a farm monitoring and mapping platform for sugarcane in Thailand.

Flight conditions that affected the quality of the reflectance map, such as the height that determines the ground sampling distance, traveling speed, front overlapping, and side overlapping, were tested and verified. The acquired images were generated using five bands of reflectance maps—blue, green, red, NIR, and red edge. Then, these reflectance maps were used to calculate the potential vegetation indices and paired with averaged references values to create simple linear regression models.

Chea et al. reported calibration performance of cane properties using the UAV system [14]. The best R^2 values for different vegetation indices for Brix, Pol, CCS, and fiber were 0.84, 0.77, 0.68, and 0.50, respectively. They also showed a seasonal trend and a distribution of different varieties in the fields. Because the physiological characteristics of different varieties affect the vegetation indices chosen for use in the prediction models, feasibility tests are required to confirm whether these prediction equations can be applied to varieties which possess other noticeable physiological characteristics. This platform can be useful for optimizing harvest schedules and supply chain management and for planning cultivation in the next season based on the cane quality distribution in the field (Fig. 17.6).

Fig. 17.6 Distribution of Brix of sugarcane in the field predicted by multispectral reflectance map from UAV imagery reported by farm mapping and monitoring service (Khon Kaen University, Khon Kaen, Thailand)

17.5   Advanced Unique Applications

Soil sensors are already available for analyzing soil properties in real time. Shibusawa suggested that soil components can be estimated using an NIR sensor attached to a tractor [15, 16]. Currently, an automated tractor controlling system based on the global navigation satellite system (GNSS) is under development. This system stores results measured by real-time soil sensors in the field in a database. These data can be used for fertilization management and understanding farmland characteristics.

Recent years have seen rapid developments in microcomputers and device miniaturization. Microcomputers, like the Raspberry Pi, enable user-specific customization for realizing various applications [17, 18].

Overall, the findings suggest that the combination of NIR and ICT can serve as a powerful tool in the field of agriculture for optimizing limited resources and supporting farming operations.

References

1. P. Tillmann, T.-C. Reinhardt, C. Paul, Networking of near infrared spectroscopy instruments for rapeseed analysis: a comparison of different procedures. J. Near Infrared Spectrosc. **8**, 101–107 (2000)
2. E. Taira, M. Ueno, N. Furukawa, A. Tasaki, Y. Komaki, J.-I. Nagai, K. Saengprachatanarug, Networking system employing near infrared spectroscopy for sugarcane payment in Japan. J. Near Infrared Spectrosc. **21**, 477 (2013)
3. K. Watanabe, M. Nakabaru, E. Taira, M. Ueno, Y. Kawamitsu, Relationships between nutrients and sucrose concentrations in sugarcane juice and use of juice analysis for nutrient diagnosis in Japan. Plant Prod. Sci. **19**, 215–222 (2016)
4. M. Kumi, Feedback of the fruit quality data measured by N1R spectropbotometer to production control using GIS and taking new turn of agricultural research. Horticult. Res. (Japan) **3**, 245–250 (2004)
5. S. Morimoto, W.F. McClure, B. Crowell, D.L. Stanfield, Near infrared technology for precision environmental measurements: part 2. Determination of carbon in green grass tissue. J. Near Infrared Spectrosc. **11**, 257–267 (2003)
6. W.F. McClure, B. Crowell, D.L. Stanfield, S. Mohapatra, S. Morimoto, G. Batten, Near infrared technology for precision environmental measurements: part 1. Determination of nitrogen in green- and dry-grass tissue. J. Near Infrared Spectrosc. **10**, 177–185 (2002)
7. E. Taira, M. Ueno, Y. Kawamitsu, R. Matsukawa, High efficient diagnosis of sugarcane farm land using NIR Spectroscopy networking system, in *Proceedings of International Society of Sugar Cane Technologists* (2007), pp. 143–149
8. M. Ueno, Y. Kawamitsu, L. Sun, E. Taira, K. Maeda, Combined applications of NIR, RS and GIS for sustainable sugarcane production. Sugar Cane Inter. **23**, 8–11 (2005)
9. A. Phuphaphud, K. Saengprachatanarug, J. Posom, K. Maraphum, E. Taira, Prediction of the fibre content of sugarcane stalk by direct scanning using visible-shortwave near infrared spectroscopy. Vib. Spectrosc. **101**, 71–80 (2019)
10. K. Maraphum, S. Chuan-Udom, K. Saengprachatanarug, S. Wongpichet, J. Posom, A. Phupha-phud, E. Taira, Effect of waxy material and measurement position of a sugarcane stalk on the rapid determination of Pol value using a portable near infrared instrument. J. Near Infrared Spectrosc. **26**, 287–296 (2018)
11. E. Taira, M. Ueno, K. Saengprachatanarug, Y. Kawamitsu, Direct sugar content analysis for whole stalk sugarcane using a portable near infrared instrument. J. Near Infrared Spectrosc. **21**, 281–287 (2013)
12. R. Colombo, M. Meroni, A. Marchesi, L. Busetto, M. Rossini, C. Giardino, C. Panigada, Estimation of leaf and canopy water content in poplar plantations by means of hyperspectral indices and inverse modeling. Remote Sens. Environ. **112**, 1820–1834 (2008)
13. M.F.d. Souza, L.R.d. Amaral, S.R.d.M. Oliveira, M.A.N. Coutinho, C. Ferreira Netto, Spectral differentiation of sugarcane from weeds. Biosyst. Eng. **190**, 41–46 (2020)
14. C. Chea, K. Saengprachatanarug, J. Posom, M. Wongphati, E. Taira, Sugar *Yield Parameters and Fiber Prediction in Sugarcane Fields Using a Multispectral Camera Mounted on a Small Unmanned Aerial System (UAS)* (Sugar Tech, 2020), pp. 1–17
15. S. Shibusawa, Soil sensors for precision farming, *Handbook of Precision Agriculture* (CRC Press, 2006), pp. 87–120
16. M. Kodaira, S. Shibusawa, Using a mobile real-time soil visible-near infrared sensor for high resolution soil property mapping. Geoderma **199**, 64–79 (2013)
17. K. Bougot-Robin, J. Paget, S.C. Atkins, J.B. Edel, Optimization and design of an absorbance spectrometer controlled using a raspberry Pi to improve analytical skills. J. Chem. Educ. **93**, 1232–1240 (2016)

18. L.M. Kumar, B. Pavan, P. Kalyan, N.S. Paul, R. Prakruth, T. Chinnu, Design of an embedded based control system for efficient sorting of waste plastics using Near Infrared Spectroscopy, in *2014 IEEE International Conference on Electronics, Computing and Communication Technologies (CONECCT)* (IEEE, 2014), pp. 1–6

Chapter 18
Near-Infrared Spectroscopy in the Pharmaceutical Industry

Benoît Igne⃝ and Emil W. Ciurczak

Abstract Since the first applications of near-infrared spectroscopy in the pharmaceutical industry to today's in-line and in real-time monitoring and control of manufacturing processes, the technology has come a long way in sensitivity, robustness, and deployability. The pharmaceutical industry is now able to rely on the technology to release medicine to patients without having to sample for off-line analyses. The sensitivity of NIR light to the matrix, which makes it a tool of choice for raw material identification and counterfeit detection, can be an issue for quantitative and qualitative methods as much of the variance has to be captured by the model prior to validation to avoid repeated method updates. Nevertheless, its flexibility of implementation, hardware ruggedness, and wide range of applicability makes it a tool of choice for process understanding, monitoring, and control. In this chapter, an overview of the current usage is provided for the development, understanding, and control of pharmaceutical processes for small molecule drug substances, drug products, and biopharmaceutical materials. A discussion of the regulatory environment and available guidance documents is provided. Finally, the intended method use and the associated method validation requirements are discussed in the context of building fit for purpose methods.

Keywords Near-infrared spectroscopy · Pharmaceutical · Drug substance · Drug product · Biopharmaceutical

B. Igne (✉)
Vertex Pharmaceuticals Inc, 50 Northern Avenue, Boston, MA 02210, USA
e-mail: Benoit_Igne@vrtx.com

E. W. Ciurczak
Doramaxx Consulting, 77 Park Road, Goldens Bridge, NY 10526, USA
e-mail: emil@ciurczak.com

© Springer Nature Singapore Pte Ltd. 2021
Y. Ozaki et al. (eds.), *Near-Infrared Spectroscopy*,
https://doi.org/10.1007/978-981-15-8648-4_18

18.1 Introduction

The pharmaceutical industry aims to discover, develop, and manufacture chemical and biological compounds with a therapeutic activity that benefits patients. These active pharmaceutical ingredients (APIs) can be synthetized by synthetic chemical reactions (small molecules) or natural processes through the use of microorganisms (large molecules). After an entity has been found to have a desired therapeutic effect and is approved for sale by health authorities, it needs to be manufactured in sufficient quantity to ensure patient access.

Since the first applications of near-infrared spectroscopy (NIRS) in the pharmaceutical industry [1] to today's in-line and in real-time monitoring and control of manufacturing processes, the technology has come a long way in sensitivity, robustness, and deployability. While that evolution was inevitable, it still took several decades to demonstrate the value the technology can bring to ensure product quality. In this chapter, an overview of the current uses of near-infrared spectroscopy will be provided for the development, understanding, and control of pharmaceutical processes for small molecule drug substances, drug products, and biopharmaceutical materials.

While an active field of research and development in the 1980s and 1990s, it is the US FDA that put NIRS and other rapid, non-destructive analytical tools on the map of senior stakeholders with the Process Analytical Technology (PAT) guidance of 2004 [2]. It defined Process Analytical Technologies as "a system for designing, analyzing, and controlling manufacturing through timely measurements (i.e. during processing) of critical quality and performance attributes of raw and in-process materials and processes, with the goal of ensuring final product quality" [2]. While NIRS is certainly not the only process analytical tool, it is one of the most robust, the safest, and most widely used tool by pharmaceutical scientists. The reasons are multiple (i.e., good sample penetration depth, no sample preparation, rapid acquisition, and good sensitivity), but the combination of spectroscopy and chemometrics led to the development of NIRS as a technology of choice and to the training of a generation of process analytical scientists, skilled in spectroscopy, chemometrics, sampling, and engineering. It is nevertheless important to note that much work has been done with IR, Raman, fluorescence, and UV–Vis spectroscopy. Readers can get an idea of the current state of the process analytics field in several reviews [3, 4].

In this chapter, an overview of the current usages of NIR Spectroscopy in the pharmaceutical industry for small and large molecules will be presented. For a historical perspective, readers should refer to other books and book chapters [5–7].

18.2   ICH Guidance, Validation Principles, and Lifecycle Management

The industry is highly regulated to ensure that patients are provided with safe and effective active pharmaceutical ingredients (APIs). Health authorities publish guidelines that describe the clinical and manufacturing principles that companies should consider to potentially gain market approval in their respective jurisdictions. In an effort to harmonize expectations and requirements across the industry, the International Council for Harmonization of Technical Requirements for Pharmaceuticals for Human Use (ICH) has published since 1990 several guidelines on manufacturing, quality control and assurance and some specific countries have also published guidelines on the use of PAT [2], describing a framework designed to encourage companies to use process analytics tools to support development and innovation in the manufacturing of medicines.

The development of PAT applications typically requires a multidisciplinary approach, combining the use of simple process sensors and analyzers, chemometric tools (multivariate tools for experimental design and analysis), process supervision and control, and continuous improvement tools.

While the majority of spectroscopic PAT applications are for process understanding, an increasing number of submissions are specifically for assuring quality or releasing product. Because spectroscopic PAT methods often rely on a multivariate model, the impact of these models to the patient dictates the required level of model verification and validation. Three levels were discussed by ICH [8]:

- Low-Impact Models: typically used to support product and/or process development
- Medium-Impact Models: assuring quality of the product but are not the sole indicators of product quality
- High-Impact Models: prediction from the model is a significant indicator of quality of the product.

ICH-Q2(R1) [9] is usually the framework followed by pharmaceutical scientists to validate analytical methods. It references the requirements for the various figures of merit an analytical method must include (i.e., accuracy, precision, linearity, range, robustness, …). A number of articles describing the validation of near-infrared analytical methods have been published [10, 11]. In addition, frameworks of analytical development practices have been published such as the Analytical Quality by Design [12]. Several standards discuss these in detail [13, 14] and regulatory agencies have also published their expectations [15, 16].

Finally, it is necessary to state that the current challenge in the use of PAT methods, including NIRS, is their lifecycle management. Method validation is now well understood but as these tools get used in commercial settings and need updates as the instruments, raw materials, and processes change, more experience needs to be developed in how models can be made more robust, reducing the need for updates, or stream-lining the updates, or applying re-calibration/re-validation approaches to ensure these methods are ready with minimal to no downtime [17].

18.3 Large Molecules

The manufacturing of biologics (i.e., monoclonal antibodies, recombinant proteins and DNA, vaccines, etc.) relies on complex cellular systems with high sensitivity to their environment and feeding regimen. The production of large molecules by microbes and mammalian cells requires the control of numerous processing param-eters such as nutrient concentration, temperature, pH, gases, agitation, etc. The host cells, the product, the by-products (lactate, ammonium, CO_2, etc.), and the growth medium constitute a complex mixture with many of the chemical species present in a bioreactor at levels undetectable by many analytical tools, including NIR spectroscopy.

The production of large molecules typically follows a two-step process: first the microorganisms produce the molecules of interest, then the molecule is purified from the growth medium, cells, viruses, and other impurities. However, to date, much of the published work involving NIRS has been to produce large molecule in bioreactors.

The manufacturing process heavily relies on the in-line and in real-time measure-ments and control of processing parameters such as pH, dissolved oxygen, dissolved CO_2, and other elements impacting cell health. Depending on the desired feeding strategy, nutrients (i.e., glucose) may need to be measured and controlled throughout the duration of the batch and by-products (i.e., lactate, ammonia) may also need to be monitored. However, their measurements have been and remain a challenge. As a consequence, manual sampling and off-line measurements with fundamental primary analytical methods are still predominant. Nevertheless, the use of in-line spectroscopy as a process analytical tool to monitor and control these bioreactors has seen a significant increase over the last decade. While Raman spectroscopy may appear to be better suited to a water rich process, near-infrared spectroscopy has been widely utilized [18].

18.3.1 Bioreactor Monitoring and Control

Initially employed for the analysis of pulled samples at-line or off-line, it is now commonly used on-line through a recirculation loop or in-line with a probe directly introduced in the bioreactor.

The first report of the use of off-line spectroscopy was performed on a fermentation process for *A. awamori* and *P. oxalicum* [19] at 1L scale.

Further work was performed for the analysis of the yeast fermentation processes for the production of mammalian proteins. A significant body of literature exists on *Pichia pastoris*. The monitoring of product, methanol, glycerol, and biomass using both transmittance and reflectance (when sample was too optically dense) NIR spectroscopic modes was demonstrated [20]. On-line monitoring through a recirculation loop showed the ability to monitor methanol and developing feed-back control to maintain its level at various set points [21]. In-line monitoring of a fermentation process with *Streptomyces coelicolor* was published for the prediction of glucose and ammonium using a fiber-based system. Authors compared the results with off-line samples and showed that signal attenuation above 2000 nm resulted in lower quality models for ammonium [22]. Further in-line examples of fermentation monitoring demonstrated prediction errors relevant for considering the reduction or elimination of manual sampling while maintaining a higher level of process monitoring, control, and fault detection [23].

Fermentations are progressively being replaced by mammalian cell-based bioreactors such as the Chinese Hamster Ovary (CHO) cells for the production of complex monoclonal antibodies. For mammalian cell reactors, the monitoring of glucose, lactate, and ammonia has been investigated [24, 25].

An example of deployment of several transflectance probes using a multiplexed FT-NIR spectrometer was published. Author monitored 12,500 L bioreactors using partial least-squares (PLS) regression models for seven parameters: glucose concentration, osmolality, packed cell volume, product titer, viable cell density, integrated viable cell count, and integrated viable packed cell volume [26]. Figure 18.1 presents an example of a bolus fed-batch glucose profile as a function of time.

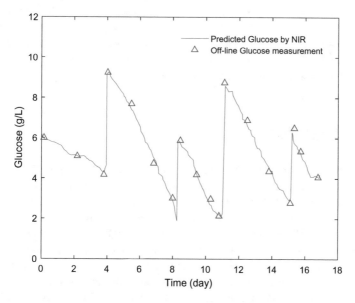

Fig. 18.1 Predicted glucose content as a function of growth time. The red triangles correspond to off-line measurements

Partial least-squares models have routinely been employed to develop the models discussed above, but other quantitative approaches and variable selection methods have been utilized to improve method performance. An example of fermentation monitoring that employed variable selection was demonstrated with improved performance over full-scale/ selected-region models through the use of interval PLS [27]. Interval PLS approaches divide the full spectra in sub-regions and find which combination of variables give the best results. Models with the full spectra or selected regions were also developed, but with higher errors than the interval PLS [27].

All the examples provided above show how NIRS has be used during a bioreactor run. But a large amount of information can be obtained from retrospective analyzes to compare batch-to-batch variability. A principal component analysis (PCA) applied on NIR spectra collected in real-time during five batches allowed the characterization of the batches based on cell densities (scores on the first principal component) and batch-to-batch variations (scores on the second principal component) [28].

But despite the reported work, a review of on-line monitoring and control tools for mammalian cell cultures challenged the applicability of NIR for bioprocesses [29]: "[...], the molar absorptivity in the NIR range is typically quite small; thus, the method is not ideal for diluted or minor components. This can be a limiting factor for the application to mammalian cell cultures, since a key process control objective during fed-batch operation is to keep glucose and glutamine at very low concentrations to prevent the accumulation of the toxic by-products ammonia and lactate. Furthermore, the overlapping signals seen in the NIR range results in very broad peaks, leading to complex spectra and hindering the assignment of specific features to individual compounds. These characteristics require a chemometric data-mining step to relate spectral information with the target compounds". For these reasons, NIRS remains underutilized when compared with Raman for the monitoring of pharmaceutical bioprocesses. But, innovation in sensitivity and chemometrics will help support the technology in the long term.

18.3.2 Lyophilization

Another area of application of NIRS in bioprocesses is lyophilization. The removal of water from the final drug product (after purification of the growth medium containing the molecule of interest) is necessary to ensure that the monoclonal antibodies remain stable and can be stored and shipped without affecting their therapeutic effects. Lyophilization, or freeze drying, can be used to achieve that. Since water has a strong molecular absorptivity in the NIR region, it is a tool of choice for the monitoring and control of these processes. Publications have shown the suitability of the technique for the analysis of water content through the container during water removal [30, 31]. The spatial distribution of moisture within vials was also investigated with NIR chemical imaging [32].

To ensure that the proteins will maintain their therapeutic effect after water removal, it is necessary to understand whether the water removal process affects the protein structures by creating aggregates or irreversible modifications. Several studies have looked at the protein conformation during lyophilization and the interactions with the lyoprotectant compounds [33].

18.3.3 Summary

While biological processes use water as a medium and water has a strong NIR absorptivity, a large body of work has been established, demonstrating the suitability of the technique. Simplification of the modeling and sensitivity enhancement could help NIRS support the development of robust processes and, when relevant, be employed to monitor or control processes during manufacturing operations.

18.4   Small Molecules

Engineered to generate a particular therapeutic response, small molecules have been the workhorse of the pharmaceutical industry. While more and more biological products are being brought to market, small molecules will remain a significant part of the portfolio of pharmaceutical companies for the foreseeable future.

There are usually two main manufacturing steps in bringing a synthetic active pharmaceutical ingredient to patients: the first consists in producing the small molecule in large quantity with the desired properties; the second will take the API and transform it in a form that can be supplied to the patients (tablets, capsules, transdermals, creams) or used by medical professionals (injectables). The remainder of the chapter will discuss how NIRS has been used in the support of these activities.

18.4.1 Drug Substance Manufacturing

The synthesis routes leading to the discovery of APIs are usually not optimized for purity and manufacturing efficiency. Process chemists and engineers will often develop a simplified and robust process while ensuring that the molecule retains its safety, efficacy, and physico-chemical characteristics. The synthetic route must ensure that all the raw material properties and unit operations that lead to the formation of the final drug substance are understood so that properties such as API particle size and shape, form, impurity level, and yield are controlled during production.

A synthetic synthesis will involve combining raw materials, reactions with the formation of intermediates, isolation of intermediates, and final product isolation. For each step, the impact of temperature, pressure, steering rate (amongst other parameters) on the solubility, and reaction kinetics will need to be carefully studied to design a process that maximizes the yield while minimizing impurities and ensuring the targeted bioavailability for the API.

Process analytical technologies have been used in the development of these routes to ensure that the chemists and engineers can identify and optimize critical parameters to develop robust processes [3, 34, 35]. In this section, examples of the use of NIR spectroscopy for reaction and purification monitoring and control are presented.

18.4.1.1 Reaction Monitoring and Control

Gaining understanding of the reaction under consideration is the primary value proposition that NIRS provides to the development of drug substance manufacturing. Chemists and engineers want to develop the most robust processes and will often sample the reaction for analysis by HPLC or NMR, which can take several hours and potentially be affected by sampling and quenching. Having the ability to track the reactions in-line and in real-time can provide additional insights into the state of the reaction in the vessel, its dynamics, and end-point. This is particularly relevant for the monitoring of intermediates and final product in the early phases of development or when sampling is difficult (such as for hydrogenation reactions performed at high temperature and pressure). When relevant, NIRS may be used for reaction control, making decisions on the process to ensure the quality of the product for the patient.

An example of reaction monitoring was published by Blanco et al. [36] Authors followed the esterification of myristic acid by isopropanol using multivariate curve resolution. Figure 18.2 shows the relative trends of the component parameters. In simple reactions like this, PCA may also be used but when intermediates are formed, the variance described by the principal components may be distorted by the appearance and disappearance of species. This could affect the score trends and impair the ability of PCA to adequately track the reaction. Multivariate curve resolution, with its constraints for specificity can allow a better understanding of the reaction [37].

An example of the deployment of NIRS to reduce the safety risks to operators was published by Wiss et al. [38]. The authors used NIRS with a transmittance immersion probe to monitor and control a highly exothermic reaction during the formation of a Grignard reagent. After building calibration models, the system was used to quantitatively track the reagents as a function of reaction time. Figure 18.3 shows the evolution of the reaction components. The authors subsequently used NIRS to automatically control the feed rate of a reagent and limit the safety risks associated with the highly reactive process.

A very similar example was described for the monitoring and control of a distillation process [4]. The authors built quantitative models for the API concentration and the solvent composition and fed-back that information to the distillation system controlling the temperature and reflux ratio of the column when product of a variable extraction was continuously fed for distillation. The results showed far better control of the process when the distillation was optimized to account for the variability of the incoming material. Figure 18.4 shows the process variability of API %w/w, with

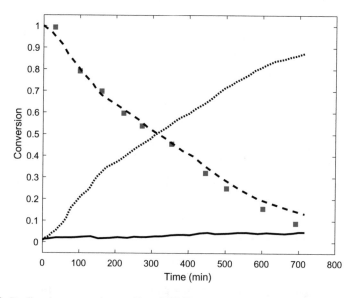

Fig. 18.2 Predicted concentration profiles. (■ ■ ■) myristic acid; (- -) isopropyl myristate; (—) isopropanol; (■): acid concentrations obtained by titration. (Reproduced with permission from [36])

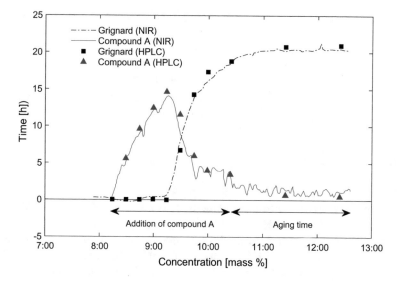

Fig. 18.3 Compound formation and consumption during a highly exothermic reaction (Reproduced with permission from [38])

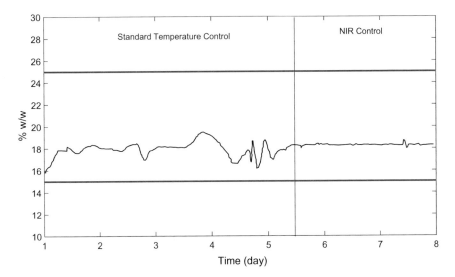

Fig. 18.4 Active ingredient production with and without temperature control by NIRS (Reproduced with permission from [4])

and without temperature control. Other examples of solvent composition monitoring and control were described [3].

The enhanced understanding of a drying reaction was demonstrated by tracking the OH- absorption band as a function of drying and rehydration time [3]. Form monitoring during drying was also demonstrated using a reflectance probe. The monitoring allowed in real-time determination of the conversion efficiency, allowing processing of the batch once the conversion to the desired form was achieved.

18.4.1.2 Purification

To remove impurities and control their form, APIs are usually crystalized by using temperature and solubility to precipitate the molecules out of solution. NIRS has been used to monitor the yield (how much API is removed from the solution) and the form of the resulting crystals. As the instrumentation evolved, form analysis moved from at-line [39] to in-line [40]. A study of crystallization kinetics as a function of processing parameters (temperature, habit, size of seeds, and solvent) allowed the optimization of the process [40]. The scale up of a crystallization from pilot to industrial scale was aided by a quantitative NIR model for the prediction of the form composition. Probe fouling was mitigated through the use of a nitrogen purge in front of the sapphire window [41].

18.4.2   Drug Product Manufacturing

Once a drug substance has been manufactured in sufficient quantity, the optimization of the process that manufactures the drug product delivered to patients can start. The form of the drug product can vary significantly based on the intended use and the desired pharmaco-kinetic profile. Medicines for delivery to patients by health-care providers will often be manufactured in the form of sterile products for intravenous delivery. However, other means of drug delivery are necessary for patients to take at home. Tablets are the most commonly used oral solid dosage forms because of their convenience (which improve treatment compliance) and ability to change their delivery profiles (immediate release or sustained release). But a visit to the pharmacy will highlight the variety of delivery forms available, from creams and ointments to inhalers, capsules, and transdermal patches to name a few.

Active pharmaceutical ingredients are often poorly flowing materials and so merely pouring them directly in a capsule filling machine or tablet press would result in variable dosage forms (not providing the patient with the right dose for the intended therapeutic effect). To ensure the final drug product has the characteristics advertised on the label, the API is usually combined with inactive ingredients (excipients) that will result in a better flowing material that can be processed at high speed with the desired product attributes. When no other processing step than mixing is necessary to ensure homogeneity and weight consistency, the process is called direct compression. However, when the API is present in large quantity in the formulation or has poor flowing properties, it may be necessary to granulate the material (through physical means (dry granulation or extrusion) or using a solvent and binder (wet granulation)) to obtain a free-flowing powder that can be processed at high speed on a tablet press or encapsulator.

Because of the difficulty of working with powders and their ability to aggregate or segregate, particular attention has to be given to the intermediate and final quality attributes of the drug product. Near-infrared spectroscopy has been a tool of choice for use in the manufacturing of oral solid dosage forms. Near-infrared light penetrates deeper than UV, Raman, or infrared light and rugged equipment is available to measure the drug content during the manufacturing process. In addition, with the transition of the pharmaceutical industry from batch to continuous manufacturing, NIRS has become a tool of choice for in-line and in real-time measurements and control of the manufacturing of drug products. In this section, examples of the use of NIRS for the main unit operations involved in manufacturing tablets are provided.

18.4.2.1   Powder Homogeneity

This particular unit operation consists in mixing active and inactive pharmaceutical ingredients to obtain a homogeneous mix that can be used for tablet/capsule manufacturing or for input to a granulation operation. The most important parameters associated with blend quality are homogeneity and stability. Blending is used in all drug product manufacturing operations and has thus been the subject of extensive research.

Batch blending

The traditional approach to batch blend quality evaluation relies on powder sampling at various locations of the blender bin after a given time of mixing and analysis by HPLC for the API content. It is a discrete operation that provides an evaluation of an entire bin at a particular time point. But because it requires manual sampling, it has been shown that it can induce errors [42]. Alternatively, a wireless NIR unit can be mounted on a blender. While the amount of powder analyzed at each rotation is usually lower than what would be considered by manual sampling (between 10 and 50 mg based on the collection optics and powder density), it allows the understanding of blend kinetics for not only the active ingredient but also the major excipients, the comparison of blend trends across batches without having to manually sample, and the possibility to determine a blend end-point.

Blend analysis by NIRS has been performed off or at-line on sampled powders by single point spectrometer [43] or imaging [44] but most of the work has been done on-line with qualitative and quantitative methodologies. Hailey et al. and Sekulic et al. first reported the use of on-line NIR with qualitative approaches to evaluate the end-point [45]. They used a moving block standard deviation to determine when the blend was done evolving. In that approach, the pooled standard deviation across the variables is calculated for a set block of spectra and compared with previous and subsequent blocks. Figure 18.5 shows the application of the moving block standard deviation approach with a block size of 25 spectra onto NIR spectral data collected on-line for blend comprised of acetaminophen (34.5%w/w), lactose (34.7%/w), microcrystalline cellulose (24.8%/w), croscarmellose sodium (5.5%/w), and magnesium stearate (0.5%/w) [46]. The spectra were preprocessed with Standard Normal Variate to reduce the effects of physical properties. The approach determined that the blend reaches a plateau of variability after about 40 rotations with no long-term trend. A threshold could be set using historical information to identify when a blend has reached homogeneity. However, this approach requires historical information of what should be considered homogeneous. To address this limitation, an F-Test-based method was proposed [47]. The spectral variance of two sequential blocks is calculated and compared with an F-critical value calculated based on the size of the blocks and the confidence limit. This approach avoids having to set a value of standard deviation corresponding to blend stability but rather uses the block-to-block information to determine when the two populations have a similar variance, indicating that the blend is no longer changing given the confidence limit considered.

These two methods use the spectral variance in the form of a spectrum standard deviation. It could be envisioned that for a same resulting standard deviation, different regions of the spectra could be changing; thus, indicating different phenomena taking

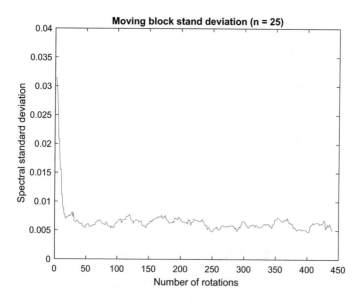

Fig. 18.5 Moving block standard deviation with a block of 25 spectra applied to on-line NIR spectra of an acetaminophen blend

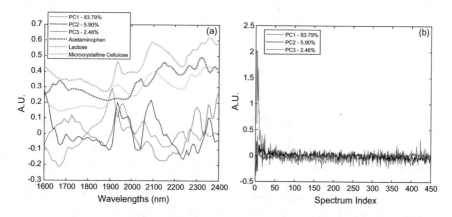

Fig. 18.6 Principal components and overlaid pure components (**a**) scores as a function of mixing time (**b**)

place in the blend. Principal component analysis has been used to help with the analysis of the origin of the variance as well as blend monitoring and end-point. Soft Independent Modeling of Class Analogy (SIMCA) was employed to identify when a spectrum belonged to the class corresponding to homogeneous spectra [48]. Figure 18.6 shows the PCA scores and loadings plots for the blend discussed above.

The loading plot allows the understanding of the origin of the change in the spectral data as a function of mixing while the scores provide the variance trend as

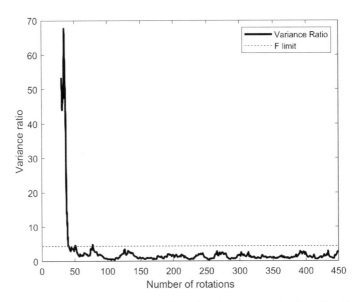

Fig. 18.7 Variance ratio trend as a function of blending time compared with the F-critical value

a function of time. Compared to the moving block standard deviation method, the PCA approach shows that the information described by the first principal component (PC1) and representing 84% of the available variance would describe the blend to reach stability after approximately 75 rotations. This was not a piece of information available by looking only at the pooled spectral standard deviation. However, this type of analysis is rather difficult to use to determine an end-point as historical knowledge would be necessary.

A hybrid approach called the caterpillar, that performs a variance comparison on blocks of spectra after local PCA analyses, was proposed [49]. An F-test is calculated to compare the variance in each block and an F-critical can be calculated to determine when the variance across blocks is no longer significantly different. In addition, the shape of the components for each local model can be compared, providing specificity to the components of interest. Figure 18.7 shows an example of a caterpillar output with the calculated F-value trend as a function of the number of rotations.

The methods discussed above provide an idea of the blend kinetics, but not directly blend content. It could be inferred however with a PCA model. If a PCA model is built on spectra proven to correspond to homogeneous powders (through sampling and HPLC analysis), new spectra could be projected onto that model and diagnostics (Hotelling's T^2 and Q-residuals) could be used to determine whether they present the same variability (thus the same content) [46]. Other authors have used PLS regression [50]. The development of these methods is however very resource intensive with samples needed at various concentration levels to build the model.

When a process is scaled up from laboratory to plant or manufacturing scale, differences can be observed in the spectra due to changes in powder density against the sensor's window, resulting in the need for data collection at large scale which can be expensive and time consuming.

Continuous blending

In situations where the blend is manufactured continuously, approaches have been developed to monitor the homogeneity as a function of time. However, a few additional considerations need to be taken into account when discussing continuous blending. First, the scale of scrutiny (or the amount of sample analyzed at a given time) becomes more important. In batch blend monitoring, the same powder will be analyzed rotation after rotation until the entirety of the powder becomes homogeneous and stable. In continuous blending, the powder may never be re-analyzed and so the sampling must be representative of the mass of powder that patients will take. Thus, spectra representing at least one-unit dosage form should be collected so a relevant determination of homogeneity can be achieved. A discussion of scale of scrutiny and spectral collection geometry on model performance showed the interdependence between process design and sampling representativity [51]. Probe fouling can also be a significant issue in continuous blending. While in batch, it would be as simple as stopping the bin, wiping the window, and restarting the batch if coating on the window is detected, it is not possible to do that often in continuous blending as manufacturing time dictates the quantity produced. When sufficient shear is present, and the window self-cleans, little to no control may be necessary. This may be the case in a tablet press feed-frame for instance. But where the shear is low or the material sticky, it may be necessary to have an active control that will wipe or clean the probe at set frequencies. Commercial solutions are available to perform these frequent probe cleanings.

   The monitoring of blend homogeneity in a continuous process has been reported at the exit of a blender or in the feed-frame of a tablet press. Similarly to batch blending, continuous blend monitoring can utilize qualitative or quantitative approaches, depending on the intended purpose of the method. A qualitative approach based on the F-test was proposed by Fonteyne et al. using the same principles of the block F-test discussed above for the monitoring of blends [52]. A number of articles discuss the prediction of blend uniformity for continuous systems. A triangle interface was used by Vargas et al. to monitor the homogeneity of powder at the discharge of a blender [53]. Quantitative models were developed for the active ingredient after a spectral quality evaluation ensured that the powder bed was representative and did not contain air pockets potentially formed as the blend travels from the blender to downstream unit operations. Other articles report the monitoring of powder uniformity in feed-tube to a tablet press and other powder interfaces [54].

   A significant amount of work has been performed on the analysis of powder in the feed-frame. Initially presented by Liu and Blackwood [55], an example of method development and validation was published by De Leersnyder [11]. Authors investigated the relationship between press parameters (turret speed, paddle speed and the distance between the paddle and the sensing window) and the quality of spectral signal.

18.4.2.2   Granulation and Drying

When a powder flows poorly and is not amenable to high speed processing, powder consolidation can be used to improve its flow characteristics using a solvent and binder, compaction, or heat to create granules. In wet granulation, a solvent (usually water) and binder (in solid or liquid form) are added to the powder along with shear to create aggregates. The solvent is then removed through drying. A granulation process will result in particles of much larger sizes and density than the initial powders. NIRS has been used to monitor the particle growth, homogeneity, and solvent level after drying.

High-shear wet granulators and fluid bed dryers are the most common types of batch granulator and dryers. Since NIR has a very strong water signal, it has been used to monitor the water level of the granules during granulation and drying. Frake et al. used the technology to monitor the granulation end-point in a fluid bed dryer [56]. Another study looked at how the NIR signal could be used to monitor granule growth in a fluid bed dryer [57]. Corresponding work in a high-shear wet granulator was performed using the change in slope of the NIR spectra as a function of physical differences during the process [58].

The removal of the granulation solvent can also be monitored and potentially controlled by NIRS. Fluid bed drying is most commonly used. An example of method development and validation for the monitoring of water was published [59]. Alcalà et al. showed the suitability of NIRS for the prediction of granule bulk density [60].

Figure 18.8 shows an example of fluid bed granulation and drying process trajectory. After a brief phase of mixing without solvent (dry mixing), the water is added and then removed. The first two principal components clearly show a combination of water and particle size information with the water content increasing and decreasing to a similar level, but with a powder physical property shifting significantly.

Twin-screw granulation is a continuous wet granulation method used in continuous manufacturing lines. Granule quality and the impact of processing parameters such as liquid feed rate, moisture content, and screw configuration were investigated by NIR chemical imaging [61]. Authors showed that the visualization capability of the technology allowed the characterization of the unit operation as the entirety of the produced materials can be analyzed in-line and in real time, accelerating process development and improving understanding.

Dry compression does not use a solvent but physical force to consolidate powder into granules of larger particle size. The density of the ribbons and particle size distribution of the resulting granules are parameters that have been investigated by NIRS [62].

Hot melt extrusion uses heat and shear to embed the API with polymers creating an extrudate with high homogeneity and increased bioavailability for API with low solubility. A probe can be attached to the die of the extruder to monitor the properties

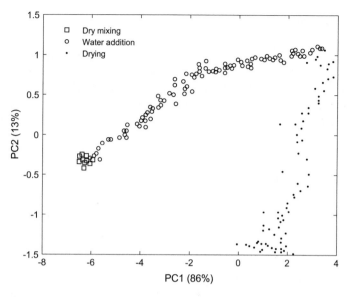

Fig. 18.8 PCA score trends for a fluid bed granulation and drying process. (Reproduced with permission from [60])

of the extrudate [63]. Drug and polymer concentrations as well as the solid state of the API have been studied.

18.4.2.3   Dosage Form Analysis

Since analyzing a tablet or capsule by NIR takes only a few seconds with no sample preparation, numerous companies and academic groups have investigated the suitability of the technology. Some deployments have been proposed to replace final product testing as an alternative release method. A well-documented example was published by Goodwin et al. [64] In their work, authors proposed a control strategy for a product based on compaction force weight control, periodic at-line weight measurements, and NIRS analysis of individual tablet content. The authors developed a NIR spectroscopic method with a validation error of 1.12% of label claim and demonstrated its equivalency with the reference HPLC method using a two one-sided t-Test.

A demonstration of high-speed tablet analysis was published by Boiret et al. [65] The authors used a reflectance measurement based on a spatially resolved spectroscopic probe to investigate (using a conveyor belt and acquisition times as low as 1 ms) the homogeneity and distribution of API in tablets.

However, while much work has been done to show that intact dosage forms can be successfully analyzed, the industry appears to be moving toward in-line and in real-time measurements of content.

18.4.3 Summary

The variety of applications in drug substance and drug product manufacturing shows the versatility and usefulness of the technology for the development of robust processes to deliver patients with safe and efficacious medicines. It also shows the breadth and depth that a PAT scientist must have to successfully deliver a NIRS analytical method.

18.5 Raw Material Identification

Raw material identification has been made possible by the flexibility of the hardware, particularly the handheld systems. In essence, developing a raw material identification method relies on the creation of a relevant spectral library and the determination of a set of criteria that can determine the class of a new sample. A comprehensive review on the topic is available [66].

The development of a spectral library is not a simple task. Similar to developing a quantitative method, the spectral library must include spectra representative of the variability that will be encountered during the use of the method. If future samples are expected to vary in physical properties or have a range of chemical properties, they should be included in the library. Failure to do so will falsely identify valid samples as not belonging to the expected population, thus requiring the need for investigation and updating the library and model. While it is never possible to foresee all the sources of variability that will be encountered, it is recommended to consider various lots and chemical/physical property ranges to define each class as precisely as possible.

Once a representative library is created, an algorithm is used to discriminate between classes. A large variety of methods is available [67].

Control over the raw material characteristics has also been investigated for biopharmaceutical products. The performance of basal powders on process performance and product quality was used to identify classes amongst spectra [68]. The grouping of the samples was attributed to the blend uniformity, impurities, and heat sensitivity during the milling process.

18.6 Summary

From the first usage of NIRS in the field to its now prominent position for process monitoring and control, near-infrared spectroscopy has come a long way in hardware, software, and modeling. The pharmaceutical industry is now able to rely on the technology to release medicine to patients without having to sample for off-line analyses. The sensitivity of NIR light to the matrix, which makes it a tool of choice for raw material identification and counterfeit detection [69], can be an issue for quantitative and qualitative methods as much of the variance has to be captured by the model prior to validation to avoid repeated method updates. Other techniques (i.e., Raman and Infrared spectroscopy) are also complementing and sometimes replacing NIRS when better suited. Nevertheless, its flexibility of implementation, hardware ruggedness, and wide range of applicability make it a tool of choice for process understanding, monitoring, and control.

References

1. E.W. Ciurczak, Uses of near-infrared spectroscopy in pharmaceutical analysis. Appl. Spectros. Rev. **23**(1–2), 147–163 (1987)
2. USFDA, PAT—A framework for innovative pharmaceutical development, manufacturing, and quality assurance, 2004
3. A. Chanda, A.M. Daly, D.A. Foley, M.A. Lapack, S. Mukherjee, J.D. Orr, G.L. Reid, D.R. Thompson, H.W. Ward, Industry perspectives on process analytical technology: Tools and applications in API development. Org. Process Res. Dev. **19**(1), 63–83 (2015)
4. L.L. Simon, H. Pataki, G. Marosi, F. Meemken, K. Hungerbühler, A. Baiker, S. Tummala, B. Glennon, M. Kuentz, G. Steele, H.J.M. Kramer, J.W. Rydzak, Z. Chen, J. Morris, F. Kjell, R. Singh, R. Gani, K.V. Gernaey, M. Louhi-Kultanen, J. Oreilly, N. Sandler, O. Antikainen, J. Yliruusi, P. Frohberg, J. Ulrich, R.D. Braatz, T. Leyssens, M. Von Stosch, R. Oliveira, R.B.H. Tan, H. Wu, M. Khan, D. Ogrady, A. Pandey, R. Westra, E. Delle-Case, D. Pape, D. Angelosante, Y. Maret, O. Steiger, M. Lenner, K. Abbou-Oucherif, Z.K. Nagy, J.D. Litster, V.K. Kamaraju, M.S. Chiu, Assessment of recent process analytical technology (PAT) trends: a multiauthor review. Org. Process Res. Dev. **19**(1), 3–62 (2015)
5. E. Ciurczak, B. Igne, *Pharmaceutical and Medical Applications of Near-Infrared Spectroscopy* (CRC Press, Boca Raton, 2015)
6. C.C. Corredor, D. Bu, G. McGeorge, Chapter 9—Applications of MVDA and PAT for drug product development and manufacturing, in *Multivariate Analysis in the Pharmaceutical Industry*, ed. by A.P. Ferreira, J.C. Menezes, M. Tobyn (Academic Press, 2018), pp. 211–234
7. B. Igne, R.W. Bondi, C. Airiau, Chapter 8—Multivariate data analysis for enhancing process understanding, monitoring, and control—active pharmaceutical ingredient manufacturing case studies, in *Multivariate Analysis in the Pharmaceutical Industry*, ed. by A.P. Ferreira, J.C. Menezes, M. Tobyn (Academic Press, 2018), pp. 185–210
8. ICH, Points to Consider for ICH Q8/Q9/Q10 Implementation. 2011
9. ICH, Validation of Analytical Procedures Q2(R1). 1996
10. G.E. Ritchie, R.W. Roller, E.W. Ciurczak, H. Mark, C. Tso, S.A. MacDonald, Validation of a near-infrared transmission spectroscopic procedure: part B: application to alternate content uniformity and release assay methods for pharmaceutical solid dosage forms. J. Pharm. Biomed. Anal. **29**(1–2), 159–171 (2002)
11. F. De Leersnyder, E. Peeters, H. Djalabi, V. Vanhoorne, B. Van Snick, K. Hong, S. Hammond, A.Y. Liu, E. Ziemons, C. Vervaet, T. De Beer, Development and validation of an in-line NIR spectroscopic method for continuous blend potency determination in the feed frame of a tablet press. J. Pharm. Biomed. Anal. **151**, 274–283 (2018)
12. G.L. Reid, J. Morgado, K. Barnett, B. Harrington, J. Wang, J. Harwood, D. Fortin, Analytical quality by design (AQbD) in pharmaceutical development. Am. Pharm. Rev. 2013
13. American Society for Testing and Materials, E2476–16—Guide for risk assessment and risk control as it impacts the design, development, and operation of PAT processes for pharmaceutical manufacture. 2016
14. United States Pharmacopoeia, USP 1119—Near-infrared spectrophotometry. 2003

15. European Medicine Agency, Guideline on the Use of Near Infrared Spectroscopy by the Pharmaceutical Industry and the Data Requirements for New Submissions and Variations. 2014
16. United States Food and Drug Administration, Draft Guidance—Development and Submission of Near Infrared Analytical Procedures Guidance for Industry. 2015
17. T. Miyano, H. Nakagawa, T. Watanabe, H. Minami, H. Sugiyama, Operationalizing maintenance of calibration models based on near-infrared spectroscopy by knowledge integration. J. Pharm. Innov. **10**(4), 287–301 (2015)
18. M. Hoehse, J. Alves-Rausch, A. Prediger, P. Roch, C. Grimm, Near-infrared spectroscopy in upstream bioprocesses. Pharm. Bioprocess **3**, 153–172 (2015)
19. R.W. Silman, L.T. Black, K. Norris, Assay of solid–substrate fermentation by means of reflectance infrared analysis. Biotechnol. Bioeng. **25**(2), 603–607 (1983)
20. J. Crowley, S.A. Arnold, N. Wood, L.M. Harvey, B. McNeil, Monitoring a high cell density recombinant Pichia pastoris fed-batch bioprocess using transmission and reflectance near infrared spectroscopy. Enzyme Microb. Technol. **36**(5), 621–628 (2005)
21. M. Goldfeld, J. Christensen, D. Pollard, E.R. Gibson, J.T. Olesberg, E.J. Koerperick, K. Lanz, G.W. Small, M.A. Arnold, C.E. Evans, Advanced near-infrared monitor for stable real-time measurement and control of Pichia pastoris bioprocesses. Biotechnol. Prog. **30**(3), 749–759 (2014)
22. N. Petersen, P. Odman, A.E. Padrell, S. Stocks, A.E. Lantz, K.V. Gernaey, In situ near infrared spectroscopy for analyte-specific monitoring of glucose and ammonium in streptomyces coelicolor fermentations. Biotechnol. Prog. **26**(1), 263–271 (2010)
23. L.O. Rodrigues, L. Vieira, J.P. Cardoso, J.C. Menezes, The use of NIR as a multi-parametric in situ monitoring technique in filamentous fermentation systems. Talanta **75**(5), 1356–1361 (2008)
24. M.J. McShane, G.L. Coté, Near-infrared spectroscopy for determination of glucose, lactate, and ammonia in cell culture media. Appl. Spectrosc. **52**(8), 1073–1078 (1998)
25. C.B. Lewis, R.J. McNichols, A. Gowda, G.L. Coté, Investigation of near-infrared spectroscopy for periodic determination of glucose in cell culture media in situ. Appl. Spectrosc. **54**(10), 1453–1457 (2000)
26. M. Clavaud, Y. Roggo, R. Von Daeniken, A. Liebler, J.-O. Schwabe, Chemometrics and in-line near infrared spectroscopic monitoring of a biopharmaceutical Chinese hamster ovary cell culture: Prediction of multiple cultivation variables. Talanta **111**, 28–38 (2013)
27. Z. Sun, C. Li, L. Li, L. Nie, Q. Dong, D. Li, L. Gao, H. Zang, Study on feasibility of determination of glucosamine content of fermentation process using a micro NIR spectrometer. Spectrochim. Acta Part A Mol. Biomol. Spectrosc. **201**, 153–160 (2018)
28. J.G. Henriques, S. Buziol, E. Stocker, A. Voogd, J.C. Menezes, Monitoring mammalian cell cultivations for monoclonal antibody production using near-infrared spectroscopy, in *Optical Sensor Systems in Biotechnology*, ed. by G. Rao (Springer, Berlin, Heidelberg, 2009), pp. 29–72
29. A.P. Teixeira, R. Oliveira, P.M. Alves, M.J.T. Carrondo, Advances in on-line monitoring and control of mammalian cell cultures: supporting the PAT initiative. Biotechnol. Adv. **27**(6), 726–732 (2009)
30. M.S. Kamat, R.A. Lodder, P.P. DeLuca, Near-infrared spectroscopic determination of residual moisture in lyophilized sucrose through intact glass vials. Pharm. Res. **6**(11), 961–965 (1989)
31. A. Funke, R. Gross, S. Tosch, A. Tulke, Recent achievements in NIR-based on-line monitoring of lyophilisation processes. Eur. Pharm. Rev. **21**(3), 50–53 (2016)
32. D. Brouckaert, L. De Meyer, B. Vanbillemont, P.J. Van Bockstal, J. Lammens, S. Mortier, J. Corver, C. Vervaet, I. Nopens, T. De Beer, Potential of near-infrared chemical imaging as process analytical technology tool for continuous freeze-drying. Anal. Chem. **90**(7), 4354–4362 (2018)
33. S. Pieters, T. De Beer, Y. Vander Heyden, Near-infrared and Raman spectroscopy: potential tools for monitoring of protein conformational instability during freeze-drying processes. Am. Pharm. Rev. **15**(1), 2012

34. A. Kandelbauer, W. Kessler, R.W. Kessler, Online UV–visible spectroscopy and multivariate curve resolution as powerful tool for model-free investigation of laccase-catalysed oxidation. Anal. Bioanal. Chem. **390**(5), 1303–1315 (2008)
35. P. Hamilton, M.J. Sanganee, J.P. Graham, T. Hartwig, A. Ironmonger, C. Priestley, L.A. Senior, D.R. Thompson, M.R. Webb, Using PAT to understand, control, and rapidly scale up the production of a hydrogenation reaction and isolation of pharmaceutical intermediate. Org. Process Res. Dev. **19**(1), 236–243 (2015)
36. M. Blanco, M. Castillo, R. Beneyto, Study of reaction processes by in-line near-infrared spectroscopy in combination with multivariate curve resolution. Esterification of myristic acid with isopropanol. Talanta. **72**(2), 519–525 (2007)
37. M.C. Antunes, J.J. Simão, A.C. Duarte, R. Tauler, Multivariate curve resolution of overlapping voltammetric peaks: quantitative analysis of binary and quaternary metal mixtures. Analyst. **127**(6), 809–817, (2002)
38. J. Wiss, M. Länzlinger, M. Wermuth, Safety improvement of a grignard reaction using on-line NIR monitoring. Org. Process Res. Dev. **9**(3), 365–371 (2005)
39. P.K. Aldridge, C.L. Evans, H.W. Ward Ii, S.T. Colgan, N. Boyer, P.J. Gemperline, Near-IR detection of polymorphism and process-related substances. Anal. Chem. **68**(6), 997–1002 (1996)
40. G. Févotte, J. Calas, F. Puel, C. Hoff, Applications of NIR spectroscopy to monitoring and analyzing the solid state during industrial crystallization processes. Int. J. Pharm. **273**(1–2), 159–169 (2004)
41. S.E. Barnes, T. Thurston, J.A. Coleman, A. Diederich, D. Ertl, J. Rydzak, P. Ng, K. Bakeev, D. Bhanushali, NIR diffuse reflectance for on-scale monitoring of the polymorphic form transformation of pazopanib hydrochloride (GW786034); model development and method transfer. Anal. Methods **2**(12), 1890–1899 (2010)
42. M. Poux, P. Fayolle, J. Bertrand, D. Bridoux, J. Bousquet, Powder mixing: Some practical rules applied to agitated systems. Powder Technol. **68**(3), 213–234 (1991)
43. A.S. El-Hagrasy, J.K. Drennen, A process analytical technology approach to near-infrared process control of pharmaceutical powder blending. Part III: quantitative near-infrared calibration for prediction of blend homogeneity and characterization of powder mixing kinetics. J. Pharm. Sci. **95**(2), 422–434 (2006)
44. J.G. Osorio, G. Stuessy, G.J. Kemeny, F.J. Muzzio, Characterization of pharmaceutical powder blends using in situ near-infrared chemical imaging. Chem. Eng. Sci. **108**, 244–257 (2014)
45. P.A. Hailey, P. Doherty, P. Tapsell, T. Oliver, P.K. Aldridge, Automated system for the on-line monitoring of powder blending processes using near-infrared spectroscopy Part I. System development and control. J. Pharm. Biomed. Anal. **14**(5), 551–559 (1996)
46. B. Igne, A.D. Juan, J. Jaumot, J. Lallemand, S. Preys, J.K. Drennen, C.A. Anderson, Modeling strategies for pharmaceutical blend monitoring and end-point determination by near-infrared spectroscopy. Int. J. Pharm. **473**(1–2), 219–231 (2014)
47. R. Besseling, M. Damen, T. Tran, T. Nguyen, K. van den Dries, W. Oostra, A. Gerich, An efficient, maintenance free and approved method for spectroscopic control and monitoring of blend uniformity: the moving F-test. J. Pharm. Biomed. Anal. **114**, 471–481 (2015)
48. S.S. Sekulic, J. Wakeman, P. Doherty, P.A. Hailey, Automated system for the on-line monitoring of powder blending processes using near-infrared spectroscopy. Part II. Qualitative approaches to blend evaluation. J. Phar. Biomed. Anal. **17**(8), 1285–1309 (1998)
49. G.R. Flåten, R. Belchamber, M. Collins, A.D. Walmsley, Caterpillar—an adaptive algorithm for detecting process changes from acoustic emission signals. Anal. Chim. Acta. **544**(1–2 SPEC. ISS.), 280–291 (2005)
50. O. Berntsson, L.G. Danielsson, B. Lagerholm, S. Folestad, Quantitative in-line monitoring of powder blending by near infrared reflection spectroscopy. Powder Technol. **123**(2–3), 185–193 (2002)
51. M.A. Alam, Z. Shi, J.K. Drennen, C.A. Anderson, In-line monitoring and optimization of powder flow in a simulated continuous process using transmission near infrared spectroscopy. Int. J. Pharm. **526**(1–2), 199–208 (2017)

52. M. Fonteyne, J. Vercruysse, F. De Leersnyder, R. Besseling, A. Gerich, W. Oostra, J.P. Remon, C. Vervaet, T. De Beer, Blend uniformity evaluation during continuous mixing in a twin screw granulator by in-line NIR using a moving F-test. Anal. Chim. Acta **935**, 213–223 (2016)

53. J.M. Vargas, S. Nielsen, V. Cárdenas, A. Gonzalez, E.Y. Aymat, E. Almodovar, G. Classe, Y. Colón, E. Sanchez, R.J. Romañach, Process analytical technology in continuous manufacturing of a commercial pharmaceutical product. Int. J. Pharm. **538**(1), 167–178 (2018)

54. Y.M. Colón, M.A. Florian, D. Acevedo, R. Méndez, R.J. Romañach, Near infrared method development for a continuous manufacturing blending process. J. Pharm. Innov. **9**(4), 291–301 (2014)

55. Y. Liu, D. Blackwood, Sample presentation in rotary tablet press feed frame monitoring by near infrared spectroscopy. Am. Pharm. Rev. (2012)

56. P. Frake, D. Greenhalgh, S.M. Grierson, J.M. Hempenstall, D.R. Rudd, Process control and end-point determination of a fluid bed granulation by application of near infra-red spectroscopy. Int. J. Pharm. **151**(1), 75–80 (1997)

57. W. Li, J. Cunningham, H. Rasmussen, D. Winstead, A qualitative method for monitoring of nucleation and granule growth in fluid bed wet granulation by reflectance near-infrared spectroscopy. J. Pharm. Sci. **96**(12), 3470–3477 (2007)

58. A.C. Jorgensen, P. Luukkonen, J. Rantanen, T. Schaefer, A.M. Juppo, J. Yliruusi, Comparison of torque measurements and near-infrared spectroscopy in characterization of a wet granulation process. J. Pharm. Sci. **93**(9), 2232–2243 (2004)

59. A. Peinado, J. Hammond, A. Scott, Development, validation and transfer of a Near Infrared method to determine in-line the end point of a fluidised drying process for commercial production batches of an approved oral solid dose pharmaceutical product. J. Pharm. Biomed. Anal. **54**(1), 13–20 (2011)

60. M. Alcalà, M. Blanco, M. Bautista, J.M. González, On-line monitoring of a granulation process by NIR spectroscopy. J. Pharm. Sci. **99**(1), 336–345 (2010)

61. J. Vercruysse, M. Toiviainen, M. Fonteyne, N. Helkimo, J. Ketolainen, M. Juuti, U. Delaet, I. Van Assche, J.P. Remon, C. Vervaet, T. De Beer, Visualization and understanding of the granulation liquid mixing and distribution during continuous twin screw granulation using NIR chemical imaging. Eur. J. Pharm. Biopharm. **86**(3), 383–392 (2014)

62. R.W. Miller, K.R. Morris, A.G., Roller compaction scale-up, in *Pharmaceutical Process Scale-Up*, 2nd Ed., ed. by M.Levin (Taylor and Francis, Boca Raton, 2006), pp. 237–266

63. B. Smith-Goettler, C.M. Gendron, R.F. Meyer, Understanding hot melt extrusion via near infrared spectroscopy. NIR News **25**(7), 10–12 (2014)

64. D.J. Goodwin, S. van den Ban, M. Denham, I. Barylski, Real time release testing of tablet content and content uniformity. Int. J. Pharm. **537**(1–2), 183–192 (2018)

65. M. Boiret, F. Chauchard, Use of near-infrared spectroscopy and multipoint measurements for quality control of pharmaceutical drug products. Anal. Bioanal. Chem. **409**(3), 683–691 (2017)

66. J. Luypaert, D.L. Massart, Y. Vander Heyden, Near-infrared spectroscopy applications in pharmaceutical analysis. Talanta **72**(3), 865–883 (2007)

67. M. Blanco, M.A. Romero, Near-infrared libraries in the pharmaceutical industry: a solution for identity confirmation. Analyst **126**(12), 2212–2217 (2001)

68. A.O. Kirdar, G. Chen, J. Weidner, A.S. Rathore, Application of near-infrared (NIR) spectroscopy for screening of raw materials used in the cell culture medium for the production of a recombinant therapeutic protein. Biotechnol. Prog. **26**(2), 527–531 (2010)

69. B. Krakowska, D. Custers, E. Deconinck, M. Daszykowski, Chemometrics and the identification of counterfeit medicines—a review. J. Pharm. Biomed. Anal. **127**, 112–122 (2016)

Chapter 19
Bio-applications of NIR Spectroscopy

Christian W. Huck

Abstract Near-infrared (NIR) spectroscopy occupies a distinct spot as an investigation tool in bioscience. It gained an ultimate value in several areas of application, e.g., in characterization of plant material, examination of body fluids, exploration of the structure and properties of water and biomolecules in aqueous environment. On the other hand, certain limitations of this technique have been apparent and its full potential seems yet to be unveiled. In recent years, key advancements in technology and methods have pushed the frontier of NIR spectroscopy in bio-applications. Trend-setting studies demonstrated the capacity of NIR spectroscopy to excel in previously unattainable scenarios such as in vivo examination of entire organisms. The advent of miniaturized instrumentation enabled a new spectrum of applications in plant-related research. Advancements in data analytical methods decisively pushed the limit in interpretability of NIR spectra, enabling better understanding of NIR spectral features of biomolecules. These advancements were accompanied by a continuous refinement of established approaches. This chapter discussed the established applications, current developments and future prospects of NIR spectroscopy in broadly understood bio-applications.

Keywords Bio-applications · NIR biospectroscopy · Plant analysis · Biomolecules

19.1 Introduction

Near-infrared (NIR) spectroscopy occupies a particular spot across the field of bioscience. On the one hand, it has become the tool-of-choice in various applications concerning the assessment of bio-related samples, e.g., in medicinal plant analysis or quality control of natural products. On the other hand, in several fields such as in-laboratory bioanalytical research and biomedical diagnosis, it steadily gains in importance and challenges the related techniques such as IR or Raman spectroscopies that are well-established tools therein. Because of several reasons, in the

C. W. Huck (✉)
Institute of Analytical Chemistry and Radiochemistry, CCB-Center for Chemistry and
Biomedicine, Leopold-Franzens University, Innrain 80-82, 6020 Innsbruck, Austria
e-mail: Christian.W.Huck@uibk.ac.at

© Springer Nature Singapore Pte Ltd. 2021   413
Y. Ozaki et al. (eds.), *Near-Infrared Spectroscopy*,
https://doi.org/10.1007/978-981-15-8648-4_19

past decades, a relatively modest attention has been paid to the potential that NIR spectroscopy is able to bring to the latter field. Recent years have demonstrated that NIR spectroscopy may be used in a number of similar applications concurrently with IR or Raman spectroscopy. To better present the strengths and limitations of the eponymous method, it is useful to briefly summarize the essential similarities and differences existing between these approaches. The chemical specificity of IR or Raman spectra is relatively superior to that of NIR spectra, and they are more straightforward in a direct interpretation. The observed bands are broader because of strong overlapping, reducing the potential for linking the spectra with the structural information. This becomes particularly significant in the case of chemically complex samples of biological origin. However, recent years have witnessed rapid progress in our ability to understand NIR spectra of such samples, with combined use of spectral imaging, novel chemometrics or theoretical methods of spectra calculation [1, 2]. Relevant examples will be discussed in this chapter that demonstrate the progress recently achieved at this direction.

IR and NIR spectroscopy differ significantly in the typical sampling depth, or in other words, the information on the sample is collected from distinctively different sample volumes. This fact has a notable influence, as it is common in bioscience to investigate highly inhomogeneous, often micro-structured samples, such as cells and tissues of either plant or animal origin. For example, the consequences of that difference is well exemplified in medical application, in which the optimal sample thickness for IR transmission measurements conveniently matches the typical configuration of microtomed tissue specimen used in conventional medical diagnosis. However, the absorptivity of organic matter is up to two orders of magnitude lower in NIR than in IR region. Consequently, no useful NIR signal can be obtained from such specimen. On the other hand, this permits NIR radiation to reach deeper and measure the spectrum of the sample beneath its surface. This enables, for instance, sensing the information from beneath human skin or examining entire organisms such as fish embryo. Deep tissue sampling is a key advantage for bioanalytical applications. Moreover, for the same reason, a larger sample volume is permissible in NIR spectroscopy, making it better suited for the analysis of bulk materials essential for bio-related studies (e.g., natural products). Moreover, NIR spectroscopy is relatively better suited for examination of samples with high water content. Measurements in transmission or diffuse reflection mode without sample preparation are more feasible than in IR spectroscopy, which requires attenuated total reflection (ATR) approach is such cases. Compared with Raman, which is suitable for examining moist samples, NIR technique is applicable to specimen with high content of fluorophores; those most typically encountered in bioscience are, e.g., chlorophyll or proteins. Because of that, NIR spectroscopy is easily applicable for examination of plants and plant-related materials, as well as protein-rich samples. Better suitability of the principal features of NIR instrumentation in certain applications may be mentioned. Availability of fiber probes makes in vivo diagnosis easier. Miniaturized instrumentation is readily available for NIR spectroscopy. In contrast, such IR sensors are practically limited to ATR, and their application faces difficulties, e.g., because of the stability of sampling conditions. Worth mentioning is a relatively well-established functional NIR spectroscopy in medical diagnosis, where it serves the purpose of functional neuroimaging.

19.2 Medicinal Plant Analysis

Most often, the therapeutic and medicinal properties of herbs and plants are related to individual bio-active compounds, and their content affects the general usefulness of a given natural product. The chemical composition and the concentration of the bio-active compounds can be analyzed by NIR spectroscopy. The worldwide trend in using medicinal plant products is permanently increasing. In 2018, the turnover with freely available products from pharmacies was more than 1 billion Euro and that from other sources including online business more than another 1 billion Euro. This trend creates high demand for high throughput, in situ analytical methods capable of fast, non-invasive, simultaneous analysis of chemical and physical parameters in order to ensure quality of the natural medicine. Analytical methodologies based on portable, miniaturized NIR instrumentation are essentially favored for this purpose, as direct assessment and optimization of the cultivation conditions and parameters become possible, e.g., the harvest time. However, this application field remains quite new, and the applicability and performance profiles of handheld NIR devices remain continuously investigated.

Kirchler et al. described in their comprehensive examination the capability of NIR spectroscopy supported with various tools to determine the antioxidative potential and related properties of plant medicine. This trend-setting study proposed a new analytical strategy [3]. The performances of one benchtop and two different types of miniaturized NIR spectrometers were tested and compared for the first time by the determination of the rosmarinic acid (RA) content of dried and powdered *Rosmarinus officinalis*, folium (i.e., Rosmarini folium). The recorded NIR spectra (Fig. 19.1) were utilized in hyphenation with multivariate data analysis (MVA) to calculate partial least squares regression (PLSR) models (Table 19.1). Quality parameters obtained from cross-validation (CV) revealed that the benchtop spectrometer achieved the best result with a R^2 of 0.91 and a RPD of 3.27. Miniaturized NIR spectrometer MicroNIR 2200 showed a satisfying calibration value R^2 of 0.84 and a RPD of 2.46. The analysis performed by miniaturized microPHAZIR, with a R^2 of 0.73 and a RPD of 1.88, was less precise and revealed room for improvements. All recorded spectra of the different devices were additionally studied by two-dimensional correlation spectroscopy (2D-COS; details on this technique are available in Chapter 6 Two-dimensional correlation spectroscopy) analysis; in order to support the performed PLS regression models (Fig. 19.2). Differences in the sensitivity of the spectrometers were visualized by 2D hetero-correlation plots as well. These approaches were found to be helpful to identify discrepancies between microPHAZIR and MicroNIR 2200 compared to the benchtop instrument. With the aim to obtain a better understanding of the factors which determine the analyzed PLS regression models, in this study,

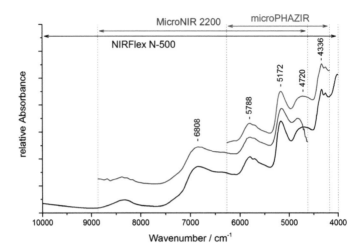

Fig. 19.1 NIR spectra of 60 *Rosmarini folium* samples measured on benchtop (NIRFlex N-500) and two handheld (microPHAZIR and MicroNIR 2200) NIR spectrometers. Reproduced from Ref. [3] with permission from The Royal Society of Chemistry

Table 19.1 Results of all PLSR models for the quantification of rosmarinic acid in *Rosmarinus officinalis, folium*

Spectrometer		NIRFlex N-500		microPHAZIR		MicroNIR 2200	
Samples		60		60		60	
Outliers		6		8		4	
C (w/w) range/%		1.138–2.425		1.138–2.425		1.138–2.425	
Validation method		*CV*	TSV	*CV*	TSV	*CV*	TSV
R^2		*0.91*	0.91	*0.73*	0.73	*0.84*	0.85
SECV/%	SEP/%	*0.072*	0.069	*0.12*	0.11	*0.091*	0.11
SECV/SEC	SEP/SEC	*1.46*	1.43	*1.28*	1.24	*1.55*	2.09
Factors		*8*	8	5	5	*11*	12
RPD		*3.27*	3.41	*1.88*	2.06	*2.46*	2.14

quantum chemical calculation of NIR spectrum of RA was carried out. In the process, this approach enabled us to understand, interpret and attribute the main influences in the regression coefficients plots; further information on spectra calculation is given in Sect. 10.8 and in a recent review article [1]. The study by Kirchler et al. demonstrated that the performance of NIR spectroscopy with benchtop and miniaturized devices as a fast and non-invasive technique is able to replace time- and resource-consuming analytical tools [3].

Pezzei et al. compared the suitability of benchtop and portable NIR spectrometers for predicting the optimum harvest time of *Verbena officinalis* [4]. In this project, NIR analyses were performed non-invasively on the fresh plant material based on

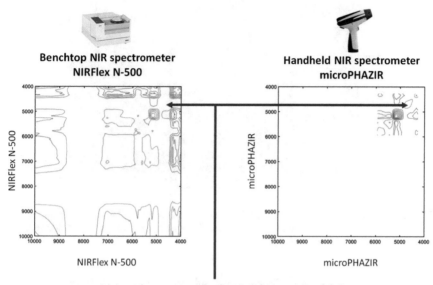

Fig. 19.2 Application of 2D correlation spectroscopy (2D-COS) for the visual assessment and comparison of the sensitivity profiles between different NIR spectrometers; reference benchtop Büchi NIRFlex N-500 vs. handheld Thermo Fisher Scientific microPHAZIR. The comparison of sensitivity: microPHAZIR shows an autopeak located at ca. 4000–5000 cm^{-1} (not observed on the benchtop instrument); this highlights an additional source of spectral variability due to working characteristics of the handheld spectrometer (instrumental nature)

the quantification of the key secondary metabolite ingredients verbenalin and verbascoside. NIR spectroscopic measurements were performed applying a conventional NIR benchtop device as well as a laboratory independent portable NIR spectrometer. A high performance liquid chromatography (HPLC) method served as the reference method. For both instruments, PLSR models were established performing cross-validations (CV) and test-set validations (TSV). Quality parameters obtained for the benchtop device revealed that the newly established NIR method allows sufficient quantifications of the main bio-active compounds of the medicinal plant, verbenalin and verbascoside. Results obtained with the miniaturized NIR spectrometer confirmed that accurate quantitative calibration models could be developed for verbascoside achieving a comparable prediction power to the benchtop instrument. PLS models for verbenalin were less precise suggesting that the application of portable devices may be limited by its spectral range, resolution and sensitivity (Fig. 19.3). Finally, this work demonstrated the suitability of NIR vibrational spectroscopy performing direct measurements on pharmaceutically relevant fresh plant material enabling a quick and easy determination of the ideal harvest time.

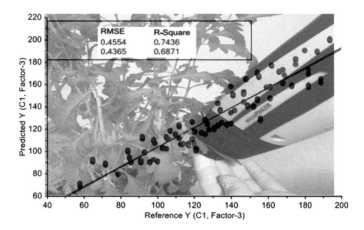

Fig. 19.3 On-site application of portable NIR spectrometer (microPHAZIR) for the optimization of harvest time of medicinal plant *Verbena officinalis*. This is accomplished through PLSR prediction of the content of bio-active compound, verbenalin

Delueg et al. described the online monitoring of the extraction process of Rosmarini folium by an analytical combination of wet chemical assays, UHPLC analysis and a newly developed NIR spectroscopic analysis method [5]. In the stage of experimental design, three different specimen-taking/sampling plans were chosen. At first, monitoring was carried out using three common analytical methods: (a) total hydroxycinnamic derivatives according to the European Pharmacopoeia, (b) total phenolic content according to Folin–Ciocalteu and (c) RA content measured by UHPLC-UV analysis. The combination of the recorded NIR spectra and the previously obtained analytical reference values in conjunction with multivariate data analysis enabled the successful establishment of PLSR models. Coefficients of determination (R^2) were: (a) 0.94, (b) 0.96 and (c) 0.93 (obtained by test-set validation), respectively. Since Pearson correlation analysis revealed that the reference analyses correlated with each other, just one of the PSLR models was required. Therefore, it was suggested that PLSR model (b) be used for monitoring the extraction process of Rosmarini folium. This example demonstrated the potential of NIR spectroscopy in providing a fast and non-invasive alternative analysis method, which can subsequently be implemented for on- or in-line process control in phytopharmaceutical industry.

NIR spectroscopy is a potent tool in qualitative analysis of plants and related samples. For example, the classification, discrimination or authentication of different natural products by the use of NIR spectroscopy is feasible. In particular, classifying the origin of natural products, detecting of adulteration and verifying authenticity of natural products are commonly performed. A considerable utilization this technique has found on the market of traditional Chinese medicines (TCMs). As an example of the applied methodology and achieved accuracy and applicability, Huck-Pezzei et al. established a procedure to discriminate between pharmaceutical formulations prepared from either *Hypericum perforatum* or *Hypericium hirsutum* originating from China [6]. It has been demonstrated that NIR spectroscopy is capable of discriminating between different plant species, varieties as well as cultivars, plant grown in different conditions or locations. For example, a rapid and accurate discrimination of *Chrysanthemum* varieties using NIR hyperspectral imaging technique (for further details of this technique, reader is referred to the chapter discussing NIR hyperspectral imaging) operating in 11,442–5767 cm^{-1} region (874–1734 nm) was demonstrated by Wu et al. [7]. The examination of the spectral images obtained from 11,038 samples was carried out by means of deep convolutional neural network (DCNN). The study indicated that NIR hyperspectral imaging combined with DCNN is a potent tool for rapid and accurate discrimination of plant varieties. These accomplishments may advance the qualitative analysis useful for producers, consumers and market regulators. Analysis of chemical compositions and various other properties of plants is often the main aim of various applications in agro-food sector. For further information, reader is referred to the chapters focused on this field of application.

19.3 Cell Analysis

IR and Raman spectroscopy have become greatly matured techniques in medical diagnosis of tissues, with prime importance for carcinoma diagnosis. In this field, NIR spectroscopy is still under development, with recent few years marking its significant progress in these applications. For example, its applicability to characterizing breast cancer cells was studied. The behavior of gold nanorods (AuNRs) in metastatic breast cancer cells was investigated by Zhang et al. [8]. The study used absorption spectroscopy in a broad ultraviolet–visible-NIR (UV-Vis-NIR) region (25,000–10,000 cm^{-1}; 400–1000 nm). That case serves an interesting example of how electronic absorption bands that extend to NIR region can be investigated in practical bio-applications (Fig. 19.4). UV-Vis-NIR absorption spectroscopy was employed in combination with inductively coupled plasma mass spectrometry (ICP-MS), transmission electron microscopy (TEM) and dark-field microscopic observation as reference methodologies for examination of the positively charged AuNRs in the highly metastatic tumor cell line MDA-MB-231. Absorption spectra of AuNRs in the living cells were acquired in that study; Fig. 19.4b presents the effects of serum on absorption spectra of AuNRs dispersed in SCM. It was described that characteristic surface plasmon resonance (SPR) peaks of AuNRs can be detected using spectroscopic method with living cells that have taken up the nanorods. The peak area of transverse SPR band was shown to be proportionally related to the amount of AuNRs in the cells determined with ICP-MS. The established spectroscopic analysis method can be used to monitor the behaviors of AuNR. Zhang et al. have demonstrated how successful monitoring the behaviors of AuNRs in the cells can be accomplished through an easy-tu-use UV-Vis-NIR absorption spectroscopic method [8].

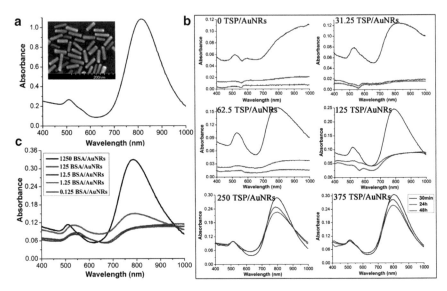

Fig. 19.4 Characterization of AuNRs dispersed in water and culture media. **a** The UV-Vis-NIR absorption spectrum of AuNRs dispersed in water. The inserted SEM image shows the morphology of AuNRs deposited from the aqueous solution. **b** Absorption spectra of AuNRs dispersed in serum containing media (SCM) with 0, 2.5, 5, 10, 20 and 30% of fetal bovine serum (FBS) as a function of incubation time. Concentration of AuNRs in the media was 120 pM. The corresponding ratios of total serum proteins (TSP) to AuNRs (TSP/AuNRs) were 0, 31.25, 62.5, 125, 250 and 375. **c** The absorption spectra of AuNRs dispersed in basic media containing different content of only bovine serum albumin (BSA) for 30 min. The ratios of BSA to AuNRs (BSA/AuNRs) were 1250, 125, 12.5, 1.25 and 0.125. Reproduced in compliance with CC-BY 4.0 license, Ref. [8]

Other examples include similar approaches to prostate cancer cells. In correspondence to carcinoma markers, NIR calibration models for the analysis of glucose, lactate, glutamine and ammonia were established by Rhiel et al., respectively [9]. For the calibration, an adaptive procedure was developed aimed at selective removing metabolism-induced covariance between these analytes arising in the cultivar of PC3-human prostate cancer cells. PLSR models were generated from single-beam NIR spectra recorded between 4800 and 4200 cm^{-1}. Calibration models were in the first attempt developed with both the full spectral range and also with selected optimized spectral ranges. The lowest standard errors of prediction were 0.82, 0.94, 0.55 and 0.76 mM, respectively, for glucose, lactate, glutamine and ammonia. It was demonstrated that NIR spectroscopy can be used effectively in the off-line analysis of glucose and glutamine, as the important nutrients, and lactate and ammonia, as the byproducts, present in a serum-based cell culture medium. Two years later in 2004, the same authors reported about the online monitoring of human prostate cancer cells in a perfusion rotating wall vessel by NIRS [10]. In that study, a perfusion vessel volume, equipped with a silicone membrane oxygenator, peristaltic pump and liquid handling manifold, was installed. For retaining the cells, a 100-μm polypropylene filter was used and separation of cells from other parts was achieved by rotation.

Recording of spectra and establishment of calibration and validation procedures was carried out in the same manner as in the previous contribution published by the same authors [9].

Interestingly, the problem of biological cells appearance in the sample was investigated as a factor potentially influencing the performance of NIR spectroscopy in food analytical applications. For example, Tsenkova et al. examined the influence of high somatic cell count (SCC) in non-homogenized cow milk on the accuracy of NIR spectroscopic determination of fat, protein and lactose content [11]. Transmittance spectra of 258 milk samples were analyzed in SW-NIR, $14,285–9090$ cm^{-1} (700–1100 nm) region. The most accurate calibrations, evaluated through analyzing the standard error of prediction and the correlation coefficient, were obtained for the samples with low SCC. The accuracy decreased notably in the scenario, where calibration models constructed on the basis of low SCC milk were used to predict the content of the examined components in samples with high SCC, and vice versa. Therefore, SCC factor is meaningful and highly influences the accuracy of fat, protein and lactose determination. This dependence strongly affects robustness of analysis and needs to be taken into consideration during the determination of milk chemical composition by NIR spectroscopy.

19.4   Serum Analysis

Nioka et al. employed NIR spectroscopy in their approach to test breast tumor-bearing patients who are undergoing a biopsy [12]. The aim was to see if angiogenesis and hypoxia can be used as meaningful factors in detecting cancer. In that attempt, continuous short-wave NIR (SW-NIR) spectroscopy was employed to measure blood hemoglobin concentration and to obtain blood volume. This would allow answering the question, whether the correlated parameters, the total hemoglobin content and oxygen saturation, can serve as the biomarkers for the angiogenesis and hypoxia. Through monitoring these two parameters, high total hemoglobin and hypoxia score, the sensitivity and specificity of cancer detection could be achieved at 60.3% and 85.3% levels, respectively. It was concluded that smaller-size tumors prove to be more challenging for detection by NIR spectroscopy, whereas ductal carcinoma in situ (DCIS) can be detected using configuration assumed in the discussed study. It was noted that in larger-size tumors, there is significantly higher deoxygenation in invasive and ductal carcinoma in situ DCIS than in that of benign tumors [12].

Blood-oxygen-level-dependent contrast functional magnetic resonance imaging (BOLD-fMRI) is a favored tool for detection of brain cancer. However, this technique faces some limitations. The BOLD-fMRI diagnosis in brain disorders such as stroke and brain had been shown in the previous studies to be prone to yield incorrect image activation areas correctly in such cases. Sakatani et al. performed an investigation upon the application of NIR spectroscopy for this purpose [13]. To clarify the characteristics of the cerebral blood oxygenation (CBO) changes occurring in stroke and brain tumors, the authors have been comparing

NIR spectroscopy and BOLD-fMRI recording during functional brain activation in patients. Noteworthy, NIR spectroscopy delivered good diagnostic performances in the cases, in which BOLD-fMRI performed poorly. It was, therefore, concluded that a combined use of both techniques could lead to a higher level of accuracy in the functional imaging of diseased brains [13].

Kasemsumran et al. performed a series of investigations of the analytical capability of NIR spectroscopy in the analysis of human serum albumin (HSA), γ-globulin and glucose for the needs of biomedical purposes. These studies demonstrated the potential for simultaneous determination of HSA, γ-globulin, and glucose by NIR spectroscopy in a model phosphate buffer solution and in a control serum solution that represents a complicated biological fluid [14, 15]. In the study using phosphate buffer solution, five levels of full factorial design were used to prepare a sample set consisting of 125 samples of three component mixtures with various concentrations and examined at 37 °C. The spectral dataset was analyzed using moving-window partial least squares regression (MW-PLSR), which determined the spectral ranges of 4648–4323, 4647–4255 and 4912–4304 cm^{-1} as the most informative and correlated with the content of the targeted molecules (Fig. 19.5) [14]. Subsequently, the analysis of HSA, γ-globulin and glucose in a more complex matrix, the control serum solution, was attempted using an evolutionary chemometric method, searching combination moving window partial least squares (SCMW-PLS). In that study, the control serum IIB (CS IIB) solutions with various concentrations were prepared, and NIR spectroscopy supported by SCMW-PLS was able to successfully determine simultaneously the concentrations of HSA, γ-globulin and glucose in a complex biological fluid [14].

19.5 Saliva Analysis

A diagnostic method based on NIR spectroscopy has been proposed by Murayama et al. for oral cancer detection from one drop of saliva without any specific diagnosis marker [16]. In that study, the NIR spectra of one drop of saliva were measured using a capillary tube method. Principal component analysis with the second and third factors calculated with the second-derivative NIR spectra clearly discriminated between the two groups.

Application of NIR spectroscopy to measurement of hemodynamic signals accompanying stimulated saliva secretion was demonstrated by Sato et al. [17]. That study aimed to explore the feasibility of indirect measurement of human saliva secretion in response to taste stimuli for potential application to organoleptic testing. NIR spectroscopy was used to monitor extracranial hemodynamics, through Hb signals around the temples, of healthy participants upon application of taste stimuli. Functional magnetic resonance imaging (fMRI) was used to provide reference. Statistical analysis indicated that the Hb response and saliva volume are greater upon giving sucrose solution than distilled water to the test group. It was concluded that NIR

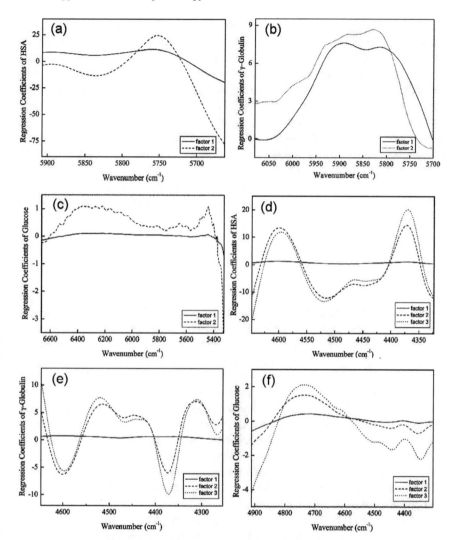

Fig. 19.5 Regression coefficients for the PLS model predicting the analytes concentrations in the study reported by Kasemsumran et al. [Ref. 14]. **a** Factor 1 and factor 2 of HSA in the 5907–5663 cm^{-1} region; **b** factor 1 and factor 2 of γ-globulin in the 6080–5700 cm^{-1} region and **c** factor 1 and factor 2 of glucose in the 6665–5325 cm^{-1}; **d** factor 1, factor 2 and factor 3 of HSA in the 4648–4323 cm^{-1} region; **e** factor 1, factor 2 and factor 3 of γ-globulin in the 4647–4255 cm^{-1} region; **f** factor 1, factor 2 and factor 3 of glucose in the 4912–4304 cm^{-1} region. Reproduced from Ref. [14] with permission from The Royal Chemical Society

spectroscopy is effective in assessing hemodynamic signals accompanying stimu-lated saliva secretion [17]. NIR spectroscopy appears to be a promising tool for investigation of saliva and processes accompanying swallowing, as this research lane is continuously explored. For instance, recently, Kober and Wood compared the hemodynamic response observed during swallowing of water or saliva using NIR spectroscopy [18]. Relative concentration changes were evidenced in oxygenated and deoxygenated hemoglobin during swallowing. NIR spectroscopy demonstrated high sensitivity to topographical distribution and time course of the hemodynamic response between distinct swallowing tasks. The authors concluded that this approach shows potential for application in diagnostic practice and in supporting therapy for swallowing difficulties [18]. Further examples of the potential that NIR spectroscopy bears in analysis of body fluids are provided in the chapter discussing medicinal applications.

19.6   Tissue Analysis

Tissue analysis is one of the major applications fields of spectroscopy in biomed-ical sciences. This topic is exhaustively discussed in another chapter of this book that discusses medicinal applications; therefore, only basic information is presented here. For tissue analysis, NIR spectroscopy should be partitioned according to short-wave (SW; 750–1100 nm) and long-wave (LW; 1100–2500 nm) NIR wavelength intervals. At short NIR wavelengths, the hem proteins (hemoglobin, myoglobin and oxy-derivatives) and cytochrome of the tissue dominate the spectra and provide infor-mation concerning tissue blood flow, oxygen saturation and consumption, and the redox status of the enzymes. In the LW-NIR region, the observed absorptions are caused by combinations and overtones of vibrations involving hydrogen-containing molecular substructures. Valuable information concerning the chemical composition of the tissue with its main components of lipids, proteins, carbohydrates and water can be gathered from LW-NIR region.

Most of the NIR investigations dealing with human tissues were on breast cancer. Quantitative chemical information from breast tissue based on oxy-hemoglobin and deoxy-hemoglobin, water and lipids has been reported [19]. From these parame-ters, total hemoglobin concentration and tissue hemoglobin oxygen saturation were calculated and are expected to provide information on tumor angiogenesis and hyper-metabolism.

19.7   Hemodialysis Analysis

Henn et al. described how hemodialysis monitoring can be performed using MIR and NIR spectroscopy with PLSR as the data analytical algorithm [20]. The study aimed to evaluate the feasibility of both techniques and compares their performances. Blood constituents such as urea, glucose, lactate, phosphate and creatinine are important for monitoring the process of detoxification, especially in ambulant dialysis treatment. Henn et al. compared these two different vibrational spectroscopic techniques to determine the targeted molecules quantitatively in artificial dialysate solutions. The goal of the study was to compare the definitive suitability of NIR and MIR spectroscopy for this purpose. These methods were compared directly by means of statistical errors determined in PLSR analysis, while using the same sample set. Interestingly, Henn et al. presented a detailed analysis of the structure of the PLSR vector developed for quantification of the target analytes in the sample on the basis of MIR and NIR spectra (Fig. 19.6). This comparison demonstrates that relatively

Fig. 19.6 NIR (Panel I) and MIR (Panel II) spectra of the calibration samples used by Henn et al. [Ref. 20]. In Panel I are presented: raw NIR absorbance spectra (**A**); regression vector intensity in percent referenced to the maximum for glucose (**B**) and urea (**C**). In Panel II are presented: raw difference MIR spectra (**A**); the regression vectors in % intensities for urea (**B**), glucose (**C**), lactate (**D**), phosphate (**E**) and creatinine (**F**). Reproduced in compliance with CC-BY 4.0 license, Ref. [20]

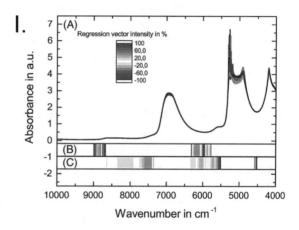

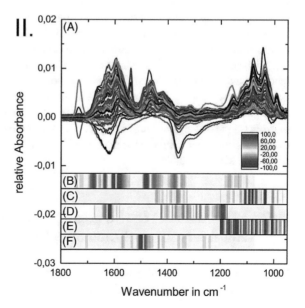

few NIR wavenumbers meaningful for regression of glucose and urea concentration in artificial dialysate solutions are found, and these wavenumbers are located in the regions free from a strong absorption of water (Fig. 19.6-I). Multilevel/multifactor design was employed to cover the relevant concentration variations during dialysis. The results demonstrated that MIR spectroscopy is better suited to analyze the molecules of interest. When employed in a multi-reflection ATR mode, it enables reliable prediction of all target analytes. In contrast, the NIR spectroscopic method did not give access to all five components but only to urea and glucose. However, it offered advantages of practical nature, such as easy sampling. For both methods, coefficients of determination R^2 are greater or equal to 0.86, as elaborated in the test-set validation process for urea and glucose. The method applied to the analysis of lactate, phosphate and creatinine performed well in the MIR with $R^2 \geq 0.95$ using test-set validation (Table 19.2). This study indicates that there exists room for improvement in the performance levels of NIR spectroscopy applied to hemodialysis analysis (Table 19.2).

19.8 Examination of Entire Organisms

As mentioned, the underlying physical principles of NIR spectroscopy make it relatively more suitable for interrogation of high-volume samples such as entire biological organisms. As a good example, the properties of fish embryo have recently been comprehensively examined in vivo at the molecular level by Ishigaki et al. [21]. The authors used NIR spectroscopy and imaging for monitoring of the growth of fertilized eggs of Japanese medaka fish. This approach enabled non-destructive examination of the inner components such as proteins, lipids and water, over the 6200–4000 cm^{-1} region. Changes in chemical structure of oil droplets and egg yolk over the time period from the first day after fertilization to the day before hatching were monitored (Fig. 19.7). The study demonstrated that NIR spectroscopy can decipher signs of hatching and metabolic changes in the egg non-invasively. It was revealed that the percentage of strongly hydrogen-bonded water in the oil droplets is larger than in other parts and that yolk has quite different water environments from those found in embryo parts. Furthermore, insights into secondary structure of proteins were obtained. From characteristic bands at 5756 and 4530 cm^{-1} appearance of membrane structures was concluded.

19.9 NIR Studies of the Structure, Properties and Interactions of Biomolecules

Exhaustive presentation of NIR spectroscopy in elucidating information on the molecular structure, properties and interactions is included in Chapter 13 Overview of

Table 19.2 PLSR results for the five-component mixture in dialysate derived from MIR and NIR spectra

Model			Factor	R^2	RMSECV in mg/dL	RMSEP in mg/dL	LOD_{min} in mg/dL	LOD_{max} in mg/dL	LOQ_{min} in mg/dL	LOQ_{max} in mg/dL
Urea	CV	NIR	4	0.97	12	–	10	24	29	72
		MIR	4	0.99	7.9	–	10	18	31	55
	TV	NIR	4	0.98	–	19	–	–	–	–
		MIR	5	0.99	–	6.6	–	–	–	–
Glucose	CV	NIR	4	0.89	37	–	36	73	108	218
		MIR	3	0.96	22	–	47	142	140	428
	TV	NIR	4	0.86	–	54	–	–	–	–
		MIR	2	0.99	–	11	–	–	–	–
Lactate	CV	NIR	–	–	–	–	–	–	–	–
		MIR	5	0.95	8.2	–	28	90	84	271
	TV	NIR	–	–	–	–	–	–	–	–
		MIR	8	0.99	3.0	–	–	–	–	–
Phosphate	CV	NIR	–	–	–	–	–	–	–	–
		MIR	8	0.99	1.1	–	1.0	2.6	3.0	7.9
	TV	NIR	–	–	–	–	–	–	–	–
		MIR	8	0.95	–	2.0	–	–	–	–
Creatinine	CV	NIR	–	–	–	–	–	–	–	–
		MIR	5	0.98	1.8	–	2.6	4.5	7.9	13
	TV	NIR	–	–	–	–	–	–	–	–
		MIR	4	0.96	–	2.1	–	–	–	–

– ... value not available

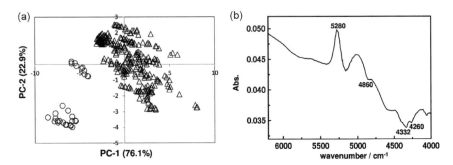

Fig. 19.7 **a** PC-1 and PC-2 PCA scores plot of all NIR spectra of yolk measured over the period from the first day after fertilization until the day before hatching. △ indicates data from the first to the tenth day and ① denotes data from the day before hatching. **b** Loading plot of PC-1. Reproduced from Ref. [21] in compliance with CC-BY 4.0 license.

application of NIR spectroscopy to physical chemistry; therefore, only brief overview of this field in the context of selected biomolecules is discussed here. Biomolecules are typically complex molecules and they tend to interact with their chemical neighborhood which further complicated their NIR spectra. However, by applying sophisticated methodology, one can obtain valuable information on the behavior of biomolecules. For example, Watanabe et al. employed perturbation-correlation moving-window two-dimensional correlation analysis (PCMW2D) method to monitor the temperature-dependent structural changes in hydrogen bonds occurring in microcrystalline cellulose (MCC) [22]. This approach allowed deducting from NIR and IR spectra that in the temperature range of 25–130 °C, structural changes occur gradually in the strong hydrogen bonds in MCC; the extent of these changes becomes greater above 130 °C. It was concluded that intermediate strength and weak hydrogen bonds arise from the structural changes between 40–90 °C, whereas the appearance of very weak hydrogen bonds becomes dominant above 90 °C. Additionally, PCMW2D correlation analysis enabled band assignments for the first overtone region, and OH groups of MCC exemplifying different hydrogen bonding strength could have been identified. The results of that study enabled further investigations into water adsorption onto MCC [23]. NIR spectroscopy combined with PCMW2D and PCA methods was applied to interrogate a sample set of MCC with the moisture content ranging in 0.2–13.4 wt%. The chosen data analytical methods helped to distinguish OH stretching bands, which heavily overlap in the NIR region due to contributions from MCC and water. Nonetheless, it could have been concluded that a decrease in the free or weakly hydrogen-bonded and an increase in the strong hydrogen-bonded OH groups of MCC occur, with the increase of moisture content. At the same time, an increase of the water adsorbed on MCC was observed. These results suggest that the inter- and intrachain hydrogen bonds of MCC are formed by monomeric water molecule adsorption. The study revealed that ca. 3–7 wt% of adsorbed water is responsible for the stabilization of the hydrogen-bond network in MCC at the cellulose–water surface [23].

NIR spectroscopy is a potent tool in exploring the complex properties of proteins. Protein research by NIR spectroscopy includes several significant contributions, e.g., analysis of the secondary structure [24]. Furthermore, this technique finds unique usefulness in investigating hydration process of proteins. Monitoring changes that occur in hydration as well as that in the protein secondary structure at the same time is possible by NIR spectroscopy; in contrast, IR and Raman spectroscopies can hardly investigate these properties simultaneously. The potential of NIR spectroscopy to investigate proteins in aqueous environment and the appropriate methodology can be presented on the example from literature [25, 26]. Murayama et al. [25] performed a comprehensive comparison of the methods for analyzing NIR spectra that are suitable for investigating proteins in aqueous solution. Conventional spectral analysis methods, chemometrics (PCA) and 2D-COS spectroscopy were evaluated in that case. The study was based on the NIR spectra of human serum albumin (HSA) in aqueous solutions within the concentration range of 0.5–5.0 wt%. It was concluded that basic conventional methods of spectra pretreatment and analysis, such as second-derivative and difference spectra, remain critically important for analysis of protein in relatively low concentration in water. For example, the difference spectra unveiled that various species of water are responsible for the observed gradual concentration-dependent changes in the broad feature in the 7100–6500 cm^{-1}. PCA is more resistant against spectral noise; however, 2D-COS is more informative on the correlations between individual bands and also elucidates sequences of spectral changes. Therefore, the best approach is to combine various methods, as this yields highest potential for interpretation of spectral variability.

With a similar aim, this study has been continued by Yuan et al. who compared different methods for treating NIR spectra of bovine serum albumin (BSA) [26]. However, this time the source of spectral variability was the temperature perturbation (45–85 °C), while concentration of protein was constant at 5.0 wt% and the pH of the sample was 6.8. The evaluated methods were extended by the addition of chemometric algorithm of evolving factor analysis (EFA). That study confirmed the previous conclusions about the usefulness of conventional methods of spectral analysis and the significance of combined use of various approaches. Namely, the difference spectra were essential in finding the change in protein hydration that occurs in the temperature range of 61–65 °C. However, this finding was supported by analyzing the temperature profile through three-factor EFA in the 7400–6400 cm^{-1} region. The investigation has also revealed that the structural variation of BSA in the aqueous solution just precedes the change in the protein hydration, indicating that the change in the hydration is initiated by the structural modifications in the protein itself. For yielding these deeper insights from NIR spectra, application of EFA combined together with the other methods was essential. Note, the spectral variability observed in NIR region upon concentration change of HSA protein in water differs from the one observed upon temperature change in aqueous solution of BSA.

NIR spectral bands of proteins can be used to follow complex biological processes in vivo, such as embryonic development, [27] for example. The current state of the art of protein research by NIR spectroscopy is exhaustively covered in a book chapter by Ishigaki and Ozaki [28]. As mentioned in introduction and in the chapter referred above, NIR bands can deliver unique information on the properties of such molecules in their native environment. The above examples demonstrate that interpretability of NIR spectra of complex molecules such as biomolecules remains a challenge, and often, only speculative NIR band assignments are available, with selected ones being resolved, e.g., the intense and strongly affected by intermolecular interaction OH stretching bands. Recent advances in quantum chemical calculation of NIR spectra of biomolecules should be highlighted, [1, 29–32] which bring significant progress in the interpretability of their NIR absorption, as well as the insight into how it is influenced by intermolecular interactions. Rapid development of the applicability of these new methods for improving our understanding of NIR spectra of biomolecules should be expected in the near future [1].

19.10   Selected Other Applications

The attributes of NIR spectroscopy have been recognized in ecology and environmental studies. Most often, in such applications, samples of biological origin are investigated. Moreover, spectroscopic analysis is in its essence chemical reagent-free, and as such, spectroscopy itself is environmental friendly. A focused review of the role of NIR spectroscopy in modern research in the field of ecology and environment is available in the recent literature [33].

Over the last few years, entirely, new possibilities have emerged as the result of the progress in unmanned aerial vehicle (UAV, i.e., airborne drones) technology. Such accomplishment became possible on the basis of breakthroughs made in the past decade, with sensor miniaturization, new low-power technology and progress in spectral data processing. There appears an enormous potential from deploying spectroscopic sensors mounted on drones. Applications of airborne NIR sensors installed on UAV are recently strongly advancing in agriculture and environmental studies. Such configuration enables unparalleled high-throughput capability and remote sensing of large Earth surface areas. The current evolution trends are aiming toward real-time monitoring and imaging by using airborne NIR spectrometers. A comprehensive overview of the current state-of-the-art airborne spectroscopy, including NIR, is available in the recent literature [34].

It is worthwhile to mention a narrow field of bio-significant research at which NIR spectroscopy has remarkable accomplishments; investigation of water structure and properties. While not an organic matter, water is an essential biological environment and its properties as well as interactions with other molecules, and biomolecules in particular, are critical for our understanding of biological processes. At this direction, NIR spectroscopy delivered unique insights. Noteworthy is the pioneering research by Segtnan et al. on the structure of water (Fig. 19.8), [35] and combined studies of NIR and IR spectra have also been carried out [36]. Notably, water absorption is comparatively weaker in NIR region than IR region, and simultaneous observation of

Fig. 19.8 Effect of temperature perturbation of water observed in the NIR spectra by Segtnan et al. [35]. Thirty-eight NIR spectra of water measured over a temperature range of 6–80 °C at 2 °C increments: **a** untreated spectra; **b** mean normalized spectra; **c** selected second-derivative spectra derived from B. Adopted with permission from Ref. [35]. Copyright (2001) American Chemical Society

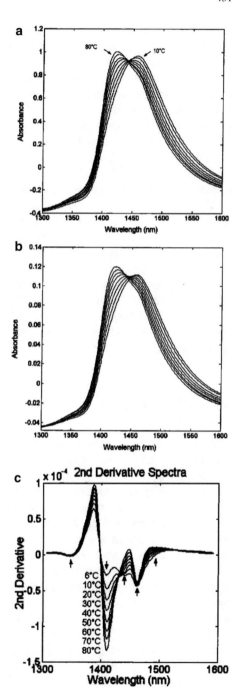

the characteristic bands of the solvent and soluted molecules is possible. A comprehensive review of the NIR research on water structure and properties is available in the recent literature [37].

A novel way of interpretation of the complex NIR spectra of biomolecules is available through quantum chemical calculation [1]. A comprehensive presentation of this topic is available in Chapter 5 of this book, Introduction to Quantum Vibrational Spectroscopy; therefore, accomplishments essential to further development of bio-applications of NIR spectroscopy are highlighted here. Recently, NIR spectra of several biomolecules such as short-, [29] medium-, [30] and long-chain [31] fatty acids, as well as nucleic acid bases, [32] were successfully reproduced with these methods and their absorption bands could have been comprehensively explained by Beć, Grabska and co-workers [1, 29–32]. This approach was also helpful in interpreting the meaningful wavenumbers in PLSR model of bio-active compounds in plant medicines. Few important bio-active constituents of medicinal plants and natural products have been examined by this approach, e.g., thymol [38] and RA [3]. The studies of thymol supported by spectra simulation yielded fundamental findings about the relationship between the specific vibrational modes and the features of PLS regression coefficients vector [38]. This approach is essential for improving the inherently inferior chemical specificity, which is one of the few properties of NIR spectroscopy at which it exemplifies a great room for improvements in comparison with IR or Raman spectroscopy.

19.11 Conclusions

NIR spectroscopy in the bio-fields offers a huge potential in various applications following its advantages: wide applicability to variety of samples, capability of examining moist samples, flexible instrumentation including miniaturized sensors. Accompanied by advanced chemometric data analytical tools, NIR spectroscopy has been proved to be of great value in various bio-scientific investigations. On the other hand, in certain other fields, it is still a developing discipline, with room for improvement as compared with other techniques. For example, in bioanalytical research and medical diagnosis, it still faces strong competition from IR and Raman spectroscopy. However, the recent literature indicates that NIR spectroscopy steadily conquers this demanding field of application. In the near future, additional support might come by quantum chemical simulation of spectra. New achievements accomplished at this field enable improving the interpretability of NIR spectra; shortening the gap between this technique and highly chemical specific IR or Raman spectroscopy. Novel hand-held NIR spectrometers are indispensable in on-site examination of medicinal plants with aim to optimize the cultivation conditions and ensure highest quality of natural drugs. Progress in the instrumentation enabled engineering remote NIR sensors as well. Airborne, UAV-mounted NIR spectrometers become increasingly important in environmental monitoring, where, e.g., large amount of data on is collected on flora and fauna from wide areas. In summary, bio-applications of NIR spectroscopy can be expected to thrive in the forthcoming decade, with continuation of those strongly advancing lanes of research and possibly an appearance of entirely new ones.

References

1. K.B. Beć, C.W. Huck, Breakthrough potential in near-infrared spectroscopy: spectra simulation. A review of recent developments. Front. Chem. **7**, 48 (2019)
2. W. Huck, K.B. Beć (eds.), *Advances in Near Infrared Spectroscopy and Related Computational Methods* (MDPI, Basel, 2020)
3. G. Kirchler, C.K. Pezzei, K.B. Beć, S. Mayr, M. Ishigaki, Y. Ozaki, C. W. Huck, Critical evaluation of spectral information of benchtop vs. portable near-infrared spectrometers: quantum chemistry and two-dimensional correlation spectroscopy for a better understanding of PLS regression models of the rosmarinic acid content in Rosmarini folium, Analyst **142**, 455–464 (2017)
4. K. Pezzei, S.A. Schönbichler, C.G. Kirchler, J. Schmelzer, S. Hussain, V.A. Huck-Pezzei, M. Popp, J. Krolitzek, G.K. Bonn, C.W. Huck, Application of benchtop and portable near-infrared spectrometers for predicting the optimum harvest time of Verbena officinalis. Talanta **169**, 70–76 (2017)
5. S. Delueg, C.G. Kirchler, M. Meischl, Y. Ozaki, M.A. Popp, G.K. Bonn, C.W. Huck, At-line monitoring of the extraction process of rosmarini folium via wet chemical assays, UHPLC analysis, and newly developed near-infrared spectroscopic analysis methods. Molecules **24**, 2480 (2019)
6. V.A. Huck-Pezzei, L.K. Bittner, J.D. Pallua, H. Sonderegger, G. Abel, M. Popp, G.K. Bonn, C.W. Huck, A chromatographic and spectroscopic analytical platform for the characterization of St John's wort extract adulterations. Anal. Methods **5**, 616–628 (2013)
7. N. Wu, C. Zhang, X. Bai, X. Du, X., Y. He, Discrimination of chrysanthemum varieties using hyperspectral imaging combined with a deep convolutional neural network. Molecules **23**, 2831 (2018)
8. W. Zhang, Y. Ji, J. Meng, X. Wu, H. Xu, Probing the behaviors of gold nanorods in metastatic breast cancer cells based on UV-vis-nir absorption spectroscopy. PLoS ONE **7**, e31957 (2012)
9. M.H. Rhiel, M.B. Cohen, D.W. Murhammer, M.A. Arnold, Nondestructive near-infrared spectroscopic measurement of multiple analytes in undiluted samples of serum-based cell culture media. Biotechnol. Bioeng. **77**, 73–82 (2002)
10. M.H. Rhiel, M.B. Cohen, M.A. Arnold, D.W. Murhammer, On-line monitoring of human prostate cancer cells in a perfusion rotating wall vessel by near-infrared spectroscopy. Biotechnol. Bioeng. **86**, 852–861 (2004)
11. R. Tsenkova, S. Atanassova, Y. Ozaki, K. Toyoda, K. Itoh, Near-infrared spectroscopy for biomonitoring: influence of somatic cell count on cow's milk composition analysis. Int. Dairy J. **11**, 779 (2001)
12. S. Nioka, M. Shnall, E. Conant, S.C. Wang, V.B. Reynolds, B.C. Ching, J.H.T. Swan, P.C. Chung, L. Cheng, D. Shieh, Y. Lin, C. Chung, S.H. Tseng, B. Chance, Breast cancer detection of large size to DCIS by hypoxia and angiogenesis using NIRS. Adv. Exp. Med. Biol. **789**, 211–219 (2013)
13. K. Sakatani, Y. Murata, N. Fujiwara, T. Hoshino, S. Nakamura, T. Kano, Y. Katayama, Comparison of blood-oxygen-level–dependent functional magnetic resonance imaging and near-infrared spectroscopy recording during functional brain activation in patients with stroke and brain tumors. J. Biomed Opt. **12**, 062110 (2007)
14. S. Kasemsumran, Y.P. Du, K. Murayama, M. Huehne, Y. Ozaki, Simultaneous determination of human serum albumin, γ-globulin, and glucose in a phosphate buffer solution by near-infrared

spectroscopy with moving window partial least-squares regression. Analyst **128**, 1471–1477 (2003)

15. S. Kasemsumran, Y.P. Du, K. Murayama, M. Huehne, Y. Ozaki, Near-infrared spectroscopic determination of human serum albumin, γ-globulin, and glucose in a control serum solution with searching combination moving window partial least squares. Anal. Chim. Acta **512**, 223–230 (2004)

16. K. Murayama, M. Tomida, Y. Ootake, T. Mizuno, J.-I. Ishimaru, Principal component analysis for diagnosis of oral cancer using capillary near-infrared spectroscopy of onedrop of human saliva. ITE Lett. Batter. New Technol. Med. **6**, 603–606 (2005)

17. H. Sato, A. Obata, Y. Yamamoto, M. Kiguchi, K. Kubota, H. Koizumi, I. Moda, K. Ozaki, T. Yasuhara, A. Maki, Application of near-infrared spectroscopy to measurement of hemodynamic signals accompanying stimulated saliva secretion. J. Biomed. Opt. **16**, 047002 (2011)

18. E. Kober, G. Wood, Hemodynamic signal changes during saliva and water swallowing: a near-infrared spectroscopy study. J. Biomed. Opt. **23**, 015009 (2018)

19. B.J. Tromberg, A. Cerussi, N. Shah, M. Compton, A. Durkin, D. Hsiang, J. Butler, R. Mehta, Imaging in breast cancer: Diffuse optics in breast cancer: detecting tumors in pre-menopausal women and monitoring neoadjuvant chemotherapy. Breast Cancer Res. **7**, 279 (2005)

20. R. Henn, C.G. Kirchler, Z.L. Schirmeister, A. Roth, W. Mäntele, C.W. Huck, Hemodialysis monitoring using mid- and near-infrared spectroscopy with partial least squares regression. J. Biophotonics **11**, 201700365 (2018)

21. M. Ishigaki, S. Kawasaki, D. Ishikawa, Y. Ozaki, Near-infrared spectroscopy and imaging studies of fertilized fish eggs: in vivo monitoring of egg growth at the molecular level. Sci. Rep. **6**, 20066 (2016)

22. Watanabe, S. Morita, Y. Ozaki, Temperature-dependent structural changes in hydrogen bonds in microcrystalline cellulose studied by infrared and near-infrared spectroscopy with perturbation-correlation moving-window two-dimensional correlation analysis. Appl. Spectrosc. **60**, 611–618, (2006)

23. Watanabe, S. Morita, Y. Ozaki, A study on water adsorption onto microcrystalline cellulose by near-infrared spectroscopy with two-dimensional correlation spectroscopy and principal component analysis. Appl. Spectro. **60**, 1054–1061 (2006)

24. K.I. Izutsu, Y. Fujimaki, A. Kuwabara, Y. Hiyama, C. Yomota, N. Aoyagi, Near-infrared analysis of protein secondary structure in aqueous solutions and freeze-dried solids. J. Pharm. Sci. **95**, 781–789 (2006)

25. K. Murayama, B. Czarnik-Matusewicz, Y. Wu, R. Tsenkova, Y. Ozaki, Comparison between conventional spectral analysis methods, chemometrics, and two-dimensional correlation spectroscopy in the analysis of near-infrared spectra of protein. Appl. Spectro. **54**, 978–985 (2000)

26. Yuan, K. Murayama, Y. Wu, R. Tsenkova, X. Dou, S. Era, Y. Ozaki, Temperature-dependent near-infrared spectra of bovine serum albumin in aqueous solutions: spectral analysis by principal component analysis and evolving factor analysis. Appl. Spectro. **57**, 1223–1229 (2003)

27. M. Ishigaki, T. Nishii, P. Puangchit, Y. Yasui, C.W. Huck, Y. Ozaki, Noninvasive, high-speed, near-infrared imaging of the biomolecular distribution and molecular mechanism of embryonic development in fertilized fish eggs. J. Biophotonics **11**, e201700115 (2018)

28. M. Ishigaki, Y. Ozaki, *Near-infrared spectroscopy and imaging in protein research*, in *Vibrational Spectroscopy in Protein Research*, eds. by Y. Ozaki, M. Baranska, B.R. Wood, I. Lednev (Elsevier, 2020)

29. J. Grabska, M. Ishigaki, K.B. Beć, M.J. Wójcik, Y. Ozaki, Structure and near-infrared spectra of saturated and unsaturated carboxylic acids. An insight from anharmonic DFT calculations. J. Phys. Chem. A, **121**, 3437–3451 (2017)

30. J. Grabska, K.B. Beć, M. Ishigaki, M.J. Wójcik, Y. Ozaki, Spectra-structure correlations of saturated and unsaturated medium-chain fatty acids. Near-infrared and anharmonic DFT study of hexanoic acid and sorbic acid. Spectrochim. Acta A **185**, 35–44 (2017)

31. J. Grabska, K.B. Beć, M. Ishigaki, C.W. Huck, Y. Ozaki, NIR spectra simulations by anharmonic DFT-saturated and unsaturated long-chain fatty acids. J. Phys. Chem. B **122**, 6931–6944 (2018)
32. K.B. Beć, J. Grabska, Y. Ozaki, M.A. Czarnecki, C.W. Huck, Simulated NIR spectra as sensitive markers of the structure and interactions in nucleobases. Sci. Rep. **9**, 17398 (2019)
33. K.R. Counsell, C.K. Vance, Recent advances of near infrared spectroscopy in wildlife and ecology studies. NIR News **27**, 29–32 (2016)
34. K.C. Elliot, R. Montgomery, D.B. Resnik, R. Godvin, T. Mudumba, J. Booth, K. Whyte, Drone use for environmental research. IEEE Geosc. Rem. Sen. M. **7**, 106–111 (2019)
35. V.H. Segtnan, S. Šašić, T. Isaksson, Y. Ozaki, Studies on the structure of water using two-dimensional near-infrared correlation spectroscopy and principal component analysis. Anal. Chem. **73**, 3153–3161 (2001)
36. S. Czarnik-Matusewicz, J.P. Pilorz, Hawranek, Temperature-dependent water structural transitions examined by near-IR and mid-IR spectra analyzed by multivariate curve resolution and two-dimensional correlation spectroscopy. Anal. Chim. Acta **544**, 15–25 (2005)
37. J. Muncan, R. Tsenkova, Aquaphotomics—From innovative knowledge to integrative platform in science and technology. Molecules **24**, 2742 (2019)
38. K.B. Beć, J. Grabska, C.G. Kirchler, C.W. Huck, NIR spectra simulation of thymol for better understanding of the spectra forming factors, phase and concentration effects and PLS regression features. J. Mol. Liq. **268**, 895–902 (2018)

Chapter 20
Medical Applications of NIR Spectroscopy

Herbert Michael Heise

Abstract In recent years, near-infrared (NIR) spectroscopy has seen much progress in instrumentation and measurement techniques. It has been used for monitoring in many fields of analytical spectroscopy. Examples are the characterization of materials from processes of the chemical and pharmaceutical industry in addition to the broad field of food industrial and biotechnological applications. Another important field for NIR spectroscopy is found within the medical sciences with topics such as clinical chemistry, sensing and monitoring of changes of homeostasis of the body, with biofluids and tissues from many organs involved. Here, *in vitro* laboratory work and *in vivo* monitoring must be mentioned. Regarding instrumentation, laboratory analyzers are kept firmly in our view, but point-of-care (POC) applications need also to be taken into account. Sensing devices for non-invasive measurements on special parameters such as blood glucose and hemoglobin or information on the redox status of tissues is another broad area with oxygenation of hemoglobin and myoglobin as most important parameters. Absorption measurements are most often carried out with transmission and reflection techniques, but due to the synthesis of new marker substances, also fluorescence measurements in the NIR spectral range become more advanced, especially for imaging and immunoassay developments.

Keywords Near-infrared spectroscopy · Medical applications · Clinical chemistry · Spectral histopathology · Oximetry · Monitoring of physiological processes · Optical tomography

20.1 Introduction

Near-infrared spectroscopy has been established as a versatile analytical method for characterizing liquid and solid chemicals as available in various fields, which include for example the food and forage industry, petrochemistry, polymer chemistry and biotechnology. Analytical spectroscopy can provide important information on

H. M. Heise (✉)
Interdisciplinary Center for Life Sciences, South-Westphalia University of Applied Sciences, Frauenstuhlweg 31, 58644 Iserlohn, Germany
e-mail: heise.h@fh-swf.de

© Springer Nature Singapore Pte Ltd. 2021
Y. Ozaki et al. (eds.), *Near-Infrared Spectroscopy*,
https://doi.org/10.1007/978-981-15-8648-4_20

437

composition and properties of materials and products such as polymers, pharmaceutical, agricultural products or beverages to list a few. The use of optical spectroscopy within the near-infrared regime in particular is widespread in analytical chemistry, and chemical process monitoring has received much attention in the past for the aim of optimization and quality control in the aforementioned application fields, especially in industrial processes.

Vibrational spectroscopy is based on molecular vibrations, which are localized in molecular substructures or larger molecular entities. The quantized energies for exciting vibrations within molecules start with longest wavelengths in the far-infrared range, with wavelengths of up to 1000 μm, and reach down to the short-wave NIR (SW-NIR) with a minimum wavelength of 780 nm by definition. NIR spectroscopy is mostly restricted to the study of substances with OH, NH and CH group absorptions, providing spectral patterns with much band overlap, but multivariate chemometric techniques render these best suited for quantitative and qualitative analysis as for raw material identification. The NIR bands arise from combination and overtone vibrations and for reaching reasonable absorbances in transmission measurements or for $-\log$ (reflectance) values of powders and tissues, the optical path in samples can be millimeters or with higher overtones even centimeters.

This part of NIR spectroscopy was for a long time considered a "sleeping giant" before its rapid implementation in applications for the forage and food industry, for petrochemicals, polymers and pharmaceuticals took place especially for quantitative assays. In the last few years, process analytical chemistry, also called process analytical technology (PAT), has been implemented with robust process NIR spectrometers, especially when online data were made available for reaction controlling and product quality monitoring. Other areas and applications within life sciences have also been dealt with in earlier chapters of this handbook.

Similarly, clinical chemistry applications also evolved for the analysis of blood and other body fluid constituents. The late emergence of near-infrared spectroscopy into the clinical chemistry field is mainly due to the complexity of the body fluids under investigation. These contain many analytes of rather low concentrations in the per mille and even lower range apart from total protein or albumin quantification with lower percentage values. Advantageously has been the high reproducibility for the recording of spectra and their high signal-to-noise ratios, which could be accomplished with sensitive photodetectors within Fourier transform or excellent dispersive spectrometers. Early clinical chemistry applications focused on the analysis of blood and blood plasma with blood glucose as the most promising analyte. This was also due to expectations that this technique could finally be used for non-invasive blood glucose monitoring.

Tissue spectroscopy for skin cancer detection or wound healing has also been investigated using the so-called combination band region, showing most spectral information for discrimination and identification. With shorter wavelengths band, half-widths become broader and with less structure, so that the selectivity for analytical assays is usually reduced compared to the combination band region. Within the short-wave near-infrared (SW-NIR) regime, also other phenomena give rise to absorptions: these are biological molecules such as hemoglobin, myoglobin and cytochromes to mention a few compounds of great interest especially for medical practice. Some applications have been established in form of valuable sensors for patient monitoring with the pulse oximeter, representing the most prominent device for measuring arterial oxygen saturation through oxy and deoxyhemoglobin quantification. Further *in vivo* medical spectroscopy was initiated by Frans Jöbsis in the 1970ies, who reported on the high transparency of brain tissue in the optical window to be used for non-invasive measurements on vital physiological parameters [1]. The groups around David Delpy and Brittan Chance had brought much progress into the field of analyte quantification in near-infrared tissue spectroscopy, exploiting different spectral signatures of redox-active substances for brain and muscle oximetry [2, 3]. Finally, different techniques were developed for functional imaging used in monitoring brain activities.

All aforementioned near-infrared spectroscopic applications are based on absorption measurements. Quite different are applications with fluorescence spectroscopy, which is finding more interest in biomedical applications such as for sensitive analyte detection using immunoassays with according fluorescent labels and for bio-imaging by achieving better contrast.

The importance of this chapter is to provide an overview of the different areas with applications of NIR spectroscopy in the medical sciences, including clinical chemistry with its aim for diagnostics using mainly bodyfluids, non-invasive sensing for monitoring dynamic physiological states, and processes and imaging of organs such as skin, brain, heart and others.

20.2 Applications in Clinical Chemistry

Early clinical chemistry applications using near-infrared spectroscopy have followed research work from several groups worldwide, who had started with mid-infrared (MIR) spectroscopy. Due to the high information content especially of the fingerprint region, biofluid analysis could be achieved for analytes of "high" concentrations as mentioned above. The fascination about the spectroscopic methods came from the fact that these worked fast and without reagents and could be exploited for multi-analyte applications. Therefore, it had attracted also companies involved in medical diagnostics and sensing, mainly for the development of analyzer instrumentation.

The rationale behind medical diagnosis by spectroscopy is based on the fact that diseases are accompanied by changes in the biochemistry of the cells and tissues within the organs in our body. Deviations from homeostasis can thus be monitored by analytical spectroscopy of body fluids with physiological parameters usually fluctuating for the healthy state around a normal range. Since spectroscopic methods, in particular vibrational spectroscopy with infrared, near-infrared and Raman, can provide information on biological molecules like proteins and peptides, nucleic acids, carbohydrates, lipids and others, it is a useful tool for medical diagnostics. There are some restrictions for the near-infrared, as mid-infrared and Raman spectroscopy make use of fingerprint region information, which cover many low-wavenumber fundamental and combination bands from molecular entities, which are not visible in the near-infrared spectral range. Despite this, the quantification of individual compounds is still possible and the first step in information gathering for medical diagnosis.

Another goal for disease diagnostics was not only based on blood analyte concentrations but exploiting the information from whole spectrum analysis aiming at "disease pattern recognition." A straightforward classification of a bodyfluid sample with linking to a disease state was then made possible by using multivariate chemometrics. However, thus applications have mostly been observed for mid-infrared and Raman spectroscopy.

20.2.1 Analysis of Blood and Other Bodyfluids

The most analyzed body fluid is certainly whole blood of which several fluids can be derived from such as plasma and serum. This is easily accessible by punctuation or the use of syringes. An important problem with the analysis of biofluids is certainly associated with the strong absorptions of water and its temperature and solute dependency due to the hydrogen bonded network. In Fig. 20.1, absorbance spectra of water are presented for a transmission cell of 1 mm pathlength, showing also the temperature sensitivity in the shorter NIR wavelength region. For the longwave region with combination bands, usually cell pathlengths around 0.5 mm are used, whereas for the short-wave NIR range even 10 mm are required to reach optimal

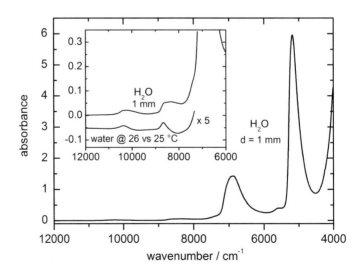

Fig. 20.1 NIR spectra of water for a layer thickness of 1 mm (calculated from data available at http://www.ualberta.ca/~jbertie/JBDownload.htm [4]); the subplot shows the SW-NIR region with a difference spectrum illustrating the temperature sensitivity of the water spectrum

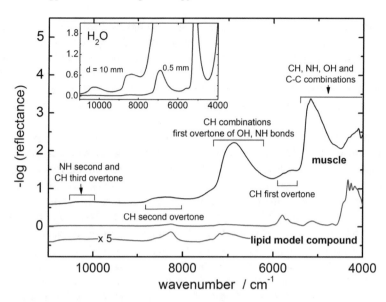

Fig. 20.2 NIR spectrum of muscle tissue as measured in diffuse reflection with band assignments including a spectrum of lecithin as model compound for lipids, whereas the subplot provides absorbance values for water at different layer thicknesses

absorbances for quantitative analysis. For the analysis of soft tissues, e.g., from biopsies or in vivo measurements, the same statement for water as the main constituent is valid, although such spectra, when recorded with diffuse reflection techniques, also show significant impact from tissue scattering. In Fig. 20.2, a model tissue spectrum is shown, elucidating also the different intervals with overtone and combination bands from molecular sub-groups of the main substance classes, i.e., proteins, lipids, carbohydrates and genetic materials, but also glycoproteins and lipoproteins and many others, as found in biomaterials.

One important parameter is certainly blood glucose, since this parameter is one of the most frequently measured analytes within the clinical chemistry laboratory. Water is the major constituent of biomedical samples, and in particular of biofluids, while dry-film preparations have not been popular as seen in mid-infrared spectroscopy, for which often only microliter samples have been prepared by water evaporation. With larger bodyfluid volumes, evaporation needs much more time.

The information content of different spectral regions for clinical chemistry applications has been the subject of many investigations, and one early study was undertaken by the research group of Henry Mantsch. They studied serum dry films prepared on simple glass slides for multi-analyte quantifications, restricting the analysis to absorption bands above 2000 cm^{-1}, thus avoiding fingerprint information. It was shown that several blood parameters such as albumin, cholesterol, glucose, total protein, triglycerides and urea could be analyzed with standard errors allowing even routine clinical analysis. Practical aspects of such an approach were also discussed

[5]. In Fig. 20.3, the drying process of a blood sample is illustrated, for which as substrate gold-coated abrasive paper had been used allowing for a diffuse reflection measurement. Whereas in the beginning of the drying process the spectral features are dominated by the water spectrum, absorption features of blood proteins with albumin and hemoglobin as important representatives, become evident after some time lapses. The similarity of NIR spectra of different proteins will be discussed later.

However, another aspect of NIR spectroscopy of different substances is presented by showing some model compound spectra. In Fig. 20.4, several carbohydrates of special medical importance are shown. As examples of sugars, the monosaccharide glucose and the disaccharide sucrose are shown, all measured as crystalline powders. In addition, two polysaccharides, i.e., cellulose and starch, are also presented. While the crystalline substances show much fine structure due to the crystalline state, the other two have rather broad absorption band signatures. With Fig. 20.5, even quantitative data are provided for glucose as measured in aqueous solution and in a glassy state from preparations starting from high-concentrated syrup samples, showing similar band structures as the polysaccharides with glucose as structural subunits. Starch, for example, consists of two types of polysaccharides, which are the linear and helical amylose and branched amylopectin. Glycogen, as the glucose store of humans, is a more highly branched amylopectin, but with a similar spectrum. In addition to glucose, also a fructose spectrum from a glassy state preparation is shown, illustrating the spectral differences within the combination band region above 4000 cm^{-1}, which is sufficient for quantitative discrimination. This aspect will later be discussed again

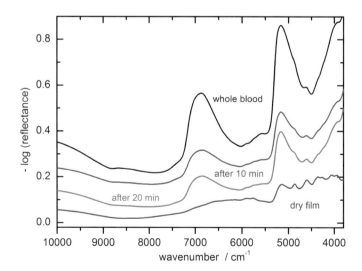

Fig. 20.3 Diffuse reflection spectra of blood during dry-film preparation by water evaporation. Blood samples had been placed on diffusely reflecting substrates (gold-coated abrasive paper) and were measured after different time lapses

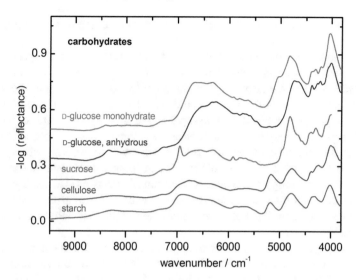

Fig. 20.4 Diffuse reflection spectra of crystalline sugars and two polysaccharides with glucose as structural subunits

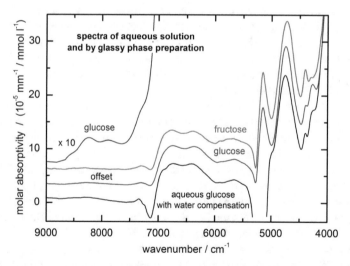

Fig. 20.5 Absorptivities of glucose and fructose obtained from measurements of aqueous solutions and of glassy monosaccharides from syrup after water evaporation (scaled to aqueous phase absorptivities)

for further clinical applications. Such quantitative data for glucose, alanine, ascorbate, lactate, triacetin and urea have been used for evaluating the selectivity of NIR spectroscopic clinical assays, using the combination band (5000–4000 cm^{-1}) and first overtone region (6450–5400 cm^{-1}) or for the development of non-invasive blood glucose assays with different measurement techniques based on skin spectroscopy [6, 7].

Some more information is allowed that the aqueous glucose spectrum was obtained with scaled water absorbance subtraction, which leads to an incomplete compensation due to the disturbance of the water hydrogen-bonding network by the sugar molecules, evident from the spectral dips. Interestingly, the spectral features of glassy sugars, as obtained from syrup samples after careful water evaporation, show the same wavelength dependencies as the aqueous phase spectrum. Using this technique, the otherwise opaque spectral intervals with strong water absorption are now accessible, despite of some still existing uncompensated water absorptions in the regions of the strong water bands. A spectrum of fructose is also displayed, which may be relevant for diabetics in non-invasive glucose measurement after consumption of the other sugar, illustrating discrimination limitations based on the spectral interval used.

For many years, our group has been involved in the development of *in vitro* reagent-free multi-analyte systems for blood analysis based on vibrational spectroscopy and two selected papers, shedding light on the potential of near-infrared spectroscopy for clinical chemistry, need to be discussed [8, 9]. In this context, also various important chemometric aspects such as spectral variable selection or different validation strategies for multivariate calibration modeling have been investigated. In nearly all studies, partial least squares (PLS) calibration has been utilized, but also science-based calibration (SBC) modeling was successfully tested for *in vitro* analysis. The latter approach uses explicitly the quantitative analyte spectral signatures, while other spectral contributions are estimated statistically. For more details, the reader is referred to our book chapter [9].

The usual strategy, applied by many research groups, has been used to demonstrate the spectroscopic assay capability for *in vitro* analysis by having aqueous mixtures of a few compounds such as albumin, glucose, lactate and others. It is certainly advantageous that an elaborate experimental design can be followed to cover an adequate physiological composition with a reasonable spread of differing concentrations, which can be most accurately prepared, for example, by gravimetry. Samples from patients are more complex in composition and with a distribution of analyte concentrations that is usually defined by the natural spread. Such specimens can be selected, for example, from a population of healthy and/or diabetic subjects if blood glucose assays have to be tested and the assay will focus on such kinds of subjects. In some cases, also spiking of samples has been taken into account for increasing the concentration levels or the distribution of concentration values. The assays for obtaining reference concentrations are often limited in precision and accuracy, unless very special effort is undertaken for improved clinical analytics. Usually, such testing scenarios are carried out under special experimental conditions with a tightly thermostated sample cell of constant thickness, which is important especially for near-infrared spectroscopy of aqueous biofluids because of the hydrogen-bonding network of the water molecules that is very sensitive to temperature changes (see again Fig. 20.1).

Table 20.1 Analytical cross-validation results for blood plasma compounds obtained from PLS calibration models based on different spectral ranges and logarithmized single beam data (same plasma sample population from 124 patients) [8]

Analyte	Standard error of prediction			
	SW-NIR[a]	NIR[b]	MIR[c]	Units
Total protein	1.08	1.07	0.90	$g\,l^{-1}$
Total cholesterol	15.4	7.7	8.2	$mg\,dl^{-1}$
Triglycerides	23.4	12.1	10.3	$mg\,dl^{-1}$
Glucose	47.3	16.2	9.8	$mg\,dl^{-1}$
Urea	–	4.7	2.6	$mg\,dl^{-1}$

[a]Spectral calibration interval for all substrates 11,015–7620 cm^{-1}; transmission cell of 10 mm pathlength
[b]Spectral range for protein 6000–5510 cm^{-1}, for cholesterol and triglycerides with additional interval of 4520–4212 cm^{-1}, for glucose and urea 6790–5460 and 4735–4210 cm^{-1}; transmission cell of 1 mm pathlength
[c]Spectral range for protein 1700–1350 cm^{-1}, for cholesterol 3000–2800, 1800–1700 and 1500–1100 cm^{-1}, for triglycerides same intervals except the upper one, for glucose 1200–950 cm^{-1}, and urea 1800–1130 cm^{-1}; ATR micro-circle cell

As a result of a project performed with the German Diabetes Research Institute, a large population of blood plasma samples had been analyzed by mid-infrared spectroscopy using the attenuated total reflection technique, by near-infrared spectroscopy using a 1 mm transmission quartz cell and by short-wave near-infrared spectroscopy using a 10 mm cell. Reference analyses had been carried out in triplicate using well-calibrated clinical analyzers as appropriate for selected individual parameters. In our publication [8], we summarized the performances of the different spectroscopic assays and results are reported in Table 20.1.

In principle, results from these studies were milestones for the development of clinical spectroscopic assays. Taking into account mean analyte concentrations of the studied plasma sample population, variation coefficients were usually lower than 5%, which is acceptable for clinical routine work. Our analytical results have often been challenged in recent years, but the same picture has been obtained by other groups; see, for example, the review by Perez-Guaita et al. [10]. One can state that lipidic parameters such as total cholesterol, triglycerides, high-density lipoprotein (HDL) and low-density lipoprotein (LDL) can be analyzed with similar accuracy [11]; a similar statement is valid for total protein, even with application of SW-NIR spectral data. On the other side, parameters such as glucose and urea, which also suffer from lower concentrations in blood derived fluids, will see deterioration in analytical performance compared to mid-infrared spectroscopy. Another extensive study on the use of NIR spectroscopy for serum analysis had been carried out for total protein, albumin, cholesterol, triglycerides, urea and lactic acid by Hazen et al., who studied a total of 242 serum samples with spectra recorded within the interval of 5000–4000 cm^{-1}. However, their results had been worse compared with our study apart from SEP values for urea [12].

There have been studies by other groups by comparing the performance of NIR versus Raman spectroscopy assays by using solutions of glucose, lactate and urea in aqueous phosphate buffer. Surprisingly, the NIR-assay outperformed that by using Raman spectroscopy. Standard errors of prediction were 0.24, 0.11 and 0.14 mmol l^{-1} for glucose, lactate and urea, respectively, from near-IR spectra, while SEP values of 0.40, 0.42 and 0.36 mmol l^{-1} were achieved with Raman calibration modeling for glucose, lactate and urea, respectively. Differences between instrumental signal-to-noise ratios were responsible for the better performance of the near-IR spectrometer. It is certainly an advantage that NIR spectra can be measured by transmission cells with quartz or glass windows or inside glass vials or less often by total reflectance using NIR spectrometers that is more affordable than MIR or Raman instrumentation [13].

Assay selectivity has been and is still in the focus of many analytical spectroscopists, and one valuable approach has been developed by using the net analyte signal for investigating the non-overlapping spectral features of individual analytes. Arnold et al. [14] investigated the selectivity of near-infrared spectroscopy for the independent measurement of glucose, glucose-6-phosphate and pyruvate in ternary mixtures under physiological pH conditions. Spectral data from multiple measurements within the combination band region had been collected and multivariate calibration modeling was carried out using the PLS method. Despite the high similarity of the individual spectra, the principal conclusion was that selective analytical measurements were possible for the three analytes, based on high quality long-wave near-infrared transmission spectra.

Another important analyte in clinical chemistry is hemoglobin, whose concentration is commonly used in clinical medicine to diagnose anemia, identify bleeding, and for managing red blood cell transfusions. Automated hematology analyzers are usually applied for quantitative analysis. Usual reference methods for hemoglobin are not reagent-free in contrast to an NIR spectroscopic assay. Besides the use of traditional laboratory equipment, anemia diagnosis can also be accomplished by quantifying the hemoglobin (Hb) concentration with point-of-care testing (POCT) devices such as the HemoCue test systems (Ängelholm, Sweden) [15]. In the past, several papers have been published on hemoglobin quantification with NIR spectrometry; see for example [16]. Here, the combination band region was considered also for wavelength variable selection using multiple linear regression (MLR) with linear summation equations based on three and four characteristic wavelengths. The cross-validated standard error of prediction (SEP) for hemoglobin was 1.25 g dl^{-1} with a four term model over the concentration range from 5.9 to 20 g dl^{-1}. A recently presented development is based on using a laptop-controlled NIR spectrometer (via USB interface, spectral range 900–1700 nm [11,100–5900 cm^{-1}]). The device had been connected to a supercontinuum broadband white light laser source by using an optical fiber bundle. Transmission cuvettes were of 5 mm optical pathlength. Prediction errors from a validation sample population were around 0.44 g dl^{-1}, either with full spectrum use or with different variable selection methods, using three or two wavelengths only. An informative literature overview on past *in vitro* and non-invasive approaches is also given [17].

An extension to the last-mentioned approaches was presented by Han et al., describing a novel near-infrared-spectroscopy-based quantification method for glycated hemoglobin (HbA1c), which is an important clinical diagnosis indicator of diabetes. The analytical method was developed on the basis of simultaneous determination of hemoglobin (Hb) and the absolute HbA1c content in human hemolysate samples [18]. Several wavelength selection algorithms were tested for a search range covered by spectral intervals not saturated by water absorption (780–1880 nm [12,800–5320 cm^{-1}], 2090–2330 nm [4785 – 4290 cm^{-1}]). For Hb and total HbA1c, only 6 and 14 wavelengths were selected with equidistant combination partial least squares (EC-PLS), respectively. The so-called competitive adaptive reweighted sampling PLS (CARS-PLS) and a Monte Carlo uninformative variable elimination PLS (MC-UVE-PLS) required 23 and 30 wavelengths, as well as 100 and 120 wavelengths, respectively. For details, the reader is referred to the publication.

Another area as mentioned above deals with disease diagnosis based on multivariate spectral classification algorithms, which has been more related to the domain of mid-infrared and Raman spectroscopy. Recently, also blood analysis for Alzheimer's disease (AD) diagnosis has been presented, using multivariate classifiers based on NIR spectral signatures [19]. Robust and early diagnosis may be a first step toward tackling this disease of mainly elderly people by allowing timely intervention with novel synthesized pharmaceuticals. In the presented study, blood plasma samples were analyzed with NIR spectroscopy as a minimally invasive method to distinguish patients with AD from non-demented volunteers. Dry-film spectra were recorded from 50 μL samples of blood plasma on IR-reflective glass slides after overnight drying. Spectra were truncated to the biochemical fingerprint region (1850–2150 nm [5400–4650 cm^{-1}]). By means of a multivariate classifier (principal component analysis with quadratic discriminant analysis – PCA-QDA), AD individuals were correctly identified with 93% accuracy, 87.5% sensitivity and 96% specificity. The results show the potential of NIR spectroscopy as a simple and cost-effective diagnostic tool for AD.

Another bodyfluid is urine, which also contains valuable biomarkers for disease diagnosis. Concerning urine analysis, not many publications are available. There is a paper from the Mantsch group who analyzed protein, creatinine and urea in urine [20], but other notable work has been carried out by the group around Mark Arnold who investigated the use of NIR spectrometers for monitoring the dialysate during hemodialysis of patients with renal failure, which is the pathway of a number of kidney diseases [21]. The treatment choices for patients with impaired renal function are dialysis (hemo- or peritoneal dialysis) or renal transplantation. Accumulation of toxic end products of the nitrogen metabolism such as urea, creatinine and uric acid in blood and tissue, a disturbed homeostasis of water and different salt minerals are the complications resulting from renal failure. Whereas for their feasibility studies, an FT-NIR spectrometer had been used; they used a spectrometer with a temperature-controlled acousto optical tunable filter (AOFT) in conjunction with a thermoelectrically cooled extended wavelength Indium Gallium Arsenide (InGaAs) photodetector, providing spectral measurements with a 20 cm^{-1} resolution over the combination band region (4000–5000 cm^{-1}) of the near-infrared spectrum. Their best PLS calibration models led to SEP values between 0.30 and 0.52 mmol/l for the removal of urea during hemodialysis sessions [21].

Just recently, an interesting study was published on urine analysis using a measurement system based on light-emitting diodes (LEDs) by considering future transition to LED light sources. In their study, LEDs with ten standard wavelengths (1400–2300 nm, in 100 nm increments [7150–4350 cm^{-1}]) were employed. The aim was to estimate multiple components such as urea and creatinine in spot urine samples [22]. A multiple regression analysis using all combinations of 10 wavelengths was performed. NIR spectra from glucose-spiked urine samples from 10 healthy adults were used for an appropriate wavelength selection. For estimating urinary urea and creatinine levels, they obtained SEP values of 42 mg/dl and 7.34 mg/dl, respectively, using four wavelengths for urea and five wavelengths for creatinine. This optical, reagent-free method is suitable for practical determination of the urea-to-creatinine ratio, which allows assessing protein intake in chronic kidney disease (CKD) patients.

A forensic application of NIR imaging is also presented, for which the identification and location of body fluid stains was investigated. For this application, the potential of near-infrared hyperspectral imaging (NIR-HSI) was investigated [23]. The authors used a hyperspectral camera (Specim, Oulu, Finland), which was equipped with a mercury cadmium telluride (MCT) cryogenically cooled detector, providing access to the spectral range from 1000–2500 nm [10,000–4000 cm^{-1}]. With the combination of the NIR-HSI data and simple chemometric techniques such as principal component analysis (PCA) and classical least squares (CLS) regression, the evidence and location of semen, vaginal fluid and urine in bodily fluid stained fabrics could be revealed.

Another clinical application with an intended purpose for diabetes screening has recently been reported. This indication is accessible due to the analysis of glycated hemoglobin, especially of the HbA1c fraction, which is usually quantified by HPLC methodology. An interesting review on optical methods for studying glycation of proteins including hemoglobin, also with a focus on Raman spectroscopy, has been given by Pandey et al. [24]. The HbA1c fraction is usually reported as percentage against total hemoglobin (for healthy people < 6.0%). In Fig. 20.6, NIR spectra of several proteins are given, which have been considered for glycation testing as there are albumin, collagen, keratin and hemoglobin. Without blood testing, another assay for a non-invasive assay for assessing the glycation rate in people with diabetes has been suggested based on glycated keratin as available from finger nail clippings [25]. For testing their hypothesis, the researchers used artificial non-enzymatic glycation of keratin in aqueous glucose solution. For the *in vitro* glycation samples, spectral changes versus the spectra of untreated keratin were observed in the range between 4300 and 7500 cm^{-1}. Spectral classification based on people with diabetes mellitus versus healthy subjects was performed using partial least square discriminant analysis (PLS-DA). Using standard normal variate normalization and Savitzky-Golay smoothing with first derivative preprocessing, a prediction rate with 100% correct assignments was achieved for a test set. This result is seen with some skepticism based on our study of *stratum corneum* with spectra measured non-invasively by attenuated total reflection mid-infrared spectroscopy. The data base included healthy subjects

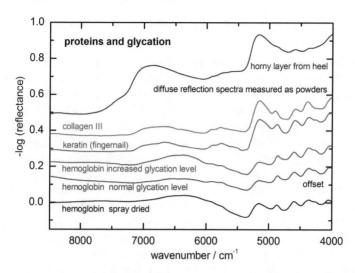

Fig. 20.6 Diffuse reflection spectra of different proteins considered for glycation measurements in diabetes screening and therapy monitoring (measured as powders); normal glycation levels for hemoglobin as for healthy people are HbA1c values < 6%; increased pathological glycation levels of HbA1c are around 14%. The horny layer from heel was from a thick *stratum corneum* layer consisting of keratin and water

and people with diabetes, including patients experiencing poor insulin therapy. The correlation based on multivariate PLS calibration with a regression against HbA1c values was rather poor (R = 0.303, standard deviation for a linear regression of cross-validated predictions versus HbA1c reference values was 0.49%) [26]. Non-invasive diabetes screening by diffuse reflection NIR skin spectroscopy has also been evaluated by my group with linear discriminant analysis (LDA) using the optimal spectral interval of 9780–4500 cm^{-1} (accuracy of 87.8% with leave-5-out cross-validation) and with spectral variable selection of eight wavelengths within the same interval, providing a prediction accuracy of 85.4% [27].

20.3 Applications of Non-invasive Technology in Clinical Chemistry

Non-invasive transcutaneous measurements for blood analysis have been the dream of our generation, and with today's multitude of body-sensing equipment for heart rate, blood pressure, activity measurement and others, certainly an important vital parameter such as blood glucose is high on the list. Other prominent analytes are blood ethanol and hemoglobin, for which some developments have been reported.

20.3.1 Non-invasive Technology for Glucose Monitoring

The benefits of tight glycemic control in people with diabetes have been repeatedly well described in the past. Previous studies had indicated that intensive insulin therapy in diabetic patients can dramatically delay the onset of serious micro- and macrovascular complications affecting eye sight, kidney, perfusion of the extremities and others. Most diabetic patients are using blood glucose self-monitoring (SMBG) devices for monitoring their glucose levels and adjustments of their insulin dosage to achieve normoglycemia. When undergoing intensive insulin therapy, current glucose monitoring requires people with diabetes to prick their fingers for blood sampling several times a day. Besides that also needle-type wearable sensors based on enzyme-mediated electrochemistry are nowadays available for continuous monitoring of interstitial glucose within the subcutaneous skin tissue (see also our recent review [28]), but suffer from systematic deviations from the capillary or arterial blood glucose level as gold standard for treatment with insulin. A non-invasive measurement device certainly eliminates the inconvenience and pain of frequent finger pricking or, as in the case of continuously sensing systems, avoids the invasiveness of needle-type sensors, allowing also a much higher frequency of readings than using SMBG invasive techniques.

A wide range of optical techniques, such as fluorescence and polarization measurements, Raman spectroscopy and optical coherence tomography (OCT) and others, has been designed for the development of robust non-invasive methods for glucose sensing; see, for example, [28]; see also Fig. 20.7 with an overview on optical methods with most promising methods highlighted. NIR spectroscopy of the skin in combination with diffuse reflection techniques also seemed to be a valid candidate for achieving such goal, as transmission spectroscopy needs rather thin web-like skin sections. Figure 20.8 provides information about different measuring techniques

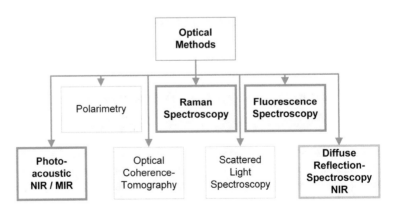

Fig. 20.7 Overview on optical methods suggested for non-invasive blood glucose measurements; highlighted methods are currently most promising. Fluorescence-based sensors are implants or use implanted modified substrates to be interrogated by *ex-vivo* optical devices (with some invasiveness)

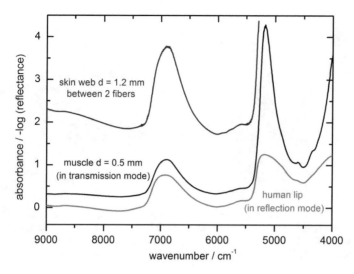

Fig. 20.8 NIR spectra of body tissues at different thicknesses as measured in transmission and reflection mode, illustrating the opportunities and limitations for non-invasive whole skin measurements

involved in skin spectroscopy, which is mainly based on diffuse reflection measurements. The most useful spectral intervals, containing important fingerprint signatures of the analyte, include the aforementioned combination and overtone NIR-regions. The successful implementation of NIR spectroscopic glucose assays using serum, blood plasma or whole blood samples, as described in the previous section, raised hope for the realization of non-invasive assays based on skin spectroscopy. A statement is allowed that shorter wavelengths, as chosen for experiments within the SW-NIR, will suffer from reduced selectivity due to broader absorption bands at the same tissue complexity; for a discussion on such assays, see our recent review [28]. Due to the small spectral signatures of the glucose, hidden among a largely variable background, multivariate calibration techniques based on wide spectral intervals are required to provide the selectivity and precision for quantification of blood glucose as outlined above.

Important is certainly the optical accessory for obtaining optimal signal-to-noise ratios. Fiber-optics, as used for infrared wave guiding, illumination and photon collection, are suffering from the small acceptance angle of backscattered photons due to the numerical aperture of the optical fibers. A much larger solid angle can be realized by using mirror-based optics for collecting backscattered NIR radiation. Figure 20.9 provides details on different optical set-ups used for diffuse reflection measurements of skin. Another improving feature is the so-called optical clearing by applying a non-absorbing and skin-friendly fluid to the skin surface for reducing the scattering from the *stratum corneum*, which can be used especially for optics allowing a larger solid angle for photon collection. Other designs for fiber arrangements for illumination and detection have been published; important is certainly the distance between emitting

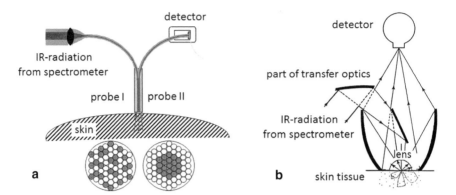

Fig. 20.9 Diffuse reflection accessories used for skin measurements: fiber-optic probes with different fiber arrangements for illumination and photon detection (**a**) and a rotational ellipsoidal mirror-based device (**b**), providing tissue spectra with different probing depths; reprinted from [27] with permission of SPIE - The International Society for Optics and Photonics

and detecting fibers for reaching different penetration depths within skin tissue. The optical accessory is significantly shaping the experimental diffuse reflection spectra and some examples for different skin sites are given with Fig. 20.10.

The non-invasive sensing of glucose is limited due to high background absorption of water, spectral baseline shifts, instrumental drift, lack of analyte sensitivity, analytical overfitting, poor precision and incorporation of spuriously correlated spectral variance into a calibration model. Noteworthy are the impressive experiments by Olesberg and coworkers, using a rat animal model and transmission measurements through a skin fold [29], as well as a recent publication from the group around K. Maruo in Japan, using a fiber-optic probe for diffuse reflection measurements in human subjects [30], for which the scattering optical parameters of skin are of importance.

Multivariate calibrations, i.e., exploiting broad spectral intervals with many spectral wavelengths to reach the required selectivity for glucose concentration determination, should have their footing on real wavelength-dependent absorption by glucose. This has been requested by using the net analyte signal as the chemometric approach for proving the spectrometric model. In the past, several papers claimed the realization of non-invasive assays, but the spectroscopic basis for the non-invasive determination of glucose and the development of appropriate calibration models, separating the glucose signal from the complex biological matrix spectral variance, needs to be proven; see also our review [28]. Spectroscopic assay pitfalls owing to overfitting, when calibration is based on too many variables and unsound model validation, have been illustrated earlier by us [9].

The success of the traditional approach using statistical PLS calibrations, even with sophisticated data pretreatment, is rather limited, but alternative calibration methods apart from artificial neural network (ANN) approaches had not been available [31]. For improving the model robustness, spectral variable selection on the

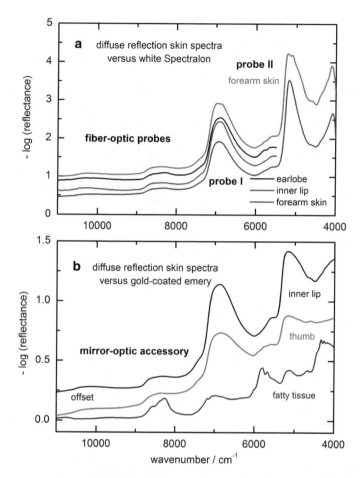

Fig. 20.10 Skin spectra obtained with different accessories (fiber-optic and optical mirror-based devices) illustrating the shape and intensities affected by their different solid angles for backscattered photon collection. Water band intensities allow for an estimate of the average photon pathlength within the tissue. Most protein signatures are from the *stratum corneum* layer; however, with larger penetration depth also subcutaneous fatty tissue can be probed

basis of the PLS regression vector weights and more strict validation with day-to-day testing had been taken into account instead of cross-validation based on different validation package sizes. Another approach, the so-called science-based calibration method, has been developed by us, which combines *a priori* information such as the spectral absorptivities of the component searched for with statistical estimates of the variance of the population of steady state samples with negligible glucose concentration dynamics [9]. This method has been successfully tested with transmission NIR spectra of plasma samples, which can be compared with results from previous PLS calibration models [8]. The situation with transcutaneous skin spectra, obtained by diffuse reflection, is more difficult because of the wavelength-dependent photon penetration depths, which requires a scaling of the aqueous glucose absorptivity spectrum ("response spectrum").

The problems with the repeatability of *in vivo* spectroscopic measurements are described by various aspects as follows. Glucose in different compartments, e.g., interstitial and vascular space, blood flow and blood volume, skin temperature gradients, comparability of reference venous or capillary blood glucose concentration with tissue probed concentration and others must be taken into account and experiments have to be carefully planned for reliable results. For the complexity of *in vivo* measurements, several extensive investigations have been carried out with results published earlier, but the challenge of achieving reliable non-invasive skin measurement repeatability is still in the focus of researchers; for particular challenges observed for the developments, see for example [31, 32].

Some information on the variability of tissue spectra can be gained from Fig. 20.11, which provides spectra as measured at different temperatures. For a tissue without blood perfusion, the temperature dependency of water as the main constituent is readily observable; for *in vivo* measurements of skin at different temperatures, also other effects come into play due to the response of the vasculature. Thus, increasing temperature causes changing perfusion, blood volume and also diminishing arterial-venous concentration differences due to higher blood flow.

Another more advanced approach arises from the opportunities, which exist in photoplethysmography (PPG) [33]. This measurement technique enables the sensing of information from the arterial vascular compartment. It can detect the periodic blood volume changes due to the systolic and diastolic blood pressures from the beating heart throughout the cardiac cycle based on NIRspectra, which has been routinely implemented in pulse oximetry. A difference to the latter is that broad spectral intervals need to be analyzed for determining the arterial blood glucose concentration. This strategy was followed by Yamakoshi and coworkers, starting with spectra between 900 and 1700 nm and coining their measurement technique as "pulse glucometry." After several improvements, an advanced set-up was designed for side-scattered finger photoplethysmography that was presented in 2017 [34]. First, PPG signals with three wavelengths: 808 nm, 1160 nm and 1600 nm (coinciding with nearly peak glucose and strong water absorptions) were compared, while the source-detector spacing was successively increased circumferentially around the fingertip. A second experiment was performed with six wavelengths from 1550 to 1749 nm for accessing glucose absorption bands. The pulsatile signal-to-noise ratios were claimed to be measured around 15 dB, giving hope for its potential for realizing a practical measurement of arterial blood constituents including glucose, but awaiting further developments.

It is interesting to take a look at early experiments performed by Nahm and Gehring, who studied the pulsatile spectrum from time-resolved near-infrared spectroscopic experiments with an injected bolus of indocyanine green (ICG) during transillumination of a fingertip [35]. Their results showed that the pulsatile component was by a factor up to 300 smaller than that of the static vasculature component, underlining the challenges of pulse glucometry versus integral tissue measurements.

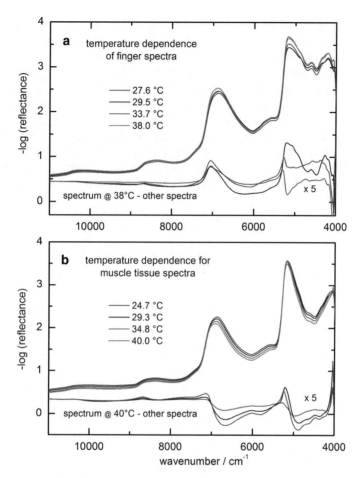

Fig. 20.11 Temperature dependency of *in vivo* skin diffuse reflection NIR spectra measured by a tip-thermostated fiber-optic probe of type I (**a**) and for a tissue phantom without blood micro-circulation effects, showing clearly the temperature dependency of the water constituent (**b**); reproduced in part from [27] with permission of SPIE—The International Society for Optics and Photonics

For further information on these developments using photoplethysmography with regard to blood glucose sensing, the reader is referred to our recent review [28].

20.3.2 NIR Spectroscopy of Skin–Optical Data for Photon Migration Modeling

The diffuse radiation transport in biological tissues can be modeled by using different mathematical tools providing an understanding, e.g., of the probed tissue volume and photon penetration depths. An interesting tutorial has recently been published to which the readers are referred [36]. The mean optical pathlength for radiation within mucosa tissue, as given for diffuse reflectance accessories, is wavenumber dependent as illustrated above (see Fig. 20.10). Tissue optical properties are the absorption and the scattering coefficients, μ_a and μ_s (in units of mm^{-1}), respectively, and the anisotropy of scattering g (dimensionless). From the latter two parameters, the reduced scattering coefficient $\mu_s' = \mu_s (1 - g)$ can be calculated. From diffusion theory for photon transport in tissues, an optical penetration depth can be estimated based on these optical constants with $\delta = (3 \, \mu_a \, (\mu_a + \mu_s'))^{-1/2}$ [36].

In Fig. 20.12, experimentally derived optical constants from the upper skin layers are shown, i.e., absorption and scattering coefficients; the inset in panel B shows also

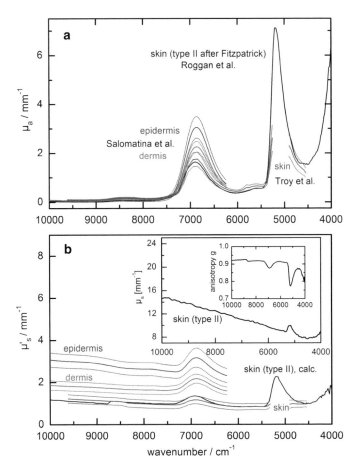

Fig. 20.12 Optical constants of dermis and epidermis in the NIR spectral range including uncertainties (μ_a absorption coefficient, μ_s' reduced scattering coefficient with $\mu_s' = (1 - g) \, \mu_s$, which is a property incorporating the scattering coefficient μ_s and the anisotropy g; all data were from Refs. [37–39]; note the significant differences in the scattering coefficients for the NIR region

Fig. 20.13 Average
wavelength-dependent
photon tissue penetration
from calculations based on
optical constants and optics
geometry [39]

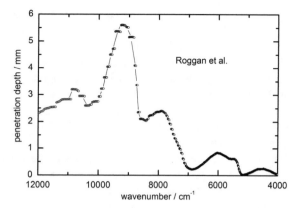

the anisotropy factor g, which describes the mainly forward scattering characteristics
of NIR photons. The optical constants were compiled from three publications [37–
39]. The optical data for epidermis and dermis (including standard deviations—
dashed curves- available down to $6250\,cm^{-1}$) from Salomatina et al. [38] are different
from the other compilations, which is certainly understandable for the thin epidermis
layer.

From the optical constants for skin, also the mean optical penetration depth can
be calculated, which was done by Roggan et al. [39], using the above mentioned
equation as derived from photon diffusion theory. Other methods include Monte
Carlo simulations of the "photon random walk" in scattering tissue, which are based
on optical constants for absorption and scattering. In Fig. 20.13, the wavelength-
dependent penetration depth estimates are provided for the SW-NIR spectral range
as published in [39]. From such calculations, also information on anatomical side
conditions can be derived, for example, for transillumination feasibility of a fingertip
or for estimating the fraction of photons reaching the skin dermis. For underlining
the success of theoretical simulations, an early published study by Qu and Wilson
[40] must be mentioned. The authors used extensive Monte Carlo modeling calculations for evaluating the effect of physiological factors and other analytes on the
in vivo determination of glucose concentrations by near-infrared optical absorption and scattering measurements. By this, much valuable insight can be obtained
for the minimum requirements of the experimental set-up with spectrometers and
accessories for the detection of minute glucose signatures within a varying complex
spectral background.

The quantification of glucose in complex multi-component systems requires a
unique absorption pattern and a significant contribution to the spectrum from this
component above the existing spectral noise level. As the NIR-assay mainly relies
on the absorption effects of glucose inside the aqueous intravascular and interstitial compartments in the skin tissue, glucose absorptivities obtained from aqueous
solution measurements have been already presented in Fig. 20.5, providing also
quantitative data on another monosaccharide of fructose and information on the

spectral quality. As already discussed, best discrimination between both monosaccharides can be obtained by using spectral data from the combination band region, whereas in higher wavenumber, regions absorptivities are nearly indistinguishable. For *in vivo* diffuse reflection measurements, the photon penetration depth in tissue is wavenumber dependent, much different from *in - vitro* measurements in cuvettes, which must be taken into account. A classical least squares (CLS) approach had been chosen by Maruo and Yamada [41] by evaluating diffuse reflection spectra of human forearm skin between 1350 and 1850 nm [7400 – 5400 cm^{-1}]. It is based on a modified Beer's law, assuming that absorbance difference spectra, as measured versus a first series spectrum, can be modeled by a linear combination of water, protein, glucose and fat spectra, as well as a baseline for scattering equivalent absorption. Another approach is SBC calibration modeling [9], which requires data input such as quantitative analyte absorptivity, as well as estimates of the instrumental noise. Based on the analyte spectral signals and of those tissue constituents showing cross-sensitivities in combination with radiation penetration depth, valuable analytical method parameters such as limit of detection and method selectivity can be derived also for diffuse reflection measurement scenarios. Due to known penetration depths, wavenumber-dependent scaling of the component spectral signatures can thus be realized.

20.3.3   Non-invasive Technology for Hemoglobin and Blood Ethanol Monitoring

The importance of hemoglobin (Hb) measurements has already been highlighted when presenting *in vitro* assays. However, there are scenarios, where it is desirable to monitor hemoglobin continuously as during transfusion, for patients under intensive care or with a postoperative follow-up. In principle, the time-resolved signals from photoplethysmography are evaluated with the observed alternating current (AC) and direct current (DC) signals measured. The selected radiation is found within the visible and SW-NIR spectral range with wavelengths between 600 and 1000 nm, realized, e.g., by a light-emitting diode (LED) array with center wavelengths of 569, 660, 805, 940 and 975 nm (selected wavelengths represent two isosbestic points and three for compensation of tissue scattering). For example, different physical models were considered, and with a "finger model," the ratio of AC and DC signals at different wavelengths (at least two) is taken into account. A more sophisticated model considers also the scattering of the arterial blood arising from the red blood cells. In principle, one parameter can be established and linked with a wavelength-dependent variable thickness due to scattering. Multiple linear regression analysis was applied for the prediction of total hemoglobin concentration of 129 different patients. The relative percentage error and standard deviation of the prediction set were 8.5% and 1.14 g/dl, respectively [42]. Further improvements were achieved by designing a special finger probe with optimizations of the detector area, the emission area of a light source and the distance between the light source and the detector. Such an optimally designed finger probe provided a correlation coefficient of 0.869 and a standard deviation of 0.81 g/dl in predicting total hemoglobin [43].

Another development of a non-invasive hemoglobin analyzer is presented in [44]. The instrument uses eight laser diodes with wavelengths between 600 nm to 1100 nm for a synchronous recording of photoplethysmographic signals. For a simplification of the optical assembly, the light sources were modulated with orthogonal square waves, and together with the design of a corresponding demodulation algorithm, a beam-splitting system could be avoided. A newly designed algorithm improved the accuracy of the dynamic spectrum extraction. A population of 220 subjects was involved in the clinical testing. A machine learning calibration model, regressing the plethysmographic data, deriving from the arterial pulse cycle, versus the hemoglobin concentration, was developed. The correlation coefficient and SEP values were 0.8645 and 0.85 g/dl, respectively. The results indicated that the hemoglobin concentration values could be obtained with acceptable precision and accuracy to allow future clinical translation [44].

A different wavelength regime is again exploited for non-invasive blood ethanol quantification, which is similar to the development of glucose NIR-sensors based on the spectral interval of 8000–4000 cm^{-1}. The spectral signatures of ethanol are even more pronounced as those of glucose, but the concentration interval of interest is quite similar, i.e., in the per mille range. Standard methods for blood alcohol determinations include the analysis of breath from subjects by using electrochemical sensors. More accurate are gas chromatographic methods with head-space analysis, requiring the sampling of blood by a syringe for laboratory analysis. A non-invasive *in vivo* spectroscopic method certainly offers a promising alternative to the other established assays.

Several companies were interested in the development of fast and reagent-free analyzers, and a few reports on progress have been published nearly a decade ago. With the last published paper, an impressive device with appropriate spectral data, as taken from measurements within the NIR combination band region, had been presented [45]. Using data from Monte Carlo simulations of the photon transport and experiments, it could be shown that the tissue fiber-optic probe design had an expected substantial impact on the effective photon random pathlength through the skin and the signal-to-noise ratio of the spectroscopic measurements. Spectral data were recorded within 8000–4000 cm^{-1}.

The so-called alternate-site phenomenon is known from blood glucose sensors, and several clinical studies have shown that interstitial finger and forearm glucose concentrations can exhibit significant concentration differences over time, with concentrations in fingers much less delayed compared to blood glucose values. A similar behavior was observed for non-invasive blood ethanol when finger and forearm tissues were tested. Results of a 26-subject clinical study with controlled drinking were reported that were designed to evaluate the spectroscopic technique preferentially at finger measurement sites in comparison to spectroscopic readings of alternate volar forearm venous blood, and breath measurements. Comparisons with results from breath, blood and tissue assays demonstrated significant differences in ethanol concentration, attributable to both assay accuracy and pharmacokinetics in the elimination phase of alcohol. With error analysis, a significant fraction of the concentration variance could be explained by alcohol pharmacokinetics using a first order kinetic model. It is interesting to note that the PLS calibration models gave different prediction errors when using either data from the entire monitoring session, or data from the alcohol elimination phase only; minimum SEP values were for regressions of finger tissue spectra versus venous blood concentration with a value of 6.6 mg/dl. As summary, the statement is valid that their work provided a first investigation of the relationships between breath alcohol, venous blood alcohol and interstitial tissue alcohol concentration, as measured at multiple skin sites [45].

20.4   NIR Spectroscopy for Tissue Analysis

20.4.1   Applications for Spectral Histopathology

A few tissue NIR spectra have already been presented in previous sections, in particular when skin spectroscopy for non-invasive assays is involved. Here, different skin areas have been studied such as oral mucosa of the inner lip, finger and forearm skin sites and most spectra, apart from a few transmission measurements, were presented as measured with fiber-optic and mirror-based accessories in diffuse reflection mode. Muscle and fatty tissue spectra have also been introduced and it is shown for all that water features are dominating the spectra. NIR microscopic studies are rare; see, for example, Ref. [32], where the authors were studying the heterogeneity of skin with its impact on non-invasive glucose measurements by skin spectroscopy. The histopathology of biopsy tissues on the microscopic level is a domain of mid-infrared and Raman spectroscopy, where imaging applications are supporting the pathologist on tissue assessment after, for example, a colonoscopy or bronchoscopy using endoscopic techniques. However, at the end of this section, such a NIR spectroscopic application with an inverted microscope is discussed. Larger tissue volumes can be easily accessed by fiber-optic probes attached to FT-NIR or dispersive spectrometers as described in one of the previous subchapters. For spectroscopy, the tissues or organs must be accessible, although also endoscopic techniques have been described in the past with more sophisticated instrumentation.

There have been different tissues studied for cancer, which comprise breast tumors, cervical dysplasia and cancer, glioma, human melanoma xenografts, head and neck tumors or cancerous bronchial mucosa; for the literature collection, see our review on NIR spectroscopy in cancer diagnosis and therapy [46]. Further studies have been performed earlier on pancreatic and colorectal tissues by my group, which has also been covered by our review. For dermatological studies on different skin cancers, the paper by McIntosh et al. should be consulted [47]. When looking at the physiological relevance of NIR spectroscopy for cancer studies, one has to look at marker substances that are providing differences in absorption and scattering characteristics. Hemoglobin concentrations and its redox status, described by oxy- and deoxygenated forms for providing oxygenation saturation, water and lipid content have been listed for discrimination from healthy tissue. Tumor tissues were found with significantly high levels of water and of hemoglobin because of higher vascularization, but also with less lipids and lower oxygen saturation due to cell hypermetabolism. Commonly, in tumorous tissue of breast cancer, abnormal proliferation of cells will result in an angiogenesis with an increased number of blood vessels, eventually increasing the local blood Hb concentration. A recent breast cancer study has been reported by Mehnati et al., who differentiated between normal and tumorous breast tissue by local hemoglobin concentrations [48].

Cancerous tissues from pancreatic and colorectal tumors have been classified by studying, in particular, first and second overtone bands of the lipid CH_2 stretching vibrations, which could be intensified by first spectral derivative calculations; for the characteristic wavenumber intervals and band assignments, see also Fig. 20.2. Besides diagnosis, also margin location of cancerous tissue, especially intraoperative malign tumor margin assessment is important for supporting the surgeon's decisions during operations. Furthermore, NIR spectroscopy is useful for therapeutics monitoring, especially after chemotherapy, radiation treatment, neoadjuvant and photodynamic therapy. For more details, the reader is again referred to our review [46]. A recent instrumental development of an LED-based NIR sensor for kidney tumor diagnostics has also been published [49]. Four LEDs with emission band maxima at 940, 1170, 1300 and 1440 nm were used in combination with a fiber-optic reflection probe for an improved sensor prototype as compared with the established application of a tungsten halogen lamp. A dispersive monochromator-based spectrometer equipped with an InGaAs linear array detector was employed. Spectra of renal biopsies with diameters of one to two cm were investigated, applying principle component analysis (PCA) and partial least squares discriminant analysis (PLS-DA) for kidney tumor detection.

Novel imaging instrumentation has been developed, e.g., on the basis of a broadband NIR source such as a tungsten halogen lamp and a liquid-crystal tunable filter (LCTF) for wavelength dispersion, realizing bandpass filtering without mechanical movement. A recent review on NIR imaging for biological tissues, including chemometric tools for spectral data analysis, has been published by Ozaki et al. [50], so that the reader is referred to that chapter for more information.

A very recent application dealing with multispectral imaging has been reported for the histopathology of skin biopsies. This technique combines spectral resolution of spectroscopy with spatial resolution of imaging to show several merits for biomedical applications. As mentioned above, infrared and Raman microscopy have been employed for spectral histopathology, also in combination with conventional tissue staining for tumor diagnostics. The biofingerprint interval within the low-wavenumber region has been mentioned already, but instrumentation for imaging will require different photonics allowing only shorter wavelengths. Spreinat et al. [51] described an imaging set-up using a Xenon lamp for NIR radiation, a monochromator

for wavelength dispersion, optics for homogenous wide-field transmission illumination, an inverted microscope and further detection by a thermoelectrically cooled
InGaAs camera for wavelengths between 900 and 1500 nm. Abnormal skin samples
including melanoma, nodular basal-cell carcinoma and squamous-cell carcinoma
were acquired from dermatology and studied to distinguish healthy from diseased
tissue regions, illustrating the potential for cancer diagnostics.

There are also developments dealing with molecular beacons (dyes), which
have been suggested as endogenous chromophores for providing more contrast in
NIR imaging applications. Such substances have recently received much attention
and actual developments are described in the last section on NIR-fluorescence in
biomedicine.

20.4.2   Monitoring of Blood-Tissue Oxygenation and Cytochrome Redox Status

Within the NIR region, the wavelength interval of 650–1100 nm, also called "optical
window" in tissue, is rather transparent for NIR radiation in scattering soft tissues.
When discussing the photon penetration depth for non-invasive blood glucose monitoring (see also Fig. 20.13), this information is clearly displayed. The penetration
depth was calculated by using optical absorption and scattering parameters of skin.
Besides the determination of blood glucose, there are further special biomolecules
of great interest for monitoring deviations from homeostasis in the body. Natural
pigments such as hemoglobin as major component in the red blood cells, myoglobin
in muscle, as well as cytochromes generally found in tissues, have characteristic
absorbance spectra within the visible and the SW-NIR spectral range. Their spectra
are dependent on the degree of oxygenation of the hemoproteins and the redox state of
the cytochromes. Therefore, spectroscopy can provide information about the in vivo
state of tissue oxygen supply.

In Fig. 20.14, absorptivities of hemoglobin in its oxy- and deoxygenated state
are shown, illustrating the possibilities for quantitative spectroscopy within several
spectral ranges from the visible up to the SW-NIR regime. Figure 20.15 provides a
blow-up of the SW-NIR region with a linear scale and different oxygenation degrees
of hemoglobin with a spectral isosbestic point around 805 nm. In particular, non-
invasive continuous measuring techniques are of special interest. Many applications
are found in cerebral monitoring with its successful advent in special clinics for
neonatal intensive care or for monitoring during labor. In the past, great interest had
also been in muscle oxygenation studies.

NIR spectroscopy of skin has been subject of many investigations, e.g., in skin
transplantation as in plastic and reconstructive surgery or burn injury treatment and
wound healing [52, 53]. Regional and temporal variations in skin tissue oxygenation
are of special interest and can be assessed continuously using NIR spectroscopy.
Especially, the research group around Michael Sowa has a long-time engagement

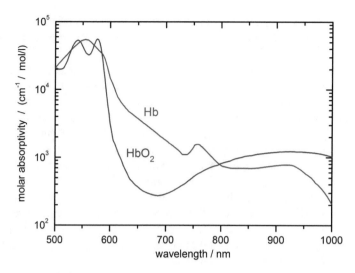

Fig. 20.14 Fundamental optical data for oxygenated and deoxygenated hemoglobins as blood constituents in the visible and short-wave near-infrared spectral range (downloaded from https://omlc.org/spectra/)

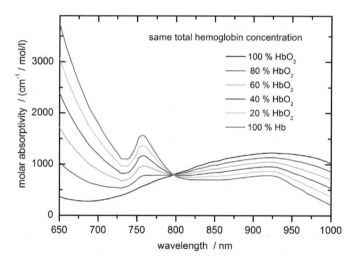

Fig. 20.15 Measurement scenario with different hemoglobin oxygenation using the so-called isosbestic wavelength with same absorption for both species, i.e., oxy- and deoxyhemoglobin, and neighboring wavelengths for species quantification. At the isosbestic wavelength at 800 cm^{-1}, the total absorbance of a sample does not change upon differences in oxygenation

in developing technology for skin viability testing, using also instrumentation with liquid-crystal tunable filters for spectroscopic imaging in the visible and near-IR spectral range.

Some delicate wound healing processes need intensive management, so that people have to consult hospitals with experienced dermatologists. This may be the case, for example, for people with insufficiently treated *diabetes mellitus* or for some elderly immunocompromised patients. Such treatment can be challenging and costly as the healing process may require longer times. Phases of wound healing are described in the review by Sowa et al. [53], providing an overview on NIR spectroscopic methods including alternative technology. Among the applied methods so far, including laser speckle contrast imaging, NIR imaging is still one of the most promising technologies, but the established indocyanine green fluorescence angiography method is competing for a successful objective wound assessment during the entire healing process. The evaluation of the pulsatile signals from NIR tissue oximetry, as described above for photoplethysmography, can provide further information, for example, by assessing arterial sufficiency.

Oximetry in the visible spectral range had found many applications and a study on the evaluation of reflection spectra from hemoglobin-free perfused heart tissue is given as an example of such investigations [54]. Visible spectra of myoglobin (oxy- and deoxygenated form) and several cytochromes with their different redox states can be found in the latter publication. Another study, focusing on the development of instrumentation for breast cancer screening based on hemoglobin spectroscopy, has been carried out by us, using two or three specially selected laser wavelengths mainly within the SW-NIR range for optical mammography [55]. In a recently published study, the method of opto-acoustic imaging of relative blood oxygen saturation and total hemoglobin has been applied for breast cancer diagnosis. Here, the measurements were done at a wavelength of 757 nm and 1064 nm, respectively, with prepared phantoms that were mimicking breast tissue for mapping oxygen saturation differences in vessels with a depth reaching down to about 50 mm [56]. A more complex assessment of tissue compounds was performed by Chen et al., who studied an NIR tomographic imaging system with several wavelengths between 633 and 980 nm in continuous wave mode for a quantification of four analytes, i.e., oxy- and deoxyhemoglobin, water and lipids, in addition to a scattering factor [57]. They came up with an optimum of seven wavelengths, which were adapted to chromophore concentrations as found for breast tissues of young and elderly women; the recommended laser diodes had wavelengths of 650, 690, 705, 730, 870/880, 915 and 937 nm.

As pointed out in the previous sections, NIR spectroscopy (NIRS) is able to measure oxygenation in human tissues, but statements can be found that it suffers lack of quantification. Methods had been developed for measuring optical pathlengths in tissue, initially enabling the detection of changes in concentrations to be quantified, and subsequently, methods for absolute quantification of HbO_2 and Hb were developed; for details, see also Ref. [58]. Commercially available monitors such as the NIRO 200 NX, which is used in clinics, can measure the following parameters: Tissue oxygenation index TOI (%), normalized tissue hemoglobin index nTHI (absolute value in arbitrary unit), oxygenated hemoglobin change, ΔHbO_2 (μmol/l), deoxygenated hemoglobin change ΔHb (μmol/l) and total hemoglobin change ΔcHb (μmol/l). For this device, three LEDs (735 nm, 810 nm, 850 nm: nominal values) are used. As spectral evaluation methods, the so-called SRS method (spatially resolved spectroscopy) and MBL method (modified Beer-Lambert) are applied; for details, see NIRO 200 NX brochure (Hamamatsu, Japan). A wide range of applications has been reported, reaching from the management of brain oxygenation status during surgery, over clinical studies related to muscle function [59], kidney transplant perfusion [60], and in particular to brain function and metabolism [61]. The latter review provides a comprehensive and actual overview on different optical methods employed in brain monitoring, with special focus on functional near-infrared spectroscopy (fNIRS), diffuse correlation spectroscopy (DCS), photoacoustic imaging (PAI) and optical coherence tomography (OCT). In addition, also wearable devices have been discussed. For details on the individual measurement techniques, the reader must be referred to the review with its extensive literature list, as otherwise this would go beyond the scope of this chapter.

Nevertheless, a few special applications will be mentioned. Non-invasive monitoring in preterm infants in intensive care units is another important activity in hospital care. Meanwhile, there is undeniably significant progress, although much work remains mostly on an experimental level due to lack of standardization and use of various algorithms for the spectral assessment. The state of the art and an overview of past studies concerning NIRS in preterm infants were presented by Korček et al. [62].

Human functional brain mapping applications have gained a new dimension when looking back at the latest developments of functional near-infrared spectroscopy (fNIRS). This technology usually refers to a point measurement of the hemodynamics that is providing information on the tissue perfusion via blood flow as well as blood volume changes, and oxygenation changes, marking oxygen supply. Schematics for the instrumentation and basic principle of fNIRS are presented with Fig. 20.16. In contrast, measurements can also be performed simultaneously at different locations for imaging purposes of the probed area. Other descriptors such as optical topography, near-infrared imaging and diffuse optical imaging have also been in use. Its aim is to detect simultaneous changes in the optical properties of the brain from multiple measurement sites with according results displayed as a map or image of the monitored area. In case of depth-resolved processing of the spectral information, one speaks of NIR tomography, also known as diffuse optical imaging (DOI) or NIR optical tomography.

The history and developments of fNIRS using continuous wave (CW) measurements, as well as time and frequency resolved instrumentation, have been presented several times, but a review on the state of the art and recent advances in hardware and signal processing from Yücel et al. [63] is worth to be mentioned. CW-based devices, which take the great advantage of being simple and more cost-efficient, compared with time and frequency domain instrumentation, have been applied for measurement of brain activation induced hemoglobin concentration changes that was reviewed by Scholkmann et al. [64]. Details on the different involved technologies, their advantages and disadvantages, cannot be given here, so that the reader is referred to the

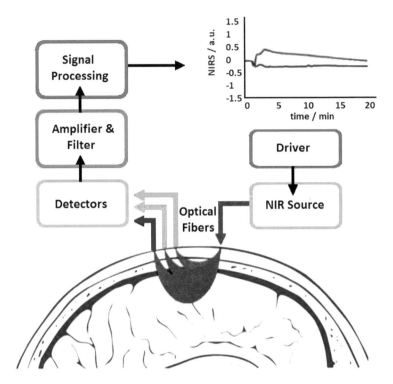

Fig. 20.16 Instrumentation and basic principle of functional NIR spectroscopy (fNIRS), for which the penetration depth is dependent on the distance between emitting and detecting fibers (from: Optics based label-free techniques and applications in brain monitoring 2020 [61]); licensed under CC-BY (http://creativecommons.org/licenses/by/4.0/)

original publications listed in the above mentioned reviews. Application areas can be listed as for neuro-development in newborns, infants and children, perception and cognition, motor control psychiatric disorders, neurology and anesthesia and further topics, which are made possible with wearable instrumentation [63].

In this context, another substance of great interest is the cytochrome-c-oxidase (COX) enzyme, which is present in all cellular mitochondria and is involved in more than 95% of oxygen consumption. There is much interest in monitoring COX in the brain, as it is a metabolic marker, especially for detecting brain injuries. Sudakou and coworkers looked at time-resolved near-infrared spectroscopy for estimating the uncertainty in the determination of cytochrome-c-oxidase concentration changes by Monte Carlo simulations with depth-resolved assessment [65].

20.4.3 Non-invasive Pulsatile NIR Spectroscopy

In most presented applications of short-wave near-IR biospectroscopy so far, the focus was on integral tissue probing apart from glucose assays using photoplethismography, allowing the probing of the intravascular space through dynamic monitoring of tissue absorption based on subsecond measurements. This technique can be employed for a spectroscopic measurement of the cardio-vascular pulse wave that is correlated to periodic changes in blood volume, since the blood is maximally diffused through the vascular system during the heart systole, whereas at the diastole blood pressure is minimal.

The major clinical application with exploitation of the cardiac blood volume modulation is in pulse oximetry providing values of the arterial hemoglobin oxygen saturation. Meanwhile, it is incorporated into the generally accepted standard of care and fundamental in the support of critical-care medicine, applied for adult and neonatal monitoring. An early review on theory and applications of this technique including practical limitations was published years ago by Mendelson [66]. Two different wavelengths had been used to measure the actual difference in the absorption spectra of oxygenated and deoxygenated hemoglobins, i.e., for example, at 660 nm (red light) and 940 nm in the SW-NIR, using the ratio of the pulsatile (AC) to the non-pulsatile (DC) signals at each wavelength for further calculations. This effective scaling process results in a normalized red/near-infrared ratio that is largely independent from the incident intensities. Problems at that time were concerned with low peripheral vascular perfusion, motion artifacts and systematic errors induced by different hemoglobin variants and derivatives (e.g., HbCO). Over the intervening years, many patents have been published on advanced pulse oximetry sensors with the consequence that problems attributed to motion artifacts or straylight have been diminished.

The commonly used fingertip-type pulse oximeter is taking measurements in transmission, thus limiting its application to the fingertip or earlobe to provide physiological parameters. However, their inconvenience for long-term monitoring in daily life has some shortcomings so that other types of wearable pulse oximeters, measuring in reflection mode, had been envisaged. For the purpose of developing reflection pulse oximetry, the light propagation in tissue was simulated to estimate the measured intensities of reflected light using analytical and numerical solutions of the diffusion approximation equation for photon migration in the visible and near-infrared region [67]. A comparison between reflection and transmission modes was investigated with experimental data and the research results showed that it is possible to model a reflectance pulse oximeter by simulating the random walk of photons by diffusion theory.

20.5   Applications of NIR-Fluorescence in Biomedicine

Near-infrared fluorescence probes (NIFPs) have often been used in immunoassays, bio-imaging and medical diagnosis. These fluorescent compounds have a characteristic molecular structure with highly conjugated polyene systems for enabling long emission wavelengths between 650 nm up to 900 nm. In this region, lower tissue autofluorescence exists with deep tissue penetration and minimal background interference. Its high sensitivity and selectivity is remarkable with the consequence of fluorescence spectroscopy, finding a broad range of applications also in bioanalytical chemistry and, in particular, for imaging applications [68]. As most biomolecules show no or only weak fluorescence, their detection sensitivity is rather low. For improvement, fluorescent labeling has been used in the past, and near-infrared (NIR) fluorescence detection shows obvious advantages in biological material analysis. The potential of cancer NIR imaging can certainly be realized with NIR dyes in conjugation with tumor specific ligands [69]. A recent review on near-infrared fluorescent dyes and their classification, providing a fine overview on the current state of the art with applications, has been published [70]. Besides several organic fluorescent dyes such as cyanine dyes, rhodamine, thiazine and oxazine dyes from past developments (indocyanine green must especially be noticed for measurements with the vascular system involved), also the synthesis of new classes has been introduced such as fluorescent quantum dots or rare earth complexes and even single-walled nanotubes (SWNT) must be mentioned. One wide area is the use within immunoassays, where these dyes are used as marker molecules in medical diagnostics. For further details, the reader is referred to the recent review [69]. Several studies have been carried out for indocyanine green enhanced near-infrared optical imaging of acutely damaged muscle, for which often animal models have been applied. For imaging applications, nanofluorophores have also been presented that are applicable within the so-called second near-infrared window (1000–1700 nm), thus providing high spatial resolution, low background and deep tissue penetration. There are further recent studies, e.g., for tumor imaging, which could be achieved by using a hypoxia-triggered single molecule probe for background-free NIR II fluorescent imaging with deep tissue penetration at the centimeter level, providing in addition possibilities in photothermal therapy for curative tumor treatment [71].

20.6   Concluding Remarks

There are many interdisciplinary research activities worldwide, searching for new and more efficient analytical methods and techniques for expanding the field of medical diagnostics. Upgraded technology for clinical chemistry should be based on reagent-free and automation-capable NIR spectroscopy and the recent advent of hand-held instrumentation will be widening the application range. For example, personal measurements on cholesterol and triglycerides in blood have been suggested for screening in pharmacies, as life style management will create a corresponding demand. Such portable NIR spectrometers could have an enormous potential in the area of medical technology for performing routine diagnostic testing. Clinical chemistry will certainly profit from less expensive miniaturized spectrometers that hover just over the horizon.

Non-invasive monitoring technologies of blood glucose and blood ethanol are candidates for another market, but there are also competing optical methods such as Raman and MIR spectroscopy. Spectral histopathology is another area, but the medical community is sometimes skeptical and quite slow in accepting novel developments. However, interdisciplinary collaboration will advance the introduction into clinics. Another promising area is optical tomography for breast cancer screening despite the image blurring owing to photon scattering in tissue. Functional NIR spectroscopy seems to be unchallenged so far, but will need further standardization. There is still a plethora of conventional or classical assays to be replaced by faster and more accurate spectroscopic methodology. Furthermore, direct and fast diagnostics and classification of diseases may be derived from the spectroscopic fingerprints of biomedical samples.

Acknowledgements I am tremendously grateful for the collaboration and enormous support from my collaborator Sven Delbeck over many years at the Interdisciplinary Center for Life Sciences in Iserlohn during research and for teaching. Some earlier work was done at the Leibniz-Institute for Analytical Sciences-ISAS, my former research institute in Dortmund. I am grateful for the productive working atmosphere, which had been very much appreciated.

References

1. F.F. Jöbsis, Non-invasive infrared monitoring of cerebral and myocardial oxygen sufficiency and circulatory parameters. Science **198**, 1264–1267 (1977)
2. D.T. Delpy, M. Cope, Quantification in tissue near-infrared spectroscopy. Philos. Trans. R. Soc. Lond. B. **352**, 649–659 (1997). https://doi.org/10.1098/rstb.1997.0046
3. B. Chance, E. Anday, S. Nioka, S. Zhou, L. Hong, K. Worden, C. Li, T. Murray, Y. Ovetsky, D. Pidikiti, R. Thomas, A novel method for fast imaging of brain function, non-invasively, with light. Opt. Express **2**(10), 412–423 (1998)
4. J.E. Bertie, Z. Lan, Infrared intensities of liquids XX: The intensity of the OH stretching band of liquid water revisited, and the best current values of the optical constants of $H_2O(l)$ at 25°C. Appl. Spectrosc. **50**, 1047–1057 (1996)
5. R.A. Shaw, H.H. Mantsch, Multianalyte serum assays from mid-IR spectra of dry films on glass slides. Appl. Spectrosc. **54**(6), 885–889 (2000)
6. A.K. Amerov, J. Chen, M.A. Arnold, Molar absorptivities of glucose and other biological molecules in aqueous solutions over the first overtone and combination regions of the near-infrared spectrum. Appl. Spectrosc. **58**(10), 1195–1204 (2004)
7. R. Marbach, T. Koschinsky, F.A. Gries, H.M. Heise, Noninvasive blood glucose assay by near-infrared diffuse reflectance spectroscopy of the human inner lip. Appl. Spectrosc. **47**, 875–881 (1993)
8. H.M. Heise, R. Marbach, A. Bittner, T. Koschinsky, Clinical chemistry and near-infrared spectroscopy: multicomponent assay for human plasma and its evaluation for the determination of blood substrates. J. Near. Infrared. Spectroscopy. **6**, 361–374 (1998)

9. Heise HM, Lampen P, Marbach R (2009) Near-infrared reflection spectroscopy for non-invasive monitoring of glucose – Established and novel strategies for multivariate calibration. In: Handbook of Optical Sensing of Glucose in Biological Fluids and Tissues, Tuchin VV (ed.), CRC Press, Chapter 5, 115–156

10. Perez-Guaita D, Garrigues S, de la Guardia M (2014) Infrared-based quantification of clinical parameters, Trends. Anal. Chem. 62 (2014) 93–105]

11. K.Z. Liu, M. Shi, A. Man, T.C. Dembinski, R.A. Shaw, Quantitative determination of serum LDL cholesterol by near-infrared spectroscopy. Vib. Spectrosc. **38**, 203–208 (2005)

12. K.H. Hazen, M.A. Arnold, G.W. Small, Measurement of glucose and other analytes in undiluted human serum with near-infrared transmission spectroscopy. Anal. Chim. Acta **371**, 255–267 (1998)

13. M. Ren, M.A. Arnold, Comparison of multivariate calibration models for glucose, urea, and lactate from near-infrared and Raman spectra. Anal. Bioanal. Chem. **387**, 879–888 (2007)

14. L. Liu, M.A. Arnold, Selectivity for glucose, glucose-6-phosphate, and pyruvate in ternary mixtures from the multivariate analysis of near-infrared spectra. Anal. Bioanal. Chem. **393**, 669–677 (2009)

15. R.D. Whitehead Jr., Z. Mei, C. Mapango, M.E.D. Jefferds, Methods and analyzers for hemoglobin measurement in clinical laboratories and field settings. Ann. N. Y. Acad. Sci. **1450**(1), 147–171 (2019)

16. I. Vályi-Nagy, K.J. Kaffka, J.M. Jákó, É. Gönczöl, G. Domján, Application of near infrared spectroscopy to the determination of haemoglobin. Clin. Chim. Acta **264**(1), 117–125 (1997)

17. H. Tian, M. Li, Y. Wang, D. Sheng, J. Liu, L. Zhang, Optical wavelength selection for portable hemoglobin determination by near-infrared spectroscopy method. Infrared Phys. Technol. **86**, 98–102 (2017)

18. Y. Han, J. Chen, T. Pan, G. Liu, Determination of glycated hemoglobin using near-infrared spectroscopy combined with equidistant combination partial least squares. Chem. Intell. Lab. Syst. **145**, 84–92 (2015)

19. M. Paraskevaidi, C.L.M. Morais, D.L.D. Freitas, K.M.G. Lima, D.M.A. Mann, D. Allsop, P.L. Martin-Hirsch, F.L. Martin, Blood-based near-infrared spectroscopy for the rapid low-cost detection of Alzheimer's disease. Analyst **143**, 5959–5964 (2018)

20. R.A. Shaw, S. Kotowich, H.H. Mantsch, M. Leroux, Quantitation of protein, creatinine, and urea in urine by near-infrared spectroscopy. Clin. Biochem. **29**, 11–19 (1996)

21. D.S. Cho, J.T. Olesberg, M.J. Flanigan, M.A. Arnold, On-line near-infrared spectrometer to monitor urea removal in real time during hemodialysis. Appl. Spectrosc. **62**(8), 866–872 (2008)

22. R. Suzuki, M. Ogawa, K. Seino, M. Nogawa, H. Naito, K. Yamakoshi, S. Tanaka, Reagentless estimation of urea and creatinine concentrations using near-infrared spectroscopy for spot urine test of urea-to-creatinine. Adv. Biomed. Eng. **7**, 72–81 (2018)

23. F. Zapata, F.E. Ortega-Ojeda, C. García-Ruiz, Revealing the location of semen, vaginal fluid and urine in stained evidence through near infrared chemical imaging. Talanta **166**, 292–299 (2017)

24. R. Pandey, N.C. Dingari, N. Spegazzini, R.R. Dasari, G.L. Horowitz, I. Barman, Emerging trends in optical sensing of glycemic markers for diabetes monitoring. Trends. Anal. Chem. **64**, 100–108 (2015)

25. T. Monteyne, R. Coopman, A.S. Kishabongo, J. Himpe, B. Lapauw, S. Shadid, E.H. Van Aken, D. Berenson, M.M. Speeckaert, T. De Beer, J.R. Delanghe, Analysis of protein glycation in human fingernail clippings with near-infrared (NIR) spectroscopy as an alternative technique for the diagnosis of diabetes mellitus. Clin. Chem. Lab. Med. **56**(9), 1551–1558 (2018)

26. Heise HM, Delbeck S, Küpper L (2018) Recent advances in sensor developments based on silver halide fibers for mid-infrared spectrometric analysis. In: Gupta VP (ed). Molecular and Laser Spectroscopy: Adv. Appl. Elsevier, San Diego, Chapter 3, 39–63

27. H.M. Heise, S. Haiber, M. Licht, D.F. Ihrig, C. Moll, M. Stücker, Recent progress in non-invasive diabetes screening by diffuse reflectance near-infrared skin spectroscopy. Proc. SPIE **6093**, 609310 (2006)

28. S. Delbeck, T. Vahlsing, S. Leonhardt, G. Steiner, H.M. Heise, Non-invasive monitoring of blood glucose using optical methods for skin spectroscopy – opportunities and recent advances. Anal. Bioanal. Chem. **411**, 63–77 (2019)

29. J.T. Olesberg, L. Liu, V. Van Zee, M.A. Arnold, In vivo near-infrared spectroscopy of rat skin tissue with varying blood glucose levels. Anal. Chem. **78**, 215–223 (2006)

30. Kessoku S, Maruo K, Okawa S, Masamoto K, Yamada Y (2011) Influence of blood glucose level on the scattering coefficient of the skin in near-infrared spectroscopy. Proc. AJTEC2011, paper No AJTEC2011-44471; https://doi.org/10.1115/ajtec2011–44471

31. Heise HM (2000) *In vivo* Assay of glucose. In: Encyclopedia of Analytical Chemistry: Instrum. Appl. Meyers RA (ed.), Wiley, Chichester Vol. I, 56–83

32. N.V. Alexeeva, M.A. Arnold, Impact of tissue heterogeneity on noninvasive near-infrared glucose measurements in interstitial fluid on rat skin. J. Diabe. Sci. Technol. **4**(5), 1041–1054 (2010)

33. J. Allen, Photoplethysmography and its application in clinical physiological measurement. Physiol. Meas. **28**, R1–R39 (2007)

34. Y. Yamakoshi, K. Matsumura, T. Yamakoshi, J. Lee, P. Rolfe, Y. Kato et al., Side-scattered finger-photoplethysmography: experimental investigations toward practical noninvasive measurement of blood glucose. J. Biomed. Opt. **22**(6), 67001 (2017)

35. W. Nahm, H. Gehring, Non invasive in vivo measurement of blood spectrum by time resolved near-infrared spectroscopy. Sens. Actuators. B. **29**, 174–179 (1995)

36. S.L. Jacques, B.W. Pogue, Tutorial on diffuse light transport. J. Biomed. Opt. **13**(4), 041303 (2008)

37. T.L. Troy, S.N. Thennadiel, Optical properties of human skin in the near infrared wavelength range of 1000 to 2200 nm. J. Biomed. Opt. **6**, 167–176 (2001)

38. E. Salomatina, B. Jiang, J. Novak, A.N. Yaroslavsky, Optical properties of normal and cancerous human skin in the visible and near-infrared spectral range. J. Biomed. Opt. **11**, 064026 (2006)

39. A. Roggan, J. Beuthan, S. Schründer, G. Müller, Diagnostik und Therapie mit dem Laser. Phys. Blätter. **55**, 25–30 (1999)

40. J. Qu, B.C. Wilson, Monte Carlo modeling studies of the effect of physiological factors and other analytes on the determination of glucose concentration in vivo by near infrared optical absorption and scattering measurements. J. Biomed. Opt. **2**(3), 319–25 (1997)

41. K. Maruo, Y. Yamada, Near-infrared noninvasive blood glucose prediction without using multivariate analyses: introduction of imaginary spectra due to scattering change in the skin. J. Biomed. Opt. **20**(4), 047003 (2015)

42. K.J. Jeon, S.J. Kim, K.K. Park, J.W. Kim, G. Yoon, Noninvasive total hemoglobin measurement. J. Biomed. Opt. **7**(1), 45–50 (2002)

43. G. Yoon, S.J. Kim, K.J. Jeon, Robust design of finger probe in non-invasive total haemoglobin monitor. Med. Biol. Eng. Compu. **43**, 121–125 (2005)

44. X. Yi, G. Li, L. Lin, Noninvasive hemoglobin measurement using dynamic spectrum. Rev. Sci. Instrum. **88**, 083109 (2017)

45. T.D. Ridder, B.J. Ver Steeg, B.D. Laaksonen, E.L. Hull, Comparison of spectroscopically measured finger and forearm tissue ethanol concentration to blood and breath ethanol measurements. J. Biomed. Opt. **16**(2), 028003 (2011)

46. V.R. Kondepati, H.M. Heise, Recent applications of near-infrared spectroscopy in cancer diagnosis and therapy. Anal. Bioanal. Chem. **390**, 125–139 (2008)

47. L.M. McIntosh, M. Jackson, H.H. Mantsch, R. Mansfield, A.N. Crowson, J.W.P. Toole, Near-infrared spectroscopy for dermatological applications. Vib. Spectrosc. **28**, 53–58 (2002)

48. P. Mehnati, S. Khorram, M.S. Zakerhamidi, F. Fahima, Near-infrared visual differentiation in normal and abnormal breast using hemoglobin concentrations. J. Lasers. Med. Sci. **9**(1), 50–57 (2018)

49. A. Bogomolov, U. Zabarylo, D. Kirsano, V. Belikova, V. Ageev, I. Usenov, V. Galyanin, O. Minet, T. Sakharova, G. Danielyan, E. Feliksberger, V. Artyushenko, Development and testing of an LED-based near-infrared sensor for human kidney tumor diagnostics. Sensors **17**, 1914 (2017)

50. Y. Ozaki, C.W. Huck, M. Ishigaki, D. Ishikawa, A. Ikehata, H. Shinzawa, Near-infrared spectroscopy in biological molecules and tissues, in *Encyclopedia of Biophysics*, ed. by G.C.K. Roberts, A. Watts (Springer, Berlin, 2018)
51. A. Spreinat, G. Selvaggio, L. Erpenbeck, S. Kruss, Multispectral near infrared absorption imaging for histology of skin cancer. J. Biophotonics. **13**, e201960080 (2020)
52. M.G. Sowa, L. Leonardi, J.R. Payette, Karen M. Cross, K.M. Gomez, J.S. Fish, Classification of burn injuries using near-infrared spectroscopy. J. Biomed. Opt. **11**(5), 054002 (2006)
53. M.G. Sowa, W. Chuan Kuo, A.C.T. Ko, D.G. Armstrong, Review of near-infrared methods for wound assessment. J. Biomed. Opt. **21**(9), 091304 (2016)
54. J. Hoffmann, D.W. Lübbers, H.M. Heise, Applicability of the Kubelka-Munk theory for the evaluation of reflectance spectra demonstrated for haemoglobin-free perfused heart tissue. Phys. Med. Biol. **43**, 3571–3587 (1998)
55. A. Zybin, V. Liger, R. Souchon, H.M. Heise, K. Niemax, Examination of the oxygenation state of hemoglobin in a phantom and in-vivo tissue applying absorption balancing with two and three laser wavelengths. Appl. Phys. B. **83**, 141–148 (2006)
56. J. Zalev, L.M. Richards, B.A. Clingman, J. Harris, E. Cantu, G.L.G. Menezes, C. Avila, A. Bertrand, X. Saenz, S. Miller, A.A. Oraevsky, M.C. Kolios, Opto-acoustic imaging of relative blood oxygen saturation and total hemoglobin for breast cancer diagnosis. J. Biomed. Opt. **24**(12), 121915 (2019)
57. L.Y. Chen, M.C. Pan, C.C. Yan, M.C. Pan, Wavelength optimization using available laser diodes in spectral near-infrared optical tomography. Appl. Opt. **55**(21), 5729–5737 (2016)
58. D.T. Delpy, M. Cope, Quantification in tissue near-infrared spectroscopy. Phil. Trans. R. Soc. Lond. B. **352**, 649–659 (1997)
59. A. Curra, R. Gasbarrone, A. Cardillo, C. Trompetto, F. Fattaposta, F. Pierelli, P. Missori, G. Bonifazi, S. Serrant, Near-infrared spectroscopy as a tool for in vivo analysis of human muscles. Sci. Rep. **9**, 8623 (2019)
60. G. Malakasioti, S.D. Marks, T. Watson, F. Williams, M. Taylor-Allkins, N. Mamode, J. Morgan, W.N. Hayes, Continuous montoring of kidney transplant perfusion with near-infrared spectroscopy. Nephrol. Dial. Transplant. **33**, 1863–1869 (2018)
61. P. Karthikeyan, S. Moradi, H. Ferdinando, Z. Zhao, T. Myllylä, Optics based label-free techniques and applications in brain monitoring. Appl. Sci. **10**(6), 2196 (2020)
62. P. Korček, G. Straňák, J. Širc, G. Naulaers, The role of near-infrared spectroscopy monitoring in preterm infants. J. Perinato. **37**, 1070–1077 (2017)
63. M.A. Yücel, J.J. Selb, T.J. Huppert, M.A. Franceschini, D.A. Boas, Functional near infrared spectroscopy: enabling routine functional brain imaging. Curr. Opin. Biomed. Eng. **4**, 78–86 (2017)
64. F. Scholkmann, S. Kleiser, A.J. Metz, R. Zimmermann, J. Mata Pavia, U. Wolf, M. Wolf, An review on continuous wave functional near-infrared spectroscopy and imaging instrumentation and methodology. Neuroimage **85**, 6–27 (2014)
65. A. Sudakou, S. Wojtkiewicz, F. Lange, A. Gerega, P. Sawosz, I. Tachtsidis, A. Liebert, Depth-resolved assessment of changes in concentration of chromophores using time-resolved near-infrared spectroscopy: estimation of cytochrome-c-oxidase uncertainty by Monte Carlo simulations. Biomed. Opt. Express. **10**, 4621–4634 (2019)
66. Y. Mendelson, Pulse Oximetry: Theory and Applications for Noninvasive Monitoring. Clin. Chem. **38**, 1601–1607 (1992)
67. M. Mehrabi, S. Setayeshi, S.H. Ardehali, H. Arabalibeik, Modeling of diffuse reflectance of light in heterogeneous biological tissue to analysis of the effects of multiple scattering on reflectance pulse oximetry. J. Biomed. Opt. **22**(1), 015004 (2017)
68. G.S. Hong, A.L. Antaris, H. Dai, Near-infrared fluorophores for biomedical imaging. Nat. Biomed. Eng. **1**, 0010 (2017)
69. S. Luo, E. Zhang, Y. Su, T. Cheng, C. Shi, A review of NIR dyes in cancer targeting and imaging. Biomaterials **32**, 7127–7138 (2011)
70. X.H. Chang, J. Zhang, L.H. Wu, Y.K. Peng, X.Y. Yang, X.L. Li, A.J. Ma, J.C. Ma, G.Q. Chen, Research progress of Near-Infrared Fluorescence Immunoassay. Micromachines **10**, 422 (2019)

71. X. Meng, J. Zhang, Z. Sun, L. Zhou, G. Deng, S. Li, W. Li, P. Gong, L. Cai, Hypoxia-triggered single molecule probe for high-contrast NIR II/PA tumor imaging and robust photothermal therapy. Theranostics **8**(21), 6025–6034 (2018)

Chapter 21
Applications of NIR Techniques in Polymer Coatings and Synthetic Textiles

Tom Scherzer

Abstract This chapter provides a survey on the current state of the art of in-line analysis by various NIR techniques for process control of two very specialized categories of polymer materials: polymer coatings and textiles from synthetic fibers. In case of coatings, monitoring of the conversion of radiation-curable monomers such as acrylates, methacrylates, cycloaliphatic epoxies and vinyl ethers that is achieved during irradiation is primarily discussed, since the conversion strongly determines application and processing properties of such coatings. Moreover, in-line measurement of the coating thickness (from only a few up to several hundreds of micrometers), the spatial distribution of various parameters of interest across the coatings as well as the characterization of thin printed layers in the printing press are further subjects of the first part. The second part deals with the application of NIR methods for process monitoring and quality control in textile converting. Technical textiles are often subject of special treatment and finishing steps such as impregnation, coating, lamination etc. which have to be controlled in order to ensure adequate processing. NIR techniques have been shown to be an appropriate tool for this problem. In particular, hyperspectral imaging can help to retain the required homogeneity of textile webs or laminates after finishing, e.g., with respect to the application weight of functional finishes or adhesive layers. Furthermore, NIR spectroscopy is used for identification and sorting of textiles with the objective of recycling of the materials. Hence, an overview of the current status of the use of NIR spectroscopic techniques in textile technology is given.

Keywords NIR spectroscopy · Hyperspectral imaging · Monitoring of UV curing reactions · Characterization of coatings · Textiles · Process control

T. Scherzer (✉)
Department of Functional Coatings, Leibniz-Institut für Oberflächenmodifizierung e.V. (IOM), Permoserstr. 15, D-04318 Leipzig, Germany
e-mail: tom.scherzer@iom-leipzig.de

© Springer Nature Singapore Pte Ltd. 2021
Y. Ozaki et al. (eds.), *Near-Infrared Spectroscopy*,
https://doi.org/10.1007/978-981-15-8648-4_21

21.1 Introduction

Polymers have become one of the most important classes of materials today, which is mainly due to the amazingly broad range of properties that may be achieved with this kind of materials. Accordingly, there is almost no range in our daily life, which goes without polymers. A multitude of different polymer materials has been developed so far and is commercially available. The widespread use and commercial importance of polymers makes also great demands on the analytics of such materials. With respect to the production of polymers, their converting, finishing, recovery or disposal, powerful and versatile analytical methods are required for characterization, identification, process and quality control etc. In particular, techniques used for sorting and control applications have to comply with special requirements such as robustness, reliability, durability, safety, significance of the data etc. [1–3]. Furthermore, in most cases they should be able to record data in a non-contact mode.

Near-infrared spectroscopy satisfies these specifications to a great extend. Moreover, due to its specific measurement principle NIR spectroscopy has high sensitivity to typical molecular structures in organic chemistry that are relevant for synthetic polymers. Consequently, NIR spectroscopy is particularly well suited for applications in polymer production and processing and related areas [4–9].

Apart from the underlying measurement principle, the great potential of NIR spectroscopy for its use in science and technology is mainly related to

(i)   The high transmission of quartz and several glasses for radiation in this spectral range, which allows the use of conventional windows, lenses, prisms and last but not least optical fibers in experimental setups, which provides easy access to the region of interest,

(ii) The possibility to measure in reflection or transflection mode, which is advantageous in case of intransparent materials and

(iii) The rather low extinction coefficients of most organic matter in the near-infrared in comparison to other spectral ranges used in analytics such as mid-infrared or UV, which allows transmission measurements on samples with rather high thickness or optical path lengths in the range of several millimeters or even more without the need of dilution, cutting, pressing or other special sample preparation steps.

A quite recent application of NIR spectroscopy in polymer technology is the in-line analysis of polymer coatings. In particular, the conversion in coatings of radiation-curable monomers and oligomers such as acrylates, methacrylates, vinyl esters, thiol-ene systems, cycloaliphatic epoxies and vinyl ethers that is achieved after UV or electron beam (EB) irradiation is of great interest, since it strongly determines both the application and processing properties of such coatings. However, in-line analysis of rather thin coatings with thicknesses between a few up to several hundreds of micrometers is a considerable challenge for several reasons such as the required sensitivity and time resolution. Consequently, only very few attempts have been made in the past to monitor technical coating and curing processes by analytical methods. NIR spectroscopy does not appear to be an obvious method for in-line analysis of thin coatings due to the rather low extinction coefficients in this spectral range. Nevertheless, it has been shown during the last two decades that NIR spectroscopy has great potential for monitoring of the conversion, the thickness and other properties of polymer coatings if spectroscopic equipment with high sensitivity is combined with powerful chemometric approaches. Furthermore, the distribution of the parameters of interest across the coatings may be monitored as well by means of hyperspectral imaging.

An even greater challenge are printed layers since their thickness is even lower than those of most coatings. UV-curable printing inks applied for example by offset printing have typical thicknesses in the range between 0.5 and 3 g/m^2 (which very roughly corresponds to their thickness in micrometers). Moreover, printing speeds are at least one order of magnitude higher than in coating technology, which further increases the specific requirements. Despite these difficulties it has been demonstrated that the conversion in printed layers as well as their thickness may be predicted with surprisingly high precision from the NIR spectra. The first part of this chapter will provide a survey on the current state of the art of in-line monitoring of the properties of polymer coatings and printed layers by NIR spectroscopy.

Synthetic fibers and textiles play an important role in our daily life. They are widely used not only for clothes, but also for interior design of living rooms, offices, cars and other means of transportation as well as in numerous technical applications. Nevertheless, despite their widespread use, the diversified production and processing technologies and the high quality requirements, synthetic textiles have been rarely the subject of specific spectroscopic investigations dealing with their characterization or the monitoring of production and finishing processes. The majority of analytic studies is dealing with textiles made of natural fibers such as cotton, wool, flax, silk etc. However, technical textiles, which make particularly high demands on quality and compliance with the specification, are mostly based on fibers and textiles of synthetic polymers such as polyethylene terephthalate (PET), polyamide (PA), polypropylene (PP) or more special materials such as poly(p-phenylene terephthalamide) (para-aramid). Moreover, they are often subject of special treatment and finishing steps such as impregnation, coating, lamination etc. which have to be controlled in order to ensure adequate processing. Recently, NIR spectroscopy has been discovered as an appropriate tool for process monitoring and quality control in textile technology. Moreover, sorting of textiles with the objective of recycling the materials becomes a more and more established practice, and NIR spectroscopy is used for identification. The second part of this chapter will give an overview of the current status of the use of NIR spectroscopy in textile technology for the characterization of materials and in-line control of production and converting processes.

21.2   Polymer Coatings and Printed Layers

21.2.1   Specific Challenges of the Analysis of Coatings and Other Thin Layers by NIR Spectroscopy

21.2.1.1   Spectroscopic Characterization of Polymer Coatings

Polymer coatings are present everywhere in our daily life. They are applied to numerous materials for an extremely broad spectrum of functions ranging from decorative purposes only via protection against various external influences up to highly specialized coatings that provide the material with certain functional properties. Coatings can be made from a multitude of polymers, and also the methods of their application show a wide variety. Except for spray coating with solvent-containing varnishes or polymer dispersions one of the most widespread methods for the preparation of coatings is the application of resinous materials such as viscous monomer or oligomer formulations, which may be cross-linked by thermosetting (e.g., epoxy resins), UV photopolymerization (e.g., acrylates/methacrylates) or as reactive two-component system (e.g., PUR varnishes, polyester resins). Such coatings resulting from cross-linking reactions are characterized by high durability, excellent resistance to various impacts (mechanical, chemical, moisture, weathering, etc.), good adhesion and many other beneficial properties. However, it is obvious that the properties of such systems strongly depend on the degree of cross-linking that is achieved during curing. In particular, this applies for many protective and functional properties. If such coatings are applied in continuous processes such as roll coating, insufficient cross-linking may have fatal consequences because the high production speed will rapidly lead to large amounts of rejects. Consequently, monitoring the application process and the actual state of the applied coating would be highly useful for an efficient process control because it enables rapid intervention in case of serious deviations from the specification, e.g., the degree of cure.

Surprisingly, only very few attempts have been made so far to control technical coating and curing processes by analytical methods. Undoubtedly, this is related to the high experimental requirements. Appropriate measuring methods must have very high sensitivity since the conversion has to be determined in thin layers with a thickness in the range of some microns only. They must be able to record data at high sampling rates because of the usually high web speeds in coating technology. Moreover, high reproducibility and reliability of the data as well as robustness of the instrumentation to withstand the conditions in technical environments (dust, variations in temperature, vibrations, etc.) are required. The experimental method must not damage the material, hence measurement in a non-contact mode is desired.

The most promising analytical method for in-line monitoring of curing reactions in coating technology is NIR spectroscopy. It is non-destructive and possesses sufficient time resolution and sensitivity as well as comprehensive analytical potential for quantitative monitoring of chemical reactions. Measurements can be easily carried out in reflectance mode, which strongly expands the range of samples that are accessible to this technique. However, there is one significant potential obstacle of NIR-based techniques with respect to the characterization of coatings: the extinction coefficients in the near-infrared are much lower than in the mid-infrared range. Whereas this characteristic is useful in most other application areas of NIR spectroscopy in process control, where samples are typically rather compact objects with thicknesses of several millimeters or even more, it is rather unfavorable in case of the investigation of thin layers with a thickness in the range of some micrometers only. This constellation makes very high demands on the sensitivity of the spectroscopic equipment used for in-line monitoring as well as to the efficiency of the chemometric methods applied for quantitative analysis of the data.

Possibly, it is attributed to these experimental challenges that only very few studies dealing with the spectroscopic monitoring of coating processes can be found in the scientific literature. Most papers on the analytics of "coatings" by NIR spectroscopy are related to coatings on tablets or microspheres in pharmaceutics, which will not be discussed in this chapter. Studies on NIR-based monitoring of coating and curing processes with respect to the application of technical coatings are primarily focused on UV-cured coatings and printed layers. Therefore, this paragraph will mainly give an outline about the characterization of such coatings and layers.

21.2.1.2 Coatings Made by UV Photopolymerization

UV-curable coatings are advantageous with respect to saving of energy, environmental protection and waste reduction since UV curing consumes less energy than thermal drying or curing. Furthermore, solvent-free varnishes are widely used in UV curing technology, which may reduce environmental pollution. Due to the almost instantaneous cross-linking of the coating during the short irradiation with UV light (typically some tens of milliseconds), UV curing is a highly efficient coating technology. Moreover, the wide variety of acrylic and other monomers grants access to coatings with a broad spectrum of functional properties.

The most important parameter of UV-cured coatings is the conversion, which determines mechanical properties such as abrasion and scratch resistance or hardness, but also the content of extractables as well as their migration, chemical stability, weathering resistance, etc. Furthermore, sufficient conversion is also required for further processing of the coatings. For instance, wipe resistance must be achieved before stacking or winding. The conversion depends on a large number of chemical and technical parameters. Apart from the chemistry of the specific monomers (radical or cationic polymerization), the most important factor is the applied irradiation dose, which is given by the line speed and the irradiance of the incident UV light. Variations in the composition of the reactive formulation or the ambient conditions (e.g., temperature, humidity, inertization) may affect the conversion as well. Unfortunately, only some of these parameters can be easily controlled under the conditions of technical UV curing, whereas it is too complex for other influences. Consequently, monitoring of UV-induced photopolymerization reactions is an important issue.

In stationary applications of NIR spectroscopy, the conversion may be obtained from the ratio of a specific band of the corresponding functional group before and after irradiation. In case of acrylates, methacrylates, vinyl ethers, etc. the band of the first overtone of the C-H stretching vibration of the carbon double bond at 1620 nm can be used for quantification. This band is sufficiently separated from the corresponding absorptions of other C-H bonds (~1670–1750 nm), which usually allows integration without complex pretreatments such as band separation. In contrast, no specific band is available for the epoxide groups in cylcoaliphatic epoxy resins [10]. However, this approach comes to its limits in most in-line process control applications. For example, variations of the thickness during the process due to changes of the line speed [11] or the inherent polymerization shrinkage (e.g., of acrylic coatings [12]) distort the band ratio and prevent a correct determination of the conversion. Moreover, the precision of the integration method decreases at very high conversions. Therefore, almost all applications of NIR spectroscopy in process control are based on efficient chemometric methods.

The preparation of photopolymerized samples with a predefined conversion, which are required for calibration, is hardly possible. Therefore, the conversion resulting from the application of different UV doses and/or the use of different amounts of photoinitiator in the samples has to be characterized independently. Usually, the preferred reference method for the conversion is FTIR spectroscopy in the mid-infrared range. In case of (meth)acrylates, quantification is obtained by band integration of the peak of the $= CH_2$ twisting vibration of the acrylic bond at 810 cm^{-1}. However, infrared spectra obtained by FTIR-ATR spectroscopy reflect the chemical state close to the surface only due to the limited penetration depth of IR radiation into the layer, which is typically a few microns if a diamond is used as ATR crystal (~ 1-3 μm) [13]. However, many coatings in UV curing technology are much thicker (tens of μm). Moreover, UV-cured coatings often show a strong gradient of the conversion with increasing depth due to the limited penetration of UV light. Thus, ATR spectra do not provide precise information about the conversion across the profile of a coating.

In contrast, it is well known that NIR radiation penetrates to much greater depths of organic coatings as a consequence of the lower molar absorptivity in this spectral region, in particular in transparent systems. Thus, NIR spectra give an average response of the whole depth profile of the coating. Therefore, conversion data obtained from the ratio of band integrals before and after irradiation represent an average value of the conversion within the coating. Consequently, they are better suited for the calibration of chemometric models [14].

21.2.1.3 Interference Suppression in the Spectra of Thin Polymer Films

In order to be able to monitor the properties of coatings on both transparent and intransparent substrates with the same equipment, NIR spectra are taken almost exclusively in reflection in process control. Due to the low reflectance of transparent substrates such as thin polymer films, the spectra of coatings on those materials are

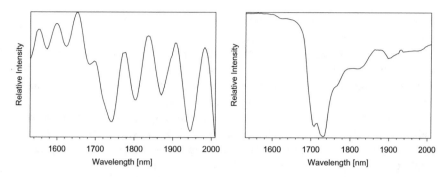

Fig. 21.1 NIR reflection spectra of a 10 g/m² acrylate coating on 20 μm PP film after UV irradiation. **a** without diffuser plate (left), and **b** with diffuser plate (right). Moreover, the probe head was tilted against the surface normal of the polymer film. Reprinted with permission by Wiley-VCH from Ref. [15]

recorded in transflection mode using a ceramic or metallic reflector that is placed underneath the coated film. However, in the spectra of thin transparent films of optically high-grade polymers such as polypropylene (PP), polyester (PET), polycarbonate (PC), etc. that have a thickness in the range of or only little higher than the wavelength of the probe light (i.e., up to ~30 μm) interference fringes may appear, which result from the superposition of the incident probe light with the light reflected at the front and back surfaces of the polymer film. Such interference patterns may completely mask the spectrum and accordingly prevent any analysis of the properties of the coating, which is applied to the film (see Fig. 21.1a). The removal of the interferences by mathematical means, i.e., Fourier transformation of the spectrum and cut-out of the sharp peak resulting from the sine wave, fails due to inevitable vibrations of the film web in the roll coating machine. Moreover, the calculation of the Fourier transform is too time-consuming for process control applications. Instead, the interference problem can be overcome by experimental means, i.e., (i) by use of a diffuser plate, which is mounted between probe head and sample, and (ii) by a tilt of the optical path of the incident light against the surface normal of the polymer film [15]. The combination of both approaches suppresses the interferences very effectively (see Fig. 21.1b) and thus enables quantitative analysis of coatings on thin transparent polymer films.

21.2.2 Monitoring of the Thickness of Coatings by NIR Spectroscopy

Although there are several measuring methods and specific sensors for in-line monitoring of the thickness of extensive laminar materials such as paper, polymer films and textile webs, there is no established method to monitor the thickness or the surface weight of UV-cured or other polymer coatings applied to such substrates. Currently, control of the thickness of the applied layers in technical coating processes is mostly carried out off-line by gravimetric determination of the coating weight. However, it is apparent that this approach is not able to respond to sudden changes of the thickness, and thus it is poorly suited for process control. However, it was shown that thickness or coating weight can be followed by NIR reflection spectroscopy under the conditions of technical coating processes [16, 17].

Clear acrylate coatings with thicknesses in the range from 5 to 100 μm were studied [16]. Although calibration samples were prepared with different thicknesses for PLS modeling with a set of various Baker applicators with well-defined gaps, the resulting thickness had to be measured after UV curing mainly due to the well-known shrinkage of acrylate formulations during the cross-linking reaction and spreading of the applied wet coating. Shrinkage and other effects affecting the thickness may lead to a decrease of the thickness up to 30%. Instead of direct thickness measurements (e.g., with a thickness gauge), gravimetry may be used alternatively to provide reference data. Depending on the range covered by the specific PLS model, prediction errors were found to be in the order between less than one and about 4 μm [16, 17]. Similar results were also found for white-pigmented coatings containing 10 wt% titanium dioxide [18]. For example, the thickness of such coatings with a thickness from 5 to 60 μm was predicted with a precision of about 1 μm.

Moreover, in-line monitoring of the thickness was carried out at a pilot-scale roll coating machine. Quantitative data were recorded at line speeds up to 50 m/min. Very close correlation between data predicted from NIR spectra and reference data determined off-line after the end of the coating trials was observed. Prediction errors were found to be very similar to those obtained in the external validation tests although the experimental conditions at the coating machine were less optimal than in the lab. In particular, in-line measurements were exposed to disturbing factors such as vibrations of the coating machine including the mount of the probe head and flutters of the moving film web. In order to simulate abrupt changes of the thickness during a real coating process, the nip between the applicator rolls was stepwise reduced and increased. An example for clear coats is shown in Fig. 21.2 [17]. Similar investigations were also carried out for pigmented coatings [18]. It was demonstrated that changes in layer thickness lower than 1 μm can be clearly detected by NIR spectroscopy.

The thickness of clear and pigmented UV-cured coatings was also monitored at different line speeds up to 100 m/min [16–18]. A marked increase of the thickness with increasing speed was observed up to about 50 m/min, which was followed by a decrease at even higher web speeds. Although this behavior seems to be somewhat surprising, it is well-known in coating technology [19, 20]. In fact, it is due to a very complex interaction of several physical and technological parameters of the coating process. In particular, acrylate formulations are typical non-Newtonian fluids, i.e., their viscosity depends on the shear rate. In-line NIR spectroscopy offered for the first time the unique possibility to monitor the thickness changes upon variation of the line speed directly during the coating process in a quantitative manner.

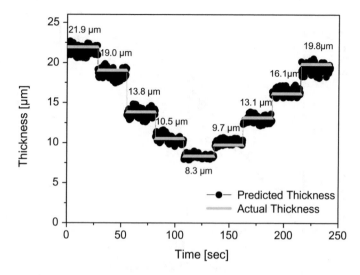

Fig. 21.2 In-line monitoring of the thickness of an acrylate coating on 20 μm PP foil upon stepwise decrease and increase of the nip between the applicator rolls of the coating machine at a line speed of 40 m/min. Actual values were obtained off-line by a thickness gauge. Reprinted with permission by Elsevier B. V. from Ref. [17]

The dependence of the applied thickness on the web speed has serious consequences for the in-line measurement of the conversion. Calibration models are typically based on calibration samples that have different conversions, but the same thickness of the coating. Variation of the thickness in the process as compared to the calibration samples may lead to significant mispredictions of the conversion during process control. Approaches to avoid such errors will be discussed paragraph 21.2.3.3.

Pigmentation of varnishes hampers the penetration of near-infrared radiation into polymer coatings. At high thicknesses, this may lead to a non-linear relationship between the thickness of the coating and its reflectance. Therefore, several investigations were carried to study the effect of increasing thickness on the total absorbance in the NIR spectrum quantitatively in order to determine the maximum thickness, up to which the thickness of pigmented coatings can be predicted from NIR spectra using PLS algorithms. Obviously, this limit depends on the kind and the percentage of pigmentation as well as on the chemistry of the specific varnish. In case of white-pigmented acrylate coatings (10 wt% TiO_2), linearity was found for thicknesses up to 300 μm at least [21]. This is possible since titanium dioxide does not significantly absorb in the near-infrared region. In contrast, investigations on a red pigmented PUR varnish applied by spray painting to steel sheets clearly showed that the integral of the absorbance in the region of the first C-H stretching overtones (1650 to 1850 nm) was linear to the thickness up to 180 μm only, which restricts the effective measurement range of NIR spectroscopy for thickness measurements (Fig. 21.3).

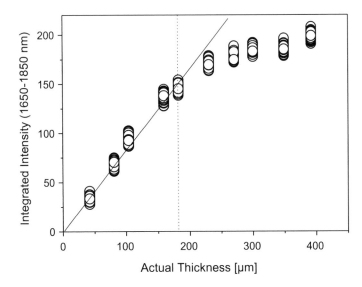

Fig. 21.3 Integrated intensity of the absorbance in the first overtone region of C-H stretching vibrations in dependence on the coating thickness of a red pigmented PUR varnish applied to steel sheets by spray painting

PUR coatings with thicknesses higher than 100 μm may be used for example for protective coatings against weathering in demanding outdoor applications.

In case of very thin coatings, the sensitivity becomes an issue that has to be considered with respect to the detection limit of the NIR method, since the low extinction coefficients in the NIR region pose a considerable challenge. For example, thin acrylate layers with a thickness of a few microns only can be photopolymerized without addition of a photoinitiator by irradiation with short-wavelength UV radiation. The initiating mechanism is based on direct excitation of the acrylate molecules by photons with high energy. An investigation by NIR in-line monitoring at a roll coating machine [22] demonstrated that a photoinitiator-free acrylate coating with an application weight of 4 g/m^2 can be cured by irradiation at 222 nm under inert conditions using a KrCl* excimer double lamp system (175 mW/cm^2). Adequate conversion was achieved at line speeds up to 30 m/min. Despite the low thickness of the coating, spectra were found to show sufficient signal-to-noise ratio to provide significant conversion data.

Ink jet printing is widely used not only in home office applications, but also for labeling in packaging technology, in electronics, nanotechnology and similar applications. Inks for such applications are aqueous systems, which poorly wet most polymer surfaces. Therefore, very thin hydrophilic ink absorption layers have to be applied to polymers to make them printable. The typical thickness of such coatings may be 1 g/m^2 or less. JIANG et al. [23] applied polyvinyl alcohol layers with nanoscale titania particles to PET films with thicknesses between 0.25 and 1.25 g/m^2. For controlling the application process and the quality of the applied layers, they developed a method based on FT-NIR spectroscopy. Several parameters such as coating weight, gloss and smoothness were predicted from appropriate PLS models. In case of the coating weight, reference data were obtained by gravimetry before and after dissolution of the layers. RMSEP was found to be 0.054 g/m^2 with a coefficient of determination (R^2) of 0.98. Convincing results were also obtained for the smoothness, whereas the method was of limited suitability only for the determination of the gloss. It was demonstrated that all three parameters could be also predicted in-line during the coatings process.

21.2.3 Conversion of UV-Cured Coatings

21.2.3.1 Acrylic Coatings

The conversion that is achieved during UV irradiation is the most important parameter of any UV-cured coating. As outlined above, it determines all functional and handling properties of the coatings, which are relevant for their application or further processing. Consequently, continuous monitoring of the conversion during a coating process would be highly desirable. However, until recently no efficient analytical method was available, which allowed direct in-line monitoring of this important parameter. At-line or off-line determination of the conversion is state of the art, which is often carried out by FTIR spectroscopy. But IR spectroscopy in the mid-infrared (MIR) region is not suited for in-line control, since non-contact reflection measurements over a rather large measuring distance are difficult or even impossible.

In contrast, vibrational spectroscopy in the near-infrared range offers this possibility, which provides an opportunity for monitoring the conversion during the coating process. The overtone band of the C-H stretching vibration of the carbon double bond at 1620 nm is well separated from the corresponding bands of other C-H structures (methyl, methylene etc.). Hence, it might be used directly for the determination of the conversion in acrylates and methacrylates using band integration. However, this requires recording of some spectra of the uncured coating before irradiation, which is often difficult to realize in complex coating processes. Moreover, this approach involves considerable dangers resulting from unintended but inevitable changes of the thickness of the coating as discussed in paragraph 21.2.1.2. For this reason, chemometric modeling is the more time-consuming, but also the more reliable approach. Depending on the specific varnish (e.g., clear or pigmented), its thickness and other parameters, MIR (in transmission or using ATR) [15, 24] or NIR spectroscopy (band integration) [14, 21], Raman spectroscopy or other analytical methods [15] can be used as reference method for calibration. Even secondary parameters that depend directly on the conversion (e.g., the hardness of the coating) may be used as reference data for calibration models. A typical PLS-based calibration model for the conversion in an acrylic clear coat is shown in Fig. 21.4.

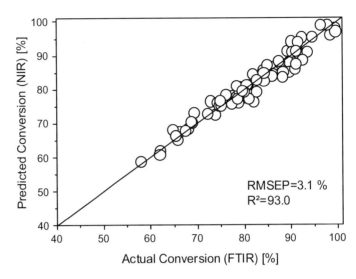

Fig. 21.4 PLS calibration model for the conversion in an acrylic clear coat. Reprinted with permission from Ref. [24]. Copyright 2010 American Chemical Society

Typically, the prediction error (RMSEP) of such models is in the order of 2–3% [14, 21, 24]. In most cases, the corresponding error of the external validation with independent test samples is only marginally larger.

When the probe head is mounted to a specific coating machine (e.g., a roll coating machine, conveyor, etc.) for in-line monitoring of the conversion after UV or electron beam (EB) irradiation, great care has to be bestowed on the correct alignment of the probe head with respect to its distance and the tilt angle relative to the film web and, if applicable, to the reflector behind the foil. It is obvious that an exact match of the measurement conditions is essential for the successful transfer of chemometric calibration models from the laboratory to process control in technical scale. It has been found, that even very minor differences between both arrangements substantially affect the precision of the predicted values.

In order to demonstrate the potential of NIR spectroscopy for monitoring changes of the conversion due to variations of the irradiation dose, both the power of the UV lamp (or the beam current of the EB accelerator) and the line speed can be varied repeatedly. An example is given in Fig. 21.5 [25]. It can be clearly seen that the conversion increases or decreases according to the applied UV dose. Evidently, changes of the line speed of the roll coating machine lead to an immediate change of the conversion. In contrast, changes of the power of the lamp appear after a delay only, what is due to the rather slow response of the mercury arc lamp, when its power is switched to a higher or lower level. Similar gradual increases or decreases of the conversion were also found for changes of the power of the EB accelerator [15, 25]. In contrast, faster response might be expected for UV LED light sources.

Analog results were obtained for acrylic coatings on both various transparent polymer films or non-transparent substrates such as paper, cardboard, opaque

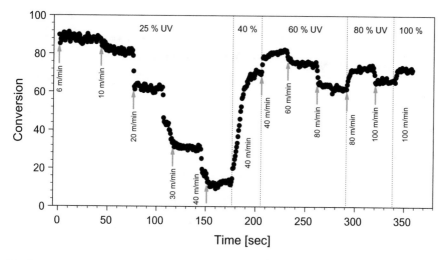

Fig. 21.5 In-line monitoring of the conversion in a 20 g m⁻² acrylic clear coat on PET film (36 μm) after UV irradiation with variable power of the mercury arc lamp and at various line speeds of the roll coating machine. The power of the UV lamp is given in percent of its maximum output. Reprinted with permission by Elsevier Ltd. from Ref. [25]

polymer films etc. [15, 25, 26]. The only difference is that a diffuse reflector has to be placed behind the film web in case of transparent materials, whereas paper serves as a reflector on its own volition. The conversion in white-pigmented varnishes was monitored in the same way as well [25].

UV curing is widely used in the wood industry, e.g., for furniture and flooring. Therefore, the conversion in acrylate coatings on rigid substrates such as fiberboard or wood was monitored. In this case, both the UV lamp and the NIR probe head were mounted above a conveyor. Since the line speed of a conveyor is usually lower than that of a roll coating machine for the application of varnishes on web-like materials, time resolution of the monitoring method is less crucial. Hence, sampling rates of 2–10 spectra/s were used. It has been demonstrated that in-line monitoring by NIR reflection spectroscopy is well suited for this kind of applications as well [15].

A special type of UV-curable formulations are UV-curable adhesives. High-molecular weight acrylic copolymers containing photoreactive groups such as benzophenone in their side chains may be used as pressure sensitive adhesives (PSA). Cross-linking of these hot melt adhesives is required in order to improve their resistance to thermal distortion. However, their adhesive properties (i.e., peel strength and shear strength) were found to be heavily dependent on the conversion. Even minor changes of the degree of cure may affect the adhesive properties adversely. Therefore, a continuous control of the UV curing process is essential for ensuring optimum adhesive properties. For monitoring the conversion in acrylic hot melts, the probe head was mounted at a slot die coating machine [27]. The adhesive layers were applied to OPP film at 90 °C with coating weights between 200 and 500 g/m² and at rather low line speeds (2 to 5 m/min). After application, they were irradiated

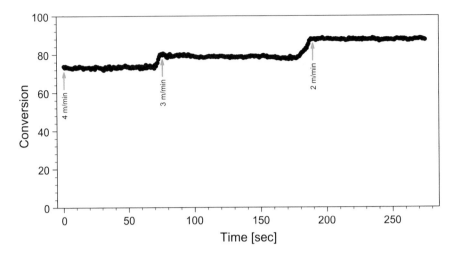

Fig. 21.6 In-line monitoring of the conversion in a 500 g m^{-2} coating of an acrylic hot melt adhesive on PP tape after 308 nm UV irradiation at various line speeds. Reprinted with permission by Elsevier Ltd. from Ref. [25]

with monochromatic UV light (308 nm) using a XeCl* excimer lamp. In spite of the very high thickness of the layers, good through cure was achieved due to the low extinction coefficient of benzophenone at the wavelength of irradiation.

In Fig. 21.6, a record of in-line monitoring of the conversion in an UV-cured acrylic hot melt after irradiation at different line speeds is shown. Similar data were also obtained upon variation of the UV intensity [25]. It is evident, that the scattering of the conversion data is distinctly lower than that of the acrylic coatings in Fig. 21.5. This is due to the much higher thickness of the adhesive layers (500 versus 20 g/m^2), which resulted in a higher signal-to-noise ratio of the recorded NIR spectra.

21.2.3.2 Epoxy Coatings

Cationic monomers and oligomers form another class of UV-curable systems. In particular, they involve cycloaliphatic epoxy resins and vinyl ethers, which may be used as photoreactive monomers and reactive diluents, respectively. Cationic photoinitiators such as diazonium, diaryliodonium or triarylsulphonium salts form Brönsted or Lewis acids as initiating species. Typically, the reaction rates of cationic photopolymerization reactions are lower than those of free-radical reactions. Consequently, the conversion in cationic formulations is mostly still rather low just after passage through the UV lamp. However, it is well-known that the cationic polymerization once initiated continues for a long time in the absence of light. This post-curing effect leads to a substantial further increase of the conversion. This time delay

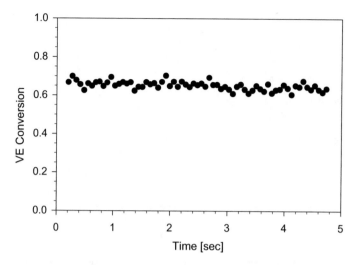

Fig. 21.7 In-line monitoring of the conversion in a coating made of 85 wt% of a cycloaliphatic epoxy resin and 15 wt% of a divinyl ether applied to OPP film at a line speed of 3 m/min. Reprinted with permission by Wiley-VCH from Ref. [10]

between irradiation and ultimate conversion implies significant consequences for in-line monitoring of such reactions. Usually, the NIR probe head is positioned immediately after the UV lamp. Consequently, the conversion that is found at this position is still pretty low. Figure 21.7 shows the conversion in a formulation consisting of a cationic epoxy resin (3,4-epoxycyclohexylmethyl-3',4'-epoxycyclohexane carboxylate; EEC) and a vinyl ether (tetraethyleneglycol divinyl ether; DVE-4) just after UV irradiation with a dose of about 2800 mJ/cm^2, which is only about 65% [10]. Although this is sufficient for winding-up the coated polymer film without blocking, it is far from the final conversion that determines the application properties of the coating. Consequently, NIR spectroscopy may provide a rough indication for the current state of the curing process, but no definite final conversion of the coating.

Evidently, the further development of the conversion during the dark reaction can be investigated only off-line. Some results are summarized in Fig. 21.8 [10]. It is apparent that most of the increase of the conversion was achieved within one hour after irradiation. Moreover, the results clearly reveal that both the initial and the ultimate conversion strongly depended on the applied irradiation dose. Whereas almost full conversion (98%) was achieved after two hours in coatings exposed to 2100 mJ/cm^2, the curing reaction leveled off at lower conversions after irradiation with lower doses.

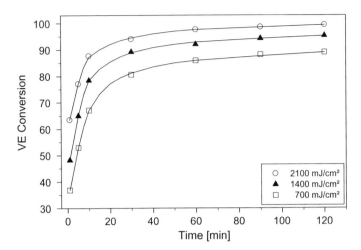

Fig. 21.8 Kinetics of the postcuring reaction of a mixture of a cationic epoxy resin (85 wt% EEC) and a divinyl ether (15 wt% DVE-4) after irradiation with various UV doses (line speed 10 m/min). Reprinted with permission by Wiley-VCH from Ref. [10]

21.2.3.3   Simultaneous Measurement of Thickness and Conversion

It was already mentioned above that variations of the coating thickness, which may be caused by accidental fluctuations or by intentional changes of the web speed, may considerably impair the quantitative monitoring of the conversion in UV-cured coatings regardless of the use of either band integration or chemometric methods. In case of the band integration method, the reference spectra defining the "zero conversion line" are no longer valid after any thickness change, whereas PLS1 models for the conversion are built up for coatings with a broad range of the conversion, but with one well-defined thickness only. However, the negative impact of variations of the thickness on the quantitative determination of the conversion during in-line monitoring can be overcome for both approaches.

If simple band integration methods are used for quantification, thickness changes may be included by a more complex analytical instrumentation. Instead of the use of a single NIR probe head, which takes spectra of the cured coatings after irradiation, two probe heads can be installed before and after the UV lamp. They may be linked to one spectrometer if a multiplexer is used to switch between both optical entrance ports. The conversion is obtained by band integration of the acrylate band at 1620 nm and calculation of the ratio of the band integrals before and after UV irradiation. In order to consider the time delay between the two probe heads, a specific offset that depends on the line speed has to be included in the evaluation scheme. Obviously, this approach corresponds to the conventional determination of the conversion in the analytical laboratory by recording spectra before and after irradiation. The efficiency of this analytic approach to compensate the effect of thickness changes on the determination of the conversion has been demonstrated by in-line monitoring experiments at a roll coating machine [11].

Alternatively, the variation of the thickness of the coating can be included as an additional parameter into the chemometric calibration models, which is certainly the most sophisticated approach. This can be achieved by use of the PLS2 algorithm, which is able to predict two or even more parameters from the same input data. Of course, this implies much higher efforts for calibration, since the calibration samples have to cover the full range of both conversion and thickness that can occur during process control. On the other hand, the higher complexity of the calibration process is offset by the fact that simultaneous prediction of both conversion and coating thickness can be achieved while using one NIR probe head only.

The suitability of this approach was demonstrated during application of clear and pigmented acrylate varnishes to PP film at a roll coating machine and their cross-linking by UV irradiation [24]. Reference samples for the setup of the PLS2 model covering a broad spectrum of combinations of conversion and thickness were characterized by FTIR spectroscopy and a thickness gauge. During roll coating, the line speed was increased in incremental steps from 20 to 100 m/min, which inevitably led to changes of the thickness as well as the conversion due to the corresponding decrease of the irradiation dose. Both parameters were predicted from the recorded NIR spectra using the PLS2 model. Results of one specific coating experiment are shown in Fig. 21.9. For quantitative evaluation of the predicted data, reference values were determined off-line for each step of the line speed after the end of the coating trial. These values are included in Fig. 21.9 as well. The data clearly prove that the conversion was predicted with a precision of 2–3% even under the conditions of changing thickness, whereas the error in the in-line measurement of the thickness was found to be about 0.5 to 1 μm. These error margins correspond to those found in independent predictions of both parameters by PLS1 models (see paragraphs 21.2.2. and 21.2.3.1.).

21.2.4 Hyperspectral Imaging of UV-Cured Coatings

In the studies reported so far, the parameters of interest (thickness, coating weight, conversion etc.) have been monitored by conventional NIR spectroscopy. Since the spot of most NIR spectrometers has a diameter of a few millimeters only, this means that these parameters can be determined in a small stripe only during monitoring a running process. However, for coatings as two-dimensional objects the spatial distribution of the parameter under investigation might be of interest for a comprehensive process control. For example, UV lamps have rod-like shape, which may lead to a decay of the intensity at their ends due to aging, pollution etc. In UV LED systems, single elements may drop out. Both effects lead to a reduced conversion in the corresponding region, which might be possibly insufficient for the intended application. Also in case of the application weight, local deviations may occur, e.g., due to running out of the varnish formulation in the roll gap or the reservoir of the coating machine,

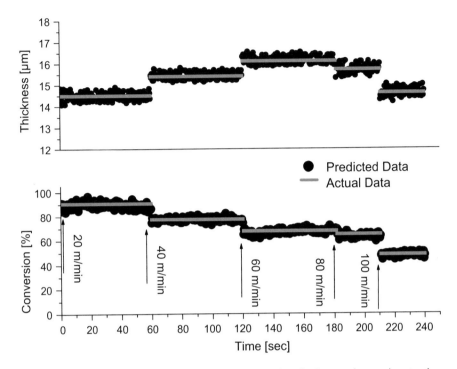

Fig. 21.9 Simultaneous monitoring of thickness and conversion of a clear acrylate coating at various line speeds of a roll coating machine. Quantitative data were predicted with a PLS2 calibration model. Gray lines represent reference values that were determined off-line by FTIR transmission spectroscopy or a thickness gauge, respectively, after the end of the coating experiment. Reprinted with permission from Ref. [24]. Copyright 2010 American Chemical Society

inclination of the web, local variation of the viscosity (e.g., due to temperature gradients) etc. In order to include the spatial distribution of the parameter of interest in process control, two-dimensional monitoring of the process is required. This is the domain of hyperspectral monitoring.

Traditionally, hyperspectral cameras have been mainly used for three-dimensional or at least rather thick objects, which resulted in applications in waste sorting, food monitoring (meat, fish, vegetables, fruits etc.) and the characterization of other objects in agriculture (e.g., logs and boards), art conservation etc. However, today's high-class hyperspectral cameras have sufficient sensitivity to monitor even thin samples such as coatings, laminates, finished textiles (see paragraph 21.3.3.), printed conductive polymer layers, pages in medieval illuminated manuscripts [28] and similar two-dimensional objects. Therefore, hyperspectral imaging was also used for monitoring thickness, conversion and other parameters of coatings. With an adequate calibration, each of these parameters can be predicted quantitatively. However, the main intention of the use of hyperspectral cameras in coating and lamination technology is monitoring the spatial distribution of these parameters across the samples as well as the detection of possible inhomogeneities.

PUR foam is widely used for cushioning and sound insulation. For faster and easier mounting in end-use applications such as automotive engineering, it may be provided with an adhesive layer forming a semifinished product. Due to the mostly coarse surface structure of such foam plastics, the adhesive cannot be applied as melted mass. Rather, melt-spun fibrous webs of thermoplastic adhesives such aliphatic polyamide or polyester are laminated to the surface of the foam mat. Higher thicknesses may be achieved by application of several adhesive webs. In this way, applications weights of the adhesive between 20 and 125 g/m^2 were achieved. The development of calibration models for the prediction of the coating weight is significantly impeded by the fact that calibration samples with well-defined homogeneous thickness can be hardly prepared due to the numerous open bubbles at the foam surface as well as due to the fibrous structure of the adhesive webs. Moreover, some of the PUR substrates are rather dark (e.g., dark gray), which leads to rather low reflectance. In order to consider a possible non-uniform distribution of the coating weight across the samples, the surface of each sample was divided into 20 rectangular regions by defining a grid of 5 columns × 4 rows. Spectra from each rectangle (several thousands in each) were averaged. The averaged spectra were allocated alternatingly, that is, according to a chessboard pattern, to the calibration and the validation set. This procedure resulted in RMSEP values of only 1.5–3.5 g/m^2 despite the heterogeneity of the adhesive web and the PUR foam substrates [29, 30].

Using these calibration models, the application weight of adhesive layers was quantitatively monitored with a hyperspectral camera mounted above a conveyor. At low coating weights, the adhesive layers shows significant local thickness variation (i.e., for samples with up to two adhesive webs), whereas it becomes more homogeneous at higher coating weights [29]. This effect mainly results from surface structure of the PUR foam. During the melting process of the web, the adhesive might partly flow into the open bubbles at the surface of the foam leading to an uneven distribution. For an overall evaluation of the quantitative prediction results, the individual values predicted from the spectra across the complete surface of each sample were averaged and compared to gravimetric values. Deviations were found to be less than 2–3 g/m^2 (the fact that some values are lower than RMSEP is related to the averaging process).

Spectral imaging was also used for the detection of inhomogeneities and coating errors, which may occur during the lamination process. During melting, the adhesives form continuous glossy transparent layers on the foam substrates, which usually prevents the detection of coating errors by visual inspection. In contrast, spectral imaging is able to highlight them. As an example, Fig. 21.10 shows the image of a coated foam sample, where the corner of one adhesive web was fold over during the lamination process [30].

Thickness and conversion of UV-cured acrylate coatings were studied by hyperspectral imaging as well. In particular, white-pigmented coatings were investigated. Calibration to the thickness was carried out for coatings with thicknesses up to 200 μm, although linearity between thickness and NIR signal was even found for much thicker coatings [21]. Calibrations to the conversion resulted in RMSEP values between 2 and 3% [14, 21]. However, specific calibration models were required for

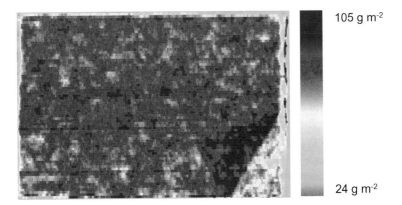

Fig. 21.10 Hyperspectral image of an adhesive layer of an aliphatic polyamide laminated on dark gray PUR foam. The lower right corner of one of the adhesive webs was fold over. Reprinted with permission by Elsevier B.V. from Ref. [30]

each varnish and each substrate. In order to reduce the considerable efforts required for the preparation of samples with different conversions, procedures for the transfer of calibration models to other substrates (e.g., glass, steel, glass fiber reinforced plastic plates) were developed. It was shown that only a marginal increase of the prediction error is caused by such transfers.

After the development of calibration models, both thickness and acrylate conversion were monitored with a hyperspectral camera installed above a conveyor belt. Similar to investigations by conventional NIR spectroscopy, quantitative data were determined for both parameters. It is well-known that curing of thick white coatings is one of the most difficult tasks in UV curing technology, since the excellent scattering power of white particles such as titanium dioxide strongly impede the penetration of UV light into the coating. This leads to strong conversion gradients within the coating. In fact, a conversion of less than 70% was found for coatings with a thickness of ~ 75 μm, which represents an average across the profile of the coating with almost complete conversion at the surface and rather low conversion in the depth. Aging of such samples led to some "self-structuring" of their originally homogeneous surface within a few days due to shrinkage and subsequent relaxation of the internal stress, which finally resulted in some kind of orange peel effect. The local thickness and conversion differences developed during this self-structuring process were clearly reflected in hyperspectral images taken from the relaxed samples [21].

Conversion differences within a coated sample can be also intentionally induced by generation of thickness differences. Although the degree of cure at the surface is certainly the same across the sample, the averaged conversions differ upon variation of the thickness due to the conversion gradient. Examples of a uniformly cured coating and a sample with conversion differences resulting from thickness changes are shown in Fig. 21.11 [14].

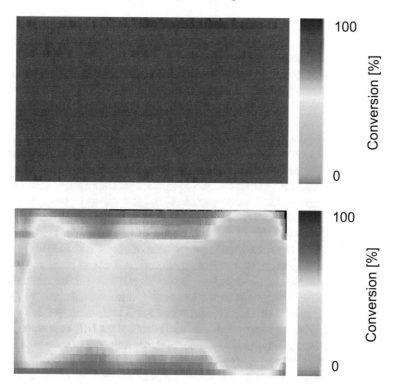

Fig. 21.11 Hyperspectral image of white-pigmented acrylate coatings: with homogeneous conversion (87%; top) or with inhomogeneous distribution of the conversion due to differences in coating thickness (bottom). Reprinted with permission by Elsevier B.V. from Ref. [14]

21.2.5 Spectroscopic Techniques in Printing Technology

21.2.5.1 Thickness of Layers in Offset Printing

Beside its application in coating and lamination technology, in dentistry, microelectronics and 3D modeling, for adhesives and composites, etc., UV curing is also widely used in printing technology and graphic arts. UV-curable inks and varnishes complement conventional oil- and water-based systems. Generally, they dry almost instantly, which results in higher printing speeds and hence in higher productivity of the printing process. Viscous UV inks and varnishes show less penetration into porous materials such as paper than other ink systems. Moreover, they result in printed layers with a high-grade optical appearance (e.g., high gloss), which makes them interesting for demanding printing applications, e.g., for packagings in cosmetics or for advertising materials.

The thickness of ink layers determines their color intensity as well as the color shade. Therefore, the exact observance of the specified thickness of the various colored layers is essential for the quality of the printed product. On the other hand, since offset printing is a complex process, numerous factors may influence the thickness of the layers. Consequently, it is indispensable to control the thickness continuously during the printing process.

The thickness of colored printed layers can be monitored in-line by use of reflectance densitometry, which relates the intensity of the reflected light to the color density. However, this method is limited to standard colors (cyan, magenta, yellow and black) only. It does not allow measurements of the thickness of layers with other color shades. Moreover, it is not suited for transparent top coats, which are widely used for the protection of printed layers, for upgrading their optical appearance, for achieving special design effects or for finishing the printed layers with special functional properties. It is obvious that the compliance of the required thickness (or coating weight) is crucial for those properties. Up to now, the surface weight of clear coats can be determined only off-line by gravimetry, which is a labor-intensive and time-consuming procedure. The typical thickness of printed layers is in the order around 1 g/m^2; it might be somewhat higher in case of clear coats. Due to the rather low extinction coefficients, NIR spectroscopy is certainly not an obvious approach to characterize such layers. Nevertheless, it was shown that similar methods like those described above for coatings with thicknesses in the range of about 5–100 μm (paragraph 21.2.2.) can be also used for process control in printing technology.

For monitoring of the coating weight, the probe head of the NIR spectrometer was mounted either above the impression cylinder of the coating unit of a sheet-fed offset printing press or in its delivery system behind the UV lamps. The first position allows easy access and enough space for mounting, but enables analysis of the wet layers only. In contrast, in the delivery unit the probe head has to be installed between the moving elements of the sheet transport system, which strongly limits the available space and requires special probe heads with reduced size and increased focal length. However, this position allows monitoring the thickness of cured layers.

Reference data of the coating weight for calibration models were determined by gravimetry. Test samples with different thickness of the layers were prepared with a printability tester. Gravimetry was also used for crosschecking the data predicted in-line. Blank sheets with known weights were numbered before laying them into the feeding system of the press. After printing, the weights of the printed layers were determined. The numbering of the sheets did not only enable an exact measurement of the coating weights, but allowed also an unambiguous assignment of coating weight and spectra to specific sheets.

During in-line monitoring, samples with different surface weights of ink or varnish were obtained by systematic variation of the operating parameters of the printing press. Investigations were carried out at printing speeds between 6000 and 12,000 sheets/h corresponding to about 90–180 m/min. NIR spectra were recorded at a rate of 30 spectra/s. At 12,000 sheets/h, this corresponds to about 3 spectra per sheet. Synchronization between NIR measurement and printing was achieved by use of a sensor that detected the front edge of each sheet. The coating weight of printed layers was checked by gravimetry after the end of the trial. As an example, Fig. 21.12 shows a result of in-line monitoring of the coating weight of printed layers of a clear varnish [31]. The labels *Opening* and *Speed* refer to specific parameters of the offset

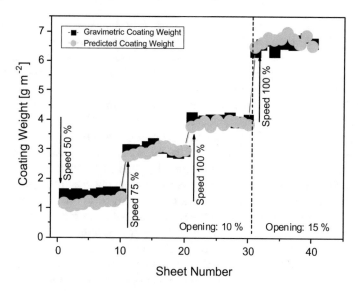

Fig. 21.12 In-line monitoring of the coating weight of a UV-curable printing varnish on paper in a sheet-fed offset printing press (printing speed 8000 sheets/h). For comparison, coating weights determined off-line by gravimetry are shown. Explanation of other labels is given in the text. Reprinted with permission by SAGE Publications Ltd. from Ref. [31]

printing press that control the amount of ink or varnish on the printing plate. An increase of these values corresponds to higher coating weights. The results demonstrate that the applied coating weight can be determined with very high precision. The prediction error was found to be about 150 mg/m^2.

Further studies on both printing inks and clear varnishes were directed toward the investigation of the influence of different printing speeds [32], the recipe of the varnish formulation [32], the color of the ink, the use of different paper and cardboard materials as substrate [33] as well as different gloss levels of the printed layers [34] on the accuracy of the prediction results. Complex calibration procedures including multistage and universal calibration models were developed, which were adapted to the requirements of each specific problem. Despite the rather low thickness of printed layers, the precision of the predictions from NIR reflection spectra recorded in-line is rather high: RMSEP values for the determination of the coating weight were found to be in the order between 120 and 200 mg/m^2.

21.2.5.2 Conversion of UV-Cured Printed Layers

Similar to UV-cured coatings, numerous factors in the printing process may influence the conversion of printed layers. This includes the characteristics of the ink or varnish (e.g., its composition and viscosity), the ambient conditions (e.g. temperature, humidity), and technical factors such as UV irradiance, printing speed and the specific setting parameters of the printing press. Any change of one of these parameters may lead to a change of the conversion of the printed layers. Generally, printing is a high-speed process, which rapidly produces large amounts of printed sheets. Therefore, sufficient conversion must be achieved before stacking in order to avoid blocking of the sheets. Moreover, adequate wipe resistance is required. It would be therefore desirable to monitor the conversion continuously during the printing process in order to ensure high quality and uniformity in this way.

However, monitoring the conversion in a printing machine is extremely difficult. So far, no in-line method for control of the conversion exists, and also its off-line determination is usually impossible in a printing plant, since no adequate analytical equipment for these measurements such as conventional IR spectroscopy is available at a printing press. Typically, indirect methods such as the determination of the coefficient of sliding friction with simple test equipment are used. NIR spectroscopy would open completely new possibilities for direct quantitative (or at least semi-quantitative) monitoring of the conversion during the printing process.

For developing a calibration model, calibration samples with well-defined thickness have to be prepared. As outlined above for coatings (paragraph 21.2.3.3.), variations of the thickness seriously disturb measurement of the conversion and may completely prevent quantitative analysis. Therefore, great care was bestowed on the preparation of printed layers with constant thickness. For each color, a number of printed paper stripes were prepared using a printability tester. The application weight depended on the color of the ink and varied between 0.8 g/m^2 (cyan) and 1.2 g/m^2 (yellow). Moreover, samples with a clear lacquer were prepared with about 3 g/m^2. Printed layers were irradiated with different UV doses in order to obtain samples with a broad conversion range. After recording the NIR spectra, discrete calibration models were developed for each color and the printing varnish. FTIR/ATR spectroscopy was used as reference method. As an example, the PLS calibration of a yellow UV printing ink is shown in Fig. 21.13 [35].

It is apparent that the scattering (and consequently RMSEP) is higher than in case of coatings with thicknesses, that are at least one order of magnitude higher (paragraph 21.2.3.1.). Nevertheless, it can be clearly seen that there is a linear correlation between reference conversion and predicted signal, which clearly proves that the conversion can be predicted from NIR reflection spectra by adequate chemometric tools. Generally, RMSEP for all inks is in the order of 3–4.5%. Due to the somewhat higher thickness of the layers of the printing varnish, RMSEP of the corresponding model was found to be only 2.3%. External validation tests confirmed the precision of the predictions [35, 36].

Using the various PLS models, the conversion of UV-cured ink and varnish layers was monitored directly in a sheet-fed offset press. It is obvious that only the position in the delivery unit behind the UV lamps can be used for mounting the probe head (see paragraph 21.2.5.1.). Except of the spatial limitations mentioned above, the probe head is exposed to vibrations, dust (e.g. talcum powder used for oil-based printing inks handled at the same press), heat, airflows, etc. at this position. Moreover, the printed sheet, which is fixed by clamps at its front edge only, glides on an air cushion through the delivery system, which may lead to sheet flutter. Variations of the

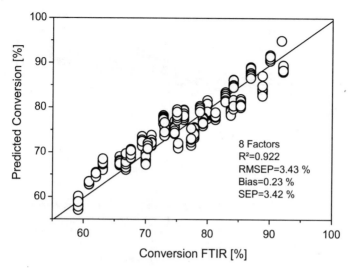

Fig. 21.13 PLS calibration model for the conversion in layers of a yellow UV-curable printing ink printed on 135 g/m^2 paper. Reprinted with permission by Elsevier B.V. from Ref. [35]

distance between probe head and paper are expected to strongly disturb quantitative NIR measurements. After several tests, the probe head was finally installed between the air nozzles in the middle of the bottom plate in order to minimize the influence of flutter on the measurement process [36].

Parameters of NIR sampling were the same as those during in-line monitoring of the thickness. Similar to in-line monitoring tests on coatings, the conversion was intentionally affected by variations of the applied UV dose, that is, by variation of either the intensity of the UV lamps or the printing speed. Figure 21.14 shows the effect of variation of the irradiance on the conversion of a cyan offset printing ink [36]. Apparently, even at irradiation with half intensity only a moderate decrease of the conversion is observed, which is due to the very high reactivity of printing inks. Although the data show somewhat stronger scattering due to the much lower thickness of the layers as compared to typical coatings, the conversion could be determined with rather high precision, which was confirmed by subsequent off-line measurements by FTIR/ATR spectroscopy. Investigations on inks with various colors as well as on clear printing varnishes in in-line monitoring tests at the printing press resulted in typical prediction errors for the conversion of 4–5%, which is only little higher than the error margins for coatings with thicknesses of 10 μm or more (2–3%). Only in case of black inks, it was found to be 5.5% due to the low reflectance of such layers.

The content of extractables in UV-cured coatings, that is, the amount of components that are not covalently bound to the cross-linked network, is directly related to the conversion that is achieved during irradiation. Extractables may lead to migration if printed materials are in contact e.g. to liquids. For this reason, the amount of extractables is an important parameter in packaging technology, in particular

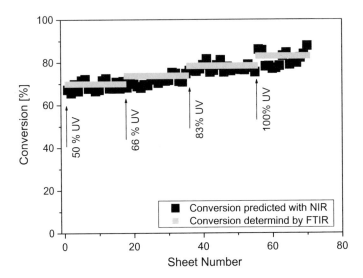

Fig. 21.14 In-line monitoring of the conversion in a cyan printing ink after irradiation with various UV doses (black). Conversions determined off-line by FTIR spectroscopy are given for comparison (gray). Reprinted with permission by Elsevier B. V. from Ref. [36]

in food packaging. The relation between conversion and the migration of acrylic compounds was studied quantitatively in dependence on the UV irradiation dose [36]. In fact, a linear relationship between the conversion and the acrylate migration was found. Using this linear relationship, the amount of acrylic extractables was directly predicted from NIR data during printing at the offset press (see. Fig. 21.15). For each irradiation dose, some random samples of printed sheets were analyzed by FTIR/ATR spectroscopy and HPLC measurements for comparison. For both the conversion and the migration, close correlation with the data predicted in-line was found. For the acrylate migration, the prediction error was estimated to be 0.03 g/m^2.

Certainly, this indirect method for the estimation of the content of extractable acrylate can neither achieve the accuracy of direct measurements by HPLC nor can replace such chromatographic analyses of the samples, which might be required e.g. for certification. However, it can provide a rough estimation of the actual state of the printed materials with respect to the migration behavior in running printing processes in real time, which allows rapid intervention, if deviations from the requested specifications occur.

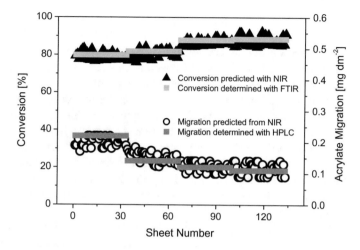

Fig. 21.15 In-line monitoring of the conversion in printed layers of a cyan printing ink and estimation of the specific acrylate migration from the NIR data. Reference values determined off-line are given for both the conversion and the migration. Reprinted with permission by Elsevier B. V. from Ref. [36]

21.3 Synthetic Fibers and Textiles

Despite their important role in daily life, the broad spectrum of different materials and forms of appearance as well as the multiplicity of production and processing technologies, textiles have been the subject of comparatively few studies in the scientific literature only that deal with their characterization or process monitoring by NIR spectroscopic methods. Moreover, the majority of the studies published so far is dealing with natural fibers such as cotton, wool, flax, silk etc. and the fabrics made there from. In the context of the present chapter on NIR investigations on special polymeric materials, we will limit this review to studies, which are only or at least mainly focused on synthetic fibers and textiles.

21.3.1 Classification of Textile Fabrics

Similar to polymers, the most important and most widely used application of NIR spectroscopy in textile technology is the sorting of used textiles. Identification of different materials (made from natural and synthetic fibers as well as blended fabrics) is indispensable for subsequent recycling processes, but typically requires extensive complicated and time-consuming analysis due to the endless broad range of different colors, patterns, specific appearances, pretreatments, construction differences, manufacturing technologies etc. In particular, the sorting of carpets became popular because of the rather large quantity of such materials resulting e.g. from the redesign of hotels, office buildings etc. [9, 37, 38]. Therefore, several special NIR-based analyzers for carpet analysis including both stationary and hand-held fiber-optic instruments are commercially available on the market.

However, identification of fibers and quantitative analysis of the composition of textiles is also beneficial in process control during production processes or for inspection purposes, e.g. in import and export of textile materials. A numbers of studies is dealing with material analysis for such applications [39–42]. A more recent application of textile classification is authentication of materials. In particular, in case of high-value materials the type and the percentage of fibers used in textiles and their blends becomes more and more important in order to prevent fraudulent supply of low-grade materials. This does not only apply to certain natural fibers (e.g. wool versus cashmere), but also to expensive synthetic materials such as aramid fibers used for example for bulletproof vests [41].

Based on the experiences with the sorting of polymer waste, NIR spectroscopy has been proven to be an extremely powerful tool for the fast and reliable identification of fibers and textiles, when it is combined with powerful chemometric algorithms such as principal component analysis (PCA) for the reduction of the dimensions of the data set and linear discrimination analysis (LDA) for classification. The majority of studies have been carried out by conventional NIR spectroscopy, but recently hyperspectral imaging has been used for data acquisition as well [41]. Most studies deal with the identification of combined sets of natural and synthetic fibers. Polymers, which are usually included in such studies, comprise typical fiber materials such as polyethylene terephthalate (PET), polyamide (PA), polypropylene (PP), polyethylene (PE), acrylic fibers such as polyacrylonitrile (PAN) [6, 9] and regenerated fibers such as viscose or tencel, but also more exotic materials such as polylactic acid (PLA) [42] and poly(p-phenylene terephthalamide) (para-aramid) [41]. Due to their different molecular structures, the differentiation between these fibers is rather easy. However, fibers with different, but very similar molecular structures of the same polymer such as PA 6 and PA 6.6 can be classified as well despite the marginal differences in their spectra [38]. More sophisticated methods such as neural networks are required to consider further properties of the fiber materials like color/dyeing or pretreatments such as heat setting.

21.3.2 Quality Control in Fiber and Textile Production

Blending is a common practice in textile manufacturing in order to combine technical and economic advantages of two or more fibers in a textile fabric. Therefore, blends make up a significant part of the textile market. In most cases, natural fibers (e.g. cotton, wool) are combined with synthetic fibers (usually polyester), but blends of synthetic fibers (e.g. PET/PA) are common as well. Typically, quality and price of textile fabrics are closely correlated with the composition of the blend. Monitoring of the actual composition is a critical issue for the control of conformity with the initial specifications, but also for process control, since certain properties that are relevant e.g. in technical applications (e.g. fire resistance, wear resistance) strongly depend on the ratio of the components [43]. Conventional analytic standard methods are time-consuming, labor-intensive and might require harmful chemicals (e.g. H_2SO_4 solution). NIR spectroscopy has been proven to be a powerful tool for the quantitative analysis of textile blends [38, 43–46]. The prediction of the content of the individual materials in such blends is often difficult due to the complex influence of a multiplicity of chemical, physical and textile parameters. Therefore, very large sample sets (e.g. about 200–300 cotton/PET samples [44–46]) are required for calibration of the models, which span not only the complete range of blend compositions (0–100% of each component), but also cover the variability of all other parameters. The immense data volume requires advanced mathematical pretreatment of the spectra as well genetic algorithms [44, 45] for variable selection in order to reduce the complexity of the multivariate models. Quantitative analysis itself is often based on the PLS regression. Moreover, support vector machine models have been tested [46]. In most cases, RMSEP for the determination of the cotton/polyester ratio was found to be 1–2% (with respect to cotton content) [43–46]. An example of a PLS model is given in Fig. 21.16 [44]. Apart from the widely studied cotton/polyester blends, wool/polyester blends were analyzed by NIR spectroscopy as well [43].

Acrylic fibers typically consist of copolymers of polyacrylonitrile (~ 85% w/w) and another vinyl monomer such as vinyl acetate, methyl methacrylate or methyl acrylate. Copolymerization is carried out in aqueous dispersion. The resulting polymer is not soluble in water. So, sodium thiocyanate has to be added to the suspension do dissolve the copolymer before fiber spinning resulting in a solution

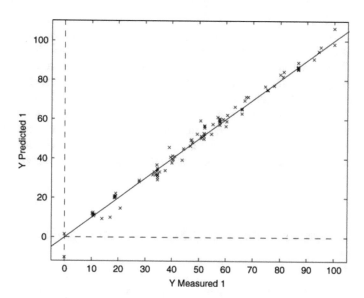

Fig. 21.16 Predicted versus measured reference values for the percentage of cotton in cotton/polyester blends using a full-spectrum PLS model. Reprinted with permission by SAGE publications Ltd. from Ref. [44]

called "dope." Extrusion of this solution through a spinneret into cold water leads to precipitation of the copolymer and the formation of fibers, which may be drawn to the desired thickness. For optimum process conditions, the actual ratio of thiocyanate and copolymer in the dope needs to be controlled continuously. NIR reflection spectroscopy was used for in-line monitoring of this ratio [47]. Difficulties in PLS modeling resulted from the fact that a variation of the NaSCN concentration led to a shift of the O-H bands of water, since the electrolyte prevents the formation of hydrogen bonds. Individual PLS1 models for NaSCN and the copolymer were developed resulting in RMSEP values of 1.6% (NaSCN) and 2.6% (copolymer), respectively. Similar results were also obtained with a PLS 2 model for simultaneous prediction of both concentration values.

Physical parameters of textiles may be determined as well. PA flocks with different degrees of fineness are used for the production of upholstery fabrics, which are indistinguishable by the human eye. Nevertheless, flocks with different degrees of fineness lead to different appearances of the final product. NIR spectroscopy was found to be an efficient tool for their discrimination [39].

Another parameter that influences the properties of textiles is the heat setting temperature applied during the production process. Heat has a significant effect on the molecular structure and the morphology of yarns. Heat setting is typically applied in order to improve textile properties with respect to shrinkage, warping, relaxation of internal stress, dye or finish fixation etc. When applied under tension, it may increase the tensile strength of fibers or fabrics and improve the behavior of the material during further manufacturing steps. For some applications (e.g. carpet manufacturing), very low variation and a high spatial homogeneity of the heat set temperature are required for the intended properties of the final products. NIR spectroscopy has been shown to be able to reveal the thermal history of the polymers in synthetic fibers [38]. In case of carpets based on PA, heat treatment is carried out at about 190–220 °C or occasionally at even higher temperatures depending on the specific method. PLS models could predict the heat set temperature of PA yarns from their NIR spectra with a precision (SEP) of about 1–2 °C for greige yarns and 2..3 °C for dyed yarns. Below 185 °C some deviation from linearity was observed, which was attributed to the Brill transition. The Brill transition in PA 6.6 occurs at about 160–180 °C and is due to the gradual transformation of the crystalline state from one triclinic structure existent at room temperature into a different triclinic structure stable at higher temperatures.

21.3.3 Finishing of Yarns and Textiles and Subsequent Drying

It is common practice in textile industry to provide fibers and yarns with special agents such as sizes (e.g. fatty acid ethoxylates, waxes, polyethylene) or lubricants (finishing oils) before processing them to textile fabrics in order (1) to make them more resistant to mechanical stress during those processes (e.g. weaving, knitting) by reinforcement and improved filament cohesion and (2) to reduce electrostatic charging as well as frictional wear and abrasion during passing them over machine parts such as needles, rolls, guides etc. The specific treatment depends on the fiber material and the intended use of the product.

Typically, very low amounts of finishing oils are deposited on the fibers (< 1 wt%). In order to achieve maximum quality of the products and optimum manufacturing conditions, the amount of finishing agents has to be controlled continuously. Conventional quantitative determination of the finishing oils is based on their removal by chlorinated solvents and analysis of the solutions by FTIR spectroscopy, which is a time-consuming and laborious procedure. BLANCO et al. have shown [48] by investigations on acrylic fibers made up of a acetonitrile–vinyl acetate copolymer that quantitative data on the amount of finishing agent can be obtained directly from the fiber during the spinning process by NIR reflection spectroscopy. The weight content of the finishing oil varied between 0.22 and 0.62 wt%. These low application weights require an excellent sensitivity of the spectroscopic method. Moreover, the properties of the acrylic fibers varied with respect to fineness, color, gloss and other possible sources of variability. The first component of the PLS model accounted for 99.99% of the spectral variance, which was related to spectral scattering due to variations in fineness of the fibers used for calibration. Generally, black samples and those dyed with dark colors were found to result in rather high absolute errors (up to 0.06 wt%). For all other samples, relative prediction errors (RMSEP) between about 6 and 7% were obtained, which roughly corresponds to the precision of data from the reference method. Similar investigations on the quantitative determination of finishing oil on the surface of PA 6.6 yarns have been reported as well [38]. Using a four-wavelength multiple linear regression (MLR) model, the amount of oil could be predicted with a SEP of 0.04 wt% for deposits of about 1 wt%.

After manufacturing of the textile fabric, the processing agents have to be removed carefully before further processing steps such as finishing with functional agents, coating, lamination, printing, etc., since the finishing oils and sizes are often hydrophobic and hence strongly affect wettability and adhesion. In particular, it has to be made sure that the oil or size is completely removed everywhere across the surface in order to avoid local wetting or adhesion problems. However, due to the invisibility of the very thin and colorless layers, the detection of local remains of the processing agents is hardly possible. Recently, NIR spectral imaging has been shown to be a powerful tool for this analytical task [49]. A size consisting of hydrocarbons and fatty acid ethoxylates was applied to polyester fabric with application weights between 0.4 and 5.5 g/m^2 in order to provide samples for calibration of a PLS model. Using this model the amount of size was determined with a precision of about 0.2 wt% by averaging the predicted individual values across the surface of the fabric. Monitoring of the distribution of the size across the textile web is of no technical relevance since it is applied to the yarn before weaving. After washing-out of the size and drying of the textile, the cleanness of the desized fabric was inspected by spectral imaging with respect to both quantification of the residual amount of size and its spatial distribution. The amount of remaining size was found to be at or below the prediction error of the external validation, which was sufficient for further processing such as ink jet printing. Moreover, traces of size applied by arbitrary spraying of a dilute solution to clean fabrics could be detected and quantified by this method as well.

Aside from the treatment with such processing agents, which have to be removed before subsequent processing, textile fabrics may be also finished with finishing agents in order to provide them permanently with certain functional properties such as flame retardancy, optical brightening, stiffening, hydrophilicity or hydrophobicity, water repellence, biocide or stain resistance, anti-static behavior and other features. Typically, such agents are applied by impregnation in aqueous solutions in a foulard, which is often followed by heat setting to improve fixation to the textile fibers. Depending on the special substrate, agent, and intended application finishes are applied with application weights in the range from less than 1 g/m^2 and to several tens of g/m^2. The finishes have to be applied according to the specified value of the application weight and with exceedingly homogenous spatial distribution across the textile fabric, which is required for both further processing of the finished textiles as well as for optimum application properties. However, similar to sizes and finishing oils most of the finishing agents form colorless layers on the surface of the textile, which usually prevents in-line detection by optical methods in the visible range. Rather, quantification is mostly carried out off-line by extraction, gravimetry etc. Recently, SCHERZER et al. [49] have demonstrated the immense potential of hyperspectral imaging for both quantitative analysis of the applied amount and monitoring of the spatial homogeneity of finishing layers on textiles fabrics. For example, it was shown that the application weight of flame retardant layers (in the order of several tens of g/m^2) applied to polyester fabric can be detected with a prediction error (RMSEP) of 2.2 g/m^2. The corresponding PLS calibration is given in Fig. 21.17. Figure 21.18 shows spectral images of two finished polyester fabrics. In one of them, a spot pattern resulting from several drops of finishing solution that were dripping down to the already dried material, can be seen. The average application weight obtained by integration over the individual values across the complete surface of the spectral image of each finished sample proved a very close correlation with the corresponding reference values obtained by gravimetry (i.e., differences < 2 g/m^2). Moreover, local inhomogeneities of the density thickness were clearly detected. In particular, the dotted pattern forming a rectangular frame close to the cut edges originates from pinning the fabric on a tentering frame for drying after impregnation and squeezing.

Similar investigations were carried out with other textile materials such as PA fabric and with other finishing systems such as stiffening agents, optical brighteners, adhesion promoters and hydrophilic layers. Depending on the specific material combination, prediction errors were found to be in the order of 1–2 g/m^2. However, there is one exception of finishing layers that cannot be monitored by NIR imaging. Water-repellent finishes are often based on fluorocarbon compounds, which cannot be detected by NIR spectroscopy.

The next step in finishing of textiles by impregnation in aqueous solutions is the drying process. Typically, this is a multistage process consisting of dewatering

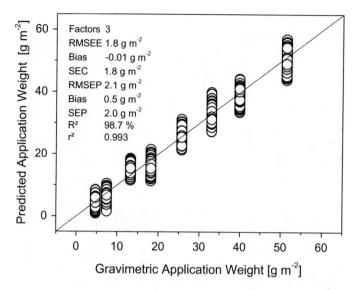

Fig. 21.17 PLS calibration model for the application weight of a flame retardant on pale beige polyester fabric (200 g/m²). Reprinted with permission by Elsevier B. V. from Ref. [49]

by squeezing, suction by low pressure and thermal drying. Obviously, drying is an energy-intensive process, which should be carried out up to the technically required level only. NIR spectroscopy or imaging are predestined for monitoring the state of the drying process due to the strong absorbance bands of water in this spectral region (e.g. the strong combination band around 1940 nm and the first overtone at about 1450 nm). Accordingly, the determination of the water content is probably the most widely used application of NIR spectroscopic methods, in particular in agriculture, food processing, pharmaceutics etc. With respect to textile processing, NIR methods are particularly suited for monitoring the content of remaining moisture at the end of the process. SCHERZER et al. [50] determined the residual damp of finished textiles by NIR hyperspectral imaging using PLS calibration models. Textile substrates and finishing systems were largely the same as those, which were studied above with respect to application weight. Figure 21.19 shows NIR spectra of a PA fabric finished with a flame retardant at different moisture levels. Figure 21.20 shows the corresponding PLS calibration model. Reference data were determined by gravimetry using a moisture analyzer. The prediction errors (RMSEP) of the PLS models of all material combinations were found to be around 0.5%, since this was the lower detection limit of the analyzer. Probably, an improved detection limit could be achieved with the NIR-based technique if a reference method with higher sensitivity would be used. Hyperspectral imaging does not only provide quantitative data on the dampness, but reveals also its spatial distribution, which can help to evaluate the homogeneity of the drying process. For example, the visualization of differences between the outer areas on both sides of broad textile webs could assist with improving the control of the drying oven.

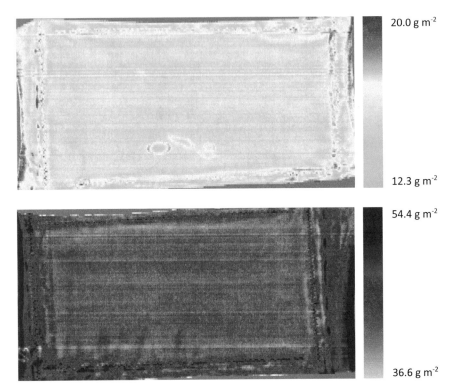

20.0 g m^{-2}

12.3 g m^{-2}

54.4 g m^{-2}

36.6 g m^{-2}

Fig. 21.18 Hyperspectral images of the distribution of a flame retardant on polyester fabric. In the upper one, a spot pattern resulting from dried drops of finishing solution can be identified. The gravimetric application weights of the samples were 13.2 (top) and 43.1 g/m^2 (bottom). Reprinted with permission by Elsevier B. V. from Ref. [49]

The moisture content of synthetic fibers and yarns has been studied by NIR spectroscopy as well. Moisture is a critical parameter with respect to both properties (e.g. morphology, many physical parameters such as glass transition temperature, tenacity etc.) and processing (e.g. dyeing or lamination) of the textile materials. On the other hand, the water content of hydrophilic polymers such as PA depends on the relative humidity of the air or the occurrence of moisture in technical processing steps. The maximum water content of PA is in the order of 2.5–3.5 wt% depending on the specific type of PA. The moisture content of spun PA fibers was determined by NIR spectroscopy with precisions (SEP) between 0.3 and 0.5% [38]. Similar investigations were reported for PAN and viscose [39].

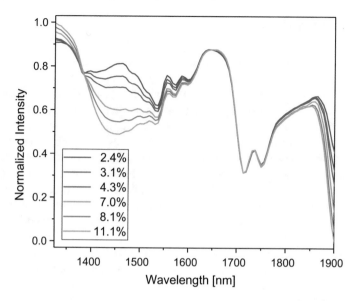

Fig. 21.19 NIR spectra of PA fabric finished with a flame retardant containing various amounts of residual moisture. Spectra were normalized to the double peak of the C-H overtone peak at 1720/1750 nm

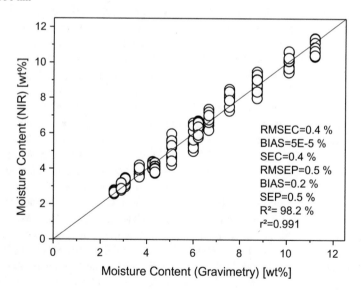

Fig. 21.20 PLS calibration to the moisture content of PA fabrics finished with a flame retardant

21.3.4   Lamination of Textiles

In several technical applications, textiles are used as laminates in order to combine the properties of different materials. For example, combinations of a fabric as top layer for the optical appearance and a hidden nonwoven material for sound insulation are widely used for interior design in automotive engineering (e.g. for door and roof linings or rear panel shelfs) and similar applications. Both textile layers are bonded to each other by laminating hot-melt adhesives. Such adhesives can be applied as powder, hot melted mass or melt-spun fibrous webs. During application of the adhesive and the subsequent lamination of the two webs, several deficiencies may occur depending on the specific adhesive system such as over/underdosage or uncoated areas for powders and melts as well as tears, holes or buckle formation in case of adhesives webs. Generally, lamination defects lead to deficient bond strength or even delamination, impairment of the visual appearance etc. Therefore, continuous monitoring of the homogeneity of the inside adhesive layer would be essential for an efficient control of the lamination process. However, visual inspection of the hidden layers or use of a corresponding camera is prevented by their invisibility and inaccessibility from outside.

Near-infrared radiation is well-known to penetrate to a certain degree into polymer materials, at least for some tens of micrometers depending on the specific material, its color and morphology. Materials for interior design in automotive engineering often contain a top layer made up of polyester piqué fabrics with weights between about 100 and 200 g/m². Colors may cover a broad range, but the majority of materials is beige, gray or black. Due to their weave structure, such fabrics appear to be slightly translucent. On the other hand, the fibrous structure of the materials may lead to diffuse scattering of incident radiation. Therefore, investigations were accomplished in order to clear up if adhesive layers hidden in textile laminates can be visualized and quantitatively analyzed by hyperspectral imaging. Generally, spectral analysis was carried out from the top side, since the nonwoven is usually thicker than the fabric and shows much stronger scattering of the probe light. In fact, it was demonstrated that the reflection spectra of such three-layer laminates show significant contributions of the adhesive layers based on aliphatic polyester [30] (see Fig. 21.21). In particular, a specific absorption was observed around 1410 nm, which is attributed to a C-H combination band in aliphatic hydrocarbons ($2\nu CH_2$ and δCH_2) [51], as well as in the range above 1700 nm. Similar results were obtained for laminates containing PA-based adhesives webs. Even reflection spectra of corresponding black textile laminates can be recorded in high quality and with explicit effect of the coating thickness of the adhesive layers [52], although the optical conditions are far from being ideal for this kind of measurements.

Reflection spectra were found to correlate with the application weight of the hot melt adhesive. A principal component analysis (PCA) of the NIR spectra of the laminates showed that the scores plot of the first two components (PC1 versus PC2) only allowed a discrimination between samples with and without adhesive layer, whereas

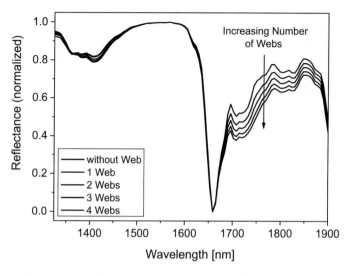

Fig. 21.21 NIR spectra of bright three-layer laminates with different numbers of adhesive webs. Reprinted with permission by Elsevier B. V. from Ref. [30]

the plot of the second and third component (PC2 versus PC 3) also enables a differentiation between adhesive layers with different thicknesses [30]. Consequently, PLS calibration models were developed. Reference data were obtained from gravimetry during sample preparation. For each sample (roughly 20 × 30 cm), several tens of thousands spectra were recorded with a hyperspectral camera. In order to consider a possible non-uniform distribution of the adhesive coating weight across the samples, their surfaces were divided into rectangular regions by defining a grid of n columns x m rows. Spectra from each region were allocated alternatingly, that is according to a chessboard pattern, to the calibration and the validation set. In case of bright laminates (beige or light gray top layer), this resulted in RMSEP values around 3.5 g/m^2 [30] for adhesive layers between about 25 and 125 g/m^2 (see for example Fig. 21.22), whereas the prediction error was higher (around 6 g/m^2) for laminates made up of black textiles only [52]. Accordingly, the color of the latter laminates somewhat impairs the precision of the predictions, but by no means it prevents the quantitative analysis of the thickness of the buried hot melt layers. External validations confirmed that the application weight of the inside adhesive layers can be actually predicted within this error limits. Moreover, averaging the predicted individual values from all NIR spectra across each sample enabled a direct comparison with gravimetric reference data. Due to this averaging process, the deviations are lower than the corresponding RMSEP (up to ~ 2 g/m^2 for bright laminates and up to ~ 5 g/m^2 for black laminates). As an example, the spectral image of a three-layer laminate consisting of black textiles is shown in Fig. 21.23 [52].

Aside from the prediction of quantitative data of the application weight of the adhesive layers, the main objective of investigations by spectral imaging is monitoring of the homogeneity of the layers and the detection of coating errors. In fact,

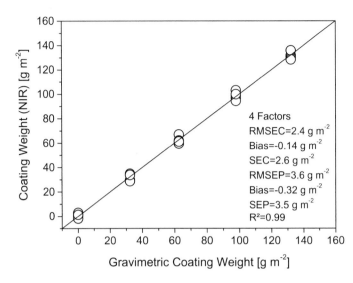

Fig. 21.22 PLS calibration model for the coating weight of the adhesive layer in bright polyester-based three-layer laminates. Reprinted with permission by Elsevier B. V. from Ref. [30]

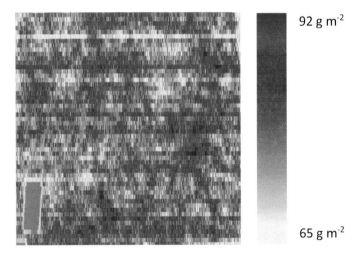

Fig. 21.23 Hyperspectral image of the distribution of the adhesive inside a black polyester-based three-layer laminate. The gravimetric application weight of the adhesive was 80.9 g m^{-2}. Reprinted with permission by Elsevier Ltd. from Ref. [52]

it was shown that the calibration and validation samples of the textile laminates had a very high homogeneity. On the other hand, laminates with specific error patterns (uncoated areas, overdosage, tears, folds, thick adhesive droplets etc.) were prepared. All defects were clearly detected in the spectral images [30, 52]. In most cases, even quantitative information about the defects were obtained. Figure 21.24 shows a repre-

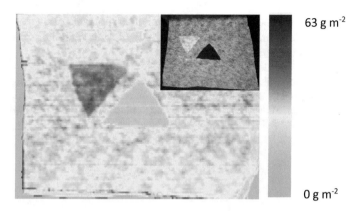

63 g m^{-2}

0 g m^{-2}

Fig. 21.24 Hyperspectral image of a bright polyester-based three-layer laminate containing an adhesive web with an intentionally prepared defect (see inset). Reprinted with permission by Elsevier B. V. from Ref. [30]

sentative example of a textile laminate with an intentionally prepared defect in the adhesive web [30]. The triangular area that was cut out of the web was placed close to the hole during lamination.

Finally, it was shown that acceptable predictions of the application weight of the adhesive can be even obtained if the thickness (or the weight per area) of the polyester top fabric varies, e.g. if laminates with different specifications have to be produced alternately [52]. The variation was included in the PLS model as well, which resulted in some increase of RMSEP (by ~ 50%), but still allowed predictions with reasonable precision.

The investigations proved the excellent potential of hyperspectral imaging for monitoring the homogeneity of hidden hot melt adhesive layers inside textile laminates, even if these laminates consist of black components only. This opens up new possibilities for continuous large-area quality and process control in technical lamination processes.

21.4 Conclusion

This chapter demonstrated that NIR spectroscopy cannot only be used for "classical" applications in polymer science and technology such as monitoring of polymerization and curing reactions. Rather, it may be also applied to a broad range of thin layers such as polymer coatings, printed layers, laminating adhesives, functional layers in textile finishing etc., although the thickness of such layers is far below that of typical subjects of investigations by NIR spectroscopy, which broadens the application range of NIR-based methods for the analysis of polymeric materials considerably. This expansion was enabled and pushed by the tremendous performance, in particular the considerably enhanced sensitivity, of modern NIR instrumentation in combination with powerful chemometric approaches.

Acknowledgements The author is deeply grateful to Dr. Olesya Daikos, Leibniz-Institut für Oberflächenmodifizierung (IOM), for careful reading of the manuscript as well for technical support in the preparation of figures etc. Moreover, he thanks all current and former members of the spectroscopy and process control group at IOM for successful longtime collaboration.

References

1. J.B. Callis, D.L. Illmann, B.R. Kowalski, Process analytical chemistry. Anal. Chem. **59**, 624A (1987)
2. D.C. Hassell, E.M. Bowman, Process analytical chemistry for spectroscopists. Appl. Spectrosc. **52**, 18A (1998)
3. K.A. Bakeev (ed.), *Process Analytical Technology* (Blackwell Publishing, Oxford, 2010)
4. L.G. Weyer, Near-infrared spectroscopy of organic substances. Appl. Spectrosc. Rev. **21**, 1 (1985)
5. C.E. Miller, Near-infrared spectroscopy of synthetic polymers. Appl. Spectrosc. Revs. **26**, 277 (1991)
6. H.E. Howell, J.R. Davis, Qualitative identification of polymeric materials using near-infrared spectroscopy, in *Structure-Property Relations in Polymers*, ed. by M.W. Urban, C.D. Craver (Am. Chem. Soc., Washington 1993), chap. 9, pp. 263–285
7. H.W. Siesler, Applications to polymers and textiles, in *Near-Infrared Spectroscopy: Principles, Instruments, Applications*, ed. by H.W. Siesler, Y. Ozaki, S. Kawata, H.M. Heise (Wiley-VCH, Weinheim, 2002), chap. 10, pp. 213–245
8. C. Krajdel, K.A. Lee, NIR analysis of polymers, in *Handbook of Near-Infrared Analysis*, ed. by D.A. Burns, E.W. Ciurczak (CRC Press, Boca Raton, 3rd ed., 2008), chap. 27, pp. 529–568
9. A.M. Brearly, S.J. Foulk, Near-infrared spectroscopy (NIR) as a PAT tool in the chemical industry: Added value and implementation challenges, in *Process Analytical Technology*, ed. by K.A. Bakeev (Blackwell Publishing, Oxford, 2010), chap. 15, pp. 493–520
10. T. Scherzer, M.R. Buchmeiser, Photoinitiated cationic polymerization of cycloaliphatic epoxide/vinyl ether systems studied by near-infrared reflection spectroscopy. Macromol. Chem. Phys. **208**, 946 (2007)
11. G. Mirschel, K. Heymann, T. Scherzer, M.R. Buchmeiser, Effect of changes of the coating thickness on the in-line monitoring of the conversion of photopolymerized acrylate coatings by near-infrared reflection spectroscopy. Polymer **50**, 1895 (2009)
12. C. Schmidt, T. Scherzer, Monitoring of the shrinkage during the photopolymerization of acrylates using hyphenated photorheometry/near-infrared spectroscopy. J. Polym. Sci., Pt. B: Polym. Phys. **53**, 729 (2015)
13. T. Scherzer, Depth profiling of the degree of cure during the photopolymerization of acrylates studied by real-time FT-IR attenuated total reflection spectroscopy. Appl. Spectrosc. **56**, 1403 (2002)
14. O. Daikos, K. Heymann, T. Scherzer, Development of a PLS approach for the determination of the conversion in UV-cured white-pigmented coatings by NIR chemical imaging and its transfer to other substrates. Progr. Org. Coat. **132**, 116 (2019)
15. T. Scherzer, R. Mehnert, H. Lucht, Online monitoring of the acrylate conversion in UV photopolymerization by near-infrared reflection spectroscopy. Macromol. Symp. **205**, 151 (2004)

16. T. Scherzer, K. Heymann, G. Mirschel, M.R. Buchmeiser, Process control in ultraviolet curing with in-line near infrared reflection spectroscopy. J. Near Infrared Spectrosc. **16**, 165 (2008)
17. K. Heymann, G. Mirschel, T. Scherzer, M.R. Buchmeiser, In-line determination of the thickness of UV-cured coatings on polymer films by NIR spectroscopy. Vibr. Spectrosc. **51**, 152 (2009)
18. K. Heymann, G. Mirschel, T. Scherzer, Monitoring of the thickness of ultraviolet-cured pigmented coatings and printed layers by near-infrared spectroscopy. Appl. Spectrosc. **64**, 419 (2010)
19. Y. Hao S. Haber, Reverse roll coating flow. Int. J. Numer. Meth. Fluids **30**, 635 (1999)
20. M.J. Gostling, M.D. Savage, A.E. Young, P.H. Gaskell, A model for deformable roll coating with negative gaps and incompressible compliant layers. J. Fluid Mechanics **489**, 155 (2003)
21. O. Daikos, K. Heymann, T. Scherzer, Monitoring of thickness and conversion of thick pigmented UV-cured coatings by NIR hyperspectral imaging. Prog. Org. Coat. **125**, 8 (2018)
22. T. Scherzer, W. Knolle, S. Naumov, L. Prager, Investigations on the photoinitiator-free photopolymerization of acrylates by vibrational spectroscopic methods. Macromol. Symp. **230**, 173 (2005)
23. B. Jiang, Y.D. Huang, Y.P. Bai, Noncontact and rapid analysis of the quality of the recording coating on ink jet printing by near-infrared spectroscopy. Analyst **136**, 5157 (2011)
24. G. Mirschel, K. Heymann, T. Scherzer, Simultaneous in-line monitoring of the conversion and the coating thickness in UV-cured acrylate coatings by near-infrared reflection spectroscopy. Anal. Chem. **82**, 8088 (2010)
25. T. Scherzer, S. Müller, R. Mehnert, A. Volland, H. Lucht, In-line monitoring of the conversion in photopolymerized acrylate coatings on polymer foils using NIR spectroscopy. Polymer **46**, 7072 (2005)
26. T. Scherzer, S. Müller, R. Mehnert, A. Volland, H. Lucht, In-line determination of the conversion in acrylate coatings after UV curing using near-infrared reflection spectroscopy. Nucl. Instr. Meth. in Phys. Res. B **236**, 123 (2005)
27. T. Scherzer, M.R. Buchmeiser, A. Volland, H. Lucht, NIR spectroscopy as powerful tool for process control in UV curing, in *Proceedings of the RadTech Europe Conference 2007* (Vincentz, Hannover, 2007)
28. L. Cséfalvayová, M. Strlič, H. Karjalainen, Quantitative NIR chemical imaging in heritage science. Anal. Chem. **83**, 5101 (2011)
29. G. Mirschel, O. Daikos, C. Steckert, T. Scherzer, Monitoring of the application of laminating adhesives to polyurethane foam by near-infrared chemical imaging, in *Proceedings of the 18th International Conference on Near Infrared Spectroscopy*, ed. by S.B. Engelsen, K.M. Sørensen, F. van den Berg (IM Publications Open, Chichester, 2019), pp. 163–168
30. G. Mirschel, O. Daikos, T. Scherzer, C. Steckert, Near-infrared chemical imaging used for in-line analysis of inside adhesive layers in textile laminates. Anal. Chim. Acta **932**, 69 (2016)
31. G. Mirschel, K. Heymann, O. Savchuk, B. Genest, T. Scherzer, In-line monitoring of the thickness of printed layers by near-infrared (NIR) spectroscopy at a printing press. Appl. Spectrosc. **66**, 765 (2012)
32. O. Daikos, G. Mirschel, B. Genest, T. Scherzer, In-line monitoring of the thickness of printed layers by NIR spectroscopy: Elimination of the effect of the varnish formulation on the prediction of the coating weight. Ind. Eng. Chem. Res. **52**, 17735 (2013)
33. G. Mirschel, O. Savchuk, T. Scherzer, B. Genest, Process control of printing processes with in-line NIR spectroscopy and elimination of the influence of the substrate on the prediction of the coating weight. Progr. Org. Coat. **76**, 86 (2013)
34. G. Mirschel, O. Savchuk, T. Scherzer, B. Genest, The effect of different gloss levels on in-line monitoring of the thickness of printed layers by NIR spectroscopy. Anal. Bioanal. Chem. **404**, 573 (2012)
35. G. Mirschel, O. Daikos, K. Heymann, T. Scherzer, B. Genest, C. Sommerer, C. Steckert, Monitoring of the conversion in UV-cured printed layers by NIR spectroscopy in an offset printing press. Progr. Org. Coat. **77**, 719 (2014)
36. G. Mirschel, O. Daikos, K. Heymann, U. Decker, T. Scherzer, C. Sommerer, B. Genest, C. Steckert, In-line monitoring of printing processes in an offset printing press by NIR

spectroscopy: Correlation between the conversion and the content of extractable acrylate in UV-cured printing inks. Prog. Org. Coat. **77**, 1682 (2014)

37. B.J. Kip, T. Berghmans, P. Palmen, A. van der Pol, M. Huys, H. Hartwig, M. Scheepers, D. Wienke, On the use of recent developments in vibrational spectroscopic instrumentation in an industrial environment: quicker, smaller and more robust. Vibr. Spectrosc. **24**, 75 (2000)

38. S. Ghosh, J. Rodgers, NIR analysis of textiles, in *Handbook of Near-Infrared Analysis*, ed. by D.A. Burns, E.W. Ciurczak (CRC Press, Boca Raton, 3rd ed., 2008), chap. 25, pp. 485–520

39. E. Cleve, E. Bach, E. Schollmeyer, Using chemometric methods and NIR spectrophotometry in the textile industry. Anal. Chim. Acta **420**, 163 (2000)

40. X.D. Sun, M.X. Zhou, Y.Z. Sun, Classification of textile fabrics by use of spectroscopy-based pattern recognition methods. Spectrosc. Lett. **49**, 96 (2016)

41. X.K. Jin, H. Memon, W. Tian, Q.L. Yin, X.F. Zhan, C.Y. Zhu, Spectral characterization and discrimination of synthetic fibers with near-infrared hyperspectral imaging system. Appl. Optics **56**, 3570 (2017)

42. J.F. Zhou, L.J. Yu, Q. Ding, R.W. Wang, Textile fiber identification using near-infrared spectroscopy and pattern recognition. Autex Res. J. **19**, 201 (2019)

43. J.S. Church, J.A. O'Neill, A.L. Woodhead, A comparison of vibrational spectroscopic methods for analyzing wool/polyester textile blends. Text. Res. J. **69**, 676 (1999)

44. C. Ruckebusch, F. Orhan, A. Durand, T. Boubellouta, J.P. Huvenne, Quantitative analysis of cotton–polyester textile blends from near-infrared spectra. Appl. Spectrosc. **60**, 539 (2006)

45. X.D. Sun, M.X. Zhou, Y.Z. Sun, Variables selection for quantitative determination of cotton content in textile blends by near infrared spectroscopy. Infrared Phys. Technol. **77**, 65 (2016)

46. Y.S. Liu, S.B. Zhou, W.X. Liu, X.H. Yang, J. Luo, Least-squares support vector machine and successive projection algorithm for quantitative analysis of cotton-polyester textile by near infrared spectroscopy. J. Near Infrared Spectrosc. **26**, 34 (2018)

47. M. Blanco, J. Coello, H. Iturriaga, S. Maspoch, J. Pagès, Use of near-infrared spectrometry in control analyses of acrylic fibre manufacturing processes. Anal. Chim. Acta **383**, 291 (1999)

48. M. Blanco, J. Coello, J.M. García Fraga, H. Iturriaga, S. Maspoch, J. Pagès, Determination of finishing oils in acrylic fibres by near-infrared reflectance spectrometry. Analyst **122**, 777 (1997)

49. G. Mirschel, O. Daikos, T. Scherzer, C. Steckert, Near-infrared chemical imaging used for in-line analysis of functional finishes on textiles. Talanta **188**, 91 (2018)

50. O. Daikos, T. Scherzer, Monitoring of the residual moisture content in finished textiles during converting by NIR hyperspectral imaging, Talanta **221**, article no. 121567 (2021)

51. J. Workman Jr., L. Weyer, *Practical Guide to Intrepretive Near-Infrared Spectrosopy* (CRC Press, Boca Raton, 2008)

52. G. Mirschel, O. Daikos, T. Scherzer, In-line monitoring of the thickness distribution of adhesive layers in black textile laminates by hyperspectral imaging. Comput. Chem. Eng. **124**, 317 (2019)

Chapter 22
NIR Imaging

Daitaro Ishikawa, Mika Ishigaki, and Aoife Ann Gowen

Abstract Visualization of the spatial distribution of surface properties is increasingly desired in many fields. Spectral imaging combines spectroscopy with imaging, implying images with three dimensions: two spatial and one spectral. NIR spectral imaging enjoys many of the useful features of NIR spectroscopy such as suitability for nondestructive measurement, in situ analysis and potential for transmission measurements. NIR imaging systems can provide high-speed monitoring and stability. These features are very attractive not only for laboratory-based studies but also for applications in a number of practical fields such as pharmaceutical, medical, engineering, biological and agricultural. NIR imaging technology is still developing by improvement of spectral analysis method; chemometrics and image analysis methods. In this section, we describe the concept of NIR spectral imaging at first and introduce the basic design of NIR imaging devices and the features of newly developed devices. Finally, the potential of NIR imaging for practical situation is demonstrated through several applications reported at recent years.

Keywords Visualization · Innovative devices · Spectral imaging · Label-free molecular imaging · Pharmaceuticals · Foods · And biological applications · Foods and biolobical applications

D. Ishikawa (✉)
Faculty of Food and Agricultural Sciences, Fukushima University, 1 Kanayagawa, Fukushima
960-1296, Japan
e-mail: daitaroishikawa@agri.fukushima-u.ac.jp

M. Ishigaki
Faculty of Life and Environmental Sciences, Shimane University, 1060 Nishikawatsu,
Matsue 690-8504, Shimane, Japan
e-mail: ishigaki@life.shimane-u.ac.jp

A. A. Gowen
School of Biosystems and Food Engineering, University College Dublin,
Belfield, Dublin 4, Ireland
e-mail: aoife.gowen@ucd.ie

© Springer Nature Singapore Pte Ltd. 2021
Y. Ozaki et al. (eds.), *Near-Infrared Spectroscopy*,
https://doi.org/10.1007/978-981-15-8648-4_22

22.1 Introduction

Spectral imaging, known also as chemical or hyperspectral imaging [1], combines imaging and spectroscopy to obtain spatial and spectral information from a sample. "Spectral imaging" is a general term covering a wide variety of spectroscopic and spectrometry techniques (e.g., fluorescence, vibrational, Raman, mass spectrometry), wavelength ranges (e.g., UV, Vis, NIR, MIR) and spatial resolutions (from the nanoscale to remote-sensing scale). Spectral images typically comprise hundreds of wavebands for each spatial position in the image. For a given spatial pixel, combining the intensity values at each waveband into a vector results in a spectrum, containing information on the interaction between light and the object or objects contained within that pixel. The resulting spectrum can be used to characterize the composition of that particular pixel. Another advantage of this technique is the ability to measure many samples simultaneously, within the same field of view, and to combine shape characteristics with spectral features.

NIR or short-wave infrared (SWIR) imaging typically refers to spectral images obtained in the wavelength region from 750 to 2500 nm. Spectral images can be arranged as three-dimensional blocks or "cubes" of data, comprising two spatial and one wavelength dimensions, as shown schematically in Fig. 22.1. The spectral image allows for the visualization of biochemical constituents of a sample, separated into particular areas of the image, since regions of a sample with similar spectral properties tend to have similar biochemical composition. Within the schematic in Fig. 22.1 is an NIR spectral image of different packaging materials. Four different materials are present in the image, identifiable by their shape. For instance, the squares are high-density polyethylene (HDPE), and the rectangles are polystyrene (PS). The reflectance spectrum of one pixel of a PS sample is plotted in the lower left-hand side of Fig. 22.1; it exhibits three main absorption features (i.e., three troughs in the reflectance spectrum) at 1146, 1209, 1412 nm, related to overtones of CH vibrations within the polymer. Due to these strong features, PS can be used as a wavelength reference material for NIR imaging. A single slice of the spectral image at 1146 nm is also shown. In this image, the PS samples appear darker than the other ones because they absorb more light than the other samples do at this particular wavelength.

An NIR spectral image of four different packaging materials (high-density polyethylene (HDPE), polystyrene (PS), cardboard and polyethylene-terephthalate (PET)) is shown. The materials were cut into different shapes to enable visual identification: squares (HDPE), rectangles (PS), circles (cardboard) and triangles (PET). The image was obtained in diffuse reflectance using a pushbroom NIR spectral imaging instrument, as described in [2]. A single pixel reflectance spectrum of the PS sample is shown on the lower left-hand side. The largest absorbance peak for PS occurs at 1146 nm. On the lower right-hand side, a single wavelength image at 1146 nm is displayed. The PS samples (rectangles in the image) appear darker than the other materials in this image due to their relatively higher absorbance of light at 1146 nm.

As is the case for conventional NIR spectroscopy, NIR spectral images can be obtained using different modalities, such as transmission, reflectance, transflectance

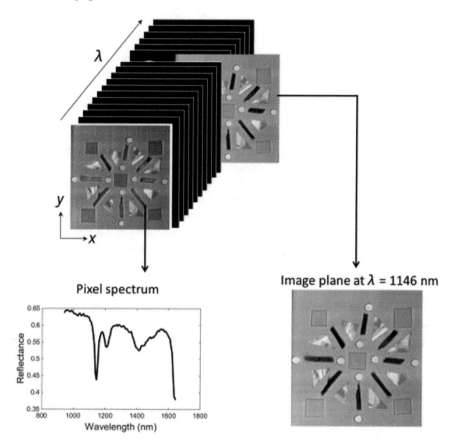

Pixel spectrum

Image plane at λ = 1146 nm

Fig. 22.1 Schematic of NIR spectral image structure: x and y represent spatial dimensions, while λ represents the wavelength dimension

or interactance. The selection of modality is largely sample dependent, and it is important to consider the optical properties of the sample and penetration depth of NIR light within. For example, transmission imaging is generally preferable to reflectance for materials that absorb high quantities of light in the NIR, such as aqueous or high water content samples. In order to enable imaging of highly absorbing samples, it is advisable to work with thin samples and/or work in transmission mode. If this is not possible, the use of a highly reflective substrate can also improve the signal-to-noise ratio achievable through transflectance measurements, although the effectiveness of this approach depends on sample thickness.

The remainder of this chapter is divided into two major sections: NIR imaging instrumentation and applications, with several illustrative examples provided throughout.

22.2 Instrumentation

22.2.1 General Features of NIR Imaging Device

Different instrumental setups exist for the acquisition of NIR images. These can vary in terms of modality (e.g., reflectance, transmission), components (e.g., light sources, detectors) and in how the spectral image is acquired. Quartz tungsten-halogen (QTH) light bulbs are often used as light sources in NIR imaging as they are low cost and cover a broad wavelength range, from 400 to 2500 nm; however, they generate significant amounts of heat, and this may alter or even burn a sample. To overcome this issue, fiber-optic lines can be used to transmit the light. The detector employed in an NIR imaging system is typically selected based on the required wavelength range, and various options exist. Systems operating in the 400–1000 nm wavelength range typically use charge-coupled device (CCD) or complementary metal-oxide semiconductor (CMOS) sensors. Such detectors are suitable for imaging in the wavelength range 300–1000 nm. Indium gallium arsenide (InGaAs) or mercury cadmium telluride (MCT) detectors are more commonly used for NIR imaging in the 1000–2500 nm range. However, these detectors are more expensive than CCDs and require cooling. Detectors used in NIR imaging often exhibit noise at extreme edges of the wavelength range, due to reduced quantum efficiency. An example of this is shown in Fig. 22.2, which shows single-waveband images extracted from the spectral image of packing materials (as previously described in Fig. 22.1). The wavelength range of the InGaAs detector used for obtaining this image is 880–1720 nm. Clearly, noise is evident in the image plane at peripheral wavebands such as 929 and 1713 nm. Typically, such noisy wavelength regions are removed from the spectra prior to further analysis. However, they can provide discriminatory information despite the presence of noise—for example the 1713 nm image shows good contrast between the HDPE/PS and other materials. For this reason, it can sometimes be useful to retain noisy waveband images and reduce the noise through the use of multivariate analysis.

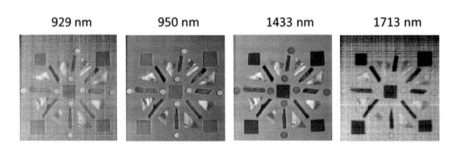

Fig. 22.2 Single-waveband images of various packaging materials (see Fig. 22.1 caption for further information on samples)

22.2.2 Spectral Image Acquisition

Currently, there are four main ways in which spectral images can be obtained: point, line, plane or snapshot imaging, as shown schematically in Fig. 22.3.

Point scanning can be regarded as an extension of conventional point spectroscopy where a spectrum is obtained at one spatial position, followed by movement of either the sample or the spectrometer to an adjacent position for spectral acquisition. This is continued until an entire image of a sample is acquired.

"Pushbroom" or line scanning refers to acquisition where the spectral image is acquired by collecting spectra of spatially contiguous lines over a sample. Typically, light from each spatial position along the line passes through a spectrograph and is split into its component wavelengths, resulting in a spectrum for each pixel on the line. The resultant two-dimensional image is captured by an array detector. The spectral image is built up either through moving the sample underneath the detector or moving the detector above the sample. "Staredown" or plane scanning spectral imaging systems acquire the hyperspectral image waveband by waveband by sequentially obtaining single-waveband images. This can be achieved through the use of a tunable filter to select only light at a specific wavelength or through the use of a tunable laser light source that only shines light of a specific wavelength on a sample. More recently, devices have been developed that can acquire an entire spectral image simultaneously in one acquisition. This so-called snapshot spectral imaging can be achieved through the use of prisms and waveplates or mirrors to project simultaneously different wavelength images of a sample onto a detector array [3] or by integrating spectral filters on top of detector elements [4]. Some recent developments in rapid and portable NIR imaging systems are described below.

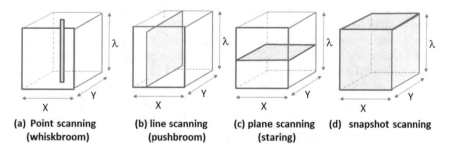

(a) Point scanning (b) line scanning (c) plane scanning (d) snapshot scanning
   (whiskbroom)   (pushbroom)   (staring)

Fig. 22.3 Schematic showing different techniques for obtaining spectral images

22.2.3 Development of Instrument

22.2.3.1 High-Speed Portable NIR Imaging System [5]

To meet the demands of diverse application fields such as pharmaceutics and agriculture, high speed and portability are essential. "D-NIRs" is a portable NIR spectral imaging system recently developed by Yokogawa Electrics Co., Ltd. D-NIRs is a point scanning-based system that enables rapid spectral imaging in the 950–1700 nm region with a maximum spatial resolution of 0.1 mm. For instance, a spectral image of a sample tablet (diameter; 8 mm) can be obtained within a few seconds. An overview and schematic diagram of D-NIRs are shown in Fig. 22.4. D-NIR consists of an NIR spectrometer ("P-NIRs"), imaging unit and light source unit. P-NIR is a polychromator-type spectrometer, developed by the group of Prof. Ishikawa for process monitoring. Although the basic system of this spectrometer is similar to a typical one, the detector of the spectrometer is more sensitive, consisting of a newly developed high-density InGaAs photodiode array detector of 640 elements with 20 μm pitch. The photodiode is sensitive to light in the 900–1700 nm wavelength range, and this spectrometer can measure spectra with a 1.25 nm interval in that range. As shown in Fig. 22.4, the diffuse reflection energy from a sample reaches the imaging unit of D-NIRs. Mapping for two dimensions (x- and y-directions) is controlled by two galvanomirrors.

22.2.3.2 Wide-Area NIR Imaging Device with High-Speed Performance [6]

Wide-area investigation of inhomogeneity for components and/or quality with high speed is important to process control of various samples, for instance, agri-food products. "Compovision" is a wide-area NIR imaging camera developed by Sumitomo Electric Industries Ltd (Fig. 22.5). This wide-area and high-speed monitoring system can measure NIR spectral images of spatial size 150×200 mm^2 with spatial resolution of 0.2 mm in less than 5 s. The high speed is achieved by a newly developed InGaAs detector which consists of InGaAs/GaAsSb type-II quantum wells (QWs) laminated on an indium phosphide (InP) substrate. This detector also contributes to the acquisition of high-quality spectral data in a wide NIR spectral region (1000–2350 nm).

22.3 Applications of NIR Imaging

NIR spectral imaging has in recent years become the topic of intense research in highly applied and industrially relevant areas such as food, pharmaceutical, polymer and biological sample analysis. The following sections provide an overview of the

(a)

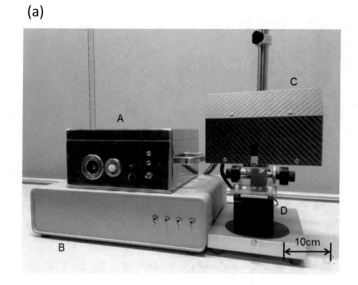

(b)

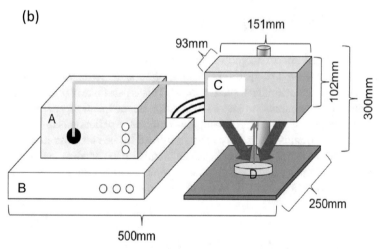

Fig. 22.4 Photo (**a**) and schematic diagram (**b**) of "D-NIRs" NIR spectral imaging system. A: NIR spectrometer (P-NIRs), B: power source unit, C: imaging unit, and D: sample tablet

range and scope of recent applications. For a more comprehensive description of these and related applications, several informative reviews have been published describing advances in spectral imaging for contaminant detection [7], food authentication [8], food quality control [9, 10], pharmaceutical quality control ([11] Gowen et al. 2008) and agricultural analysis ([12] Adao et al. 2017).

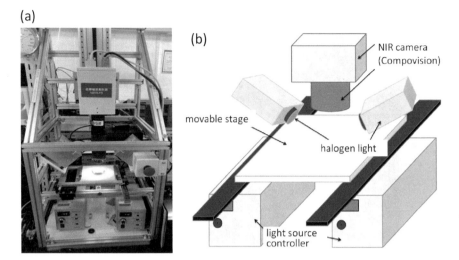

Fig. 22.5 **a** Picture of the "Compovision" NIR spectral imaging system and **b** schematic of same

22.3.1 Food-Related Applications

While conventional spectroscopic methods are useful for characterizing homogeneous products, the lack of spatial resolution often leads to an incomplete assessment of heterogeneous products, such as foods. This is particularly problematic in the case of surface contamination, where information on the location, extent and distribution of contaminants over a food sample is required. Applications of NIR spectral imaging for food quality and safety are widespread in the scientific literature. The heightened interest in this technique is driven mainly by the nondestructive and rapid nature of the technique, and the potential to replace current subjective, labor- and time-intensive analytical methods in the production process.

22.3.2 Contaminant Detection in Foods

Since spectral imaging can detect spatial variations in chemical composition, it is widely regarded as a promising tool for contaminant detection. Major food chain contaminants that can be detected in the NIR include polymers, paper, insects, soil, bones, stones and fecal matter. Diffuse reflectance is by far the most common modality of NIR spectral imaging utilized for this purpose, meaning that primarily only surface or peripheral contamination can be detected. Of particular concern in the food industry is the growth of spoilage and pathogenic microorganisms at both pre-harvest and post-harvest processing stages, since these result in economic losses and potentially threaten public health. Vis–NIR spectral imaging has been demonstrated for pre-harvest detection of symptoms of viral infection and fungal growth on plants and cereals such as maize and wheat [13].

Many studies have focused on the detection of fecal contamination on a wide variety of food products, including fresh produce, meat or poultry surfaces. For example, Vis–NIR imaging (450–851 nm) has been shown to be capable of detecting fecal contamination on apples using with high accuracy levels, suggesting that just two NIR wavebands (748 and 851 nm) could be used for detection [14]. Insect infestation is another example of contamination in the food processing chain detectable by NIR imaging. One of the earliest reported studies in this area showed the possibility of using NIR imaging to detect insects inside wheat kernels [15]. In that study, the authors demonstrated that subtraction of single wavelength images at 1300 nm from those at 1202 nm resulted in an image in which infested kernels could be distinguished from sound ones.

22.3.3 Food Authentication

In order to ensure compliance with labeling, legislation and consumer demand in an ever expanding and global food supply chain, authentication and traceability of food ingredients are critical. Due to the sensitivity of vibrational spectroscopy to molecular structure and the development of advanced multivariate data analysis techniques such as chemometrics, near- and mid-infrared spectroscopies have been successfully used in authentication of the purity and geographical origin of many foodstuffs, including honey, wine, cheese and olive oil. Spectral imaging, having the added spatial dimension, has been used to analyze nonhomogeneous samples, where spatial variation could improve information on the authentication or prior processing of the food product, for example, in the detection of fresh and frozen-thawed meat or in adulteration of flours [16].

22.3.4 Food Quality Control

NIR spectral imaging has been applied in a wide range of food quality control issues, such as bruise detection in mushrooms, apples and strawberries and in the prediction of the distribution of water, protein or fat content in heterogeneous products such as meat, fish cheese and bread [17]. Typically, when observing the NIR spectrum of high-moisture foods, the dominant spectral feature is a wide peak, centered at around 1450 nm, corresponding to the overtone and combination vibrations of fundamental - OH stretching and bending vibrations within the food matrix. The shape and intensity of this peak are sensitive to the local environment of the food matrix and can provide information on changes in the water present in food products. This is useful since many deteriorative biochemical processes such as microbial growth and nonenzymatic browning rely on the availability of free water in foods. NIR spectral imaging has also been applied to quality assessment of semi-solid foods, as reviewed in [10]. For instance, the transmittance modality has been used to nondestructively assess the interior quality of eggs [18, 19], while diffuse reflectance modality has been used to study the microstructure of yoghurt [20] and milk products [21].

22.4 Pharmaceutical-Related Applications

22.4.1 Blend Process Monitoring [22]

Pharmaceutical processes consist of several steps such as milling of materials, blending, drying, coating and tableting. The samples under processing should be monitored by the appropriate method, with measurement speed, wavelength resolution, probe type and signal-to-noise ratio being adjusted according to the application. Various studies have been performed to control pharmaceutical processes by NIR spectroscopy, such as blending. To determine a more accurate end point of the blending process, NIR spectra in the 950–1700 nm region were obtained during the blending process using the experimental setup shown in Fig. 22.6. In addition, an at-line NIR image of the sample during blending was acquired. Figure 22.7 depicts the second derivative image and binary image developed during blending period.

Fig. 22.6 Photo of experimental setup for NIR spectrometry-based process monitoring of a blending process. **A** P-NIRs for the in-line blending process monitor, **B** vessel-type blending machine. Copyright (2015) MDPI

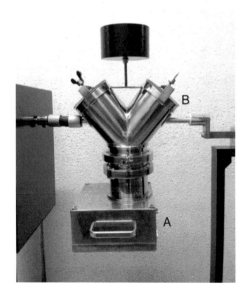

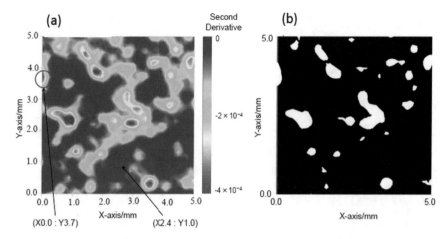

Fig. 22.7 **a** Image of the mixing sample developed by second derivative at 1458 nm for 15 min after the start of the blending. **b** A binary image of the sample at 15 min after the start of the blending obtained by thresholding (**a**) Note that the mapping was performed under special resolution of 0.1 mm. Threshold was defined by standard deviation of second derivative spectra. Copyright (2015) MDPI

The estimated concentration of ascorbic acid corresponded to actual concentration, indicating sufficient blending time, had passed to provide a homogeneous blend.

22.4.2   *Water Penetration Monitoring [23]*

In this example application, we discuss the potential of NIR spectral imaging for tablet dissolution monitoring of a tablet. A model tablet was prepared, consisting of 20% ascorbic acid (AsA) and 80% hydroxypropyl methylcellulose (HPMC). NIR spectral images of the tablet during the process of water penetration were obtained using the previously described D-NIRs system.

Absorbance and second derivative spectra during the dissolution process are depicted in Fig. 22.8. It is evident from the figure that a band at 1361 nm (assigned to the first overtone of an OH stretching vibration of AsA) and a band at 1354 nm decrease during the dissolution process. Thus, NIR ratio images were developed using ratio of second derivative at 1361 and 1354 nm (Fig. 22.9). Distribution change of AsA concentration in the tablet due to water penetration is clearly shown by using the ratio-based image.

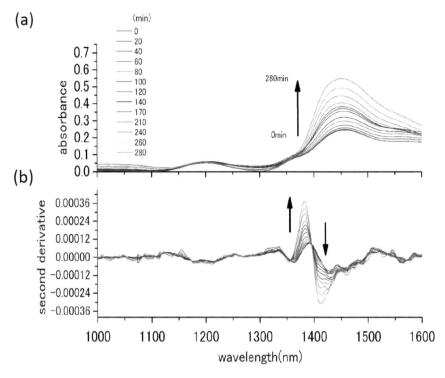

Fig. 22.8 Time-dependent changes of (**a**) NIR spectra and (**b**) their second derivative spectra of the tablet dissolution process at a point. Copyright (2013) Springer

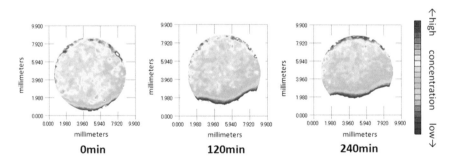

Fig. 22.9 Changes in the peak-height ratio-based image of tablet dissolution developed by using the second derivative intensities at 1361 and 1354 nm due to ascorbic acid and water, respectively. These ratio images were modified by arbitrary threshold to delete the color except the tablet part. Copyright (2013) Springer

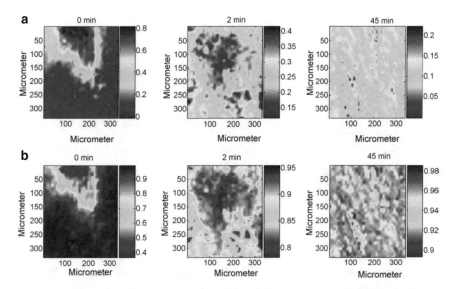

Fig. 22.10 Concentration profiles C of **a** PTX and **b** palmitic acid for each tablet ground for 0, 2 and 45 min. SMCR was applied to the NIR spectra over the spectral region of 7600–4500 cm^{-1}, and concentration profiles **c** and pure component spectra S were obtained. Copyright (2008) Elsevier

22.4.3 Investigation of Inhomogeneity During the Grinding Process [24]

In this example, NIR spectral imaging was carried out on pharmaceutical tablets containing two ingredients, a soluble active ingredient, pentoxifylline (PTX), and an insoluble excipient, palmitic acid. Figure 22.10 illustrates a series of concentration profiles C of PTX and palmitic acid, respectively, for the tablets ground for 0, 2 and 45 min. These concentration profiles were obtained by a chemometrics technique called self-modeling curve resolution (SMCR) of the tablets and can be used to estimate the distribution of PTX and palmitic acid within them. They reveal that the homogeneity of the distribution of chemical ingredients in the samples strongly depends on the grinding time and that this process plays a central role in quantitative control. Accordingly, this study clearly demonstrates that NIR imaging combined with SMCR can be a powerful tool to reveal chemical or physical mechanism induced by the manufacturing process of pharmaceutical products.

22.4.4 Identification of Defective Tablets [25]

In this section, we provide a method for detection of defective tablets using NIR spectral imaging. Normal tablets and defective tablets were prepared, and NIR spectra in

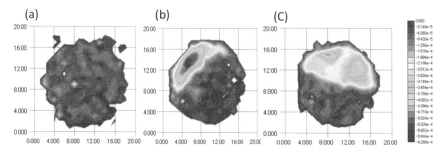

Fig. 22.11 NIR images of (a) nondefective tablet and (**b**), (**c**) defective tablets of 25 and 50% Mg-stearate developed by second derivative at 1213 nm. Note that the 25 and 50% indicates the percentage of defective area in a tablet. Copyright (2019) IM publication

the 950–1700 nm region were measured for each pixel covering the tablet surface by D-NIRs. In this study, "defective" was defined as irregular distribution of the component (in this case, Mg-stearate) in a tablet. Second derivative NIR images at 1213 nm (related to Magnesium Stearate) were produced from the original spectral images, as shown in Fig. 22.11 for nondefective and defective tablets. The defective component can be clearly identified using the standardized image. The skewness of intensity distribution in defective and nondefective tablets is plotted in Fig. 22.12. The skewness of a normal tablet was approximately 0, and it reached approximately −2.3 in the defective tablet, which had a 12.5% redundant concentration area. The skewness decreased gradually as the redundant concentration area increased. Therefore, defective tablets could be identified by the skewness of spectra.

22.5 Polymer-Related Applications

22.5.1 *Polymer Crystallinity Evaluation [26]*

Shinzawa et al. have revealed crystallinity variation in a polymer due to interactions with nanocomposites with NIR spectral imaging using a novel method: the band shift image based on shifting peak positions of the crystalline band of NIR spectra in 5000–4000 cm^{-1}. The distinctive dark pixels in Fig. 22.13 may be interpreted as massively aggregated clay particles, probably reflecting inhomogeneous distribution during the manufacturing process. Shinzawa et al. suggested that the crystallinity of the polymer was improved by the inclusion of the clay. Further studies of this group published elsewhere [27, 28] also enhanced the efficiency of NIR imaging to evaluate physicochemical properties of polymers.

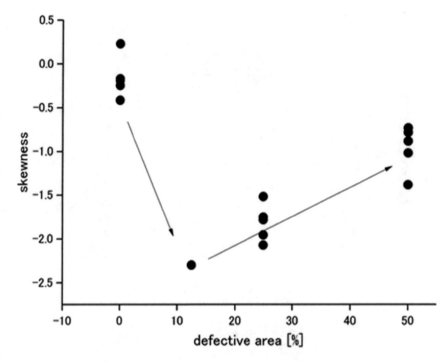

Fig. 22.12 Change in skewness according to the increase of defective area in the tablet. Copyright (2019) IM publication

22.5.2 Biodegradable Polymer Evaluation [29, 30]

As examples of quantitative analysis of crystallinity of biodegradable polymers, polylactic acid (PLA) and the effect of its concentration in PLA/poly-(R)-3-hydroxybutyrate (PHB) blends were investigated using the previously described Compovision system. PLA samples with different crystallinity and blended sample of PLA/PHB were prepared, and NIR spectral images in the 1000–2350 nm region were obtained for a spatial region of 150 × 200 mm area (with 0.25 mm spatial resolution). Application of the standard normal variate (SNV) pretreatment to collected spectra followed by partial least squares regression enabled prediction of crystallinity and concentration of PLA in the biopolymer blend. Figure 22.14 depicts NIR images of wide-area crystal evolution of PLA. Note that the PLA plate was subjected to gradual temperature slope in the range 70–105 °C. A SNV-based prediction image gave an obvious contrast of the crystallinity around the crystal growth area according to slight temperature change. Moreover, it clarified the inhomogeneity of crystal evolution over the sample area.

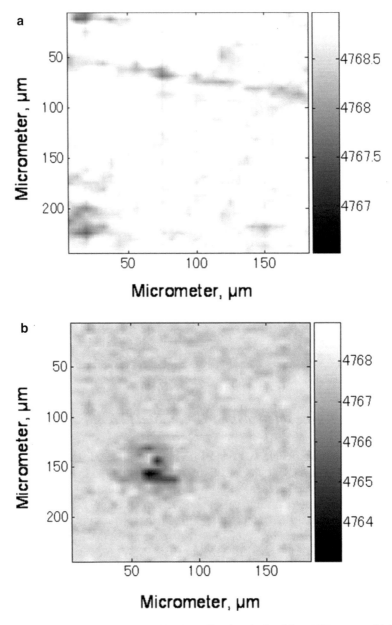

Fig. 22.13 Peak positions of the crystalline band directly calculated from NIR spectra of (**a**) neat PLA and (**b**) the nanocomposite including 15% by weight of clay. Copyright (2012) Elsevier

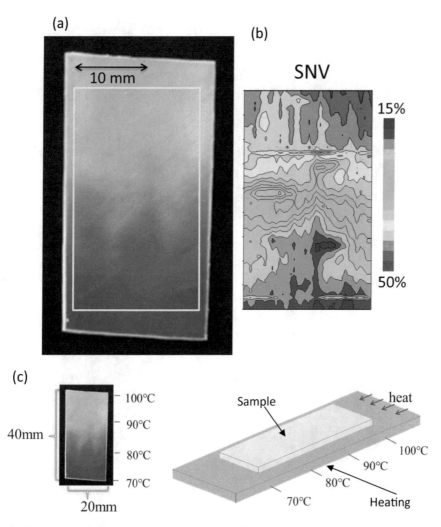

Fig. 22.14 **a** Photo of the PLA sample whose crystallinity was developed by the hot stage, **b** NIR images for the predicted crystallinity of PLA sample developed by using the SNV pretreatment in the 1600–2000 nm region. Copyright (2013) SAGE publishin. **c** Photo of sample (left) subjected to heating by developed heating stage (right).

The use of NIR spectral imaging for estimation of physical properties modified due to hydrolysis of PLA has been investigated by Muroga et al. NIR spectral images were obtained using the Compovision system, and NIR images were developed using peaks and the result of PLS modeling. Figure 22.15 shows two-dimensional distributions of the estimated response variables: (a) flexural strength, (b) flexural strain, (c) flexural modulus and (d) crystallinity. The spectra of the regression coefficients in the PLS models of the flexural strength and the flexural strain had the same

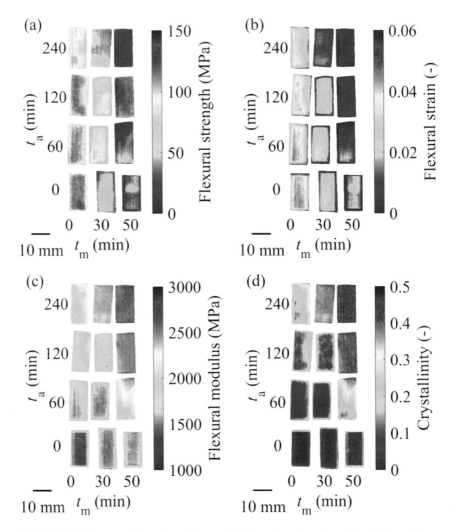

Fig. 22.15 NIR images of the **a** flexural strength, **b** flexural strain, **c** flexural modulus and **d** crystallinity of PLA estimated using PLS models. Note that t_a and t_m indicate annealing time and melting time, respectively. Copyright (2018) John Wiley & Sons, Ltd.

peaks in the range from 1400 to 1500 nm. The spectra of the regression coefficients of the flexural strength and the flexural strain largely reflected the change in the concentration of terminal hydroxyl groups induced by hydrolysis. Thus, this study successfully demonstrated the potential of NIR imaging technology to evaluate the flexural properties and crystallinity of PLA products rapidly.

22.5.3 Monitoring of Biopolymer Photodegradation [31]

Photodegradability of biopolymer is a major concern for industrial applications, such as packaging. In this study, the visualization of photodegradation is investigated during photolysis process. The evolution of crystallinity at 1917 nm in the SNV spectra of the film during photolysis is shown in Fig. 22.16. It can be noted that the behavior of SNV spectra of PLA in the 1000–2000 nm was investigated prior to imaging and the intensities of this band nm due to the second overtone of the $C = O$ stretching vibrations decreased with elapsed time. The crystallinity change in the 0-min image arises from the temperature gradient of the designed hot stage. The distribution of crystallinity is visualized: highly crystalline regions at the bottom of the sample reached to homogeneity by the saturation of crystallinity. The amorphous region at the top area of images depicted inhomogeneity of crystallinity by photolysis process, clearly. Finally, this approach successfully demonstrated that the increase in crystallinity due to photodegradation could be modeled using a logarithmic equation, and the degradation speed is slightly promoted by the initial crystallinity of the sample (Fig. 22.17).

22.6 Bioscience-Related Applications

NIR absorbance bands are mainly due to the functional groups containing a hydrogen atom such as OH, CH and NH. The major components of biological organisms are water, proteins, lipids, molecules which contain these functional groups. Therefore, NIR spectroscopy and imaging can be a useful analytical tool for biological samples to investigate structural changes of biomolecules. Furthermore, aqueous samples are easier to handle for NIR spectroscopic analysis because water absorption in the NIR wavelength region is much weaker than that in infrared (IR) region. Since NIR light

Fig. 22.16 NIR images for the predicted crystallinity of a PLA sample developed by using SNV spectra subjected to photolysis at 0 and 60 min. Copyright (2015) Springer

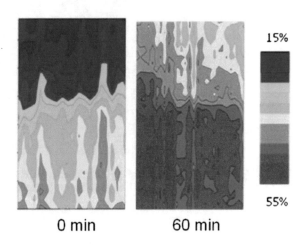

15%

55%

0 min   60 min

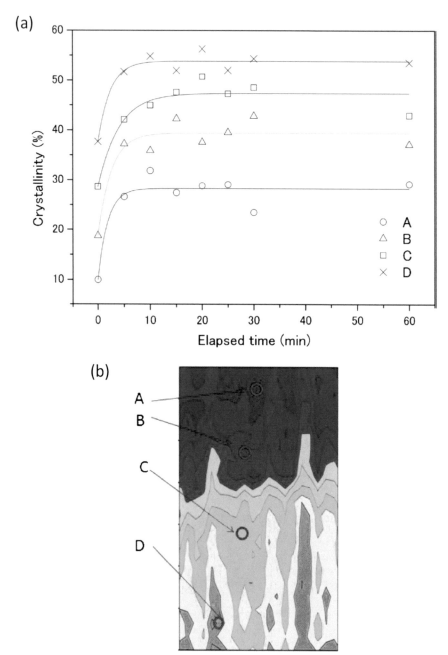

Fig. 22.17 **a** Time-series profiles of crystallinity changes in the different areas shown in **b** the image at 0 min shown in Fig. 22.16. Copyright (2015) Springer

can reach much deeper in a sample than IR light, NIR imaging can easily investigate thick samples than IR imaging without much sample preparation.

22.6.1 Application of Three Types of NIR Imaging System to Biology

Our recent studies investigate the growth of Japanese medaka fish eggs in vivo [32–36]. They are aimed at exploring embryogenesis at the molecular level by visualizing variations of biomolecule distribution and their concentration with the egg development. The size of the eggs is about 1.5 mm in a diameter, and they are almost transparent. They hatch at around two weeks after fertilization under 25 °C water temperature [37, 38]. Note that eggs on the day before hatching are termed here as "just before hatching" (JBH). The eggs consist of three parts (yolk, oil droplets, and embryo) and embryonic body can be clearly observed after the third day after fertilization (Fig. 22.18). For transmission NIR measurements both in point mode and imaging mode, eggs were fixed by two glass slides with spacers to regulate the optical path length as 0.36 or 0.5 mm. The NIR spectra were obtained using three kinds of NIR spectrometers. The first was a PerkinElmer Spectrum One FT-NIR spectrometer equipped with a HgCdTe (MCT) detector (Spectrum Spotlight 300). In the imaging mode, the spatial and wavenumber resolutions were set as 25 mμ and 4 cm^{-1}, respectively. The measurement time was about 1 min for the point mode and 20 min for the imaging mode. For comparison, the Compovision system in transmission mode was employed (Fig. 22.19). This system is equipped with an InGaAs photodiode array (T2SL SWIR focal plane array, SSW230A, Sumitomo Electric Industries Ltd) as a detector. The magnification of the objective lens was 5 × (LMPLNIR5 × , Olympus Co.). The number of pixels in the 1.5 × 1.5 mm^2 area was approximately 50,000,

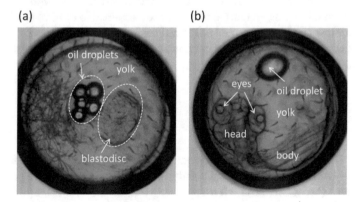

Fig. 22.18 An optical image of a fertilized medaka egg on **a** the first day and **b** the fifth day after fertilization. Reproduced from Ref. [34] with permission. Copyright (2018) John Wiley & Sons, Ltd.

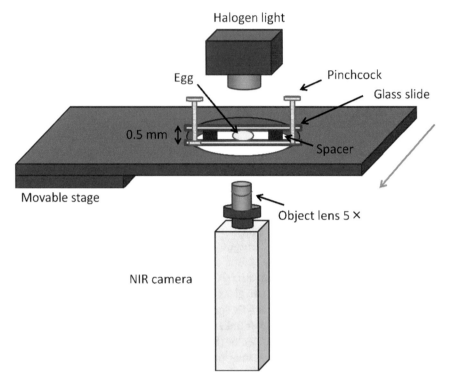

Fig. 22.19 Schematic view of the microscopic near-infrared (NIR) system with samples of medaka fish eggs. Reproduced from Ref. [34] with permission. Copyright (2018) John Wiley & Sons, Ltd.

and the pixel size corresponded to 6.8 mμ. The measurement time was about 2 s in the imaging mode. The last instrument tested was an imaging-type two-dimensional Fourier spectroscopy (ITFS) system. During NIR measurements with these three types of spectral imaging systems, embryos were confirmed to be alive by observing their blood flow and heart beat, and NIR band positions and intensities due to water or proteins did not change between the first and the end points of the measurement in the imaging mode.

22.6.2 NIR Imaging of Fish Egg Embryogenesis

Figure 22.20a and b exhibit NIR absorbance (7500–4000 cm^{-1}) and second derivative (4900–4200 cm^{-1}) spectra, respectively, obtained from different five parts of the medaka eggs (body, eye, head, oil droplet and yolk) on the fifth day after fertilization using the first device (Spectrum Spotlight 300, PerkinElmer). The broadbands at around 6950 and 5200 cm^{-1} are attributed to the combination of the antisymmetric and symmetric O–H stretching modes and of antisymmetric O–H stretching

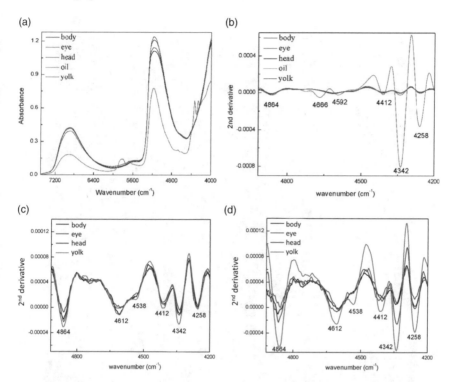

Fig. 22.20 a NIR spectra (7500–4000 cm^{-1}) and **b** their second derivatives (4900–4000 cm^{-1}) from the body, eye, head, oil droplets and yolk on the fifth day after fertilization. NIR second drivative spectra obtained from the body, eye, head, and yolk parts on **c** the 3rd day and **d** the day JBH. Reproduced from Ref. [33] with permission from The Royal Society of Chemistry

and O–H bending modes of water, respectively [39, 40]. In the oil droplet spectra, two prominent peaks at 4258 and 4342 cm^{-1} were observed, which were due to the combination of C–H stretching and bending modes [40, 41]. The second derivative spectra of the body, eye, head and yolk parts on the 3rd day and the day JBH in the 4900–4200 cm^{-1} region are shown in Fig. 22.20c and d. Some small peaks due to proteins were detected at around 4864, 4612 and 4538 cm^{-1} which are assigned to the combination modes of N–H stretching vibration and amide II (amide A + II) [42, 43], the combination modes of amide A and amide III (amide A + III) [44], and more β-sheet structure [45, 46], respectively. Since embryonic structures were in the incomplete stage where they were still being formed and they overlap with yolk parts on the third day, the second derivative spectra showed similar spectral pattern between body, eye, head and yolk (Fig. 22.20c). On JBH day, on the other hand, the differences between the yolk and other embryonic parts were made clear by reflecting the variations of chemical components associated with embryonic formation (Fig. 22.20d). That is, NIR spectra captured the molecular variations such as composition, concentration and molecular structure due to embryonic development.

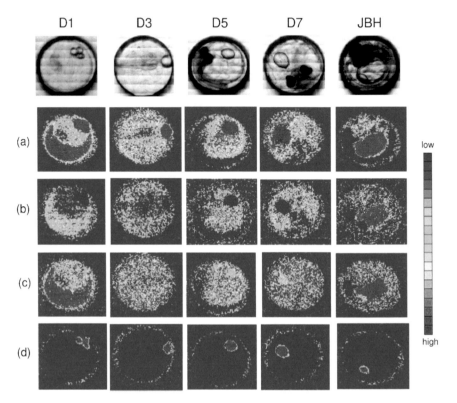

Fig. 22.21 Day-dependent variations in the NIR images of fertilized medaka eggs structured using second derivative intensities at **a** 4340, **b** 4616, **c** 4864 and **d** 4666 cm^{-1} on day 1, 3, 5, 7 and JBH using the PerkinElmer instrument. Reproduced from Ref. [33] with permission from The Royal Society of Chemistry

In imaging mode, these molecular differentiations with the embryonic development were visualized in situ without labeling. Figure 22.21 shows day-dependent variations in optical and NIR images of the eggs on the day 1, 3, 5, 7 and JBH recorded using the first device PerkinElmer. NIR images were constructed by plotting the second derivative intensities at some notable bands in two dimensions. Figure 22.21a was made by the second derivative intensity at 4340 cm^{-1} due to the combination of C–H stretching and bending modes of hydrocarbons and aliphatic compounds [40, 41], especially highlighting oil droplets and yolk parts. The NIR images in Fig. 22.21b were structured using second derivative intensity at 4616 cm^{-1} assigned as the amide mode. This band is characteristic for α–helix structure of proteins, and it has higher peak intensity with more α-helix structure [40, 45]. Therefore, proteins with α–helix structure were revealed to be included in the yolk and egg membrane, and the concentration increased with the embryonic development. The images in Fig. 22.21c were obtained from the peak at 4864 cm^{-1}, which are sensitive to the β-sheet secondary structures of proteins [45, 46]. Even though both Fig. 22.21b, c exhibit protein distributions, they showed different variation patterns depending on the protein secondary structures. The α-helix component in the egg yolk increased with egg development (Fig. 22.21b), but the protein concentration with β-sheet structure, on the other hand, became higher on day 1 and JBH than during the intermediate period (Fig. 22.21c). Monroy et al. investigated the pathway of yolk utilization during egg development using a radioactive element for imaging [47]. They revealed that the synthesis of embryonic proteins from the yolk and its utilization rate was different depending on each developmental stage [47]. That is, the yolk has different roles at each developmental stage, and NIR imaging might capture the metabolic changes depending on the phases, in situ, without labeling. Figure 22.21d shows the distribution of unsaturated fatty acids constructed by the second derivative intensity at 4666 cm^{-1} [41]. The highlighting of oil droplets and egg membrane in this image indicates the existence of the unsaturated fatty acids in these embryonic structures.

22.6.3   High-Speed NIR Imaging of Fish Egg Embryogenesis

For high-speed NIR imaging, the Compovision system was used. Figure 22.22 depicts visible images of the fish eggs in the same day series (a), and NIR images constructed using second derivative intensities and chemometrics (b)–(f) in the 1460–1767 nm (6850–5660 cm^{-1}) region. In Fig. 22.22b made by the band at 1767 nm (5660 cm^{-1}), which was assigned to the first overtone of C–H stretching mode of CH$_2$ groups in hydrocarbons and aliphatic compounds [40, 48, 49], clearly depicts the egg membrane and oil droplets. Since cell membrane generally has a phospholipid bilayer structure with many CH$_2$ groups, the structure is expected to be visualized. Furthermore, after third day, eye structures were also depicted using the band. The images of the first overtone of C–H stretching band in CH$_2$ groups at 1716 nm (5828 cm^{-1}) are shown in Fig. 22.22c [40, 41, 49]. The band position of the first overtone of C–H stretching in unsaturated fatty acids shifts from 1725 nm (5797 cm^{-1}) to 1709–1717 nm (5851–5824 cm^{-1}) [41, 50]. Fish eggs contain much polyunsaturated fatty acids such as docosahexaenoic acid and eicosapentaenoic acid. Therefore, Fig. 22.22c can be interpreted to exhibit the distribution of unsaturated fatty acids. That is, the distributions of different fatty acids depending on the degree of unsaturation can be selectively visualized. The band intensity at 1564 nm (6394 cm^{-1}) due to the first overtone of N–H stretching mode of amide groups was used to prepare Fig. 22.22d [40, 45]. The membrane proteins were expected to make clear the egg membrane. The heterogeneous highlight within oil droplets in Fig. 22.22d overlapped with the part in Fig. 22.22b. This part may show the lipoprotein distribution in which lipids present in plasma wrapped by proteins like blood. Fish eyes were also visualized in Fig. 22.22d. Figure 22.22e was developed by plotting PC1 scores expressing the weakly hydrogen bonding water species [35]. The embryonic structure, contours of oil droplets and the boundary of egg membranes were made clear. Especially, it was

542   D. Ishikawa et al.

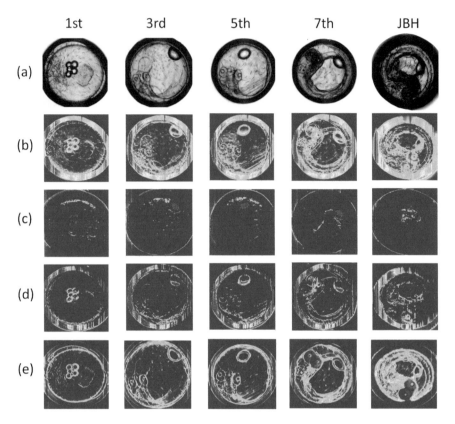

Fig. 22.22 **a** Visible images of medaka eggs from the first day after fertilization to the day JBH, and NIR images of eggs constructed by the second derivative intensities at **b** 1767 nm (aliphatic compounds), **c** 1716 nm (unsaturated fatty acids), **d** 1564 nm (proteins) and **e, f** scores calculated by projecting PC1 loadings onto imaging data to identify the contribution of weakly hydrogen-bonded water. Reproduced from Ref. [34] with permission. Copyright (2002) John Wiley & Sons, Ltd.

very interesting that the water structure at the interfaces between yolk and oil, and yolk and membrane were different. The dynamic water distributions with different structure were revealed to be visualized.

22.6.4 Blood Flow Imaging of Fish Egg Embryos

The last device used for the research was the ITFS system [51]. The characteristic feature of the system is that it has a partial movable mirror as shown in Fig. 22.23 which was developed by Ishimaru of Kagawa University and AOI ELECTRONICS Co., Ltd. in Japan. The partial movable mirror gives optical path differences to the object light and an interferogram can be obtained by continuously changing the spatial phase difference. Since interference only occurs in the rays that come from

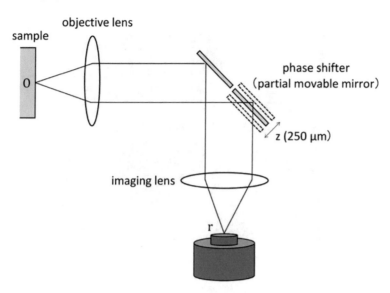

sample
objective lens
0
phase shifter
(partial movable mirror)
z (250 μm)
imaging lens
r

Fig. 22.23 Schematic view of the imaging-type two-dimensional Fourier spectroscopic system (ITFS; AOI ELECTRONICS CO., LTD., NT00-T011). Reproduced from Ref. [52] with permission from ACS Publications

the same point in the system, interference coming from outside the focal plane is not observed in the alternate current (AC) component but detected only in the direct current (DC) component. Therefore, the system can obtain confocal 3D imaging data in principle.

Using the ITFS system, a medaka fish egg was analyzed during the egg development [52]. In the reflectance mode for a medaka fish egg on the fifth day after fertilization (Fig. 22.24a), an interferogram was obtained from the yolk part, (A) as shown in Fig. 22.24b and from heart part, (B); noise-like background was additionally observed (Fig. 22.24c) due to interference with the components that had different optical frequency due to the reflection from moving objects like heart and blood cells, i.e., the "Doppler effect." Figure 22.25a and b demonstrates spectroscopic light intensity in the 1000–2500 nm and 2000–15,000 nm regions, respectively, that were extracted from the interferogram of Fig. 22.24c. The strongest peak was observed at 3768 nm, and two prominent peaks at 1884 and 1256 nm were also observed. The strong peak at 3768 nm (Fig. 22.25b) is due to the fundamental mode of the heart beat, and the two peaks (1884 and 1256 nm) in Fig. 22.25a correspond to the first and second overtones of the fundamental mode, respectively. Please refer the detailed discussion about the peak assignment in Ref. [52]. By plotting the intensities of (a) detected light and (b) absorbance at 1256 and 1884 nm in two dimensions, nonstaining blood flow images were successfully accomplished as shown in Fig. 22.26.

(a)

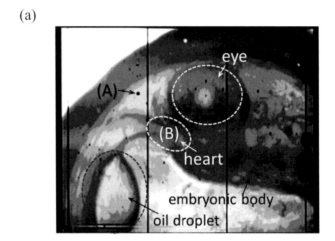

(b)

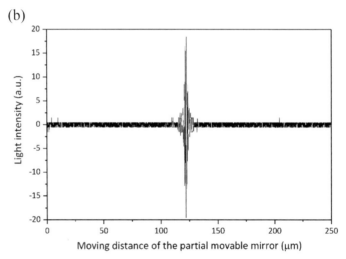

Fig. 22.24 **a** An optical image of an embryonic body of a medaka fish egg on the fifth day after fertilization. **b** Interferogram obtained from the yolk part (A) and **c** from the heart part (B) of (**a**). Reproduced from Ref. [52] with permission from ACS Publications

The measurement in transmission mode was also tried. Two peaks due to the heart beat were detected at 1880 and 2260 nm in FT spectra. In addition, absorbance peaks due to molecular vibrations were observed at 1940 and 2360 nm originating from the antisymmetric O–H stretching and O–H bending modes of water, and C–H stretching and bending modes of hydrocarbons and/or aliphatic compounds, respectively [40, 41, 49]. Figure 22.27 depicts NIR images developed by plotting the intensities of (a) detected light at 2260 nm and absorbance at (b) 1880 nm, (c) 1940 nm and (d) 2360 nm. Figure 22.27a shows the position of the heart and Fig. 22.27b exhibits the

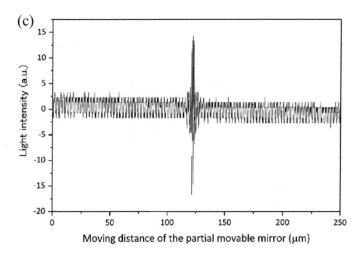

Fig. 22.24 (continued)

blood vessels spread within embryo in addition to the heart. Figure 22.27c and d reveal the detailed structure of embryo and yolk, respectively.

In this way, molecular vibration and heart beat signals can be simultaneously obtained in vivo using optical interference caused by optical path differences and light frequency shifts. The method can be applied to the detection of the early motion of cardiogenesis in the stage where the heart structure is yet not clear. For example, iPS cell differentiation into cardiomyocytes can be monitored in 3D from both pulsation and molecular composition information.

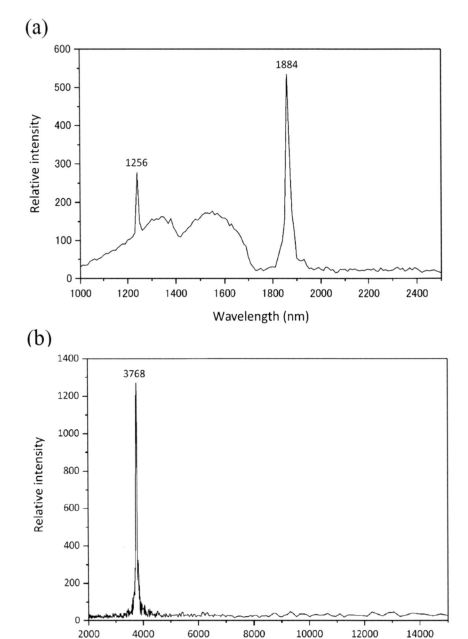

Fig. 22.25 Spectroscopic information obtained by Fourier transformation of the data in Fig. 22.24c in the **a** 1000–2500 nm and **b** 2000–15,000 nm regions. Reproduced from Ref. [52] with permission from ACS Publications

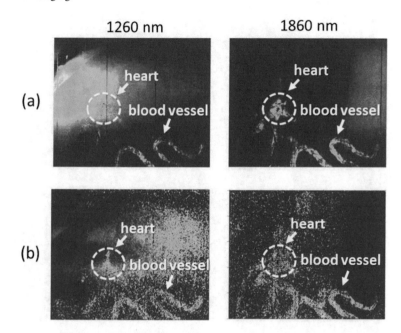

Fig. 22.26 Blood flow images of a medaka fish egg on the fifth day after fertilization obtained by plotting intensities of **a** detected light from the sample and **b** absorbance at 1260 and 1860 nm. Reproduced from Ref. [52] with permission from ACS Publications

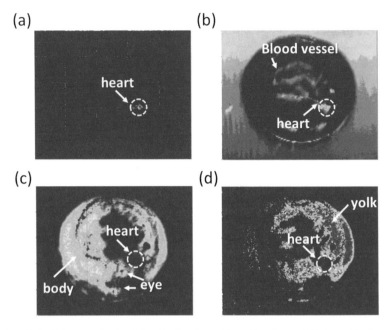

Fig. 22.27 NIR images obtained by plotting intensities of **a** detected light at 2260 nm and absorbance at **b** 1880 nm, **c** 1940 nm, and **d** 2360 nm. Reproduced from Ref. [52] with permission from ACS Publications

References

1. G. Polder, A.A. Gowen, The hype in spectral imaging. J. Spectral Imaging **9**, a4 (2020)
2. A.A. Gowen, R. Tsenkova, C. Esquerre, G. Downey, C. O'Donnell, Use of near infrared hyperspectral imaging to identify water matrix co-ordinates in mushrooms (Agaricus bisporus) subjected to mechanical vibration. J. Near Infrared Spectrosc. **17**, 363–371 (2009)
3. T. Takatani, T. Aoto, Y. Mukaigawa, One-shot hyperspectral imaging using faced reflectors, in *Proceeding of IEEE Conference on Computer Vision and Pattern Recognition (CVPR)* (2017)
4. B. Geelen, C. Blanch, P. Gonzalez, N. Tack, A. Lambrechts, A tiny VIS-NIR snapshot multispectral camera, in *Advanced Fabrication Technologies for Micro/Nano Optics and Photonics VIII* (SPIE: Bellingham, WA, USA, 2015) 937414, 9374
5. D. Ishikawa, K. Murayama, T. Genkawa, K. Awa, M. Komiyama, Y. Ozaki, Development of a component NIR imaging device with high speed and portability for pharmaceutical process monitoring. NIR News **23**(8), 19–23 (2012)
6. D. Ishikawa, T. Nishii, F. Mizuno, S.G. Kazarian, Y. Ozaki, Development of a high-speed monitoring NIR hyperspectral camera (Compovision) with wide area and its applications. NIR News **24**(5), 6–11 (2013)
7. R. Vejarano, R. Siche, W. Tesfaye, Evaluation of biological contaminants in foods by hyperspectral imaging: a review. Int. J. Food Prop. **20**(sup2), 1264–1297 (2017)
8. J. Roberts, A. Power, J. Chapman, S. Chandra, D. Cozzolino, A short update on the advantages, applications and limitations of hyperspectral and chemical imaging in food authentication. Appl. Sci. **8**, 505 (2018)
9. A.A. Gowen, C.P. O'Donnell, P.J. Cullen, G. Downey, J.M. Frias, Hyperspectral imaging—an emerging process analytical tool for food quality and safety control. Trends Food Sci. Technol. **18**, 590–598 (2007)

10. A. Baiano, Applications of hyperspectral imaging for quality assessment of liquid based and semi-liquid food products: A review. J. Food Eng. **214**, 10–15 (2017)
11. A.A. Gowen, C.P. O'Donnell, P.J. Cullen, S.J. Bell, Recent applications of chemical Imaging to pharmaceutical process monitoring and quality control. Eur. J. Pharm. Biopharm. **69**, 10–22 (2008)
12. T. Adão, J. Hruška, L. Padua, J. Bessa, E. Peres, R. Morais, J. Sousa, Hyperspectral imaging: A review on UAV-based sensors, data processing and applications for agriculture and forestry. Remote Sens. **9**, 1110 (2017)
13. E. Bauriegel, A. Giebel, M. Geyer, U. Schmidt, W.B. Herppich, Early detection of fusarium infection in wheat using hyper-spectral imaging. Comput. Electron. Agric. **75**, 304–312 (2011)
14. M.S. Kim, Y.R. Chen, B.K. Cho, K. Chao, C.C. Yang, A.M. Lefcourt, D. Chan, Hyperspectral reflectance and fluorescence line-scan imaging for online defects and fecal contamination inspection of apples. Sens. Instrum. Food Qual. Saf. **1**(3), 151–159 (2007)
15. C. Ridgway, J. Chambers, detection of insects inside wheat kernels by NIR imaging. J. Near Infrared Spectro. **6**, 115–119 (1998)
16. W.H. Su, D.W. Sun, Evaluation of spectral imaging for inspection of adulterants in terms of common wheat flour, cassava flour and corn flour in organic Avatar wheat (Triticum spp.) flour. J. Food Eng. **200**, 59–69 (2017)
17. Y. Liu, H. Pu, D.W. Sun, Hyperspectral imaging technique for evaluating food quality and safety during various processes: a review of recent applications. Trends Food Sci. Technol. **69**, 25–35 (2017)
18. N. Abdel-Nour, M. Ngadi, Detection of omega-3 fatty acid in designer eggs using hyperspectral imaging. Int. J. Food Sci. Nutr. **62**, 418–422 (2011)
19. W. Zhang, L. Pan, S. Tu, G. Zhan, K. Tu, Non-destructive internal quality assessment of eggs using a synthesis of hyperspectral imaging and multivariate analysis. J. Food Eng. **157**, 41–48 (2015)
20. J.L. Skytte, F. Møller, O.H.A. Abildgaard, A. Dahl, R. Larsen, Discriminating yogurt microstructure using diffuse reflectance images, in *Image Analysis, Proceedings of the 19th Scandinavian Conference (2015)*, (SCIA, Copenhagen, Denmark, 2015), pp. 187–198
21. O.H.A. Abildgaard, F. Kamran, A.B. Dahl, J.L. Skytte, F.D. Nielsen, C.L. Thomsen, P.E. Andersen, R. Larsen, J.R. Frisvad, Non-invasive assessment of dairy products using spatially resolved diffuse reflectance spectroscopy. Appl. Spectro. **69**, 1096–1105 (2015)
22. K. Murayama, D. Ishikawa, T. Genkawa, H. Sugino, M. Komiyama, Y. Ozaki, Image monitoring of pharmaceutical blending process and the determination of an end point by using a portable near-infrared imaging device based on a polychromator-type near-infrared spectrometer with a high-speed and high-resolution photo diode array. Molecules **20**, 4007–4019 (2015)
23. D. Ishikawa, K. Murayama, A. Kimie, T. Genkawa, M. Komiyama, SG. Kazarian, Y. Ozaki, Application of a newly developed portable NIR imaging device to dissolution process monitoring of tablets. Anal. Bioanal. Chem. 9401–9409 (2013)
24. K. Awa, T. Okumura, H. Shinzawa, M. Otsuka, Y. Ozaki, Self-modeling curve resolution (SMCR) analysis of near-infrared (NIR) imaging data of pharmaceutical tablets. Anal. Chim. Acta **619**, 81–86 (2008)
25. D. Ishikawa, K. Murayama, T. Genkawa, Y. Kitagawa, Y. Ozaki, An identification method for defective tablets by distribution analysis of NIR imaging. J. Spectral Imaging **8**, a15 (2019)
26. H. Shinzawa, M. Nishida, T. Tanaka, W. Kanematsu, Crystalline structure and mechanical property of poly(lactic acid) nanocomposite probed by near-infrared (NIR) hyperspectral imaging. Vib. Spectro. **60**, 50–53 (2012)
27. H. Shinzawa, M. Nishida, A. Tsuge, D. Ishikawa, Y. Ozaki, S. Morita, W. Kanematsu, Thermal behavior of poly(lactic acid)-nanocomposite studied by near-infrared imaging based on roundtrip temperature scan. Appl. Spectro. **68**(3), 371–378 (2014)
28. H. Shinzawa, J. Mizukado, Near-infrared (NIR) disrelation mapping analysis for poly(lactic) acid nanocomposite. Spectrochim. Acta a **181**, 1–6 (2017)
29. D. Ishikawa, T. Nishii, F. Mizuno, S.G. Kazarian, Y. Ozaki, Potential of a newly developed high speed near-infra red (NIR) camera (Compovision) in polymer industrial analyses—monitoring

of crystallinity and crystal evolution of poly lactic acid (PLA) and concentration of PLA in PLA/Poly-(R)-3-hydroxybutyrate (PHB) Blends-. Appl. Spectro. **67**(12), 1411–1416 (2013)

30. S. Muroga, Y. Hikima, M. Ohshima, Visualization of hydrolysis in polylactide using near-infrared hyperspectral imaging and chemometrics. J. Appl. Polym. Sci. **135**, 45898 (2018)

31. D. Ishikawa, D. Furukawa, T.W. Tseng, R.R. Kummetha, A. Motomura, Y. Igarashi, S.G. Kazarian, Y. Ozaki, High-speed monitoring of crystallinity change in poly lactic acid during photodegradable process by using a newly developed wide area NIR camera (Compovision). Anal. Bioanal. Chem. **407**(2), 397–403 (2015)

32. M. Ishigaki, S. Kawasaki, D. Ishikawa, Y. Ozaki, Near-infrared spectroscopy and imaging studies of fertilized fish eggs: In vivo monitoring of egg growth at the molecular level. Sci. Rep. **6**, 20066 (2016).

33. M. Ishigaki, Y. Yasui, P. Puangchit, S. Kawasaki, Y. Ozaki, In vivo monitoring of the growth of fertilized eggs of medaka fish (Oryzias latipes) by near-infrared spectroscopy and near-Infrared imaging—a marked change in the relative content of weakly hydrogen-bonded water in egg yolk just before hatching. Molecules **21**(8), 1003 (2016)

34. P. Puangchit, M. Ishigaki, Y. Yasui, M. Kajita, P. Ritthiruangdej, Y. Ozaki, Non-staining visualization of embryogenesis and energy metabolism in medaka fish eggs using near-infrared spectroscopy and imaging. Analyst **142**(24), 4765–4772 (2017)

35. M. Ishigaki, T. Nishii, P. Puangchit, Y. Yasui, C.W. Huck, Y. Ozaki, Noninvasive, high-speed, near-infrared imaging of the biomolecular distribution and molecular mechanism of embryonic development in fertilized fish eggs. J. Biophotonics **11**(4), e201700115 (2018)

36. M. Ishigaki, P. Puangchit, Y. Yasui, A. Ishida, H. Hayashi et al., Nonstaining blood flow imaging using optical interference due to Doppler shift and near-infrared imaging of molecular distribution in developing fish egg embryos. Anal. Chem. **90**(8), 5217–5223 (2018)

37. A. Shima, H. Mitani, Medaka as a research organism: Past, present and future. Mech. Dev. **121**, 599–604 (2004)

38. T. Iwamatsu, Stages of normal development in the medaka Oryzias latipes. Mech. Dev. **121**, 605–618 (2004)

39. H.W. Siesler, Y. Ozaki, S. Kawata, H.M. Heise (eds.), *Near-Infrared Spectroscopy: Principles, Instruments, Applications* (Wiley, New Yolk, 2008)

40. J. Workman Jr., L. Weyer, *Practical Guide and Spectral Atlas for Interpretive Near-Infrared Spectroscopy* (CRC Press, Boca Raton, 2012)

41. T. Sato, S. Kawano, M. Iwamoto, Near infrared spectral patterns of fatty acid analysis from fats and oils. J. Am. Oil Chem. Soc. **68**(11), 827–833 (1991)

42. S. Holly, O. Egyed, G. Jalsovszky, Assignment problems of amino acids, di-and tripeptides and proteins in the near infrared region. Spectrochim. Acta Mol. Spectro. **48**(1), 101–109 (1992)

43. W.Y. Yang, E. Larios, M. Gruebele, On the extended β-conformation propensity of polypeptides at high temperature. J. Am. Chem. Soc. **125**(52), 16220–16227 (2003)

44. P. Robert, M.F. Devaux, N. Mouhous, E. Dufour, Monitoring the secondary structure of proteins by near-infrared spectroscopy. Appl. Spectro. **53**(2), 226–232 (1999)

45. K.I. Izutsu, Y. Fujimaki, A. Kuwabara, Y. Hiyama, C. Yomota, N. Aoyagi, Near-infrared analysis of protein secondary structure in aqueous solutions and freeze-dried solids. J. Pharm. Sci. **95**(4), 781–789 (2006)

46. M. Miyazawa, M. Sonoyama, Second derivative near infrared studies on the structural characterization of proteins. J. Near Infrared Spectro. **6**, A253–A257 (1998)

47. A. Monroy, M. Ishida, E. Nakano, The pattern of transfer of the yolk material to the embryo during the development of the teleostean fish. ORYZIAS LATIPES. Embryol. **6**, 151–158 (1961)

48. F. Westad, A. Schmidt, M. Kermit, Incorporating chemical band-assignment in near infrared spectroscopy regression models. J. Near Infrared Spec. **16**(3), 265–273 (2008)

49. W. Hug, J.M. Chalmers, P.R. Griffith, *Handbook of Vibrational Spectroscopy* (Wiley, Chichester, England, 2002)

50. T. Sato, Application of principal-component analysis on near-infrared spectroscopic data of vegetable oils for their classification. J. Am. Oil Chem. Soc. **71**(3), 293–298 (1994)

51. W. Qi et al., Enhanced interference-pattern visibility using multislit optical superposition method for imaging-type two-dimensional fourier spectroscopy. Appl. Opt. **54**(20), 6254–6259 (2015)
52. M. Ishigaki, P. Puangchit, Y. Yasui, A. Ishida, H. Hayashi, Y. Nakayama, H. Taniguchi, I. Ishimaru, Y. Ozaki, Nonstaining blood flow imaging using optical interference due to doppler shift and near-infrared imaging of molecular distribution in developing fish egg embryos. Anal. Chem. **90**(8), 5217–5223 (2018)

Chapter 23
Inline and Online Process Analytical Technology with an Outlook for the Petrochemical Industry

Rudolf W. Kessler and Waltraud Kessler

Abstract The concept of process analytical technology (PAT) started around the 1970s with the advent of personal computers in combination with instrumental analytical chemistry. Over the years, increasingly sophisticated and holistic quality management concepts such as quality by design (QbD) were developed and strongly promoted, especially by the American Food and Drug Agency (FDA) around 2002. Recently, the German initiative for the Fourth Industrial Revolution "Industrie 4.0 (i40)" was introduced, which is similar to the US "Industrial Internet Consortium (IIC)" concept or the "Industrial Internet of Things (IIoT)." Another initiative in Asia is the Chinese campaign "Made in China 2025." The role of PAT in all these concepts is to develop and integrate context-sensitive intelligent sensors to enable understanding of the process at the basic mechanistic (molecular) level in order to achieve knowledge-based production in the future. This contribution starts with a short introduction into the concept of PAT/QbD and other new concepts for the next generation of spectroscopic sensors in the manufacturing industry. The fundamental limitations of spectroscopy in terms of sensitivity and selectivity are discussed, and the need to increase robustness for industrial applications is described. A critical discussion on problems and problem solutions are provided when scattering samples are investigated. An outlook on how to use NIR spectroscopy within the petrochemical industry and how to manage a PAT project complements this chapter.

Keywords Industrial Internet of Things (IIoT) · Industrie 4.0 (i40) · Near-infrared spectroscopy (NIR) · Petrochemical industry · Process analytical technology (PAT) · Process control · Process understanding · Robustness · Sensitivity and selectivity · Spectroscopy

R. W. Kessler (✉) · W. Kessler
Reutlingen University, Kessler ProData GmbH, Kaiserstr. 66, 72764 Reutlingen, Germany
e-mail: rudolf.kessler@kesslerprodata.de

W. Kessler
e-mail: waltraud.kessler@kesslerprodata.de

© Springer Nature Singapore Pte Ltd. 2021
Y. Ozaki et al. (eds.), *Near-Infrared Spectroscopy*,
https://doi.org/10.1007/978-981-15-8648-4_23

23.1 Process Analytical Technology (PAT): A Systems Approach

23.1.1 Road Map for PAT

In PAT, complex data (e.g., spectra) are obtained in real time, then processed and evaluated using the modern techniques of chemometrics, multivariate data analysis or data mining (*see previous chapters in this book*). This requires a strongly interdisciplinary and transdisciplinary approach, because process managers, chemometric data analysts, electronics engineers, process specialists, and chemists with domain knowledge, e.g., of spectroscopy have to be brought together in a business environment.

There have been dramatic changes in the manufacturing and processing industry during the past years. Increasingly sophisticated and holistic quality management systems like quality by design (QbD) have been developed over the years and strongly promoted by the US Food and Drug Agency (FDA). However, the present-day strategy focuses strongly on process optimization and the evaluation of safety risks in the chemical industry. Much less emphasis is laid on the possibility of producing a product which matches exactly the customer profile and expectations of personalized products. This process- and product-functionality design allows a precisely defined manufacture of the widest range of products with minimized costs. When computer-driven, the inverse model enables the manufacturer to develop new products and techniques within the defined process limits. Multi-objective optimization can be quickly, reliably, and cheaply achieved, because the essential steps are known from the model algorithm and are known as "knowledge-based process management." As described in the next chapter, concepts like Industrie 4.0 (i40) or the Industrial Internet of Things (IIoT) are a strategy which describes in detail the holistic approaches needed to achieve these goals [1–3].

In-process quality and process optimization will only be successful when it is based on appropriate process understanding, i.e., the analysis of the connection (cause and effect) between process parameters and the quality characteristics of the final product with its specifications. PAT within this framework means therefore understanding the causal relation between measurement and response.

Very often in chemometrics and modeling, only descriptive or statistical knowledge is produced which fits the specific dataset but cannot be used as a general or global model based on molecular information. Figure 23.1 visualizes the different levels of knowledge for process understanding [1, 4].

Complex products are usually produced in lines involving several steps. Depending on the application of the product, each step can be adjusted specifically for the quality of the incoming material and for the quality of the end product [4]. Furthermore, personalized products and materials are needed to fulfill the customers demand for smart functionalities. This is the concept of the future production.

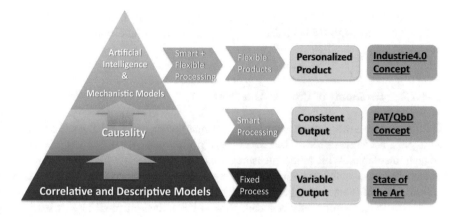

Fig. 23.1   Road map for process understanding for smart and flexible production

23.1.2   Taxonomy of Process Analyzers and Sampling

23.1.2.1   Sampling

The main concern of sampling is to get access to a representative sample. The sample set should be divided into a calibration set and a validation set as described in the European Medicine Agency (EMA) guidance for industry [5]. Each set of samples should be representative of the intended scope of the NIR spectroscopic procedure and includes samples covering the full range of potential variation in the sample population. The intensity of the analyte signal and the complexity of the sample matrix and/or interference by the matrix of the analyte signal of interest should also be taken into account. In general, the more complex and the more interference from the matrix, the more samples will be required.

After establishing a first calibration model, an external validation set of samples is strongly recommended to check the robustness of the model. This sample set should be entirely independent of those samples used to build the spectral library and should include qualitatively positive and negative samples. In principle, the external validation set should cover the calibration range of the NIR spectroscopic model, including all variation seen in the commercial process and should include pilot and production-scale batches, where possible [5].

Sampling is the most important task but presents the most significant problems. The total sampling error (TSE) will include all sampling and mass reduction error effects. The total analytical error (TAE) includes thus the total sampling error together with the handling error as well as the analysis uncertainty estimate. In most cases, sampling errors show the highest contribution to the TAE rather than the analytical error of the instrumental analysis [6–8]. Furthermore, any change in the texture and morphological structure, e.g., by milling the sample, will lead to a change in the spectral signature through (Mie-) backscattering and thus will modify the chemometric model.

23.1.2.2 Taxonomy of Process Analyzers

Offline analysis always exerts significant lag times between recognizing and counteracting against irregularities. The advantage of offline analytics is that expert knowledge is available. With atline measurements, the sample is withdrawn from the process flow and analyzed with analytical equipment that is located in the immediate environment of the industrial equipment. Hence, the reaction time for countermeasures is already significantly reduced. Due to the industrial proximity, it is often observed that atline analytical equipment is more robust and insensitive towards process environment but less sensitive or precise than laboratory-only devices [4].

In the case of online measurements, samples are not completely removed from the process flow but temporarily separated, for example, via a by-pass system which transports the sample directly through the online measurement device. Thus, the sample is analyzed in immediate proximity to the industrial machining and is afterwards either reunited with the process stream or ejected. The major advantage of this procedure is that the sample can be conditioned, e.g., filtered and measured at a constant temperature [4].

When inline devices are used, the sensor is directly immersed into the process flow and prevails in direct contact with the unmodified material flow. Sometimes the sensors are placed in front of a window to measure non-invasive. Inline measurements are sometimes also called insitu measurements [4, 9].

Figure 23.2 illustrates various sampling modes realized with inline, online, atline and offline sampling.

23.2 Future Concepts in the Process and Manufacturing Industry: Industrie 4.0, Industrial Internet of Things, and Their Impact on PAT Sensors

23.2.1 Concepts for the Next Generation of Production Systems

Process analytical technology (PAT) enables the implementation of quality by design (QbD) in industrial reality [1, 9–13]. The basic idea behind the use of spectroscopic sensors is to gain process understanding at the molecular level, which together with continuous process improvement should lead to a possible real-time release (RTL) of the product. This requires the identification of the critical process parameters (CPP) together with the critical control points (CCP). In addition, the final products and

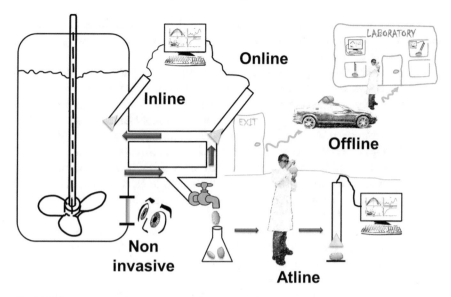

Fig. 23.2 Visualization of the taxonomy of process analyzers

intermediate products must have all critical quality characteristics defined (critical quality attributes, CQA), including the final process performance qualification (PPQ) [1, 9, 10].

Digitization as the next step will again significantly change the structures of future production. Recently, the German initiative for the Fourth Industrial Revolution "Industrie 4.0 (i40)" was presented, which is similar to the US "Industrial Internet Consortium (IIC)" concept or the "Industrial Internet of Things (IIoT)" [1, 2, 4, 13, 14]. A similar initiative is the Chinese strategy "Made in China 2025." In the meantime, there are contracts between all partners in which the same language and definitions will be used. Common to all concepts is that context-sensitive sensors will be integrated into the process and manufacturing industry [1, 15].

This shifts the focus from centrally controlled to decentralized controlled production processes even down to a lot size of 1. The resulting so-called smart products know their production history, their current and target state, and actively steer themselves through the production process by instructing machines to perform the required manufacturing tasks and ordering conveyors for transportation to the next production step [16].

Stirred-tank reactors are well-established tools in the traditional batch production with the advantage of high flexibility of the production. Major disadvantages of this technology are long setup times, difficult heat transfer rates, and inherent inefficiency. A second established production scheme is the continuous production, especially applied in the field of high-volume and low-cost materials and manufactured in highly efficient world-scale plants [16]. From a technology point of view, those plants contain, e.g., reactors with tube shape geometry which enable heat integration, good process control abilities and are highly efficient. But these world-scale plants are inflexible in the case of variations like raw material changes, reaction systems, or amount of production capacity [16].

Future production must be decentralized, modularly packed in standardized compartments, and should enable continuous production even for small- to medium-scale applications. An example is the so-called F3 factory. The objective of the F3 factory is to demonstrate the economic advantages of modular, small-scale production plants compared with multiproduct batch plants [17]. The metal framework allows for rapid module exchange and replacement as well as stepwise increase of the plant. An increasing market demand could easily be followed by just numbering up the containers, or if in cases of volatile markets, the demands decrease, individual containers could be decoupled from the production stream. No extra time for scaling up is needed [16–18].

Another core element of the new production concepts is micro- and milli-structured devices assisting continuous-flow processes due to their superior transport characteristics and small holdup. Microreactors are characterized by rapid mixing and excellent heat transfer conditions.

Figure 23.3 shows a typical modular production concept, which was developed within the European Union's project "CONSENS" [18]. The objectives of the concept are to develop standardized unit operations plant modules with miniaturized flexible equipment and local control. The "plug-and-produce" concept need no scaling up as the plants can produce in parallel according to the demand of the customers.

All these concepts need new multimodal smart sensors for PAT with an extended connectivity, internal maintenance functions, better traceability, compliance, and the ability to interact and communicate with its physical environment.

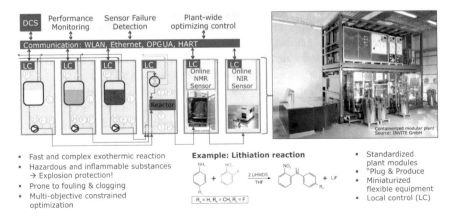

Fig. 23.3 Modular lithiation reaction setup with a multimodal spectroscopic control system (NIR and NMR) using standardized plant modules "plug and produce." Source and permission of M. Maiwald, BAM, Germany

23.2.2   Industrial Internet Reference Architectural Model of Industrie 4.0: RAMI 4.0

Figure 23.4 shows the "Reference Architecture Model Industry 4.0 (RAMI 4.0)," which was developed by a trilateral collaboration of working groups of German Industry 4.0 together with the Alliance Industrie du Futur (France) and Piano Impresa 4.0 (Italy). Harmonization was also achieved with the US Industrial Internet Consortium (IIC) in 2017 [1]. In addition, there is cooperation with China in several working groups, which led to a report on the harmonization of the Chinese counterpart IMSA as well as the Japanese counterpart IVRA and the German RAMI 4.0 [1, 2, 4, 19, 20].

In order to harmonize the requirements for digitized industrial production, common and global standards for communication structures (networks and protocols), rules for cybersecurity and data protection and the common language including characters, alphabet, vocabulary, syntax, grammar, semantics, pragmatics, and culture are being developed [21].

The concept consists of a three-dimensional coordinate system that describes all the key aspects of industry 4.0 or IIoT. In this way, complex relationships can be broken down into smaller and simpler clusters. A detailed description can be seen as well as the status of standardization [1, 20–24].

The basic idea of Fig. 23.4 and the objectives behind are summarized below and detailed in many reviews and articles [1, 2, 19–24].

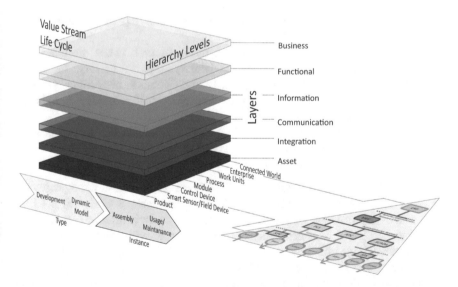

Fig. 23.4 Reference Architectural Model of Industrie 4.0 (RAMI 4.0, platform Industrie 4.0), the hierarchical structure of a PAT sensor is also illustrated, details see text below, modified from ref. [21]. *Source* and permission of M. Maiwald, BAM, Germany

The "hierarchy levels" are indicated on the right horizontal axis and are defined by the international series of standards for enterprise IT and control systems IEC 62264. These hierarchy levels represent the different functionalities within factories or plants.

The left horizontal axis represents the "Life Cycle & Value Stream" axis, which describes the life cycle of plants and products, based on IEC 62890 for life cycle management.

The "Layers" axis is the vertical axis and has six layers. They are used to describe the properties of a machine layer by layer in a structured way and are thus creating a virtual image of a machine.

In this way, RAMI 4.0 offers a common understanding for standards and use cases. This will significantly change the structure and integration of PAT tools within the factory.

23.2.3 Communication Between Cyber-Physical PAT Systems: Connected, Multimodal, Decentralized, and Secured

OPC Unified Architecture (OPC UA) is a machine-to-machine communication protocol for industrial automation developed by the OPC Foundation [21–23]. The next steps to be taken within this framework are summarized in [1, 25]:

- Identification: This is a necessary prerequisite for things to find their own way to each other within the networked production.
- Semantics: For communication between machines or between machines and workpieces, a manufacturer-independent data exchange is necessary.
- Quality of Service (QoS) for Industry 4.0 or IIoT components: Critical services such as time synchronization, real-time capability, and reliability of Industry 4.0/IIoT components must be defined.
- Communication within Industry 4.0/IIoT: There is a multitude of communication links and protocols. The most common examples are field buses based on Ethernet or OPC UA, a protocol for machine-to-machine communication.

The physical Cyber-PAT systems (CPS-PAT) are introduced in three phases [1, 21]. The first generation of CPS includes identification technologies such as RFID tags, which enable unique identification. Storage and analysis must be provided as a central service. The second generation of CPS is equipped with sensors and actuators with a limited range of functions. Third generation CPS can store and analyze data, are equipped with several sensors and actuators, and are network compatible.

The role of PAT sensors in all these concepts is therefore to develop and integrate context-sensitive information to promote understanding of the process at the basic mechanistic (molecular) level, thus enabling knowledge-based production in the future [1, 15].

The most important aspects with regard to industrial production are summarized in [1, 21]:

- Personalization: Integration of short-term customer requirements and change management requirements through digital engineering.
- Flexibilization: reduces lead time and time to market, e.g., through 3D printing and predictive analysis.
- Decentralization: enables smaller lot sizes down to a lot size of 1.

Powerful algorithms will make it possible to simulate the production route and to control and optimize the intermediate and final products in real time at all stages of the manufacturing process. The new PAT sensors are thus a basic technology for the following processes [1, 4]:

- Rationalization of processes that are not already automated.
- Pro-active process and quality management including predictive maintenance.
- Integration of quality assurance into production while increasing system availability and system reliability.
- Improved product safety and increased production efficiency.

Figure 23.5 attempts to visualize the basic requirements for a future sensor concept.

For standard sensor applications and decision systems, level 0 contains the field devices such as flow and temperature sensors (process value displays, PVR) as well as actuators (FCE), such as control valves. Level 1 contains the industrialized input/output modules (I/O modules) and the associated distributed electronic processors. Level 2 contains the monitoring computers that collect information from the

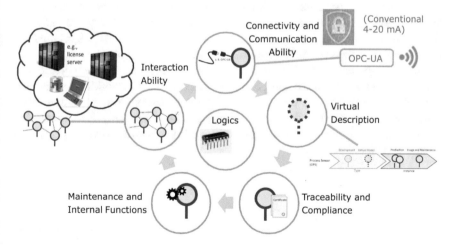

Fig. 23.5 Current and future requirements of a cyber-physical sensor system (reproduced and modified from NAMUR "Prozesssensoren 4.0" [21]). *Source* and permission of M. Maiwald, BAM, Germany

processor nodes in the system and provide the operator screens. Level 3 represents the production control level, which, however, does not directly control the process, but is concerned with monitoring production and setting targets. Level 4 is the production planning level. The future sensor levels may well be more decentralized, but the decision hierarchies will be retained (see also Fig. 23.4) [24].

For the future intelligent process analyzers, the following important goals apply in addition to miniaturization and cost reduction described in [22, 23, 26] and summarized in [1]:

- Plug-and-play capability.
- Self-configuration, self-calibration, and self-optimization.
- Higher sensitivity for inline trace analysis.
- Imaging ability.
- Multimodality with "all-in-one" functionality.

New and miniaturized instruments can drastically reduce costs and thus open up further possibilities for use. The user should also be given the possibility of "plug-and-play." Since miniaturization allows the integration of different measurement technologies, multimodal information can be accommodated in a single sensor [1, 26].

Data fusion, i.e., the combination of different data sources for classification, can be carried out at different levels. At the low level, the concatenation of data matrices at the variable level is possible. At the middle level, data matrices are concatenated using a feature selection, for example, at score level after applying a PCA. And at the high level, model predictions are combined, e.g., predictions from PLS models [27–29].

In addition to the already established multivariate data analysis and other chemometric tools in the spectroscopically based control systems, the concepts of machine learning (ML) and deep learning (DL) are also used in "Industry 4.0" applications which is described in more detail in [1, 30]. ML and DL algorithms are used to describe the system behavior instead of the previous physical–technical or chemometric-based models. This is usually referred to as predictive analytics. It must be emphasized that feature engineering is included in the ML workflow, whereas it is not available in DL. DL data acquired from the sensors ("raw measurements") can be entered directly into the deep learning algorithms. Thus, ML is more oriented towards traditional engineering methodology, while DL algorithms are largely based on artificial neural networks.

23.3 Robustness in PAT Applications with a Focus on NIR Spectroscopy: About Sensitivity, Selectivity, and Signal-to-Noise

23.3.1 General Approach to Optical Spectroscopy in PAT and Their Advantages

Optical spectroscopy describes the interaction of molecules or matter with photons [1, 4]. The inelastic interaction leads to the transition of the molecule from a lower to a higher energy level. This spectral property is called wavelength-dependent absorption. The elastic interaction, on the other hand, leads to a scattering of the photon in different directions and is also wavelength-dependent. Optical spectroscopy detects absorption and scattering simultaneously and thus provides the following information as summarized in [1]:

- The chemical composition of the molecule or matter by measuring the wavelength-dependent (inelastic) absorption of light.
- The morphological information (nanoscopic substructure) by measuring the wavelength-dependent (elastic) scattering of light (e.g., Mie scattering).
- The texture of the heterogeneous system by combining spectroscopy and image analysis. This is called hyperspectral or chemical imaging.

Optical spectroscopy is currently the workhorse in PAT. Within optical spectroscopy, near-infrared spectroscopy is the most commonly used technique [1, 31–34]. Online and inline NIR optical process spectroscopy is well-established and widely used. It is also frequently used to characterize materials or to control chemical reactions in industry at the molecular level, e.g., to monitor the concentration of a chemical component in a mixture. The spectral information is often evaluated by multivariate data analysis (MVA) and correlated with quality parameters of the material or finished product [4].

There are three key parameters that are important for the functionality of spectroscopic methods in PAT solutions: sensitivity, selectivity, and robustness of the method [1, 33].

23.3.2 Sensitivity: Definition at Molecular Level and Classification of NIRS Within the Spectroscopic Toolbox

23.3.2.1 Definition

Analytical sensitivity is the smallest amount of a substance in a sample that can be accurately determined using a specific technology. In general, this is the concentration at which the determined mean measurement signal is just above the noise limits of the zero signal. Among other things, it is defined by the absorption cross section of the molecule [1, 2, 10, 33].

The sensitivity of a molecule can be described by the quantum mechanical cross sections [1, 33, 34]. These are the effective areas that determine the probability of an event with a molecule, such as elastic scattering or absorption or emission of a photon at a certain wavelength. From a quantum mechanical point of view, the absorption cross section σ_a is usually given in cm^2/molecule and depends on the individual molecular structure of the compound and quantum mechanical selection rules. The term cross section is used in physics to quantify the probability of a certain interaction, e.g., scattering or electromagnetic absorption. For example, in the NIR range the cross section probabilities change over several orders of magnitude for the combination bands between the first and second or third overtone.

It is also important to note that in turbid media or solids the scattering cross section can be several orders of magnitude more sensitive than the absorption cross section.

Table 23.1 shows the quantum mechanical absorption cross section σ_a [cm^2] and the scattering cross sections σ_s [cm^2] including their interaction probability for selected spectroscopic methods.

The data shown here are representative of a large number of different components. The numbers are given as $-\log \sigma_a$ [cm^2/molecule]; therefore, low numbers show high sensitivity. For comparison, Rayleigh scattering figures are also shown, which is due to molecular scattering and Mie scattering, where Mie scattering is more related to particle size.

The total cross section is related to the absorption according to Lambert–Beer's law and is proportional to the concentration of the species and the path length [34]. The absorbance or absorption is given as the logarithm of the reciprocal of the transmittance.

Table 23.1 Absorption and scattering cross sections of selected spectroscopic technologies

Technology	Cross section	Extinction
Absorption	$- \log \sigma_a$ [cm^2/molecule]	ε [1/(Mol cm)]
UV–Vis/*Fluorescence*	23–**16**	*app. 10^{+4}*
Mid-Infrared	25–**18**	*app. 10^{+1}*
Near-Infrared	28–**22**	*app. 10^{-3}*
combinations	23–22	
1st overtone	25–24	
2nd overtone	27–26	
3rd overtone	28–27	
Raman	35–**28**	*app. 10^{-9}*
Scatter	$-\log \sigma_s$ [cm^2/molecule or particle]	
Rayleigh scatter	33–29–24 per molecule	
Mie scatter	20–16–10 per particle	

23.3.2.2 NIR Spectroscopic Sensitivity

Near-infrared spectroscopy (NIR) is less sensitive than MIR spectroscopy due to smaller absorption cross sections of the oscillation transitions of the higher orders. Therefore, NIR spectroscopy is not a very sensitive method and the detection limit is much lower than with MIR spectroscopy and also UV–Vis spectroscopy. It must also be emphasized that scattering of particles can significantly alter the spectral fingerprints due to the high sensitivity of the scattering cross section (see Table 23.1). Changes in morphology, e.g., during sample preparation, are often misinterpreted as changes in chemical composition. An example is the measurement of groundwood chips used for calibration for the inline measurement of macroscopic wood chips [1, 4, 13, 33].

Figure 23.6 shows as an example the absorption spectra of liquid water at room temperature in the NIR ranges measured with cuvettes of different path lengths. One can see the decrease in absorption from longer to shorter wavelengths due to the decrease in the quantum mechanical cross sections of the overtone oscillation transitions (see Table 23.1).

In general, the fundamental stretching vibrational mode in the mid-infrared range is at least 1 order of magnitude more sensitive than the combination bands in the NIR. This also applies to the transition from the 1st to the 2nd overtone and so on. From a technical point of view, the lower absorption in solution can be compensated

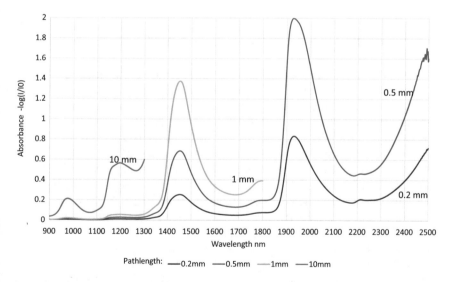

Fig. 23.6 Absorption spectra (900–2500 nm) of water for several path lengths. This illustrates the decrease of the quantum mechanical cross sections, and therefore, the decrease in absorbance from the combination bands down to higher overtone vibrational transitions at lower wavelength ranges. (details see text)

by a longer path length. This is also shown in Fig. 23.6. The figures in mm are the path lengths of the cuvettes used.

The assignment of the liquid water bands has been discussed for a long time. Usually, the peaks are assigned to the symmetric (v_s) stretching mode in combination with the asymmetric (v_a) stretching mode and often also together with the bending mode (δ) of the OH groups in water [30–33]. For example, the significant water peaks with maxima at around 970 nm (10300 cm^{-1}) are assigned to ($2v_s + v_a$), at 1200 nm (8310 cm^{-1}) to ($v_s + v_a + \delta$), and at 1450 nm (6900 cm^{-1}) to ($v_s + v_a$). The dominating strong peak at 1940 nm (5150 cm^{-1}) is suggested to be a combination of the asymmetric stretch and bending ($v_a + \delta$) mode of the OH-group. The absorbance of D_2O (*heavy water*) is lower and can be used as a standard reference material for water in solutions. There are smaller additionally peaks at approximately 770 nm (13000 cm^{-1}) and at 850 nm (11800 cm^{-1}) (both not shown in Fig. 23.6) as well as at 1780 nm (5620 cm^{-1}) which are difficult to assign correctly. The water peaks with maxima near 970, 1200, and 1450 nm shift with increasing temperature towards lower wavelengths (higher wavenumbers) and also show an isosbestic point of the optical unresolved pair of peaks [35–39].

Due to the high dipole moment change of the vibrational transitions in water, NIR and MIR spectroscopic investigations are highly sensitive to water absorption which makes spectroscopy in aqueous systems, e.g., biotechnology, quite challenging. On the other hand, water in food can easily be quantified even at very low concentrations [38, 39].

23.3.2.3 Penetration Depth of Photons in the NIR Wavelength Range

As shown in Table 23.1, the analytical sensitivity increases from the 3rd overtone to the combination bands with several orders of magnitude. The probability of photons penetrating an absorbing material is dominated by absorption, but scattering also plays an important role. This means that due to the smaller cross sections in the NIR range, a higher penetration depth of the photons can be achieved compared to MIR or UV–Vis spectroscopy. The great advantage of the NIR in PAT applications is therefore that no sample preparation (e.g., dilution) is required even at higher concentrations of an analyte. In transparent liquids, the lower sensitivity of the higher harmonics can be compensated for by longer layer thicknesses for the measurement setup. However, in solid particle systems this is not possible or difficult to achieve. On the other hand, the penetration depth in particle systems with lower absorption increases significantly. This becomes particularly clear, e.g., with the 3rd overtone [4, 13, 40]. This is an important aspect, e.g., for the quantitative determination of a pharmaceutical active substance in a tablet or for the measurement of substances, e.g., through the skin of a vegetable or a fruit. These aspects are discussed in detail in the literature [1, 32–34] and in the chapter measurement of solids.

23.3.3 Selectivity: Classification of NIRS Within the Spectroscopic Toolbox

The selectivity of a method is defined by how the method can determine certain analytes in a complex mixture without interference from other components. There-fore, if the signal of the analyte can be separated from signal of the interfering substance, the sensitivity in complex mixtures increases significantly. Several authors define a so-called net analyte signal (NAS), which describes the part of a sample spec-trum that is orthogonal to the spectra of all other components in the sample. This is also called "the unique spectral signature for the analyte of interest," i.e., the part of the instrument signal that is not lost due to spectral overlap. This underlines the importance of finding peaks of the analyte at wavelengths that are clearly separated from the residual matrix [1, 33, 41, 42].

Diode array or Fourier transform spectrometers show a fixed optical resolution over the entire wavelength range. Dispersive scanning spectrometers often compen-sate the intensity of the illumination (reference) by changing the slit width. This means that the optical resolution and thus the selectivity changes over the wave-length and depends very much on the absorption range. It is important to stress that the optical resolution is not equal to the pixel resolution.

23.3.4 Robustness, Detection Limit (DL), and Signal-To-Noise Ratio (SNR)

23.3.4.1 Definition of Robustness and SNR

The ICH quality guideline (International Council for Harmonization, ICH Q2) on the robustness of an analytical method states [10]: "Robustness of an analytical procedure is a measure of its capacity to remain unaffected by small, but deliberate variations in method parameters and provides an indication of its reliability during normal usage." This statement therefore accepts errors due to fluctuations of the method parameters, but they must be within given limits of the methodology. Furthermore, the statement for a system qualification also includes the optical setup, the wavelength-dependent detectivity of the sensors and the wavelength-dependent illumination, which determine the overall signal-to-noise ratio [33, 43, 44].

The signal-to-noise ratio (SNR) is defined as

$$SNR = (S-B)/N_{SD}$$

With S = average value of the signal, B = average value of the background, N_{SD} = standard deviation (SD) of the noise amplitude.

23.3.4.2 Absolute and Relative Measurements: Detection Limits (DL)

The different spectroscopic methods can be divided into two categories: absolute and relative measurements [33, 43]. In absolute measurements, the signal grows from zero or the background noise upwards. This means that a single photon can be detected on a dark background. Absolute technologies include NMR, fluorescence, and Raman spectroscopy, where the signal is proportional to the concentration of the analyte. For relative measurements, the measured parameter refers to a reference, which is defined as a 100% signal at the defined wavelength, as in MIR, NIR, and UV–Vis spectroscopy.

The detection limit for absolute measurements is defined as the standard deviation SD of the background noise with an SNR equals to 1:1. For absolute measurements, the SNR depends on the value of the background noise. This means that reducing this noise increases the SNR, which can be achieved simply by cooling the detector. The detection limit of relative measurements is limited by the normalized signal magnitude and thus also depends on the values that are output, e.g., as counts or as voltage. This means that all parameters such as light source (intensity, fluctuations), beam guidance, polarization of the photons, characterization of the spectrometer used (e.g., scattered light, losses), the detector used (background, sensitivity, dynamic range), and also the sample or sample chamber must be controlled and kept constant in order to compare the individual measurements [43].

23.4 Inline Spectroscopy of Liquids: Interfacing with Probes

23.4.1 Synopsis and Taxonomy of Probes

Generally, probes may be divided into two types: immersion probes which are directly integrated into the system (inline), and flow through probes (online), which are coupled to the system via a bypass. In both these categories, there is a variety of different configurations.

With the transmission probe, the liquid is illuminated directly and the wavelength-dependent absorption is measured. Derived from a transmission probe is the transflectance probe. In this case, after light has passed the medium it is reflected back by a mirror. Thus, the path length is twice the path length of the transmission setup. Apart from this, the construction of the light source and detector is identical to the transmission probe. Reflection probes belong to the type of backscattering probes and detect light which is reflected or scattered back from particles or bodies. Fluorescence probes and Raman probes are special probes for the measurement of fluorescence and Raman signals, respectively, and similar to scattering probes and will not be shown here. A particular type of probe is the attenuated total reflectance (ATR) probe. With this type of probe, the light undergoes total reflection in the probe surface. The

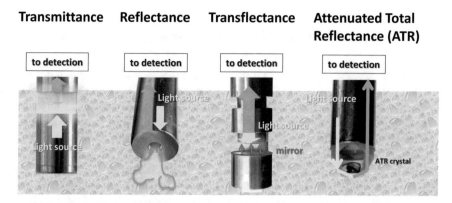

Fig. 23.7   Schematic representation of typical spectroscopic probes

evanescent light which interacts with the solution is the signal which is measured. The evanescence field penetrates only a few hundred nm up to a few microns into the solution after a single total reflectance. The penetration depth depends on the used wavelength and the refractive indices of the elements.

Figure 23.7 shows the set-up of selected probes.

According to the Beer–Lambert law, for high absorption coefficients and high concentrations in the sample, the measurement path length has to be small in order to get an optimum signal. A consequence of this is that one has to work with very thin layers when transmission probes are used. This creates many technical and mechanical problems. The ATR technique can be positioned freely, and a layer thickness in the micrometer range or less is still achieved. ATR technology is common in MIR but seldom used in NIR spectroscopy due to the lower absorption coefficients in the NIR wavelength region.

For quantitative examination, the results from transmission cells are directly usable. The Beer–Lambert law is always valid when the probe is working in the linear range (concentration not too high or not too low) and no scattering takes place. This is also the case with the ATR technique as long as the angle of incidence and detection is reproducible and within a limit. As a rule of thumb, the best working absorbance range with the lowest error of the measurement is roughly between 0.4 and 1.7 absorbance units depending on the quality (especially stray light) of the spectrometer unit [44–46].

Depending on the area of application and the wavelengths used, various probe materials are available: e.g., diamond, Suprasil, Infrasil, quartz, sapphire (synthetic monocrystalline aluminum oxide), and zirconia, which all are suited for the NIR wavelength range. Diamond, sapphire, and zirconia are very resistant to harsh environment [13].

23.4.2   Retractable Probes for Cleaning in Place (CIP) and Working with Highly Toxic or Aggressive Media

In food industry and especially in biotechnology, due to probe fouling, there is a demand for cleaning in place (CIP) which avoids contamination of the bioreactor. Highly viscous or pulpy process media form often strongly adherend deposits on the surface of an optical probe which then need aggressive cleaning procedures to re-establish a reproducible measurement. Furthermore, when working with hazardous materials, the processing unit must be reliably isolated from the environment during production even when the sensor is broken. In this case, the entire drive unit must be separated from the process stream under full process conditions. It is important, that the functionality of the optical probe is continuously under control and the probe can have a full view into the process.

Ceramically sealed sensor lock gates can withstand highly aggressive and corrosive media as well as high temperatures and pressures when stainless steel, corrosion-resistant alloys (e.g., Hastelloy), or titanium are used as process-wetted housing materials. Ceramic sealing is extremely resistant to chemical, thermal, and mechanical influences, guaranteeing maximum availability. Ball valves may not always withstand a long time the process pressure. A better, but more costly solution, is to use two planar ceramic disks that rotate toward each other to perform the sealing function and simultaneously can separate the calibration chamber from the process.

An example for such a retractable probe system is shown in Fig. 23.8, where all types of different optical sensors or any other measurement option can be integrated.

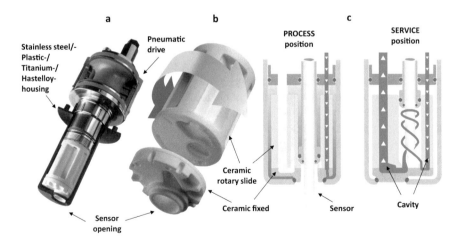

Fig. 23.8 Retractable probe for inline spectroscopic process control including cleaning in place (CIP) possibility (reproduced with permission from Knick GmbH, Berlin, Germany), **a** schematic of the housing and system integration of the retractable probe, **b** sealing and separation of the service and calibration, respectively, service position using two planar ceramic elements which rotate toward each other, **c** ceramic process gate in process position and in cleaning/calibration position

The pneumatically operated sensor lock gate allows calibrating or adjusting the measuring system as well as cleaning the sensor in the running process. The sensor can be moved into "PROCESS" position where the probe is located in the process medium. In "SERVICE" position, the probe is located in the calibration chamber. In this position, the measuring system can be calibrated or can be cleaned and dried, e.g., using compressed air. The used liquids (for cleaning or calibration) leave the calibration chamber through an outlet hose; i.e., they are displaced from the calibration chamber by following liquids or by air. The spectrometer software detects optical window contamination via the measuring signal, and the probe is cleaned as required or automatically at specific intervals.

23.5 Inline Spectroscopy of Surfaces, Thin Films, and Particulate Systems

23.5.1 Separation of Specular and Diffuse Reflectance in PAT Applications Using Polarization Spectroscopy

23.5.1.1 Specular and Diffuse Reflectance

Light scattering is an elastic interaction between the photon and the particle. The phenomena of reflection and scattering are thus closely related to particle size, angle of incident of the illuminating light source, angle of detection of the observer, respectively, sensor and the difference between the refractive index of the substance and that of the embedding medium, respectively, environment. When polarized light is used for the measurement, specular reflected light preserves its polarization, whereas diffuse reflected light is depolarized after scatter in all directions. Scatter also changes the angular distribution of the scattered light. This means, the light no longer has a preferential direction, but spreads itself in all directions. Specular reflection is always mirrored at exactly the same angle of the incident light with respect to the surface, whereas (ideal) diffuse reflected light is scattered in all directions.

For the measurement of diffuse reflection or transmission, ideally an integrating sphere is used. The sample is illuminated with directed light through an opening in the sphere. Specular reflection, where the angle of illumination and the angle of reflection is the same, is lost in a light trap (usually, an angle of around 5° is used) or measured directly with an additional detector. The diffuse reflected light is spread over the whole volume of the sphere and can be measured at one spot with a detector located where specular reflected light is avoided [47–50].

23.5.1.2   Measurement of the Film Thickness and Defect Sites by the Combination of Spectral Interference Spectroscopy and Depolarization of Light

The thickness of thin films on a surface can be measurement using, e.g., the specular reflectance arrangement where wavelength-dependent interference according to the Fresnel equations will appear. Due to refraction and reflection of the top and lower level of a film, a phase shift occurs and this phase shift can be measured by means of the interference pattern of the spectrum of the specular reflected light. Popular arrangements for inline control are, e.g., illumination at an angle of 45 degrees and detection at also 45°, which is labeled as 45R45, whereas diffuse reflected light might be measured at 0 degree, labeled then 45R0. However, if the macroscopic shapes of the sample, e.g., a shiny skin of a fruit or microscopic mirrors evenly distributed within a surface layer of a tablet, the use of polarization filters is recommended (see outline below).

Figure 23.9 visualizes the optical measurement of the diffuse and specular reflected light with an integrating sphere and the measurement of the film thickness by spectral interference spectroscopy (45R45) using Fresnel equations [47, 48]

When a surface is illuminated with parallel polarized light and measured also by a parallel polarized detector (e.g., 0R0, parallel/parallel), only specular reflected light will be measured. As diffuse reflected light has lost its polarization, diffuse reflected light can be separated from specular reflected light by measuring the intensity of the reflected light with crossed polarizer. In this case, the defect sites can be quantified by

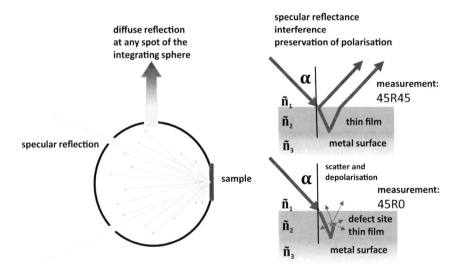

Fig. 23.9 Left: geometric arrangement for the measurement of diffuse and specular reflection with an integrating sphere. Right: Principle of the inline measurement for the layer thickness of thin films (e.g., geometry specular reflectance 45R45) and measuring the depolarization of light by defects or particles (geometry diffuse reflectance 45R0)

the decrease of the polarization. The polarization decreases with increasing surface roughness and/or the number of scattering centers formed by, e.g., defect sites such as holes or cracks in a surface layer of thin films. The proportion of depolarized light may then be a measure of the number of defect sites within a thin film caused by the particle density or roughness of the surface. Thus, besides the film thickness, the defects sites within this film can be quantified simultaneously [47, 48].

23.5.2 Penetration Depth of Specular and Diffuse Reflected Light

23.5.2.1 Photon Reflection and Photon Diffusion

Figure 23.10 top shows a sample of a pellet. White microcrystalline cellulose was mixed 1:1 with a red-dyed microcrystalline cellulose and then pressed to a pellet. The objective here is to illustrate in the visible range how photons migrate into a system as analogion to the NIR range.

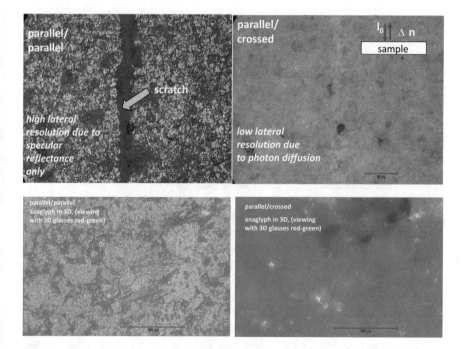

Fig. 23.10 Top—picture of the surface of a pellet with a mixture (1:1) of white and red micro-crystalline cellulose. Illuminated by parallel polarized light. Measured through a parallel polarized analyzer (left) and a crossed polarized analyzer (right). Bottom—Closer look as an anaglyph in 3D (best to view with 3D glasses red-green). Details see text

Photons which are directly (specular) reflected from the surface of the pellet can hardly uptake the chemical information "red" as no or very little penetration of the photon takes place, and the probability of absorption (= inelastic interaction) is therefore low. However, the spatial information is high as the reflected photon carries more or less the exact spatial information with high resolution. On the other hand, if the pellet is illuminated, e.g., at 0 angle degree with parallel polarization and the scattered photon is detected also at 0 degree but measured with a crossed polarizer, the probability is high to detect only diffuse reflected photons. This means, these photons have migrated deeply into the system and have thus a high chance to carry the chemical information "red." However, due to the photon diffusion into the pellet and statistical scatter several times, the photon loses its spatial coordinates and thus a blurred picture is observed from the diffusely scattered photons.

Figure 23.10 left and right side shows the result of the 2 different optical setups. A more grayish color can be seen in the setup measuring the specular reflectance (left). And a more reddish color is visible in case of the setup measuring the diffuse reflectance (right), but this view is a bit blurred. More details of the pellets and the spectra in the UV–Vis and NIR region are discussed in [51].

Due to a scratch on the surface, the photons are specular reflected off axis compared to the measurement optical axis. Thus, the intensity of the reflected light, which is captured by the detector, decreases, and the scratch appears a little bit darker. When measured with mixed polarization to identify, e.g., the active ingredient of a tablet, the lower intensity may then be misinterpreted as a region of higher absorbance and therewith an increase in concentration of the API. The bottom images can be viewed by red-green glasses in three dimensions. The specular reflectance picture (left) shows clear structures, whereas the diffuse reflectance picture (right) is blurred and shows the three-dimensional photon diffusion.

Another possibility to exclude the specular reflectance is to illuminate the sample with a so-called Lambertian illuminator (see discussion below).

23.5.2.2 Multiple Scattering and Penetration Depth

The probability of photons penetrating an absorbing material is dominated by the absorption and the scattering [1]. The prerequisite for scattering to occur is a difference in the refractive indices of the material and its surroundings of at least $\Delta n = 0.1$. In air, this difference is so great that most of the organic and inorganic powders scatter strongly. In organic media, this is not always the case, as the differences in the refractive indices become smaller.

The penetration depth depends primarily on the scattering coefficient S and the absorption coefficient K of the sample [1, 40, 47, 52–54]. The scattering is determined by the size, packing density, and relative refractive index (i.e., indirectly, e.g., also by the water content) of the scattering particles. The absorption of visible light is caused by chromophoric groups, while in the near-infrared, OH and CH harmonics are mainly responsible for the scattering.

The penetration depth of the light can be modeled with the radiative transfer equation. A detailed overview of the axial and radial scattering of photons is given in [52, 53]. Some more practical applications are described in [1, 55–60].

23.5.3 Robustness of the Inline Measurement Setup for Solids: Diffuse Illumination

Robustness of the optical setup is a key issue for successful implementation of PAT into industrial processes. The strongest "enemy" for reliable measurements of solids and surfaces is specular reflection which accounts for many perturbations. A high effort is needed especially with objects which are complex in shape like, e.g., capsules in the pharmaceutical industry, or natural variable surfaces which are found in apples, tomatoes (see Fig. 23.11) and other fruits. In hyperspectral, imaging often shades are measured and need to be excluded, e.g., by chemometric data pre-treatment which on the other side may reduce the robustness of the model [1, 40].

Ideally, the sample is illuminated by a perfect Lambertian source and the detector is also integrated into an integrating sphere. A good approximation to a Lambertian illumination can be realized in PAT as shown in Fig. 23.11. Two practical examples of an illumination setup are shown which are cheap, robust, and efficient. As material, industrially produced Teflon can be used as long as the material is easy to clean and is stable within its industrial environment. The samples with these arrangements are almost ideally and homogeneously illuminated and no shadow on the surface or on the background material is present. Furthermore, diffuse illuminated samples are very tolerant towards flutter in industrial assembly lines [13, 60, 61].

23.6 PAT in the Petrochemical Industry as an Example for Inline Process Control

23.6.1 Petrochemical Industry

23.6.1.1 Product Portfolio

The petrochemical industry is a branch of industry that produces organic intermediates such as refined products, natural gas, plastics, rubber, fiber raw materials, and many other basic chemicals using unit operations to separate and functionalize the products [62].

Natural gas processing, for example, is a complex industrial process for purifying raw natural gas, in which impurities such as various non-methane hydrocarbons and liquids are separated to produce dry natural gas in so-called pipeline quality. The inline analysis of natural gas consists primarily of methane, but also includes various

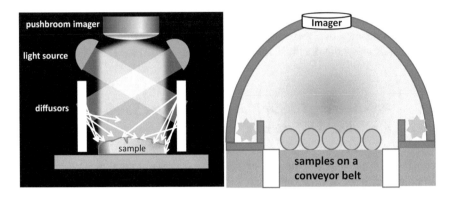

Direct illumination

- **Geometry, 3d-shape of sample**
- **Reflection of lamp**
- **Shadows**

Specular reflected light does not contain any chemical information

Indirect illumination

- **2d-shape of sample**
- **No specular reflections**
- **No shadows**

Only the diffusive reflected light contains chemical information

Fig. 23.11 Top: good approximation of a Lambertian illuminator with the advantage to be used inline on a conveyor belt. Left: Lambertian illuminator using flat illumination of diffusers. Right: Indirect illumination with a hemisphere. Bottom left: Illumination of tomatoes using 45R0 illumination, right: diffuse illumination similar to top right

hydrocarbons such as ethane, propane, butane, and pentanes as well as hydrogen sulfide and carbon dioxide. The heavier hydrocarbons are extracted to produce a purer "dry" methane for consumers. The separated natural gas liquids (NGLs) are useful as a chemical feedstock for petrochemical plants and are used for heating and cooking or are blended into automotive fuel.

The structure of an oil refinery is designed to convert crude oil, e.g., into high-octane motor gasoline (gasoline/petrol), diesel oil, liquefied petroleum gas (LPG), kerosene, fuel oil, lubricating oil, bitumen, and petroleum coke or asphalt and possibly sulfur as a by-product. Often the production of some basic chemicals such as ethylene/propylene/butadiene, aromatic hydrocarbons, hydrogen, and many others also take place there.

It would go beyond the scope of this short presentation to describe in detail all the products that can be produced. In order to understand the challenges of in-line control using NIR spectroscopy, some characteristic features of the main products are listed below.

23.6.1.2 Selected Products from a Refinery

This paragraph is just a short summary of the publications of two extended reviews [62, 63].

Crude oil:

Crude oil is a highly complex mixture of hydrocarbons and heteroatomic organic compounds with different molecular weights and polarities. Crude oil is converted into many different products through a combination of physical and chemical processes, collectively known as refining. The first stage of a refining process is the fractional distillation of crude oil, in which the crude oil is split into several components with different boiling ranges, such as gases, straight-run gasoline, naphtha, kerosene, gas oil, and atmospheric residue (AR). Straight-run gasoline, naphtha, kerosene, and gas oil, whose boiling temperatures range from 30 to about 400 °C, are clear liquids and can be easily measured by NIR spectroscopy. Atmospheric residue, which has a boiling temperature of about 350 to 800 °C, is used as the main fuel for power plants, ships, and large heating systems. It is almost black and very viscous (even solid at room temperature) and is therefore difficult to handle and measure [62, 63].

Gasoline:

The composition of petrol is a complex mixture of C4 to C9 hydrocarbons with a boiling range of 30–210 °C. It consists of more than 150 individual components of hydrocarbons. Petrol is characterized, for example, by octane number, Reid vapor pressure (RVP), aromatic content, benzene content and the content of oxygenates (e.g., MTBE, methyl tert-butyl ether) [62, 63].

Diesel:

Diesel is similar to petrol but is a mixed product with longer hydrocarbon chains. The important properties are cetane number, distillation temperature, pour point, and cold filter plugging point (CFPP). The main components of diesel are light gas oil (LGO: C13-C18, boiling point 250–350 °C) and kerosene (C9-C15, boiling point 190–250 °C). These have much higher molecular weights than the components of petrol. The pour point is controlled by adding a very small amount (100–500 ppm) of an additive, a so-called cold flow enhancer [62, 63].

Naphtha:

Naphtha is one of the most important materials in the petrochemical industry and is used as a basic material for the production of ethylene and benzene, toluene, xylene (BTX). Typically, ethylene and BTX are produced by thermal cracking or catalytic reforming of naphtha. Naphtha consists of alkanes, cycloalkanes, or naphthenes (C6–C10, boiling point 75–190 °C) and aromatic hydrocarbons present in the original crude oil. The most important property is paraffins, isoparaffins, aromatics, naphthenes, olefine (PIANO) compositions based on single carbon chain lengths with high concentrations and therefore relatively easy to calibrate [62, 63].

Heavy petroleum products, bitumen, and asphaltenes:

These samples are completely dark in color, very viscous, and difficult to handle. Solid suspensions and particles in these products also affect the reproducibility of NIR measurements. ASTM analysis methods require a long analysis time. Maintenance is costly and requires a considerable amount of work. The spectral properties of heavy hydrocarbons are much less sensitive to structural changes due to the longer chain lengths. A further hurdle for reliable spectroscopic measurement is the presence of water in crude oil, because crude oil can still contain water dispersed in the mass even after the separation of water. This changes the scattering and absorption and the measurement becomes erroneous and can only be corrected with difficulty or not at all [62, 63].

The combination of many unit operations at the same place together with the auxiliary facilities forms very large industrial complexes. Each refinery has its own unique arrangement and combination of their refining processes which are largely determined by the refinery location and used crude oil, desired products together with economic considerations. This means that every refinery has its own spectroscopic fingerprint representing the properties of the used crude oil and the settings of the unit operations.

23.6.1.3 Complexity of a Refinery: Example OMV Refinery Burghausen, Germany

OMV is an Austrian-based company and operates a total of three refineries: one in Schwechat (Austria) and one in Burghausen (South Germany), with both refineries also producing basic petrochemicals, along with the Refinery Petrobrazi (Romania). At OMV refineries, crude oil is converted into fuel, heating oil, bitumen, and petrochemical products by means of distillation, desulfurization, refining, and mixing. Special features of the refinery Burghausen, Germany, are [64]:

- Use of especially high-quality and low-sulfur crude oil types.
- Focus on petrochemistry to supply the southeast Bavarian chemical triangle.
- Residue-free crude oil processing through the complete conversion of heavy crude oil-based components into high-quality products.
- The International Airport Munich which is located approximately 125 km away is supplied with jet fuel by pipeline.
- Protection of the environment and saving of resources because of energy circulation with maximized heat recovery.

- Unique product range through the combination of special systems and procedures.

Figure 23.12 shows the flowchart of the OMV refinery as an example for a combined refinery for crude oil and simultaneously with a production of basic chemicals.

23.6.2 Objectives for the Integration of PAT Sensors in a Refinery and Future Smart Production

Inline near-infrared spectroscopy (NIR) has been increasingly used in refineries to replace hazardous manual sampling and frequent laboratory analysis. Knowledge of the current state of the crude oil being used has a major impact on the refining process design and influences all refinery operations. In addition, the availability of shale oil and other crude oils causes a greater variability of the raw material. Unknown variations of feed properties lead to instability, e.g., in the crude distillation unit.

The main objectives of introducing NIR inline measurements in a refinery for molecular characterization of components are described in detail, e.g., [65–68]. The bullets summarize the most important findings:

- Better and above all faster reaction to quality fluctuations of the crude oil mixture.
- This is associated with a reduction in the downgrading of transient products such as gas oil and residues.
- Yield increase of selected streams without affecting the quality, especially the cloud and freezing point.
- Real-time quality determination of intermediate flows to improve continuity between short-term planning, scheduling, and optimization systems (e.g., allocation of naphtha to reformers or steam crackers).
- This results in more stable downstream operations, increased profits, and reduced environmental risks.

The use of near-infrared spectroscopy (NIR) in the petroleum industry has greatly increased over the last 15 years because NIR allows fast and nondestructive online or inline multi-component analyses to be performed [62, 67]. Several suppliers of NIR instruments are now selling state-of-the-art technology for predicting and controlling important inline parameters in a refinery, as detailed, for example, in [66].

The following properties can be measured online or inline with NIR [66]:

- For research octane number (RON, ASTM D 2699), motor octane number (MON, ASTM D 2700), road octane number (RdON).
- Specific gravity of the diesel, viscosity, flash point, cold filter plugging point (CFPP), pour point, and cloud point.
- In addition, the volume percent or mole percent of individual components such as paraffins, isoparaffins, aromatics, naphthenes, and olefin (PIANO) are also measured during distillation.

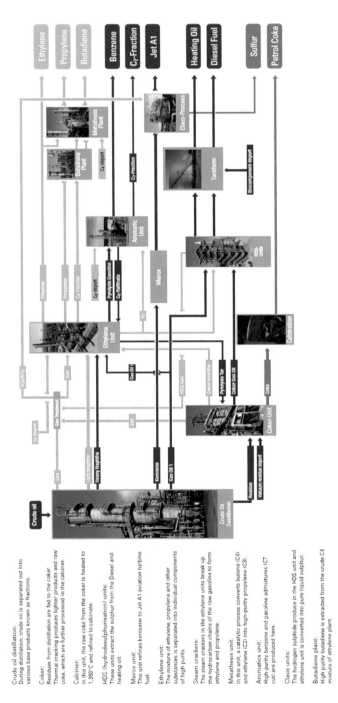

The Burghausen Refinery is a highly complex group of plants consisting of numerous columns, containers and several kilometres of pipeline.

Crude oil distillation:
During distillation, crude oil is separated out into various base products known as fractions.

Coker:
Residues from distillation are fed to the coker. Thermal cracking produces lighter products and raw coke, which are further processed in the calciner.

Calciner:
In this unit, the raw coke from the coker is heated to 1,350°C and refined to calcinate.

HDS (hydrodesulphurisation) units:
These units extract the sulphur from the Diesel and heating oil.

Merox unit:
This unit refines kerosene to Jet A1 aviation turbine fuel.

Ethylene unit:
The mixture of ethylene, propylene and other substances is separated into individual components of high purity.

Steam crackers:
The steam crackers in the ethylene units break up the hydrocarbon chains of the raw gasoline to form ethylene and propylene.

Metathesis unit:
In this unit, a catalytic process converts butene (C4) and ethylene (C2) into high-purity propylene (C3).

Aromatics unit:
High-purity benzene and gasoline admixtures (C7 cut) are produced here.

Claus units:
The hydrogen sulphide produce in the HDS unit and ethylene unit is converted into pure liquid sulphur.

Butadiene plant:
High purity butadiene is extracted from the crude C4 mixture of ethylene plant.

Fig. 23.12 Flowchart as an example of a refinery for crude oil in combination with the production of basic chemicals. Reproduced with permission and courtesy of [64]: OMV Deutschland GmbH, in: Perfect Flow: Crude oil as it makes its way through the refinery, brochure: OMV Germany, Burghausen—A refinery with great prospects, page 10; Burghausen; 2015

This allows complete control of the mixing process.

Future concepts such as Industry 4.0, Industrial Internet of Things (IIoT) and the China strategy "Made in China 2025" offer the opportunity not only to use information and measurements at the molecular level at each local plant site, but also to integrate this information for a holistic approach to the entire plant, which will enable intelligent production (see also the chapter on future production systems).

Intelligent production will transform the oil refining and petrochemical sector into a networked, information-driven environment [68]. Real-time measurement systems and the networking of all components will allow the manufacturing company to respond quickly to customer requirements and minimize energy and material consumption, while improving sustainability, productivity, innovation, and economic competitiveness.

An example of such a platform currently running at Sinopec Jiujiang Company in China is described in detail in [68] and also addressed in the chapter future production systems.

23.6.3 NIR Spectroscopy for the Petrochemical Industry

23.6.3.1 Composition of Crudes, Fuels, and Their Key Wavelengths in the NIR for Monitoring and Control

Interpretation of the Fingerprint Spectra

The CH vibrations are mainly responsible for the NIR spectral signature of the petro-chemical products. The C–H-fundamental stretching vibration in the mid-infrared region of the (aliphatic) methyl group (CH_3-) is around 2960 cm^{-1}, the methylene (CH_2−) at around 2930 cm^{-1}, methyne (CH−) at 2890 cm^{-1}, and a hydrogen atom attached to an aromatic hydrocarbon (ArCH) is around 3100 cm^{-1}. The amount of splitting and intensities between symmetric and asymmetric CH-bond stretching indicate branching (isomerization) from the linear arrangements of the chemical backbone [62, 69].

Figure 23.13 top shows the NIR-spectral fingerprint of a gasoline fuel of different research octane numbers (RON) in the full NIR spectral range. The most significant wavelength range is the combination band between 2100 and 2500 nm and is shown in more detail in Fig. 23.13 bottom. Due to the complexity of the mixtures, often model compounds are used to identify key wavelengths. An example are the spectral differences between aromatic hydrocarbons and aliphatic hydrocarbons or the spectral differences between linear, cyclic, and isomeric aliphatic hydrocarbons as overtones and the combination of the fundamental vibrations as shown and described below. The differences are discussed in detail in the cited literature and shall not be presented here [62, 69, 70]. Please also refer to the discussion in chapter sensitivity about quantum mechanical cross sections of the vibrational overtones [33].

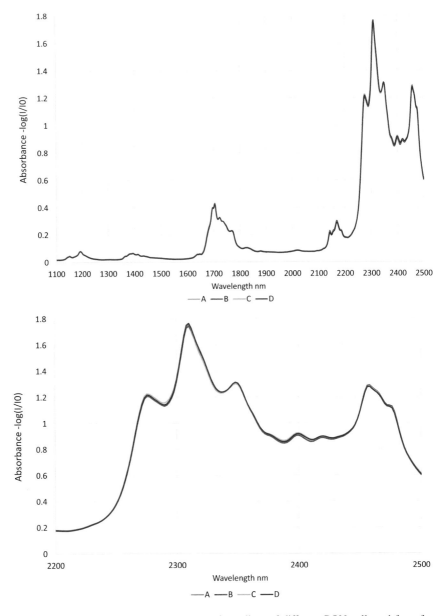

Fig. 23.13 Top: NIR spectra (whole range) of gasolines of different RON collected from four different locations in Germany in close vicinity to three refineries. A, B, and C represent the location of the three different refineries, D is a sampling point in between A and B. Bottom: spectra of the same samples in more detail within the combination band range 1 (details see cited literature and text below)

Table 23.2 Key wavelengths and interpretation of the fingerprint spectra of fuels

(0)	800–1200 nm	12,500–9090 cm^{-1}	3rd overtone
(1)	1100–1500 nm	9200–7800 cm^{-1}	2nd overtone
(2)	1300–1700 nm	7500–6400 cm^{-1}	Combination 2 (2νCH + δCH)
(3)	1600–2100 nm	6300–5200 cm^{-1}	1st overtone
(4)	2000–2500 nm	4700–4000 cm^{-1}	Combination 1 (νCH + δCH)

N = asymmetrical or symmetrical stretching vibration, Δ = bending mode vibration

The full range spectrum can be divided into five different CH-vibrational overtones (CH$_3$–, CH$_2$–, or CH–) and combinations with approximate wavenumbers and overlapping features (Table 23.2).

The aromatic ArC–H stretching vibrational bands (1st, 2nd, and 3rd) are at shorter wavelengths than the aliphatic C–H stretching vibration. The combination 1 bands of aromatic hydrocarbons are combinations of (νArC–H + νArC–C) and also at shorter wavelengths than the aliphatic C–H due to their higher bond strengths.

Tips and Tricks to Follow for Realizing Inline Control in a Refinery

From a practical point of view, the important aspects for the inline measurement of specific features to assign and control a refinery are summarized as follows [71–73]:

- Overlapping absorption peaks in the NIR together with a mixture of several hundred components make spectra too complex to interpret qualitatively. Thus, fingerprint spectra are used for the characterization together with multivariate data analysis (MVA). In future applications, artificial intelligence (machine learning, deep learning, etc.) may also be applied.
- As stated in the chapter sensitivity, NIR spectroscopy is at least 1 orders of magnitude lower in sensitivity than mid-infrared spectroscopy (MIR) of the fundamental vibrations. It can be even 10 orders of magnitude lower, when the 3rd overtone in NIR is used.
- Due to the low sensitivity, additives of a concentration of roughly below 1% are difficult to quantify.
- The advantage of the lower sensitivity of NIR in comparison with MIR is that the light penetration into the sample is much higher and thus penetration into dark-colored system is also higher. Furthermore, no sample preparation is needed before measurement for standard applications.
- As a rule of thumb, optimal path lengths with an absorbance of around 1 are achieved using at least 10 mm for the 3rd overtone, also around 10 mm for the 2nd overtone (1) and combination (2), 2 mm for the 1st overtone (3) and 0.5 mm for the combination (4).

- The difference between cyclic-, aliphatic n-, and i-hydrocarbons is mainly due to the number of -CH_2-, respectively, -CH_3 in the molecule and can be differentiated especially in the combination band range. However, at higher chain lengths, these differences diminish.
- Aromatic hydrocarbons show the most selective spectral features in the combination band area. The aromatic bands (centered around 4600 cm^{-1}) clearly increase with respect to the RON variation, since aromatic compounds usually have higher RON.
- The most commonly used oxygenates like ethanol, methyl tert-butyl ether (MTBE), and others can easily be quantified due to their high dipole moment and show thus higher sensitivity than the hydrocarbons.
- In crude oil, bitumen, and asphaltenes measurements, the design and performance of the sampling system are more important than the actual spectral collection. Since these samples are very viscous and dark and have high levels of particulates, there are high probabilities of sampling problems such as window fouling and line plugging.
- From a more practical point of view, NIR in combination with fiber optic technology for remote sensing can be used. In this case, NIR instrumentation can be located away from hazardous and explosive environments.
- NIR spectroscopy provides a lot faster and more repeatable data than, e.g., conventional online analysers. Furthermore, the data are based on molecular features rather than on macroscopic parameters. The improved harmonization of real-time analysis and process control can reach large economic benefits.
- Once correctly calibrated, NIR spectroscopy requires less maintenance than other conventional analysers used in refineries

The use of data fusion or nonlinear chemometric tools like support vector machines together with multiple analytical measurements clearly reduces the error associated with inferential property models for streams with highly complex compositions, such as crude oil, its fractions, and petrochemical products. This opens the door to a wealth of different applications and opportunities in both the laboratory and process settings [71–73].

23.6.3.2 Example: Key Spectral Features of Gasolines: Brand Name or Refinery?

In the previous section in Fig. 23.13, the spectral fingerprint of gasoline of different varieties is shown. The objective of the investigation was to show, whether the differences of the varieties are due to different brands or due to the refinery structure and the used raw material.

Figure 23.14 shows the principle component analysis (PCA) of the spectral features in the wavelength range from 2100 nm up to 2500 nm. The results of the other wavelength ranges are similar, but less pronounced.

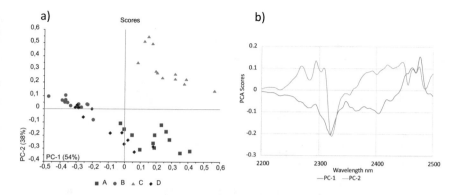

Fig. 23.14 Principal component analysis of the original spectra (left) and loading plots (right). Color labeling according to the location of the refinery. Original absorbance data are used

PC1 and PC2 explain about 90% of the variance. The spectra clearly show 3 distinct clusters which can be attributed to the locations of the 3 refineries. The variation of the brands is much smaller than the variation of the refinery's spectral signature. The different brand samples at location D are in line between the refinery A and B and therefore can be attributed to samples which were produced either in A or in B. The major distinction from C to the other two is the amount of aromatic hydrocarbons and to a certain extent also oxygenates which is higher in C than in the other 2 locations [74, 75]. It is well known, that C is specifically designed to produce high RON gasoline products.

23.6.3.3   Example Prediction: NIR Inline Measurement of the Research Octane Number RON

Figure 23.15 shows the results of a PCA and the partial least square regression (PLS-R) analysis of the samples described in the previous section. Additionally, the same sampling was repeated during summer and wintertime. The objective of this investigation was to predict the RON of the samples.

As expected, using NIR spectroscopy, it is possible to predict the RON with an error of about 0.6 RON using 3 LVs, which is higher than the reference method (0.3 RON). It is remarkable, that the predicted RON values in winter are on average 0.5 RON units lower than in summer. The model with 3 LVs only, as shown here, does not correct for summer and winter fuels. Using 5 LVs, the winter–summer difference is compensated by the 2 additional LVs and the error reduces to 0.3 RON comparable to the reference method. Most probable, different blending or additives to guarantee functionality during the winter season may be responsible for the clustering. Since a few years ago, government regulations require that oil companies offer "summer" and "winter" fuels. The latter can contain oxygenated additives in its blend that reduce the higher quantity of harmful emissions produced in cold weather. Some oil

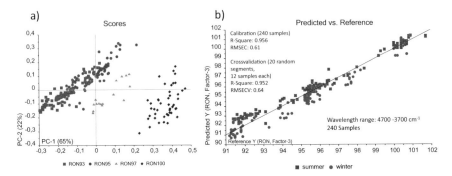

Fig. 23.15 Left: Principal component analysis of gasoline samples of RON 93–100, collected during the summer and winter season. Wavelength range analyzed: 6300–3700 cm-1 (1600–2700 nm). Labeling: different RON. Right: Regression plot predicted vs reference for the RON. RMSEC = 0.61, *R*-squared = 0.96, labeling winter and summer, model used 3 latent variable (LV). Original absorbance data are used

companies even brand this type of fuel and add ingredients that help also to prevent condensation and deposits in the fuel lines. Winter fuel is sold at the pump from fall to spring. Moreover, the summer gas blend is less volatile, which means it pollutes less, but costs slightly more to produce than winter fuel.

Selected applications of environmental issues or on adulteration are described in [76–79].

23.7 How to Run a PAT Project

23.7.1 Concept for a Knowledge-Based Production: Understanding Your Process on a Molecular Level

The importance of an understanding of the manufacturing process based on first principles is strongly emphasized. For a successful implementation of a rule-based production process, the following 3-Step Methodology is recommended [1]:

Step 1: Statistical and multivariate data analysis of historical production data, selection of the most important critical process parameters (CPPs) which have the highest impact on critical quality attributes (CQA). Define the low and high settings of the parameters and run designed experiments.

Step 2: Establish an inline monitoring based on process data and inline optical spectroscopy, use also possibly spatially resolved information, e.g., hyperspectral imaging.

Step 3: Extract knowledge from process and spectroscopic data of the designed experiments using multivariate data analysis (MVA) and machine learning (ML)

Fig. 23.16 Schematic presentation for a procedure to understand your process on a molecular level (source and permission from Kessler ProData GmbH, Germany)

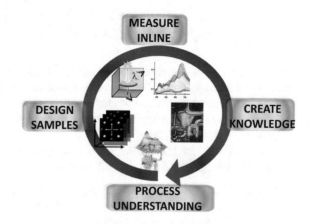

techniques. Transfer the process models into real live monitoring and control on industrial level.

Figure 23.16 illustrates the procedure.

First, robust data are the basis for all analytical and black-box model developments. In many publications, it is recommended to take as many samples as possible for the creation of chemometric models, or for process optimization to use as many datasets as possible. Using all available datasets is only meaningful when the information contained is new and not redundant. For logical reasons, it is best to use samples which are compiled with the help of a statistical experimental design (DoE).

Second, the associated measurement data of the designed samples are then suitable for modeling, which is based on empirical equations which describe the target figures with adjustable parameters. Using optical spectroscopy, information about the chemical composition of materials is obtained and in case of solid materials and particles, information about their morphology is derived simultaneously. This means information on a molecular level is achieved. If the process data are simultaneously measured, then soft sensor models may be developed. The samples can also be used to provide a reliable calibration set for online or inline process analysis. These procedures are complemented by classical methods of instrumental analysis, classical product characterization, and the traditional mathematical process description.

Third, causality is the key element to transfer data into knowledge and knowledge into process understanding. In this case, data which provide independent information should be selected and used often in combination with first principles. This procedure is important to guarantee robust process models [1, 4, 15, 35, 80, 81]. Correlated data, even a lot of them, do not provide new information.

23.7.2 Final Functionality Test of the PAT Spectroscopic System for Long-Term Operation

As described before, the concept of robustness is the capacity of a model to remain stable under small perturbations. The perturbations can be spectral in nature, chemical outliers, temperature shifts due to uncontrolled reactions or any other uncontrolled variables and data from different instruments where inter-instrument transferability is limited [33, 43–45, 82, 83].

In summary, before installation, the following sources of perturbations should be avoided and implemented into the PAT system:

- Reduced noise, especially baseline noise, for best signal-to-noise ratio.
- Selection of the optimum dynamic range of the measurement.
- Selecting the best possible reference.
- Working inside linearity and the highest precision of Lambert–Beer's law, e.g., within the Twyman–Lothian curve.
- Minimum stray light of the optical setup.
- Optimized quality of the optical setup including beam guidance, cell positioning, no instrumental wavelength variations.

The easiest way to check the system performance over time is to measure the reference once as a reference and without any change also once as a "sample." In this case, a 100% transmission straight line should be observed. This 100% line should be kept over a long time period [83]. With a suitable scaling, even small differences at the low and high ends of the usable wavelength range as well as the noise level can be judged. Running the system over a longer period, baseline drift, lamp temperature drifts and all other possible perturbation can be quantified.

23.7.3 Conclusion

A key element for a successful implementation into industrial life is the trans- and interdisciplinary collaboration of process managers, chemometric data analysts, electronics engineers, process specialists, and chemists with domain knowledge of, e.g., spectroscopy. It is important to bring them together in a business environment. But prerequisites of the project are also management support and the conviction of the staff that the project will indeed be a success. This means a "top-down" strategy of the employers and "bottom-up" strategy of the employees and a sound knowledge of the "champions" in the company.

Literature

1. R.W. Kessler, W. Kessler, Best practice and performance of hardware in Process Analytical Technology (PAT), chapter 61.1., in *Comprehensive Chemometrics 2nd edition: Chemical and Biochemical Data Analysis* (in print, Elsevier BV, 2020)
2. Industrial Internet of things (IIoT), Wikipedia: https://en.wikipedia.org/wiki/Industrial_Internet_of_Things, and references therein, Industrial Internet Consortium, in *Industrial Internet Reference Architecture Technical Report, version 1.9*, (2019). https://www.iiconsortium.org/IIRA. Accessed 01 Oct 2019
3. Industrie 4.0, https://www.plattform-i40.de/PI40/Navigation/EN/Home/home.html, online library: https://www.plattform-i40.de/PI40/Navigation/EN/InPractice/Online-Library/online-library.html. Accessed 01 Oct 2019
4. A. Kandelbauer, M. Rahe, R.W. Kessler, Process control and quality assurance–industrial perspectives, in *Handbook of Biophotonics, Vol.3: Photonics in Pharmaceutics, Bioanalysis and Environmental Research*, 1st ed., eds. by J. Popp, V.V. Tuchin, A. Chiou, A. Heinemann (Wiley-VCH Verlag GmbH & Co. KGaA, 2012), Chapter 1, pp. 1–69
5. European Medicines Agency, 2014: Guideline on the use of near infrared spectroscopy by the pharmaceutical industry and the data requirements for new submissions and variations: www.ema.europa.eu, EMEA/CHMP/CVMP/QWP/17760/2009 Rev2, Accessed 01 Oct 2019
6. K.H. Esbensen, P. Geladi, A. Larsen, The replication myth 1. NIR News **24**(1), 17–20 (2013)
7. K.H. Esbensen, P. Geladi, A. Larsen, The replication myth 2: Quantifying empirical sampling plus analysis variability. NIR News **24**(3), 15–19 (2013)
8. K.H. Esbensen, P. Geladi, A. Larsen, Myth: Process near infrared calibration spectra and reference samples can always be made to match properly. NIR News **24**(4), 23–25 (2013)
9. European Pharmacopoeia Monographs, Near Infrared spectroscopy (9th edition: 9.0, 2018): Ph. Eur. Monograph 2.2.40 Near-infrared Spectroscopy, p. 64–69, Process analytical technology (10th edition 10.0, effective from 1st January 2020): Ph. Eur. Monograph 5.25 Process analytical technology
10. The ICH International Conference on Harmonization (ICH), Quality guidelines (ICHQ): https://www.ich.org/page/quality-guidelines, Validation of analytical procedures ICH Q2 (R1, R2), Guideline on analytical procedure development Q14, Accessed 01 Oct 2019
11. US Food and drug administration FDA PAT guidance, PAT Guidance for industry: https://www.fda.gov/downloads/drugs/guidances/ucm070305.pdf, Process Validation: https://www.fda.gov/downloads/drugs/guidances/ucm070336.pdf, Accessed 01 Oct 2019
12. K.A. Bakeev (ed.), *Process Analytical Technology: Spectroscopic Tools and Implementation Strategies for the Chemical and Pharmaceutical Industries*, 2nd ed., (John Wiley and Sons, Chicester, UK, 2010)
13. R.W. Kessler (ed.), *Prozessanalytik- Strategien und Fallbeispiele aus der industriellen Praxis* (Wiley – VCH, Weinheim, 2006)
14. H. Kagermann, R. Anderl, J. Gausemeier, G. Schuh, W. Wahlster (eds.), *Industrie 4.0 in a Global Context: Strategies for Cooperating with International Partners (acatech study), Munich* (Herbert Utz Verlag, 2016). The original version of this publication is available at www.utzverlag.de or www.acatech.de
15. Kessler R. W., Perspectives in Process Analysis, J. Chemometrics 2013, 27, 369–378
16. T. Bieringer, S. Buchholz, N. Kockmann, Future production concepts in the chemical industry: Modular—Small-Scale—Continuous. Chem. Eng. Technol. **36**, 900–910 (2013)
17. T. Seifert, S. Sievers, C. Bramsiepe, G. Schenbecker, Small scale, modular and continuous: a new approach in plant design. Chem. Eng. Process. **52**, 140–150 (2012)
18. S. Kern, L. Wander, K. Meyer, S. Guhl, A.R.G. Mukkula, M. Holtkamp, M. Salge, C. Fleischer, N. Weber, R. King, S. Engell, A. Paul, M.P. Remelhe, M. Maiwald, Flexible automation with compact NMR spectroscopy for continuous production of pharmaceuticals. Anal. Bioanal. Chem. **411**, 3037–3046 (2019)
19. The Industrial Internet of Things Volume G1, Reference Architecture, Version 1.9, June 19, 2019, https://www.iiconsortium.org/pdf/IIRA-v1.9.pdf. Accessed 01 Oct 2019

20. German standardization roadmap, Industrie 4.0, version 3, 2018 https://www.din.de/blob/65354/57218767bd6da1927b181b9f2a0d5b39/roadmap-i4-0-e-data.pdf, accessed: 2019–10–01

21. J. Berthold, M. Blazek, M. Deilmann, H. Engelhardt, A. Gasch, M. Gerlach, F. Grümbel, U. Kaiser, M. Kloska, S. Löbbecke, M. Maiwald, T. Pötter, K. Rebner, E. Roos, S. Stieler, D. Stolz, M. Theuer, Technologie-Roadmap 4.0 – Voraussetzungen für die zukünftigen Automatisierungskonzepte. Ed. VDI/VDE-Gesellschaft Mess- und Automatisierungstechnik (GMA) and NAMUR – Interessengemeinschaft Automatisierungstechnik der Prozessindustrie, Düsseldorf, November 2015. Freely available at: https://www.namur.net/en/work-areas-and-project-groups/wa-3-field-devices/wg-36-analyser-systems.html or: https://www.vdi.de/ueber-uns/presse/publikationen/details/technologie-roadmap-prozesssensoren-40, or: https://www.zvei.org, Accessed 01 Oct 2019

22. K. Eisen, T. Eifert, C. Herwig, M. Maiwald, Current and future requirements to industrial analytical infrastructure—part 1: process analytical laboratories. Anal. Bioanal. Chem. **412**, 2027–2035 (2020)

23. T. Eifert, K. Eisen, M. Maiwald, C. Herwig, Current and future requirements to industrial analytical infrastructure—part 2: smart sensors. Anal. Bioanal. Chem. **412**, 2037–2045 (2020)

24. G. Engel, T. Greiner, S. Seifert, Ontology-assisted engineering of cyber-physical production systems in the field of process technology. IEEE Trans. Industr. Inf. **14**, 2792–2802 (2018)

25. J. Berthold, D. Imkamp, Looking at the future of manufacturing metrology. J. Sens. Sens. Syst. **2**, 1–7 (2013)

26. K.B. Beć, J. Grabska, H.W. Siesler, C.W. Huck, Handheld near-infrared spectrometers: where are we heading? NIR News, 1–8 (2020). 10.1177/0960336020916815

27. E. Borras, J. Ferre, R. Boque, M. Mestres, L. Acena, O. Busto, Data fusion methodologies for food and beverage authentication and quality assessment: a review. Anal. Chim. Acta **891**, 1–14 (2015)

28. B.P. Geurts, J. Engel, B. Rafii, L. Blanchet, A. Suppers, E. Szymańska, J.J. Jansen, L.M.C. Buydens, Improving high-dimensional data fusion by exploiting the multivariate advantage. Chemom. Intell. Lab. Syst. **156**, 231–240 (2016)

29. R.R. de Oliveira, C. Avila, R. Bourne et al., Data fusion strategies to combine sensor and multivariate model outputs for multivariate statistical process control. Anal Bioanal Chem **412**, 2151–2163 (2020). https://doi.org/10.1007/s00216-020-02404-2

30. Industrial Internet Reference Architecture in data analytics, The Industrial Internet of Things Volume T3: Analytics Framework, Internet of things: www.iiconsortium.org, Industrial Analytics Framework: https://www.iiconsortium.org/pdf/IIC_Industrial_Analytics_Framework_Oct_2017.pdf. Accessed 01 Oct 2019

31. J. Workman Jr., B. Lavine, R. Chrisman, M. Koch, Process analytical chemistry. Anal. Chem. **83**, 4557–4578 (2011)

32. Process analytics in Science and Industry, special edition with more than 20 papers on PAT. Anal. Bioanal. Chem. **409**, 629–857 (2017)

33. R.W. Kessler, W. Kessler, E. Zikulnig-Rusch, A critical summary of spectroscopic techniques and their robustness in industrial PAT applications. Chem. Ing. Tech. **88**, 710–721 (2016)

34. L. Rolinger, M. Rüdt, J. Hubbuch, A critical review of recent trends, and a future perspective of optical spectroscopy as PAT in biopharmaceutical downstream processing. Anal. Bioanal. Chem. **412**, 2047–2064 (2020). https://doi.org/10.1007/s00216-020-02407-z

35. P. Williams, P. Dardenne, P. Flinn, Tutorial: Items to be included in a report on a near infrared spectroscopy project. J. Near Infrared Spectro. **25**, 85–90 (2017)

36. J. Workman Jr., J.L. Weyer, *Practical Guide to Interpretive Near-Infrared Spectroscopy* (FL, CRC Press, Boca Raton, 2007)

37. H.F. Fisher, W.C. McCabe, S. Subramanian, A near-infrared spectroscopic investigation of the effect of temperature on the structure of water. J. Phys. Chem. **74**, 4360–4369 (1970)

38. K. Buijs, G.R. Choppin, Near-infrared studies of the structure of water. I. Pure Water, J. Chem. Phys. **39**, 2035–2041 (1963)

39. J.-J. Max, P. Larouche, C. Chapados, Orthogonalized H_2O and D_2O species obtained from infrared spectra of liquid water at several temperatures. J. Mol. Struct. **1149**, 457–472 (2017)
40. B. Boldrini, W. Kessler, K. Rebner, R.W. Kessler, Hyperspectral imaging: a review of best practice, performance and pitfalls for inline and online applications. J. Near Infrared Spectro. **20**, 438–508 (2012)
41. J. Vessmann, R.I. Stefan, J.F. van Staden, K. Danzer, W. Lindner, D.T. Burns, A. les Fajgelj, H. Müller, Selectivity in analytical chemistry. Pure Appl. Chem. **73**, 1381–1386 (2001)
42. S.M. Short, R.P. Cogdill, C.A. Anderson, Determination of figures of merit for near-infrared and raman spectrometry by net analyte signal analysis for a 4-component solid dosage system. AAPS Pharm. Sci. Tech. **8**(109) E1–E11 (2007). https://doi.org/10.1208/pt0804096, or: https://link.springer.com/content/pdf/10.1208%2Fpt0804096.pdf Accessed: 01 Oct 2019
43. W. Neumann, *Fundamentals of Dispersive Optical Spectroscopy Systems* (SPIE, Bellingham, WA, 2014)
44. A.G. Reule, Errors in spectrophotometry and calibration procedures to avoid them. J. Res. National Bur. Stand. Phys. Chem. **BOA**(4), 609–624 (1976)
45. H.K. Hughes, Beer's law and the optimum transmittance in absorption measurements. Appl. Opt. **2**, 937–945 (1963)
46. K.H. Esbensen, P. Geladi, A. Larsen, Concentration is a fixed, immutable property. NIR News **23**(5), 16–18 (2012)
47. R.W. Kessler, Optische Spektroskopie online und inline I: Festkörper und Oberflächen, in Prozessanalytik – Strategien und Fallbeispiele aus der industriellen Praxis, ed. by R.W. Kessler (Wiley-VCH, 2006), chapter 7, pp. 255–287
48. R.W. Kessler, W. Kessler, On line quality control of thin surface layers on metal sheets by spectral interference in combination with a neural network. J. Process Anal. Chem. **3**(4), 84–90 (1998)
49. H.H. Safwat, Effect of centrally located samples in the integrating sphere. J. Opt. Soc. Am. **60**, 534–541 (1970)
50. L.M. Hanssen, K.A. Snail, *Integrating Spheres for Mid- and Near-infrared Reflection Spectroscopy, Reproduced from: Handbook of Vibrational Spectroscopy*, ed. by J.M. Chalmers, P.R. Griffiths (Wiley, Chichester, 2002). including corrections: https://onlinelibrary.wiley.com/doi/abs/10.1002/0470027320.s2405. Accessed 01 Oct 2019
51. K. Rebner, W. Kessler, R.W. Kessler, Science based spectral imaging: combining first principles with new technologies, in *Near Infrared Spectroscopy: Proceedings of the 14th International Conference, 7.-16.11 2009, Bangkok, Thailand*, ed. by Sirinnapa Saranwong, Sumaporn Kasemsumran, Warunee Thanapase and Phil Williams, (IMPublications, Chichester, UK 2010), pp. 919–927, ISBN 978-1-906715-03-8
52. D. Oelkrug, M. Brun, K. Rebner, B. Boldrini, R. Kessler, Penetration of light into multiple scattering media: model calculations and reflectance experiments. Part I: Axial Trans. Appl. Spectro. **66**, 934–943 (2012)
53. D. Oelkrug, M. Brun, P. Hubner, K. Rebner, B. Boldrini, R. Kessler, Penetration of light into multiple scattering media: model calculations and reflectance experiments. Part II: Radial Trans. Appl. Spectro. **67**, 385–395 (2013)
54. Z. Shi, C. Anderson, Application of monte carlo simulation-based photon migration for enhanced understanding of near-infrared (NIR) diffuse reflectance. Part I: Depth Penetration Pharm. Mater. J. Pharma. Sci. **99**, 2399–2412 (2010)
55. F. Martelli, An ABC of near infrared photon migration in tissues: the diffusive regime of propagation. J. Near Infrared Spectrosc. **20**, 29–42 (2012)
56. D.J. Dahm, K.D. Dahm, Formulae for absorption spectroscopy related to idealized cases. J. Near Infrared Spectro. **22**, 249–259 (2014)
57. S.N. Thennadil, E. Dzhongova, NIR spectroscopy of turbid media: maximizing extractable information using light propagation theory. NIR News **24**(5), 12–17 (2013)
58. Y.-C. Chen, D. Foo, N. Dehanov, S.N. Thennadil, Spatially and angularly resolved spectroscopy for in-situ estimation of concentration and particle size in colloidal suspensions. Anal. Bioanal. Chem. **409**, 6975–6988 (2017)

59. M. Rey-Bayle, R. Bendoula, N. Caillol, J.-M. Roger, Multiangle near infrared spectroscopy associated with common components and specific weights analysis for in line monitoring. J. Near Infrared Spectro. **27**, 134–146 (2019)

60. W. Kessler, D. Oelkrug, R.W. Kessler, Using scattering and absorption spectra as MCR-hard model constraints for diffuse reflectance measurements of tablets. Anal. Chim. Acta **642**, 127–134 (2009)

61. E. Ostertag, M. Stefanakis, K. Rebner, R.W. Kessler, Elastic and inelastic light scattering spectroscopy and its possible use for label-free brain tumor typing. Anal. Bioanal. Chem. **409**, 6613–6623 (2017)

62. H. Chung, Applications of near-infrared spectroscopy in refineries and important issues to address. Appl. Spectrosc. Rev. **42**, 251–285 (2007)

63. K. Hidajat, S.M. Chong, Quality characterization of crude oils by partial least square calibration of NIR spectral profiles. J. Near Infrared Spectro. **8**, 53–59 (2000)

64. OMV Mediendatenbank, omv-mediadatabase.com, and https://www.omv.de/services/downlo ads/00/omv.de/1522141187364/Raffineriebroschure_DE, Accessed 31 Oct 2019

65. D. Lambert, K. Benkhelil, B. Ribero, M. Valleur, Advanced crude management by NIR spectroscopy combined with topology modelling, in *Hydrocarbon Processing, July 2019, Special Focus: The Digital Refinery*, https://www.hydrocarbonprocessing.com/magazine/ 2019/july-2019/special-focus-the-digital-refinery/advanced-crude-management-by-nir-spectr oscopy-combined-with-topology-modeling. Accessed 10 Oct 2019

66. J.C.A. Alves, C.B. Henriques, R.J. Poppi, Near infrared spectroscopy combined with support vector machines as a process analytical chemistry tool at petroleum refineries. NIR News **23**(8), 8–10 (2012)

67. J. Workman Jr., A review of process near infrared spectroscopy: 1980–1994. J. Near Infrared Spectro. **1**, 221–245 (1993)

68. Z. Yuan, W. Qin, J. Zhao, Smart manufacturing for the oil refining and petrochemical industry. Engineering **3**, 179–182 (2017)

69. G. Socrates, *Infrared and Raman Characteristic Group Frequencies: Tables and Charts*, 3 (Wiley, NY, 2004) 978-0-470-09307-8

70. L.G. Weyer, S.-C. Lo, *Handbook of Vibrational Spectroscopy*, ed. By J.M. Chalmers, P. Griffiths (John Wiley & Sons, Chichester, 2002)

71. T.I. Dearing, J.T. Wesley, C.E. Rechsteiner Jr., B.J. Marquardt, Characterization of crude oil products using data fusion of process Raman, infrared, and nuclear magnetic resonance (NMR) spectra. Appl. Spectro. **65**, 181–186 (2011)

72. Z. Xu, C.E. Bunker, P.B. de Harrington, Classification of jet fuel properties by near-infrared spectroscopy using fuzzy rule-building expert systems and support vector machines. Appl. Spectro. **11**, 1251–1258 (2010)

73. J.C.L. Alves, R.J. Poppi, Diesel oil quality parameter determinations using support vector regression and near infrared spectroscopy for hydrotreating feedstock monitoring. J. Near Infrared Spectro. **20**, 419–425 (2012)

74. M.V. Reboucas, de B.B. Netob, Near infrared spectroscopic prediction of physical properties of aromatics-rich hydrocarbon mixtures. J. Near Infrared Spectro. **9**, 263–273 (2001)

75. G. Guchardi, da P.A.C. Filho, R.J. Poppi, C. Pasquini, Determination of ethanol and methyl tert-butyl ether (MTBE) in gasoline by NIR–AOTF-based spectroscopy and multiple linear regression with variables selected by genetic algorithm, J. Near Infrared Spectro. **6**, 333–339 (1998)

76. H. Hasheminasab, Y. Gholipour, M. Kharrazi, D. Streimikiene, Life cycle approach in sustainability assessment for petroleum refinery projects with fuzzy-AHP. Energy Environ. **29**, 1208–1223 (2018)

77. H.O.M.A. Moura, A.B.F. Câmara, M.C.D. Santos, C.L.M. Morais, L.A.S. de Lima, K.M.G. Lima, L.S. de Carvalho, Advances in chemometric control of commercial diesel adulteration by kerosene using IR spectroscopy. Anal. Bioanal. Chem. **411**, 2301–2315 (2019)

78. J.C.L. Alves, R.J. Poppi, Quantification of hydrotreated vegetable oil and biodiesel contents in diesel fuel blends using near infrared spectroscopy. NIR News **27**, 4–7 (2016)

79. S. Altinpinar, D. Sorak, H.W. Siesler, Near infrared spectroscopic analysis of hydrocarbon contaminants in soil with a hand-held spectrometer. J. Near Infrared Spectro. **21**, 511–521 (2013)
80. R.W. Kessler, Projektmanagement für wissensbasierte Produkte und Verfahren, in *Prozessanalytik—Strategien und Fallbeispiele aus der industriellen Praxis Wiley-VCH 2006*, ed. by R.W. Kessler, chapter 3, pp. 49–76
81. T. Kourti, Pharmaceutical manufacturing: the role of multivariate analysis in design space, control strategy, process understanding, troubleshooting, and optimization, Chapter 15, in *Chemical Engineering in the Pharmaceutical Industry, 2nd Edition*, eds. by Mary T. am Ende David J. am Ende, (Wiley Online Library, 2019), pp. 601–629
82. S.A. Roussel, B. Igne, D.B. Funk, C.R. Hurburgh, Noise robustness comparison for near infrared prediction models. J. Near Infrared Spectro. **19**, 23–36 (2011)
83. R.W. Kessler, Optische spektroskopie: hardware für die prozessanalytik, in *Prozessanalytik – Strategien und Fallbeispiele für die industriellen Praxis*, ed. by R.W. Kessler (Wiley-VCH, 2006), chapter 6, pp. 229–252

Printed in the United States
by Baker & Taylor Publisher Services